Advances in Water Treatment and Pollution Prevention

Sanjay K. Sharma • Rashmi Sanghi
Editors

Advances in Water Treatment and Pollution Prevention

Springer

Editors
Sanjay K. Sharma
Jaipur Engineering College & Research Centre
JECRC Foundation
Jaipur
India

Rashmi Sanghi
Indian Institute of Technology Kanpur
UP
India

ISBN 978-94-007-4203-1   ISBN 978-94-007-4204-8 (eBook)
DOI 10.1007/978-94-007-4204-8
Springer Dordrecht Heidelberg New York London

Library of Congress Control Number: 2012942032

Printed on acid-free paper

Springer is part of Springer Science+Business Media (www.springer.com)

This book is for the coming generations, to make them aware about the 'cost' of water.

– Sanjay K. Sharma and Rashmi Sanghi

Preface

Anyone who can solve the problems of water will be worthy of two Nobel prizes – one for peace and one for science.

John F. Kennedy

When the well is dry, we learn the worth of water.

Benjamin Franklin

The mighty words of John F. Kennedy and Benjamin Franklin underline quite simply, and yet very powerfully, how important water is for the survival of our planet and its species and invite us to take measure against the dramatic consequences of water shortage.

Editing this book has been a very special experience for us because it has meant posing ourselves a crucial question: 'how could we live without water?'

Nowadays, the world water crisis has assumed alarming proportions and certainly two causes in particular have played a crucial role in its intensification:

- The World's growing population, and the consequent increase in water consumption and sanitation problems
- The fast-growing process of industrialization and development activities, which has led to water shortage and water pollution

The latter has been especially challenging because of its tangible and dramatic impact on ecosystems, human well-being and economies. And it is on the effects of water pollution that this book focuses its attention, examining what preventive measures can be taken against water pollution and stressing the need to implement greener water treatments.

The high quality chapters gathered in this volume make *Advances in Water Treatment and Pollution Prevention* a valuable resource to academic researchers, students, institutions, environmentalists, and anyone interested in environmental policies aimed at safeguarding both the quality and the quantity of water. We are positive that the book will provide an insightful analysis of Water Pollution and of its treatments as well as of the processes that have been studied, optimized and developed so far to sustain our environment.

We sincerely welcome feedback from our valuable readers and critics.

Jaipur, India Sanjay K. Sharma
Kanpur, India Rashmi Sanghi

About the Editors

Prof. (Dr.) Sanjay K. Sharma is a very well known author and Editor of many books, research journals and hundreds of articles from last 20 years. His recently published books are *Green Corrosion Chemistry and Engineering* (From Wiley-VCH, Germany), *Green Chemistry for Environmental Sustainability*, *Handbook on Applications of Ultrasound: Sonochemistry and Sustainability* (both from CRC Taylor & Francis Group, LLC, Florida, Boca Raton, USA) and *Handbook of Applied Biopolymer Technology: Synthesis, Degradation and Applications* (From Royal Society of Chemistry, UK). He has also been appointed as Series Editor by Springer UK for their prestigious book series *Green Chemistry for Sustainability*. His work in the field of Green Corrosion Inhibitors is very well recognized and praised by the international research community. Other than this, he is known as a person who is dedicated to educate people about environmental awareness, especially for Rain Water Harvesting.

Dr. Sharma has 13 books of Chemistry from National-International Publishers and over 48 research papers of National and International repute to his credit. Dr. Sharma is also serving as Editor-in-Chief for four international research journals *RASAYAN Journal of Chemistry*, *International Journal of Chemical, Environmental and Pharmaceutical Research*, *International Journal of Water Treatment and Green Chemistry* and *Water: Research and Development*. He is also a reviewer for many other international journals including the prestigious *Green Chemistry Letters and Reviews*.

Presently he is working as Professor of Chemistry at Jaipur Engineering College and Research Centre, JECRC Foundation, Jaipur (Rajasthan) India, where he is teaching Engineering Chemistry and Environmental Engineering Courses to B. Tech. students, Spectroscopy courses to PG students and pursuing his research interests. He is a member of American Chemical Society (USA), International

Society for Environmental Information Sciences (ISEIS, Canada) and Green Chemistry Network (Royal Society of Chemists, UK) and is also life member of various international professional societies including International Society of Analytical Scientists, Indian Council of Chemists, International Congress of Chemistry and Environment, Indian Chemical Society, etc.
E-mail: drsanjay1973@gmail.com

Dr. Rashmi Sanghi is currently working as a Research Consultant at the Indian Institute of Technology Kanpur and guest faculty at the LNM Institute of Information Technology, Jaipur. After obtaining her D.Phil degree from Chemistry Department, University of Allahabad, India, in 1994, she has been working at the Indian Institute of Technology Kanpur, India, as a Research Scientist. She was a Visiting Scientist at Chemistry Department, Rutgers University, Piscataway, NJ, USA in 1997 and worked with Prof. Alan S. Goldman.

She is passionate about environmental green chemistry and her major research interests are bioremediation, biopolymers and biosynthesis of nanomaterials using microbes and/or polysaccharides. Her research mainly focuses on the development of methods that can help in minimizing or eliminating the hazardous substances in the environment. Some of this work related to the green environmental chemistry is nothing but outstanding. She has traveled worldwide for various academic activities and/or professional talks. Her work on design and application of biopolymers in wastewater remediation shows great promise as evident by her patents *A method for preparing auto capped nano particles such as CdS in continuous flow columns* using fungus and *Poly(acryl amide) grafted Cassia grandis-silica hybrid: efficient metal ion adsorbent*. She has over 90 international journal publications to her credit. She has published three books on green chemistry: *Green Chemistry: Environment Friendly Alternatives* (2003), *Green Chemistry and Sustainable Development* (2005) and *Green Chemistry for Environmental Remediation* (2011). She is a member of many academic societies and reviewer of many international journals.
E-mail: rsanghi@iitk.ac.in; rsanghi@gmail.com

Acknowledgements

The time has come to express our sincere gratitude to all our friends, supporters and well wishers. We are heartily obliged for the support they have shown us while writing *Advances in Water Treatment and Pollution Prevention.*

First of all, we would like to thank all the esteemed contributors of this book. Without their contribution none of this would have been possible.

Professor Sharma would like to start by expressing his sincere gratitude to his teachers, Dr. R.K. Bansal, Dr. R.V. Singh, Dr. R.K. Bhardwaj, and Dr. Saraswati Mittal, the *Gurus* behind all his academic achievements and publications.

He then acknowledges Ackmez Mudhoo, Dr. Nabuk Eddy, Dr. Dong Chen, Dr. V.K. Garg, and all his friends and colleagues at the Jaipur Engineering College and Research Centre (JECRC) for their active interest and moral support.

Finally, he praises his family. His parents, Dr. M.P. Sharma and Mrs. Parmeshwari Devi, his wife Dr. Pratima Sharma, and all his family members for their never ending encouragement, moral support and patience over the months spent writing this book. A special thank you goes to his children, Kunal and Kritika: valuable moments of their lives have been missed because of his busy schedule.

Professor Sanghi would like to thank her family, who supported her all along. Her children, Surabhi and Udit, for their admirable patience and understanding that gave her the strength to face this challenging project. Her husband, Dheeraj Sanghi, whose encouragement and optimism at every stage have been precious. Her family friend, Prof. Sudhir Jain, for her support, insights and critical comments. Thank you also to her group of friends at IIT Kanpur for the relaxing walks they took together. Last but not least, a big thank you goes to Dr. Sanjay Sharma, editor of this book who 'dragged' her into this amazing project.

We are beholden to many other people whose names we have not been able to mention here but whose guidance has been very valuable. Finally, we would like to thank you our valuable readers and critics for encouraging us to do more and more research on this issue.

Save Water! Think Green!

Sanjay K. Sharma
Rashmi Sanghi
(Editors)

Abbreviations

Alum	Aluminum Sulfate
AOT	Advanced oxidation technology
AR88	Acid red 88
BDD	Boron-doped diamond
BEA-SECT	Background electrolyte-assisted sonoelectrochemical treatment
BOD	Biological oxygen demand
Cl-MBE	Mass balance error based on chlorine atoms, %
CNTs	Carbon nanotubes
COD	Chemical oxygen demand
DBS	Dodecylbenzenesulfonate
DCE	Dichloroethylene
DE	Degradation efficiency, %
DF	Diclofenac
DLVO	Derjaguin, Landau, Verwey and Overbeek
DNOC	4,6-Dinitro-o-cresol
DOE	Department of Environment
ECT	Electrochemical treatment
EDTA	Ethylenediaminetetraacetic acid
f	Frequency
F	Faraday constant, C mol^{-1}
FC	Fractional conversion, %
GC	Gaseous chromatography
GO	Graphene oxide
HPLC	High performance liquid chromatography
I_a	Ultrasound intensity
IBP	Ibuprofen
IC	Ion chromatography
MB	Methylene blue

MCP	Monocrotophos
MDB	Meldola blue
MG	Malachite green
MO	Methylene orange
MTBE	Methyl *tert*-butyl ether
MWCNT	Multi-walled carbon nanotube
NBB	Naphthol blue back
NPE	Nonylphenol ethoxylate
ODE	Ordinary differential equation
OG	Orange G
PAC	Polyaluminum chloride
PBM	Population Balance Model
p-CBA	p-Chlorobenzoic acid
PCE	Perchloroethylene
PCP	Pentachlorophenol
POMB	Palm oil mill boiler
POME	Palm oil mill effluent
PZT	Lead ziconate titanate
RB5	Reactive black 5
SCE	Saturated calomel electrode
S_{Cl}	Chloride ion selectivity
SCT	Sonochemical treatment
SECT	Sonoelectrochemical treatment without background electrolyte addition
SEF	Sonoelectro-Fenton
SERS	Surface-enhanced Raman spectroscopy
TCE	Trichloroethylene
TEM	Transmission electron microscopy
TOC	Total organic carbon
UV	Ultraviolet
$V_{A\text{-}C}$	Cathode-anode voltage
zpc	Zero point charge

Contents

Contributors

Mohammed Al-Abri Petroleum and Chemical Engineering Department, College of Engineering, Sultan Qaboos University, Muscat, Oman

M. Ashokkumar School of Chemistry, University of Melbourne, Melbourne, Australia

Jesús Beltrán-Heredia Department of Chemical Engineering and Physical Chemistry, University of Extremadura, Badajoz, Spain

P. Bonete Grupo de Nuevos Desarrollos Tecnológicos en Electroquímica: Sono-electroquímica y Bioelectroquímica, Grupo de Fotoquímicay Electroquímica de semiconductores, Departamento de Química Física e Instituto de Electroquímica, Universidad de Alicante, Alicante, Spain

Jeanette Brown Manhattan College, Riverdale, NY, USA

Fang Cai School of Chemical Engineering and Technology, Tianjin University, Tianjin, People's Republic of China

Shaoan Cheng State Key Laboratory of Clean Energy Utilization, Department of Energy Engineering, Zhejiang University, Hangzhou, China

Mei Fong Chong Department of Chemical and Environmental Engineering, Faculty of Engineering, The University of Nottingham, Semenyih, Selangor, Malaysia

Adrian Covaci Toxicological Centre, University of Antwerp, Antwerp, Belgium

Giovanni De Feo Department of Industrial Engineering, University of Salerno, Salerno, Italy

M.D. Esclapez Grupo de Nuevos Desarrollos Tecnológicos en Electroquímica: Sonoelectroquímica y Bioelectroquímica, Grupo de Fotoquímicay Electroquímica de semiconductores, Departamento de Química Física e Instituto de Electroquímica, Universidad de Alicante, Alicante, Spain

J. González-García Grupo de Nuevos Desarrollos Tecnológicos en Electroquímica: Sonoelectroquímica y Bioelectroquímica, Departamento de Química Física e Instituto de Electroquímica, Universidad de Alicante, Alicante, Spain

Tingyue Gu Department of Chemical and Biomolecular Engineering, Ohio University, Athens, OH, USA

Daniel J. Hassett Department of Molecular Genetics, Biochemistry and Microbiology, University of Cincinnati College of Medicine, Cincinnati, OH, USA

Nidal Hilal Centre for Water Advanced Technologies and Environmental Research (CWATER), College of Engineering, Swansea University, Swansea, United Kingdom

Liping Huang Key Laboratory of Industrial Ecology and Environmental Engineering, Ministry of Education (MOE), School of Environmental Science and Technology, Dalian University of Technology, Dalian, China

Yulan Ji School of Chemical Engineering and Technology, Tianjin University, Tianjin, People's Republic of China

Jing Li Engineering Department, China Tianjin Chemical Engineering Design Institute, Tianjin University, Tianjin, People's Republic of China

Giusy Lofrano Civil and Environmental Engineer, Salerno, Italy

Ling Wei Low School of Industrial Technology, Universiti Sains Malaysia, Penang, Malaysia

Tulika Malviya University of Allahabad, Allahabad, India

Ackmez Mudhoo Department of Chemical and Environmental Engineering, Faculty of Engineering, University of Mauritius, Mauritius

B. Neppolian SRM Research Institute, SRM University, Kattankulathur, Chennai, India

V. Sáez Grupo de Nuevos Desarrollos Tecnológicos en Electroquímica: Sonoelectroquímica y Bioelectroquímica, Grupo de Fotoquímicay Electroquímica de semiconductores, Departamento de Química Física e Instituto de Electroquímica, Universidad de Alicante, Alicante, Spain

Jesús Sánchez-Martín Department of Chemical Engineering and Physical Chemistry, University of Extremadura, Badajoz, Spain

Rashmi Sanghi Indian Institute of Technology Kanpur, UP, India

Sanjay K. Sharma Jaipur Engineering College & Research Centre, JECRC Foundation, Jaipur, India

Vandana Singh University of Allahabad, Allahabad, India

Thammanoon Sreethawong Baan Klangmuang Luzern, Suanluang, Bangkok, Thailand

Jinsheng Sun School of Chemical Engineering and Technology, Tianjin University, Tianjin, People's Republic of China

Tjoon Tow Teng School of Industrial Technology, Universiti Sains Malaysia, Penang, Malaysia

Chedly Tizaoui Centre for Water Advanced Technologies and Environmental Research (CWATER), College of Engineering, Swansea University, Swansea, United Kingdom

I. Tudela Grupo de Nuevos Desarrollos Tecnológicos en Electroquímica: Sonoelectroquímica y Bioelectroquímica, Departamento de Química Física e Instituto de Electroquímica, Universidad de Alicante, Alicante, Spain

Alexander L.N. van Nuijs Toxicological Centre, University of Antwerp, Antwerp, Belgium

Chapter 1
Green Practices to Save Our Precious "Water Resource"

Sanjay K. Sharma, Rashmi Sanghi, and Ackmez Mudhoo

Water is H_2O, hydrogen two parts, oxygen one but there is also a third thing, that makes it water and nobody knows what that is.

–D.H. Lawrence (1885–1930)

1.1 Introduction

Water is one of the world's most precious resources without which life is not possible on Earth. It is equally important for agriculture and industry. We cannot imagine any "crop" or any "product" without the involvement of water. Apart from its necessity for life, it has many unusual properties. Water is the only element known to man that exists naturally in all three states of matter. For example, it has exceptional ability to store heat and can modify the earth's temperature. In its solid form as ice, it has the ability to float on water allowing aquatic life to survive in winter. Water has a very significant role in chemistry as an excellent solvent that can dissolve many ionic and polar substances. That is why it is an effective medium for carrying nutrients to plants as well as to animals. Water in its liquid state can be classed as "strange" and "eccentric."

S.K. Sharma (✉)
Jaipur Engineering College & Research Centre, JECRC Foundation, Jaipur, India
e-mail: drsanjay1973@gmail.com

R. Sanghi
Indian Institute of Technology Kanpur, UP 208016, India
e-mail: rsanghi@gmail.com; rsanghi@iitk.ac.in

A. Mudhoo
Department of Chemical and Environmental Engineering, Faculty of Engineering, University of Mauritius, Mauritius
e-mail: ackmezchem@yahoo.co.uk; a.mudhoo@uom.ac.mu

S.K. Sharma and R. Sanghi (eds.), *Advances in Water Treatment and Pollution Prevention*, DOI 10.1007/978-94-007-4204-8_1,

Water makes up more than 70% of every adult human body and 50–90% of all plants and animals, and without water, we would die in a few days. The human body is comprised mostly of water: the brain is made up of 95% water, blood 82% water, and lungs 90% water, and furthermore our blood mineral content is strikingly similar to saltwater. A mere 2% drop in our body's water supply can trigger signs of dehydration. Water is also the major component of all living things, available in the form of surface water, underground water, rainwater, and seawater. However, the total quantity of water on our planet Earth is very nearly constant and it keeps circulating through the hydrologic cycle. So, the water strictly is a fixed resource and the total quantity of water on our planet is estimated at 1.36×109 km^3. Water from land and sea evaporates into the atmosphere, forms clouds, falls back to earth as rain or snow, and then into rivers and streams and back to the sea and air. The total availability of water on earth can be understood by the fact that 97.4% of earth's water is available as seawater and rest includes 1.98% ice caps and glaciers, 0.59% groundwater, 0.014% freshwater lakes and other forms of water [1]. Roughly 99% of water available on earth is not useful for us. Only 1% water is available as Earth's freshwater for all our day-to-day uses and out of this more than one-third of it is used for agricultural, industrial, and domestic uses [2].

Our earth initially had clean water. Crystal clean water bubbled forth from every spring. Since ancient times, people settled near sources of water and thus most of the cultures and communities grew along rivers and lakes in the area where some water source in the form of spring or well was available. People often drank the water from the same lake or river in which they disposed of their wastes. In the same water, they washed themselves and their animals. Consequently, they often became ill because of the water contaminated with dangerous microbes from the waste of human and animals. Centuries later, as water became more contaminated, people took steps to ensure they were drinking clean water. The ancient Greeks and Romans lined their drinking pottery with silver, believing that would ensure clean water. Later, they used boiling to ensure clean water. As recently as 1905, people used copper vessels exposed to sunlight as holding tanks that led to charcoal filters that produced fairly clean water. Although, the waterborne diseases have been controlled up to a satisfactory level, they are still very common in less developed and underdeveloped countries due to inadequate waste disposal systems. Major problems that humanity is facing in the twenty-first century are related to water quantity and/or water quality issues [2, 3] and will be more critical in future with issues such as global warming, climate change [4] resulting in floods and droughts [5]. Today, quality and pollution of surface water is a not only an issue for scientists, but also for policymakers, politicians, and the public in general.

Many reviews have appeared in the scientific literature that covers various aspects of waterborne diseases arising due to chemical pollution [6]. Pollutants have been present since time immemorial, and life on the earth has always evolved amidst them. Chemical pollutants construct a wide range of pollutants including nutrients such as nitrogen [7] and phosphorous compounds [8] and organic pollutants [9]. Industrialization and intensive use of chemical substances such as petroleum oil, hydrocarbons (e.g., aliphatic, aromatic, polycyclic aromatic hydrocarbons (PAHs),

BTEX (benzene, toluene, ethylbenzene, and xylenes), chlorinated hydrocarbons like polychlorinated biphenyls (PCBs), trichloroethylene (TCE), and perchloroethylene, nitroaromatic compounds, organophosphorus compounds), solvents, pesticides, and heavy metals are contributing to environmental pollution. Presently, large-scale pollution due to man-made chemical substances and to some extent by natural substances is of global concern. Seepage and run-offs due to the mobile nature and continuous cycling of volatilization and condensation of many organic chemicals such as pesticides have even led to their presence in rain, fog, and snow [10]. The thousands of inorganic and organic trace pollutants occurring at the nanogram to microgram per liter concentration are called "micropollutants." The sources of such micropollutants are diverse ranging from industries and municipalities [11] to agriculture [12, 13]. Though people are very much aware about their increasing concentration, designing sustainable treatment technologies is still a challenge for scientists worldwide [14–18].

> *Water is the oil of the 21st century.*
>
> –Andrew Liveris, CEO Dow Chemical, cited in "Running Dry," *The Economist*, 21 Aug 08

1.2 Some Common Pollutants

1.2.1 Persistent Organic Pollutants (POPs)

POPs have been and continue to be of greatest environmental concern worldwide. Some prominent POPs ("legacy POPs" or "the dirty dozen") have been listed in the Aarhus Protocol and the Stockhome Convention [19]. They are highly chlorinated compounds like DDT (dichlorodiphenyltrichloroethane) and polycyclic aromatic hydrocarbons (PAHs). The list of POPs is much longer [20] and allows new additions to it. The recent additions are polybrominated diphenyl ethers (PBDEs) used as flame retardants [21, 22] and perfluoroalkyl chemicals (PFCs) used in several industrial applications [23, 24]. These additional POPs are called "emerging POPs" [25]. While many POPs have been present in the atmosphere for decades, they could not be detected due to analytical limitations [26, 27]. POPs are dangerous for health because of their bioaccumulation and biomagnification potential in aquatic food web [28–31].

1.2.2 Pollutants from Agricultural Sector

Agriculture is also a common pollution source [32], because several million tons of chemicals and pesticides are being used for it globally. Use of such chemicals to increase the crop production causes numerous health hazards to biota and human beings [33–36]. Acute poisoning from direct exposure to pesticides is risky for agricultural workers, resulting in at least 20,000 deaths per year [37]. This is more frequent in developing countries [38–40] as monitoring and assessing practices are often limited [41] and enforcement of regulations are neglected [42]. The banned POPs like DDT and HCHs (hexachlorocyclohexanes) are used extensively in developing countries in agriculture and sanitation, because of their low cost and higher effectivity [43, 44].

1.2.3 *Arsenic*

Arsenic is a naturally occurring element that can be found in atmospheric, marine, mining, freshwater, groundwater [45], and terrestrial environments [46]. It is mobilized by dissolution in water and emission into the atmosphere. This is accomplished naturally through weathering of minerals and igneous rock, biological processes [47], and volcanic activities (including submarine volcanism). Human activities also introduce arsenic into the environment via agriculture, and through aerial and wastewater discharge from mining and industry.

In Bangladesh, over 35 million of 130 million inhabitants are at increased risk for cancer, cardiovascular, neurologic, and other diseases due to naturally occurring arsenic in drinking water [48]. The Bangladesh government and international aid organizations, spearheaded by United Nations Children's Fund (UNICEF), then began installing tube wells that tapped into pathogen-free aquifers as an alternative water source. The convenience and low cost of installing tube wells led millions of people to install their own private well [49]. This access to groundwater, as well as the introduction of oral rehydration therapy, was apparently effective in decreasing mortality rates. The unintended consequence, an epidemic of arsenicosis due to chronic arsenic exposure, became apparent in the 1990s. The Dhaka Community Hospital (DCH) and the School of Environmental Studies (SOES), Jadavpur University, Kolkata, India, monitored water quality from 0.1 million tube wells covering 64 districts of Bangladesh. The results showed that water samples from 53 districts had arsenic concentration above the maximum permissible limit (0.05 mg L^{-1}). The arsenic situation in the Chandipur village (7.2 km^2 and 7,430 inhabitants) for 1998–2004 showed that the situation had deteriorated. Out of 628 hand tube wells samples analyzed, 537 (85.5%) samples had arsenic in excess of 0.01 mg L^{-1} and 73.9% of the samples had arsenic in concentrations more than 0.05 mg L^{-1}. Also, the severity of the arsenic pollution in groundwater was confirmed when 93% of the 970 people surveyed on a preliminary basis showed arsenic levels more than the normally accepted levels in hair, nail, and urine samples.

Arsenic concentrations in natural water and wastewater are a worldwide problem and often referred to as a calamity of the twentieth and twenty-first centuries. Today, the chronic exposure to arsenic is also known to have an adverse effect on human health. Chronic arsenic ingestion from drinking water has been found to cause carcinogenic and noncarcinogenic health effects in humans [50]. An example of vascular effects of arsenic exposure is the so-called blackfoot disease which was found to be endemic in the southwestern coast of Taiwan associated with arsenic exposure form drinking water [51]. The primary exposure route of inorganic arsenic is ingestion from drinking water due to natural contamination in groundwater from dissolution of natural mineral deposits, industrial effluents and drainage problems [52], and of geogenic origin and caused by natural anoxic conditions in the aquifers [53]. While being a highly toxic element, arsenic has been studied to a lesser degree than other toxic elements such as lead, cadmium, or mercury. Despite its application in complete leukemia treatment, arsenic is constantly seen as a potential environmental pollution source on a large scale. As of now, ongoing research strongly evokes the polluting nature of arsenic. The pressing need for environmental protection from excessive levels of arsenic has led to the development of arsenic removal processes all over the world [54]. Existing arsenic removal technologies reported in the literature range from oxidation, precipitation, coagulation, membrane separation, ion exchange, biological treatment and removal systems, chemisorptions, filtration, to adsorption. Most of the established technologies for arsenic removal make use of several of these processes, either at the same time or in sequence. All of the removal technologies have the added benefit of removing other undesirable compounds along with arsenic – depending on the technology, bacteria, turbidity, color, odor, hardness, phosphate, fluoride, nitrate, iron, manganese, and other metals can be removed [55].

1.2.4 Heavy Metals

Rapid industrialization and urbanization have resulted in elevated emission of toxic heavy metals entering the biosphere. Various anthropogenic activities such as mining and agriculture have polluted extensive areas throughout the world [56]. The release of heavy metals in biologically available forms may damage or alter both natural and man-made ecosystems [57]. Although some heavy metals ions are essential micronutrients for plant metabolism but if present in excess concentration in soil, groundwater, and some aqueous waste streams can become extremely toxic to the receiving living micro and macro environments. Wastewater from various industries, such as metal finishing, electroplating, plastics, and pigment mining, contain several heavy metals of health and environmental concern, such as cadmium, copper, chromium, zinc, and nickel [58]. The term "heavy metals" refers to metals and metalloids having densities greater than 5 gcm^{-3} and is usually associated with pollution and toxicity although some of these elements (essential metals) are required by organisms at low concentrations [59]. Heavy metal toxicity and the danger of their bioaccumulation in the food chain represent one of the major environmental and health problems of our modern society. Primary sources of pollution is from the burning of fossil fuels, mining, and melting of metallic ferrous ores, municipal wastes, fertilizers, pesticides, and sewage sludge [60]. The most common heavy metals contaminants are: Cadmium (Cd), chromium (Cr), copper (Cu), lead (Pb), nickel (Ni), and zinc (Zn).

1.2.4.1 Cadmium

Cadmium (Cd) is one of the most toxic heavy metals and is considered nonessential for living organisms. Cadmium has been recognized for its negative effect on the environment where it accumulates throughout the food chain posing a serious threat to human health. Cadmium pollution has induced extremely severe effects on plants [61]. Cadmium, which is widely used and extremely toxic in relatively low dosages, is one of the principal heavy metals responsible for causing kidney damage, renal disorder, high blood pressure, bone fraction, and destruction of red blood cells [62].

1.2.4.2 Chromium

Chromium (Cr) is one of the most important chemical contaminants of concern. It exists in a series of oxidation states from −2 to +6 valence; the most important stable states are 0 (element metal), +3 (trivalent), and +6 (hexavalent). Cr^{3+} and Cr^{6+} are released to the environment primary from stationary point sources resulting from human activities. Acute and chronic adverse effects of chromium to warm-blooded organisms are caused mainly by Cr^{6+} compounds. Most investigators agree that chromium in biological materials is probably always in the trivalent state, that greatest exposures of Cr^{3+} in the general human population are through the diet, and that no organic trivalent chromium complexes of toxicological importance have been described. Hexavalent chromium is present in the effluents produced during the electroplating, leather tanning, cement, mining, dyeing, fertilizer, and photography industries and causes severe environmental and public health problems [63]. Its concentrations in industrial wastewater range from 0.5 to 270 mg/L and the tolerance limit for Cr^{6+} for discharge into inland surface water is 0.1 mg/L and in potable water is 0.05 mg/L.

1.2.4.3 Copper

Copper, one of the most widely used heavy metals, is mainly employed in electrical and electroplating industries, and in larger amounts is extremely toxic to living organisms. The presence of copper (II) ions causes serious toxicological concerns; it is usually known to deposit in brain, skin, liver, pancreas, and myocardium. Copper toxicity is a much overlooked contributor to many health problems, including anorexia, fatigue, premenstrual syndrome, depression, anxiety, migraine headaches, allergies, childhood hyperactivity, and learning disorders.

1.2.4.4 Nickel

Nickel is a naturally occurring element that exists in various mineral forms. It is used in a wide variety of applications including metallurgical processes and electrical components such as batteries. Nickel contamination of the environment occurs locally from emissions of metal mining, smelting, and refining operations; from combustion of fossil fuels; from industrial activities, such as nickel plating and alloy manufacturing; from land disposal of sludges, solids, and slags; and from disposal as effluents. Nickel toxicity reduces photosynthesis, growth, and nitrogenase activity of algae; fermentative activity of a mixed rumen microbiota; growth rate of marine bacteria; metabolism of soil bacteria; and mycelial growth, spore germination, and sporulation of fungi. Toxic effects of nickel to humans and laboratory mammals are documented for respiratory, cardiovascular, gastrointestinal, hematological, musculoskeletal, hepatic, renal, dermal, ocular, immunological, developmental, neurological, and reproductive systems.

1.2.4.5 Lead

Lead is one of the most dangerous contaminants which is released to the environment by various anthropogenic activities [64]. Lead poisoning is a medical condition caused by increased levels of the metallic lead in the blood. Lead may cause irreversible neurological damage as well as renal disease, cardiovascular effects, and reproductive toxicity. Humans have been mining and using this heavy metal for thousands of years, poisoning themselves in the process due to accumulation, exposure, and direct contact.

1.2.4.6 Zinc

Zinc is an essential and beneficial element for human bodies and plants. Complete exclusion of Zn is not possible due to its dual role, an essential microelement on the one hand and a toxic environmental factor on the other [65]. However, Zn can cause nonfatal fume fever, pneumonitis, and is a potential hazard as an environmental pollutant. Zinc is an essential nutrient in humans and animals that is necessary for the function of a large number of metalloenzymes.

1.2.5 Dyes

Dyes are highly dispersible aesthetic pollutants contributing to aquatic toxicity. Reminiscent or unspent dye materials are mainly responsible for the colored effluents from industries and its removal is therefore a necessary and inevitable step for the recycling and disposal of the textile industrial effluents prior to it being discharged into sewers. Dyes can be classified according to the procedures used in the dying processes or on the basis of the chromophoric groups (see Chap. 9). Synthetic dyes classified by their chromophores have different and stable chemical structures to meet various coloring requirements. Color removal from wastewater is as important as the removal of soluble colorless organic substances, which contribute the major function of chemical oxygen demand. Synthetic dyes containing various substituents like nitro and sulfonic groups are not uniformly susceptible to decomposition by activated sludge in a conventional aerobic process.

Water is the basis of life and the blue arteries of the earth! Everything in the non-marine environment depends on freshwater to survive.

–Sandra Postel, "Sandra Postel, Global Water Policy Project," *Grist Magazine* 26 Apr 04

1.3 Green Practices to Save the "Precious Water Resource"

Remediation, whether by biological, chemical, or a combination of both means, is an effective technology for combating water pollution. Remediation of the polluted sites by conventional approaches based on physicochemical methods can be both technically and economically challenging in terms of time and cost. Besides, it is very important not to cause further secondary pollution. There is a need to come up with eco-friendly and sustainable treatment technologies using renewable resources. Most of the organic pollutants are degraded or detoxificated by physical, chemical, and biological treatments before being released into the environment. Although the biological treatments are a removal process for some organic compounds, their products of biodegradation may also be hazardous. This chapter selectively examines and provides a critical view on the knowledge gaps and limitations in utilizing greener technologies as a viable solution to the problem of pollution without transferring secondary pollution to the future. It discusses the field application strategies, approaches such as biosorption, bioremediation, microwave, and ultrasonic technology for monitoring and testing the efficacy of remediation of polluted sites.

1.3.1 Biosorption of Heavy Metals

Industrial wastewater containing heavy metals is a threat to public health because of the accumulation of the heavy metals in the aquatic life, which is transferred to human bodies through the food chain. All the more, nowadays, an increasing number of hazardous organic compounds together with variable levels of heavy metals ions are also being discharged into the environment [66]. Conventional methods for the removal of the heavy metals ions from wastewater include chemical precipitation, electrofloatation, ion exchange, reverse osmosis, and adsorption onto activated carbon. But due to operational demerits, high cost of the treatment, and the generation of toxic chemical sludges, some new technologies have been tried for a long time. Recently attention has been diverted toward the biomaterials which are by-products or the wastes from large-scale industrial operations and agricultural waste materials. Among them, less expensive, nonconventional adsorbents like apple waste, peanut hull carbon, agricultural wastes, and red mud are being investigated for the removal of ions like the Cd and Ni ions. Adsorption of heavy metals onto suspended particles has been studied as a model of transportation of metals in rivers and sea. The effects of chemical composition and particle size on adsorption by suspended particles has also been studied. Attention had mostly been on carbon as adsorption materials.

Biosorption is the binding and concentration of adsorbate(s) from aqueous solutions (even very dilute ones) by certain types of inactive, dead, microbial biomass. Some biosorbents can bind and collect a wide range of heavy metals with no specific priority, whereas others are specific for certain types of metals. In the concept of biosorption, several chemical processes may be involved, such as bioaccumulation, bioadsorption, precipitation by H_2S production, ion exchange, and covalent binding with the biosorptive sites, including carboxyl, hydroxyl, sulphydryl, amino, and phosphate groups of the microorganisms. Research on biosorption shows that it is sometimes a complex phenomenon where the metallic species could be deposited in the solid biosorbent through various sorption processes such as complexation, chelation, microprecipitation, and oxidation/reduction. There is a significant research and practical interest to develop the methods that can effectively remove the heavy metals and at the same time recover them in their pure form for the potential reuse to avoid the secondary pollution problems of these heavy metals. The major advantages of biosorption over conventional treatment methods include low cost, high efficiency, minimization of chemical or biological sludge, regeneration of biosorbents, and possibility of metal recovery. Another powerful technology is adsorption of heavy metals by activated carbon for treating domestic and industrial wastewater. Active research on biosorption of heavy metals, intrinsically guided by the emerging concept of *Green Chemistry*, has led to the identification of a number of microbial biomass types that are extremely effective in concentrating metals. *Green chemistry* (environmentally benign chemistry) is the utilization of set of principles that reduces or eliminates the use or generation of hazardous substances in the design, manufacture, and application of chemical products [67]. The sections to follow highlight the pollution and toxicity characteristics of some selected heavy metals (Cd, Cr, Cu, Ni, Pb, and Zn) and summarize some selected biosorbents studied for their ability to remove heavy metal ions from synthetic and/or natural contaminated aqueous media (Table 1.1). Heavy metal removal through biosorptive pathways is also discussed in Chap. 7.

1.3.2 Bioremediation

Bioremediation, which is defined as a process that uses microorganisms, green plants, or their enzymes to treat the polluted sites for regaining their original condition has considerable strength and certain limitations. Bioremediation that involves the capabilities of microorganisms in the removal of pollutants is the most promising, relatively efficient and cost-effective, economic, versatile, and environmentally sound solution [83] technology.

Table 1.1 Biosorbents used for heavy metal removal from wastewater and synthetic solutions

Metal	Biosorbent(s)	Biosorption performance	Reference
Cd	Black gram husk (*Cicer arietinum*)	With 99.99% sorption efficiency from 10 mg/L cadmium solution, the biomass required at saturation was 0.8 g/mg cadmium. Biosorption was rapid and equilibrium was achieved in 30 min	[68]
Cd	Rice polish agricultural waste	The maximum removal of Cd^{2+} was found to be 9.72 mg/g at pH 8.6, initial Cd^{2+} concentration of 125 mg/L (20°C)	[69]
Cd	Orange wastes from orange juice production processes	Adsorption process was quick and the equilibrium was attained within 3 h. The maximum adsorption capacity of orange waste was found to be 0.40, 0.41, and 0.43 mmol/g at pH 4–6, respectively	[70]
Cd	Ethanol-treated waste baker's yeast biomass	Highest metal uptake of 15.63 mg Cd^{2+}/g	[71]
Cd	Red algae (*Ceramium virgatum*)	Biosorption capacity for Cd^{2+} ions was 39.7 mg/g	[72]
Cr	*Neurospora crassa* fungal biomass	Biosorption capacity of acetic acid-pretreated biomass was found to be 15.85 ± 0.94 mg/g biomass under optimum conditions. The adsorption constants were found from the Freundlich isotherm model at 25°C. The biosorbent was regenerated using 10 mM NaOH solution with up to 95% recovery and reused five times in biosorption–desorption cycles successively	[73]
Cr	Mucilaginous seeds of *Ocimum basilicum*	Seeds boiled in water were found to be superior in terms of mechanical stability and exhibited fairly optimal Cr^{6+} uptake kinetics. Maximum adsorption capacity from Langmuir isotherm was 205 mg Cr/g dry seeds	[74]
Cr	Palm flower (*Borassus aethiopum*)	For Cr^{3+}, maximum adsorption capacity was 6.24 mg/g by raw adsorbent and 1.41 mg/g by acid-treated adsorbent. For Cr^{6+}, raw adsorbent exhibited a maximum adsorption capacity of 4.9 mg/g, whereas the maximum adsorption capacity for acid-treated adsorbent was 7.13 mg/g. There was a significant difference in the concentrations of Cr^{6+} and total chromium removed by palm flower	[75]

Cr	Shells of walnut (WNS) *(Juglans regia)*, hazelnut (HNS) *(Corylus avellana)* and almond (AS) *(Prunus dulcis)*	Langmuir isotherm with maximum Cr^{6+} ion sorption capacities of 8.01, 8.28, and 3.40 mg g^{-1} for WNS, HNS, and AS, respectively. Percentage removal by WNS, HNS, and AS was 85.32%, 88.46%, and 55.00%, respectively, at a concentration of 0.5 mmol/L	[76]
Cr	*Turbinaria ornata* seaweed	Brown seaweed (*Turbinaria* spp.) was pretreated with sulfuric acid, calcium chloride, and magnesium chloride and tested for its ability to remove chromium from tannery wastewater. *Urbinaria* weed exhibited maximum uptake of about 31 mg of chromium for 1 g of seaweed at an initial concentration of 1,000 mg/L of chromium. Freundlich and Langmuir adsorption isotherm models were used to describe the biosorption of Cr^{3+} by *Turbinaria* spp.	[77]
Cr	*Helianthus annuus* (sunflower) stem waste	Maximum metal removal was observed at pH 2.0. The efficiencies of boiled sunflower stem absorbent and formaldehyde-treated sunflower stem absorbent for the removal of Cr^{6+} were 81.7% and 76.5%, respectively, for dilute solutions at 4.0 g/L adsorbent dose	[78]
Cu	*Neurospora crassa* fungal biomass	Biosorption capacity of acetic acid pretreated biomass was found to be 15.85 ± 0.94 mg/g biomass under optimum conditions. The adsorption constants were found from the Freundlich isotherm model at 25°C. The biosorbent was regenerated using 10 mM NaOH solution with up to 95% recovery and reused five times in biosorption–desorption cycles successively	[73]
Cu	Mucilaginous seeds of *Ocimum basilicum*	Seeds boiled in water were found to be superior in terms of mechanical stability and exhibited fairly optimal Cr^{6+} uptake kinetics. Maximum adsorption capacity from Langmuir isotherm was 205 mg Cr/g dry seeds	[74]
Cu	Palm flower (*Borassus aethiopum*)	For Cr^{3+}, maximum adsorption capacity was 6.24 mg/g by raw adsorbent and 1.41 mg/g by acid-treated adsorbent. For Cr^{6+}, raw adsorbent exhibited a maximum adsorption capacity of 4.9 mg/g, whereas the maximum adsorption capacity for acid-treated adsorbent was 7.13 mg/g. There was a significant difference in the concentrations of Cr^{6+} and total chromium removed by palm flower	[75]

(continued)

Table 1.1 (continued)

Metal	Biosorbent(s)	Biosorption performance	Reference
Cu	Shells of walnut (WNS) *(Juglans regia)*, hazelnut (HNS) *(Corylus avellana)* and almond (AS) *(Prunus dulcis)*	Langmuir isotherm with maximum Cr^{6+} ion sorption capacities of 8.01, 8.28, and 3.40 mg g^{-1} for WNS, HNS, and AS, respectively. Percentage removal by WNS, HNS, and AS was 85.32%, 88.46%, and 55.00%, respectively, at a concentration of 0.5 mmol/L	[76]
Cu	*Turbinaria ornata* seaweed	Brown seaweed (*Turbinaria* spp.) was pretreated with sulfuric acid, calcium chloride, and magnesium chloride and tested for its ability to remove chromium from tannery wastewater. T*urbinaria* weed exhibited maximum uptake of about 31 mg of chromium for one gram of seaweed at an initial concentration of 1,000 mg/L of chromium. Freundlich and Langmuir adsorption isotherm models were used to describe the biosorption of Cr^{3+} by *Turbinaria* spp.	[77]
Cu	*Helianthus annuus* (sunflower) stem waste	Maximum metal removal was observed at pH 2.0. The efficiencies of boiled sunflower stem absorbent and formaldehyde-treated sunflower stem absorbent for the removal of Cr^{6+} were 81.7% and 76.5%, respectively, for dilute solutions at 4.0 g/L adsorbent dose	[78]
Pb	Poly(2-octadecyl-butanedioic acid), prepared from polyanhydride PA-18	Adsorption capacity of this water-insoluble polymer for lead (II) was substantially higher than other heterogeneous adsorbants and is equivalent to those obtained with homogeneous sorbents	[79]
Pb	Clay/poly(methoxyethyl) acrylamide (PMEA) composite	Maximum adsorption capacity 81.02 mg/g	[80]
Pb	Diethylenetriamine–bacterial cellulose	Study provides the relatively comprehensive data for the EABC application to the removal of metal ion in the wastewater	[81]
Pb	Native and acidically modified *Cicer arientinum* pod biomass	171.28 mg/g for acid-treated biomass	[82]

1.3.2.1 Microbial Remediation

The use of microbial biomass for the removal of toxic pollutants like heavy metal ions from wastewater has emerged as an alternative to the existing conventional methods as a result of the search of low-cost, innovative methods. The disadvantages of conventional methods such as incomplete removal, high cost due to high reagent and energy requirements, and generate toxic sludge or other waste products that requires disposal can be eliminated by the microbial remediation green methods. Microorganisms, which are capable of transforming metals from one oxidation state to another, facilitate detoxification and/or the removal of metal ion, and have thus received recognition.

Because of the ubiquitous nature of microorganisms, their numbers and large biomass, wider diversity and capabilities in their catalytic mechanisms [84], and their ability to function even in the absence of oxygen and other extreme conditions, the search for pollutant-degrading microorganisms like bacteria and fungus, and developing methods for their application in the field have become an important human endeavor. Their capabilities to degrade organic chemical compounds can be used to attenuate the polluted sites. Many biomaterials such as seaweed micro-algae fungi and various other plant materials [85, 86] have been studied for their metal binding abilities. Among these, fungi have the advantage of producing small residual volume, fast removal, and easy installation of the process and possibility of valorization of fungal waste biomasses from industrial fermentations. Particularly, fungal biomass can be cheaply and easily procured in rather substantial quantities, as a by-product from established industrial fermentation processes. Live or dead microbial cells can be used as an adsorbent material for the removal of toxic metal ions from aqueous solutions. The efficiency of dead cells in biosorbing metal ions may be greater, equivalent to, or less than that of living cells and may depend on factors such as the microbe under consideration, pretreatment method used, and type of metal ion being studied.

Many azo dyes, constituting the largest dye groups may be decomposed into potential carcinogenic amines under anaerobic conditions in the environment. Attempts to develop suitable biological methods to decolorize these effluents have not been very successful.

Because synthetic dyestuffs are resistant to biological degradation, color removal by bioprocessing is difficult. Moreover, the frequently high volumetric rate of industrial effluent discharge in combination with increasingly stringent legislation makes the search for appropriate treatment technologies an important priority. By far, the class of microorganisms most efficient in breaking down synthetic dyes are the bacteria and fungus. Bacterial anaerobic reduction of azo dyes generates colorless, dead end, and aromatic amine, which can be more toxic than the parent dye. Bacterial aerobic dye degradation has been confined to chemostat-enriched cultures adapted to a single dye. Besides, decolorization by aerobic bacteria occurs mainly by adsorption of dyestuff on the cell surface rather than biodegradation; therefore, low color removal efficiencies have been obtained by conventional aerobic biological treatment systems. Expensive chemical or physical steps prior to aerobic biological treatment are currently used. Abiotic means of reduction of azo and other dyes exist but require highly expensive catalysts and reagents [87]. A number of biotechnological approaches have been suggested by recent research as of potential interest toward combating this pollution source in an eco-efficient manner, including the use of bacteria or fungi, often in combination with physicochemical processes [88–91].

The White-rot fungi (WRF) constitute a diverse ecophysiological group comprising mostly basidiomycetous (and, to a lesser extent, litter-decomposing) fungi capable of extensive aerobic lignin depolymerisation and mineralization. This property is based on the WRF's capacity to produce one or more extracellular lignin-modifying enzymes (LME), which, thanks to their lack of substrate specificity, are also capable of degrading a wide range of xenobiotics. White-rot fungi (WRF) can degrade a wide variety of recalcitrant compounds like lignin by their extracellular lignolytic enzyme system. The enzymes which allow WRF to degrade lignin also allows them to degrade a wide range of hazardous xenobiotics [92] ranging from polycyclic aromatic hydrocarbons (PAHs) [93], chlorophenols polychlorinated biphenyls [94], organochlorine pesticides [95], nitrotoluenes, fuel additives [96], and dyes. WRF offer significant advantages as the ligninolytic enzymes produced by them are substrate nonspecific, and therefore can degrade a wide variety of recalcitrant compounds, especially complex aromatic pollutants like Creosote and Aroclor all the way to carbon dioxide. WRF do not require preconditioning to particular pollutants, because enzyme secretion depends on nutrient limitation nitrogen or carbon rather than the presence of a pollutant. In addition, because the enzymes are extracellular, the substrate diffusion limitation into the cell, generally observed in bacteria, is not encountered and also they enable the WRF to tolerate higher concentration of pollutants such as cyanide. Besides, the fungi can even degrade very insoluble chemicals and hazardous environmental pollutants. WRF have a remarkable ability to degrade lignin, the complex three-dimensional structural polymer found in woody plants. The stereo irregularity of lignin makes it very resistant to attack by enzymes. The chiral carbons existing in both L and D configurations in lignin make it impossible for lignin to be absorbed and/or degraded by intracellular enzymes. White rot fungi produce oxidase enzymes that utilize glucose, glyoxal, methyl glyoxal, and other products of cellulose and lignin degradation as substrates for the production of hydrogen peroxide from molecular oxygen. Interestingly, white rot fungi do not use lignin as a carbon source for growth, but degrade the lignin to obtain the cellulose found in the wood fiber. Very inexpensive growth substrates such as corn cobs or other crop residues, wood chips, or surplus grains can be used to cultivate WRF in soil. White rot fungi can be easily grown in liquid culture too. Wastewater treatment with concomitant bioenergy production using microbial fuel cells has been discussed in more detail in Chap. 18 of this book.

1.3.2.2 Phytoremediation

Phytoremediation has been suggested as an effective and low-cost method to clean up contaminated soils and aquifers. The term phytoremediation ("phyto" meaning plant, and the Latin suffix "remedium" meaning to clean or restore) refers to a diverse collection of plant-based technologies that use either naturally occurring, or genetically engineered, plants to clean contaminated environments. Some plants which grow on metalliferous soils have developed the ability to accumulate massive amounts of indigenous metals in their tissues without symptoms of toxicity [97, 98]. The idea of using plants to extract metals from contaminated soil was reintroduced. Several comprehensive reviews have been written, summarizing many important aspects of this novel plant-based technology [99–105]. The use of plants for remediation of metals offers an attractive alternative because it is solar-driven and can be carried out in situ, minimizing cost and human exposure. Phytoremediation promotes the use of plants for environmental cleanup. Phytoremediation is a relatively new approach to the cost-effective treatment of wastewater, groundwater, and soils contaminated by organic xenobiotics, heavy metals, and radionuclides. Phytoremediation uses the green plants and their associated microbiota, soil amendments, and agronomic techniques to remove, contain, or render harmless environmental contaminants. It is an emerging technology which offers a potentially cost-effective and environmentally sound alternative to the environmentally destructive physical methods which are currently practiced for the cleanup of contaminated groundwater, terrestrial soils, sediments, and sludge.

Phytoremediation can be used to improve the quality of the secondary effluent depending on plant species. The interested plant species characteristics should be the native-salt resistance species and easy to find in the area. In addition, it will be very beneficial if these plant species can be used as raw materials for local handicraft (www.chaipat.or.th) and for enhancing landscape management of the factories for recreation area. Operation of phytoremediation is simple, uncomplicated, with low maintenancecost. The phytoremediation process consists of four steps: soil physical properties conditioning, soil chemical and biological properties conditioning, sustainable soil property improvement by using plants, and monitoring of contaminant migration. The mechanisms and effectiveness of the system can be evaluated from the effluent, groundwater discharges from the area and soil properties. In addition, a mathematical model to predict the contaminant transport in the system can be developed to enhance the theoretical knowledge of the process.

Constructed Wetland Treatment Technology

Constructed wetlands are artificial wastewater technology (CWs) consisting of shallow ponds or and energy savings, CWs have other features related to channels which have been planted with aquatic plants the environmental protection such as promoting and which rely upon natural microbial, biological, physical biodiversity, providing habitat for wetland organisms and chemical process to treat wastewater.

Phytoremediation is marginally different from the constructed wetland treatment technology, as it involves the use of living green plants for in situ risk reduction of contaminated soil, sludge, sediment, and groundwater through contaminant removal, degradation, limited to in situ cleanup areas that have been contaminated by past use. In contrast, the constructed wetland treatment technology is the involvement of living plants for ex situ cleanup of a steady flow of wastewater. In a broader sense, however, wetland treatment technology also falls under phytoremediation, since both technologies take advantage of primary producers (i.e., photosynthetic plants or other autotrophic organisms in either terrestrial or aquatic forms) to clean up and manage hazardous and nonhazardous contaminants, regardless of the fashion (i.e., in situ or ex situ) of application. Actually, the first documented plant-based system installed in Germany over 300 years ago was designed to remove contaminants from municipal wastewater. Since then, common designs such as reed-bed filters, natural and constructed wetlands, and floating plant treatment systems have been actively developed; these designs were primarily intended for purifying municipal sewage. In the past two decades, the initial concept of using plants in wastewater treatment has been expanded to remediate contaminated shallow groundwater, air, soil, and more recently, sediment and sludge. Phytoremediation integrated with Wetland treatment has been also proposed to remove total dissolved solids contaminated in the effluent discharged from the pulp and paper mill wastewater treatment plant.

Use of Macrophytes in Water Purification

Macrophytes are potent tools in the abatement of heavy metal pollution in aquatic ecosystems receiving industrial effluents and municipal wastewater. They are preferred over other bio-agents due to low cost, frequent abundance in aquatic ecosystems, and easy handling. Some aquatic plants (macrophytes) including *Eichhornia crassipes*, *Ludwigia* sp., *Salvinia*, *Hydrilla* sp., *Egeria, Typha* sp., and *Phragmtes* are very useful in the purification of the water and may be used as an eco-friendly alternate because they have excellent capacity of absorbing nutrients and other substances from the water [106]. Aquatic macrophytes in natural and constructed wetlands proved to be a potent tool for the treatment of heavy metals from industrial effluents. Several studies had been conducted globally by many researchers using the same fundamental [107–113]. Many of them experimentally proved the use of plants to remove organic contamination from water. Macrophytes are very useful in nutrient and heavy metal recycling of various ecosystems [114] and they are present in water bodies globally. Roots and leaves of aquatic plants take part in removing heavy metals and nutrients from water and wastewater. *Eichhornia crassipes* (a free-floating microphyte) is present in a great abundance throughout the world and one of the worst types of weeds. It is quite reliable for removing heavy metals and nutrients from water and wastewater [115–120] by reducing BOD, removing heavy metals like Pb, Cr, Cu, Cd, and Zn. Its roots act like filters removing suspended particles from water and decreasing turbidity [121–124].

Both live and dead biomass of macrophytes may be used in phytoremediation, though dead biomass is generally preferred in the treatment of industrial effluents due to reduced cost, easy disposal, and lack of active biochemical machinery leading to metal toxicity and death of plants. Biomass disposal problem and seasonal growth of aquatic macrophytes are some of the limitations in the transfer of phytoremediation technology from the lab to the field. Disposed biomass of macrophytes may be used for many fruitful applications. Genetic engineering, biodiversity prospecting, and X-ray diffraction spectroscopy are promising future prospects regarding the use of macrophytes in phytoremediation studies. A multidisciplinary and integrated approach may enable this embryonic technology to become the new frontier in environmental science and technology [125].

1.3.3 Photocatalytic Water Treatment Technology

Rural wastewater or treated municipal water may be used for agricultural and industrial activities [126, 127]. But, recycled wastewater is normally infected by the presence of suspended solids, dangerous coloforms, and soluble organic compounds that are expensive and tedious job to treat [128]. Current water treatment technologies like adsorption and coagulation are not useful for complete elimination of the pollutants [129], and sedimentation, filteration, membrane technology, etc. are expensive and generate toxic secondary pollutants to the water bodies [130]. On the other hand, the disinfection by-products generated in chlorination are mutagenic and carciogenic to human being [131–133]. This scenario has led to the promotion of advanced oxidation processes (AOPs) as innovative water-treatment technologies. In this connection, semiconductor photocatalytic process has proved itself a great mean as a low-cost, eco-friendly, and sustainable treatment technology producing zero waste in water/wastewater treatment science by employing semiconductor catalysts like TiO_2, ZnO, Fe_2O_3, CdS, GaP, and ZnS. Producing zero waste proves this technique as the greenest technology. To date, the most widely used photocatalyst in the water treatment is the Degussa P–25 TiO_2 catalyst [134].

1.3.4 Ultrasound Technology for Pollutant Degradation

Ultrasound is an attractive technology used in water and wastewater treatment for degradation of recalcitrant anthropogenic organic contaminants and natural organic matter, disinfection, and control of membrane fouling in water-filtration processes. The effects of ultrasound are combinations of both sonochemical and sonophysical effects. For sonochemical effects, cavitational bubble collapse produced hot spots can thermolyze contaminants along water vapor and generate •OH radicals to oxidize contaminants. For sonophysical effects, the turbulence produced by sound waves (acoustic streaming and vibration), cavitational bubbles (microstreaming and microstreamers), and bubble collapse (shock waves and microjets) can enhance mixing, cleaning, and break down particles and macromolecules. The mechanisms of ultrasound make it unique compared with other physical and chemical processes. The advantages of ultrasound include potential chemical-free and simultaneous oxidation, thermolysis, shear degradation, and enhanced mass transfer processes together.

Ultrasound is a longitudinal wave with a frequency typically between 16 kHz and 500 MHz [135, 136]. When ultrasound is introduced into liquid (e.g., water), it creates oscillating regions of positive and negative pressure. Correspondingly, the liquid molecules experience periodic compression and expansion cycles. When the pressure amplitude exceeds the tensile strength of liquid during the rarefaction of ultrasonic waves, cavitational bubbles are formed. Cavitational bubbles collapse during the compression cycle of ultrasonic wave. Localized hot spots are formed which reach temperatures and pressures around 5,000 K and 500 atm, respectively [137], depending on factors such as ultrasonic power, frequency, hydrostatic pressure, temperature, solvent property, and dissolved gas. According to the temperature profile, there are three zones associated with a cavitational bubble: (i) Thermolytic center, which is the core of the bubble with localized hot temperature ($\sim$5,000 K) and high pressure ($\sim$500 atm) during final collapse of cavitation. The high temperature results in thermolysis of volatile chemical compounds and water vapor producing radical species including •OH and •H radical. The reactions between volatile chemical compounds and •OH radicals in gaseous phase also happen in this region. (ii) Interfacial region between the cavitational bubble and bulk liquid. In this region, the thickness of the liquid is estimated about 200 nm from the bubble surface to the bulk and a lifetime is less than 2 μs. The temperature is about 2,000 K at the final cavitational collapse [138]. There are vast gradients of temperature and pressure. Thermolysis and oxidation by •OH radicals of hydrophobic, nonvolatile, and hydrophilic compounds occur in the region. In this interfacial region, hydrophobic compounds are more concentrated than bulk solution, and hydrophilic compounds have the same concentration as the bulk solution. (iii) The bulk region with ambient temperature and pressure. The self-combination of •OH radicals produces hydrogen peroxide in the solution (reaction given below) along with a small amount of •OH radicals that react with hydrophilic compounds including ionic compounds in bulk region.

$$2 \bullet OH \rightarrow H_2O_2$$

•OH radicals can oxidize broad organic pollutants in water, similar to advanced oxidation processes (AOPs). Since •OH radicals are produced as a result of cavitational collapse, they spatially concentrate inside the bubble and in the region of bubble–bulk interface during collapse. Consequently, volatile and hydrophobic compounds or parts are exposed to more •OH radicals in addition to thermolysis than hydrophilic ones and thus subject to more rapid degradation, because they tend to accumulate in these two regions.

Recently, more research has been directed toward ultrasound application in organic pollutant remediation at a laboratory scale and the results are very promising. Degradation kinetics, reaction rate, concentration range, pollution composition, reactor configuration, tested frequency, temperature, pressure, intensity, bubbling gas, chemical used, conversion path way, treatment time, and achieved removal efficiency have been reported for many pollutants. Table 1.2 presents the degradation performance of certain selected organic pollutants under ultrasonic enhanced degradation. It clearly shows that almost a large variety of organic pollutants can be remediated by using ultrasound at the laboratory scale under various experimentally imposed degradation conditions.

Perfluorinated chemicals such as perfluorooctane sulfonate (PFOS) and perfluorooctanoate (PFOA) are globally distributed, bioaccumulative, metabolically and photochemically inert, and oxidatively recalcitrant [145, 146]. In the research conducted, 200 kHz sonication cleaved the perfluorocarbon chains [145]. The half-life of PFOS and PFOA degradations was 43 and 22 min, respectively, under the atmosphere of argon. Because PFOS and PFOA molecules have both hydrophobic (perfluoroalkyl group) and hydrophilic group (acid group), they behave like an anionic surfactant: The hydrophobic groups migrate and accumulate in the bubble – bulk interfacial region and are subject to pyrolysis (or thermolysis) and •OH radical attack. Since PFOS and PFOA are nonvolatile, the pyrolysis in the gaseous phase of cavitation should be ruled out. A faster degradation rate of CCl_4 occurred at 500 kHz than 20 kHz [147]. Sonication led to almost complete mineralization of CCl_4 for a relatively short irradiation time. Radical trap addition did not change the reaction rate, since thermolysis is the mechanism of sonolytic degradation. Similarly, [148] confirmed addition of $t-$BuOH, a hydroxyl radical scavenger did not affect the degradation rate of carbon tetrachloride.

Ultrasound can also be used to degrade environmental emerging contaminants, such as pharmaceutical and personal care products, especially endocrine disruptor compounds. The conventional microbiological processes used in municipal wastewater treatment plants are not designed or capable to remove these contaminants. The sonolytic destruction of estrogen hormones in aqueous solution, including 17α-estradiol, 17β-estradiol, estrone, estriol, equilin, 17α-dihydroequilin, 17α-ethinyl estradiol, and norgestrel was investigated [149]. The results showed that 20 kHz sonolysis destructed 80–90% of individual estrogens at an initial concentration of 10 $\mu g\ L^{-1}$ within 40–60 min of reaction. The first-order degradation rate constant of individual estrogen increased with higher power intensity. However, the energy efficiency of the reactor was higher at lower power density. As a result, the choice of reactor and ultrasonic power were important to achieve optimized kinetics and energy efficiency.

The sonolysis of pharmaceutical compounds of levodopa and paracetamol in aqueous solutions was investigated [150]. In a study, the degradation kinetics and intermediate by-products of MTBE degradation were investigated with 20 kHz ultrasound [151]. The observed pseudo first-order rate constant decreased from

Table 1.2 Degradation performance of certain selected organic pollutants under ultrasonic enhanced degradation

Contaminant	Experimental conditions	Degradation efficiency/other result/remarks	Reference
4–Chlorophenol	28, 50, 100, 200, and 600 kHz disk fitted on bottom of the 1 1 50 W calorimetric power, 126.73, 126.73, 50.3, 50.3, and 50.3 cm^2 are the radiating face, respectively, reactor jacketed at 20–30°C	Removal efficiency (1) 600 kHz > 200 kHz > 50 kHz > 28 kHz (2) 100 mg/L of H_2O_2 gives higher degradation rate (3) 1 $g L^{-1}$ iron powder with smaller particle size gives good performance (4) pH = 3 with 6 μmol L^{-1} particle size and 100 mg L^{-1} H_2O_2 gives best result	[139]
Pentachlorophenol (PCP)	500 Hz, 20 kW, 3 L, water jacketed to avoid variation of temp below 3°C, 0.165 W mL^{-1}	(1) Sonication helps to dissolve more O_3 than non-sonication and stirring (2) Ozonation with sonication enhance the degradation rate of PCP (3) Audible frequency increases the rate constant than ultrasonication because audible frequency produces strong turbulence. 90% PCP of 100 mg L^{-1} degrade in 2 min under 24 mg min^{-1} ozone feed, pH 9.4 and sonication	[140]
Tetrachloride trichloroethylene (TCE) 1,2,3-tri-chloropropane	20 kHz horn, 60 W, 1.3278 cm^2 radiating area, 350 mL water-jacketed reactor	For TCE, (1) When the temp increases, rate constant reduces and half-life increases (2) At 10°C, half-life is 60 min (3) When the power intensity increases, the rate constant increases and half-life reduces (4) Ar > air > N_2 The main mechanism is pyrolysis in the bubble	[141]

$CHCl_3$ CCl_4 $CHBrCl_2$ CHBrCl	20 kHz horn, 500 W, 55 W calorimetric power, 0.2828 cm^2 vibrating area, 50 mL water-jacketed reactor at 25°C	Removal efficiency in 1 h is, component (removal efficiency), $CHCl_3$(48.2%), CCl_4(64.6%), $CHCl_2Br$(58.3) and $CHClBr_2$(54.6) degradation rate from a mixture is $CCl_4 > CHCl_2Br > CHClBr_2 > CHCl_3$	[142]
Ammonia	(a) 1.7 MHz plate transducer fitted on the bottom of the 400 mL reactor, 9.5 W, 3.1428 cm^2 vibrating area, no temperature control (18–33°C) (b) 2.4 MHz plate transducer fitted on the bottom of the 400 mL reactor, 9.5 W, 3.1428 cm^2 vibrating area, no temp control	(1) Ammonia removal efficiency was higher at 2.4 MHz than 1.7 MHz for all concentration (2) Removal efficiency reduces with liquid level increment and also with increment in ammonia concentration (3) In 2 h sonication, the maximum ammonia removal efficiency obtained was 32% by using 2.4 MHz US with 5% ammonia concentration.	[143]
DDT & PCB	20 KHz horn, 170 W, 1.23 cm^2 vibrating face, 50% sand slurry of 400 mL 100 Hz, 75 kW, 50% sand slurry	(a) 20 kHz (1) In 10 min, DDT dropped by 70% from 250 mg L^{-1} of DDT from 50% sand slurry (2) In 60 min, PCB dropped by 90% from 250 mg L^{-1} of PCB from 50% sand slurry (3) In 5 min, PAH dropped by 80% from 400 mg L^{-1} of PAH from 50% soil slurry (b) 100 Hz 700 mg L^{-1} of PCB dropped to < 2 mg L^{-1} in 105 min and the power was 0.2 kW L^{-1}	[144]

1.25×10^{-4} to 5.32×10^{-5} s^{-1} as the concentration of MTBE increased from 2.84×10^{-2} to 2.84×10^{-1} mmol L^{-1}. The rate of degradation of MTBE increased with increasing power density of ultrasound and also with the rise in system temperature. In the presence of an oxidizing agent, potassium persulpfate, the sonolytic degradation rate of MTBE was accelerated substantially. *Tert*-butyl formate and acetone were found to be the major intermediates of the degradation of MTBE. It was also found that the ultrasound coupled with Fenton reagent (Fe^{2+}/H_2O_2) effectively degraded more than 95% of MTBE (2.84×10^{-2} mmol L^{-1}) along with its intermediate products in 5 h, which was much more efficient than sonolysis alone.

Ultrasonic irradiation involves generation of intense ultraviolet wavelength light which can cause the transition of an electron from the valence band into the conduction band in the TiO_2 semiconductor, leaving a positive hole charge responsible of dissociation of water molecules to form hydroxyl free radicals. Addition of hydrogen peroxide to the US/TiO_2 system could promote further benefits to the degradation efficiency. It was proposed that under ultrasonic irradiation, hydrogen peroxide can modify the TiO_2 surface promoting the formation of titanium peroxide (TiO_3) into the solution, which is active for oxidation reactions [152]. It was demonstrated that the degradation of methylene blue by an US/TiO_2 system occurs by hydroxyl radicals generated in the process and the addition of H_2O_2 accelerated the degradation of the dye for the TiO_2-containing system up to an optimum value [153]. Recently, Abdullah and Ling [154] found that anatase TiO_2 with certain amount of rutile phase exhibited better sonocatalytic performance than pure anatase. It was attributed to the role of rutile to avoid hole–electron recombination. Authors also remarked that the physical effect of ultrasound for the disintegration of the TiO_2 particles might reduce the micro and mesoporosity leading to lower surface area. On the other hand, the above-mentioned reactivation of the sonocatalyst particles due to the cleaning effect of the ultrasonic streams enhances the efficiency of the degradation. The net effect of both is accounted for the activity of the US/TiO_2 system [154].

Most of the oxidation technologies for wastewater degradation is barely used individually for cost-effective operations, and this also includes sonocatalysis. Although photocatalysis and sonolysis have been extensively employed individually for the degradation of several organic species in water, their combined use (i.e., sonophotocatalysis) has received noticeably less attention. This combination is especially attractive as ultrasound irradiation seems to overcome some drawbacks associated to photocatalysis such as the mass transfer limitations and fouling of the solid catalyst. Promising results of synergistic effects of the integrated US–UV processes have been obtained in the degradation of model pollutant, such as salicylic acid, 2–chlorophenol, acid formic, methyl tert-butyl ether, and diazo dyes. As previously mentioned, the action of US toward UV is summarized by the increment of the catalyst surface area due to the de-aggregation action of US, the improvement of mass transfer of organic compounds between the liquid phase and the catalysis surface, and the promotion of additional hydroxyl radicals due to the residual hydrogen peroxide generated.

1.3.5 Microwave Technology for Pollutant Degradation

Microwaves (frequencies of 0.3–300 GHz and wavelengths of 1 m to 1 mm) lie between radio wave frequencies and infrared frequencies in the electromagnetic spectrum. Microwaves can be reflected, transmitted, and/or absorbed. The absorbed MW energy is converted into heat within the material, resulting in an increase in temperature. Gases, liquids, and solids can interact with microwaves and be heated. An important characteristic of microwave heating is the phenomenon of "hotspot" formation, whereby regions of very high temperature form due to nonuniform heating [155]. The formation of standing waves within the microwave cavity results in some regions being exposed to higher energy than others. This results in an increased rate of heating in these higher energy areas due to the nonlinear dependence.

Microwave energy has been widely used in several domestic, industrial, and medical applications such as food sterilization, organic/inorganic syntheses, polymerization, dehydration, analyses and extraction, and biological destruction [156]. The potential applications of microwave energy as remedial alternatives for various types of wastes and diverse contamination of soils, sludge, or wastewater have also been a growing field of research during the last decade [157]. The application of microwave energy has proven to be superior to the use of conventional heating on accelerating rates of biochemical/chemical reactions, improving yields, selectively activating or suppressing reaction pathways, allowing for greater control of the heating or drying process, and reducing equipment size and wastes and environmental remediation [158].

Microwave power has been applied widely in the remediation of contaminated soils, treatment of wastes, oil and water separation, reduction of nitrogen oxides and sulfur dioxide, polymer synthesis, organic reactions, and regeneration of activated carbon. Research carried out so far in this respect and with regard to the applicability of microwave irradiation has shown that microwave technology may be effective in treating wastewater at the various stages of the treatment process and hence improving the degradation of a variety of organic pollutants. A survey of the literature shows that there are three major types of research areas in the use of microwave irradiation for the degradation of organic pollutants. These are stand-alone microwave-assisted degradation processes, combined microwave-assisted degradation processes, and microwave-assisted catalytic degradation processes. In stand-alone processes, the microwave irradiation is the only kind of treatment conditions imposed on the contaminants. In the combined process (hybrid process), microwave irradiation is coupled or performed in the presence of other degradation processes like ozonation, electrokinetic remediation, advanced oxidation, ultraviolet illumination, or sonolysis [159]. In the microwave-assisted catalytic degradation processes, the contaminants are degraded in the presence of a catalyst, either hetereogeneous or homogeneous. Most of the studies in this category have reported the use of hetereogeous catalysts coupled with photocatalysis. Other researchers have also reported some salient findings and inferences on the advantages and possible mechanisms involved in the use of microwave irradiation for organic pollutant degradation. More elaborate discussions on the application of microwave in the degradation of pollutants in wastewater are presented in Chap. 10.

1.3.6 Nanotechnology and Wastewater Treatment

Nanotechnology, the engineering and art of manipulating matter at the nanoscale (1–100 nm), offers the potential of novel nanomaterials for treatment of surface water, groundwater, and wastewater contaminated by toxic metal ions, organic and inorganic solutes, and microorganisms. Due to their unique activity toward recalcitrant contaminants and application flexibility, many nanomaterials are under active research and development [160]. In the environmental technology industry alone, nanomaterials (nanostructured catalytic membranes, nanosorbents, nanocatalysts, and bioactive nanoparticles) will enable new means of reducing the production of industrial wastes, using resources more sparingly, and remediating industrial contamination, providing potable water, and improving the efficiency of energy production [161]. Nanoscale control of membrane architecture may yield membranes of greater selectivity and lower cost in both water treatment and water fabrication. Fullerene-based oxidant nanomaterials such as C_{60} have a high electron affinity and reactivity, and are capable of producing reactive oxygen species such as singlet oxygen and superoxides. Fullerenes might be used in engineered systems to photocatalytically oxidize organic contaminants, or inhibit or inactivate microbes. As nanochemistry emerges as an important force behind new environmental technologies, the responsibility of considering the environmental implications of an emerging technology at its inception and taking every precaution to ensure that these technologies develop as tools of sustainability rather than becoming future liabilities are of major concern.

Sorbents are widely used in water treatment and purification to remove organic and inorganic contaminants. Examples are activated carbon and ion-exchange resins. The use of nanoparticles may have advantages over conventional materials due to the much larger surface area of nanoparticles on a mass basis. In addition, the unique structure and electronic properties of some nanoparticles can make them especially powerful adsorbents. Many materials have properties that are dependent on size. Hematite particles with a diameter of 7 nm, for example, adsorbed Cu ions at lower pH values than particles of 25 or 88 nm diameter, indicating the uniqueness of surface reactivity for iron oxides particles with decreasing diameter. However, another study found that normalized to the surface area, the nanoparticles had a lower adsorption capacity than bulk TiO_2. Several types of nanoparticles have been investigated as adsorbents: metal-containing particles, mainly oxides, carbon nanotubes (CNTs) and fullerenes, organic nanomaterials, and zeolites.

1.3.6.1 Carbon Nanotubes

Carbon nanotubes (CNTs) have attracted a lot of attention as very powerful adsorbents for a wide variety of organic compounds from water. Examples include dioxin, polynuclear aromatic hydrocarbons (PAHs), DDT and its metabolites, chlorobenzenes and chlorophenols, phthalate esters, dyes, pesticides, and herbicides such as atrazine. Cross-linked nanoporous polymers that have been copolymerized with functionalized CNTs have been demonstrated to have a very high sorption capacity for a variety of organic compounds such as trichloroethylene. Carbon nanotubes and nanofibers exhibit extraordinary mechanical strength, electrical, chemical, and thermal characteristics, allowing for several potential applications [162]. Among all these applications, the use of carbon nanotubes and nanofibers as a catalyst or as a support seems to be most promising. The results published to date confirm that nanostructured carbon can represent a new class of advanced materials for catalytic applications due to specific metal support interactions given by their graphitic structure [163]. The small size of these carbon nanostructured materials, characterized by high specific surface areas of mesoporous nature, significantly contribute to the final catalytic performance of the system, since many catalytic heterogeneous reactions are governed by mass and heat transfer phenomena between the catalyst particles and the bulk medium [164]. When compared with the traditional activated carbons, characterized by very high specific surface areas, the mesoporous nature of the nanostructured materials is advantageous and encourages further exploration in applications that typically are well established with activated carbon catalysts. Multi-walled carbon nanotubes (MWCNT) can be efficiently used as support of ruthenium catalysts for the catalytic wet air oxidation of high-strength wastewater containing aniline. Garcia et al. [165] prepared catalysts using different ruthenium precursors, ruthenocene $[Ru(\eta^5–C_5H_5)_2]$, ruthenium (1,5-cyclooctadiene, 1,3,5-cyclooctatriene) [Ru(cod)(cot)], and ruthenium trichloride ($RuCl_3{\cdot}xH_2O$), different impregnation methods (excess solution and incipient wetness impregnation), and different MWCNT support surface chemistry (nitric acid-oxidized MWCNT–COOH and Na_2CO_3 ion-exchanged MWCNT–COONa). The efficiency of the aniline removal obtained with the catalysts prepared with different precursors decreases in the order [Ru(cod)(cot)] > $RuCl_3{\cdot}xH_2O$ > $[Ru(\eta^5–C_5H_5)_2]$, 100% aniline conversion being obtained after 45 min of reaction with the catalyst prepared with [Ru(cod)(cot)]. ZnO nanocrystals were prepared by precipitation method [166]. After aging, washing, and drying, solids were calcined at 300°C and examined as photocatalyst for UV-induced degradation of organophosphorus insecticide, diazinon in aqueous solution. The effects of some operational parameters such as pH value, nanocatalyst loading, and initial insecticide concentration on the degradation efficiency were discussed through the photocatalytic experiments using prepared ZnO nanocrystals as the photocatalyst. Daneshvar et al. [166] observed that the diazinon photodegradation quantum yield in UV/ZnO process increased with decreasing the diameter size of ZnO from 33 to 14 nm, and the quantum of results showed that the photocatalysis process in the presence of ZnO with mean size of 14 nm offered the best energy efficiency. Accordingly, 80% removal of the insecticide, after selecting desired operational parameters could be achieved in a relatively short time. The batch removal of dye from textile dyeing wastewater by using nanooxide-decorated multiwalled carbon nanotubes under electron beam conditions has recently been studied [167]. The effect of different nanooxide-decorated multiwalled carbon nanotubes content and irradiation dosage was also investigated. The color removal efficiency was 94.9% in dose of 17.5 kGy. The color removal efficiency with Fe_2O_3-decorated multiwalled carbon nanotubes was similar to TiO_2-decorated multiwalled carbon nanotubues. The COD removal efficiency was 52.5% in the dose of 14.0 kGy. With the irradiation dose of 17.5 kGy, the COD removal efficiency was 98.2% with TiO_2-decorated multiwalled carbon nanotubes. It was concluded that nanooxide-decorated multiwalled carbon nanotubes could effectively remove color and COD from aqueous solution under irradiation [167].

1.3.6.2 Nanocrystals

Chemically modified nanomaterials have also attracted a lot of attention, especially nanoporous materials due to their exceptionally high surface area. The particle size of such materials is, however, not in the nanorange but normally 10–100 mm. Another option is to modify chemically the nanoparticle itself. TiO_2 functionalized with ethylenediamine was, for example, tested for its ability to remove anionic metals from groundwater. Hierarchical CeO_2 nanocrystal microspheres were synthesized [168] with a nonaqueous sol-gel method at a low temperature of 120°C and tested with the batch removals of typical pollutants of Cr(VI) and rhodamine B from simulated wastewater. It was found that the nanostructured CeO_2 could effectively remove Cr(VI) without pH preadjustment. Moreover, these nonaqueous sol-gel synthesized hierarchical CeO_2 nanocrystal microspheres exhibited remarkable ability to remove rhodamine B, suggesting they are promising absorbents for wastewater treatment.

1.3.6.3 Magnetic Nanomaterials

With the latest developments in nanotechnology, various types of magnetic nanoparticles have also been successfully synthesized and have received considerable attention to solve environmental problems, such as accelerating the coagulation of sewage, removing radionuclides, adsorption of organic dyes, and remediation of contaminated soils. Apparently, the magnetic nanoparticles possess the advantages of large surface area, high number of surface active sites, and high magnetic properties, which lead to high adsorption efficiency, high removal rate of contaminants, and easy and rapid separation of adsorbent from solution via magnetic field. In addition, it is possible that after magnetic separation by the external magnetic field, the harmful components can be removed from the magnetic particles, which can be reused [169]. This study investigated the applicability of maghemite (γ-Fe_2O_3) nanoparticles for the selective removal of toxic heavy metals from electroplating wastewater. The maghemite nanoparticles of 10 nm were synthesized using a sol-gel method and characterized by X-ray diffraction and transmission electron microscopy. The surface area of the nanoparticles was determined to be 198 m^2/g using the Brunauer–Emmett–Teller method. Batch experiments were carried out to determine the adsorption kinetics and mechanisms of Cr(VI), Cu(II), and Ni(II) by maghemite nanoparticles. The adsorption process was found to be highly pH dependent, which made the nanoparticles selectively adsorb these three metals from wastewater. The adsorption of heavy metals reached equilibrium rapidly within 10 min and the adsorption data were well fitted with the Langmuir isotherm. Regeneration studies indicated that the maghemite nanoparticles undergoing successive adsorption–desorption processes retained original metal removal capacity. Mesoporous carbon capsules have been prepared and encapsulated with Fe_3O_4 nanoparticles by the successive coating of a silica layer and a subsequent mesoporous silica/carbon layer on the surface of Fe_3O_4 nanoparticles followed by chemical etching with NaOH solution [170]. These nanocomposites were superparamagnetic at room temperature with a saturation magnetization of 5.5 emu g^{-1}, which provided the prerequisite for the fast magnetic separation in wastewater treatment application. Water treatment experiments indicated that the as prepared samples exhibited higher adsorption rates and more effective removal capacity of organic pollutants compared with commercial activated carbon (AC), and their maximum adsorption capabilities for methylene blue (MB), congo red (CR), and phenol reached 608.04, 1656.9, and 108.38 mg g^{-1}, respectively. It was thus inferred that the multifunctional nanocomposites could hence be potentially used as absorbents for fast, convenient, and highly efficient removal of pollutants from the wastewater, which would play important roles in the purification or desalination of natural water and industrial effluents [170].

1.3.6.4 Nanofiltration Membranes

Nanofiltration membranes (NF membranes) are used in water treatment for drinking water production or wastewater treatment [171]. NF membranes are pressure-driven membranes with properties between those of reverse osmosis and ultrafiltration membranes and have pore sizes between 0.2 and 4 nm. NF membranes have been shown to remove turbidity, microorganisms, and inorganic ions. They are used for softening of groundwater (reduction in water hardness), for removal of dissolved organic matter and trace pollutants from surface water, for wastewater treatment (removal of organic and inorganic pollutants and organic carbon), and for pretreatment in seawater desalination.

A submerged nanofiltration membrane bioreactor (NF MBR) has been tested [172] using cellulose acetate membranes for 240 days to examine the performance of the NF membrane in domestic wastewater treatment. The workers observed that degradation of each water quality parameter over operation time could be attributed to the qualitative deterioration of the NF membrane caused by the hydrolysis of cellulose acetate, which leads to an increase in pore size and porosity, and decrease in surface charge and hydrophobicity of cellulose acetate membranes. Five different kinds of dye aqueous solutions (Direct Red 75, 80, and 81, and Direct Yellow 8 and 27) were treated [173] with nanofiltration (NF) polyamide (PA) composite membranes to obtain basic information on the dyeing wastewater reuse. The separation of dyes by the NF PA membrane appeared to be a good process for the effective removal of dyes from dyeing wastewater. The extent of the separation of dyes by the membrane was almost 100% for all of the dyes used, producing colorless water. When artificial dyeing wastewater containing Direct Red 75, PVA, NaCl, and Na_2SO_4 was used as a feed solution for the membrane separation process, it was observed [173] that separation efficiency turned out to be good, especially when one of the chemical coagulants, alum, was used for the pretreatment of the artificial dyeing wastewater. An integrated nanofiltration–membrane crystallization (NF-MC) system has been tested [174] for the removal of sodium sulfate from aqueous wastes originated by the production process of base raw materials (Ni–H) for special rechargeable batteries. Nanofiltration experiments on NF90–2540 FilmTec™ modules showed high rejection values to sodium sulfate (>99%) when operated at a pressure of 3.45×10^6 Pa; it was possible to achieve a maximum permeate recovery factor of 50% when starting from 60 g L^{-1} Na_2SO_4 feed solution. The membrane crystallizer, operated downstream to the pre-concentrated NF unit, was able to produce thenardite crystals (the anhydrous form of sodium sulfate) grown from slurry with density up to 21 kg m^{-3}; the solid product exhibited narrow size distribution (coefficient of variation 30%). Polypropylene hollow fiber membranes showed a hydrophobic character stable over 2 days of continuous operation.

1.4 Concluding Remarks

Water is the element most essential to sustain human life; therefore, it is necessary to create better understanding of water resources and foster public understanding through water education. Water education should not be only about water types, sources, uses, treatment, management, and its associated problems. It must include people's perceptions of water, the level of their consciousness toward water usage, awareness of their civic responsibilities toward water, and cultural beliefs and practices in relation to water [175]. Awareness should also be raised among people by taking the help of media and NGOs regarding conservation of water. Formation of new policies for sustainable development of water resources, regulating water prices for water conservation, conservation education, pollution prevention, as well as effective information management will help in satisfying the demand [175].

References

1. Speidel EDH, Ruedisili LC, Agnew AF (1988) Perspectives on water: uses and abuses. Oxford University Press, New York
2. Schwarzenbach RP, Escher BI, Fenner K, Hofstetter TB, Johnson CA (2006) The challenge of micropollutants in aquatic systems. Science 313:1072–1077
3. UNESCO (2009) UN Educ, Sci Cult, Organ. The United Nations World Water Development Report 3: Water in a changing world. UNESCO/Berghahn Books, Paris/New York
4. Huntington TG (2006) Evidence for intensification of the global water cycle review and synthesis. J Hydrol 319:83–95
5. Oki T, Kanae S (2006) Global hydrological cycles and world water resources. Science 313:1068–1072
6. Fenwick A (2006) Waterborne infectious diseases—could they be consigned to history? Science 313:1077–1081
7. Gruber N, Galloway JN (2008) An earth–system perspective of the global nitrogen cycle. Nature 451:293–296
8. Filippelli GM (2008) The global phosphorus cycle: past, present, future. Elements 4:89–95
9. Jorgenson AK (2009) Foreign direct investment the environment, the mitigating influence of institutional and civil society factors, and relationships between industrial pollution and human health: a panel study of less-developed countries. Organ Environ 22 (2)L:135–157
10. Dubus IG, Hollis JM, Brown CD (2000) Pesticides in rainfall in Europe. Environ Pollut 110(2):331–344
11. Cosgrove WJ, Rijsberman FR (2000) World water vision: making water everybody's business. WorldWater Council, London
12. Bockstaller C, Guichard L, Keichinger O, Girardin P, Galan MB, Gaillard G (2009) Comparison of methods to assess the sustainability of agricultural systems. A review. Agron Sustain Dev 29:223–235
13. Eliopoulou E, Papanikolaou A (2007) Casualty analysis of large tankers. J Mar Sci Technol 12:240–250
14. Larsen TA, Maurer M, Udert KM, Lienert J (2007) Nutrient cycles and resource management: implications for the choice of wastewater treatment technology. Wat Sci Technol 56:229–237
15. Lohse KA, Brooks PD, McIntosh JC, Meixner T, Huxman TE (2009) Interactions between biogeochemistry hydrologic systems. Annu Rev Environ Resour 34:65–96
16. Heisler J, Glibert PM, Burkholder JM, Anderson DM, Cochlan W (2008) Eutrophication and harmful algal blooms: a scientific consensus. Harmful Algae 8:3–13
17. Kaushal SS, Groffman PM, Likens GE, Belt KT, Stack WP (2005) Increased salinization of fresh water in the northeastern United States. Proc Natl Acad Sci U S A 102:13517–13520
18. Post VEA (2005) Fresh and saline groundwater interaction in coastal aquifers: is our technology ready for the problems ahead? Hydrogeol J 13:120–123
19. Lohmann R, Breivik K, Dachs J, Muir D (2007) Global fate of POPs current and future research directions. Environ Pollut 150:150–165
20. Muir DCG, Howard PH (2006) Are there other persistent organic pollutants? A challenge for environmental chemists. Environ Sci Technol 40:7157–7166
21. Vonderheide AP, Mueller KE, Meija J, Welsh GL (2008) Polybrominated diphenyl ethers causes for concern and knowledge gaps regarding environmental distribution, fate toxicity. Sci Total Environ 400:425–436
22. Yogui GT, Sericano JL (2009) Polybrominated diphenyl ether flame retardants in the US marine environment: a review. Environ Int 35:655–666
23. Goss KU, Bronner G (2006) What is so special about the sorption behavior of highly fluorinated compounds? J Phys Chem A 110:9518–9522
24. Prevedouros K, Cousins IT, Buck RC, Korzeniowski SH (2006) Sources fate and transport of perfluorocarboxylates. Environ Sci Technol 40:32–44

25. Schwarzenbach RP, Egli T, Hofstetter TB, Von Gunten U, Wehrli B (2010) Global water pollution and human health. Annu Rev Environ Resour 35:109–136
26. Giger W (2009) Hydrophilic and amphiphilic water pollutants: using advanced analytical methods for classic and emerging contaminants. Anal Bioanal Chem 393:37–44
27. Richardson SD (2009) Water analysis emerging contaminants and current issues. Anal Chem 81:4645–4677
28. Kelly BC, Ikonomou MG, Blair JD, Morin AE, Gobas F (2007) Food web–specific biomagnification of persistent organic pollutants. Science 317:236–239
29. Kelly BC, Ikonomou MG, Blair JD, Surridge B, Hoover D (2009) Perfluoroalkyl contaminants in an Arctic marine food web trophic magnification and wildlife exposure. Environ Sci Technol 43:4037–4043
30. Porta M, Puigdomenech E, Ballester F, Selva J, Ribas–Fito N (2008) Monitoring concentrations of persistent organic pollutants in the general population the international experience. Environ Int 34:546–561
31. Brown TN, Wania F (2008) Screening chemicals for the potential to the persistent organic pollutants: a case study of Arctic contaminants. Environ Sci Technol 42:5202–5209
32. Atapattu SS, Kodituwakku DC (2009) Agriculture in South Asia and its implications on downstream health and sustainability: a review. Agric Wat Manag 96:361–373
33. FAO (2008) UN Food Agric Organ FAOSTAT statistical database. http://faostat.fao.org/site/424/default.aspx
34. USEPA (2008) Environ Prot Agency Pesticides. http://www.epa.gov/pesticides/
35. Galt RE (2008) Beyond the circle of poison: significant shifts in the global pesticide complex 1976–2008. Glob Environ Change 18:786–799
36. Reus J, Leendertse P, Bockstaller C, Fomsgaard I, Gutsche V (2002) Comparison and evaluation of eight pesticide environmental risk indicators developed in Europe and recommendations for future use. Agric Ecosyst Environ 90:177–187
37. UNEP (2007) UN Environment Programme. Global Environment Outlook GEO4. Environment for Development
38. Tilman D, Fargione J, Wolff B, D'Antonio C, Dobson A (2001) Forecasting agriculturally driven global environmental change. Science 292:281–284
39. Oluwole O, Cheke RA (2009) Health and environmental impacts of pesticide use practices a case study of farmers in Ekiti State Nigeria. Int J Agric Sustain 7:153–163
40. Williamson S, Ball A, Pretty J (2008) Trends in pesticide use and drivers for safer pest management in four African countries. Crop Prot 27:1327–1334
41. Menezes CT, Heller L (2008) A method for prioritization of areas for pesticides surveillance on surface waters a study in Minas Gerais Brazil. Wat Sci Technol 57:1693–1698
42. Agrawal GD (1999) Diffuse agricultural water pollution in India. Wat Sci Technol 39:33–47
43. Sarkar SK, Bhattacharya BD, Bhattacharya A, Chatterjee M, Alam A (2008) Occurrence distribution and possible sources of organochlorine pesticide residues in tropical coastal environment of India an overview. Environ Int 34:1062–1071
44. Shi LL, Shan ZJ, Kong DY, Cai DJ (2006) The health and ecological impacts of organochlorine pesticide pollution in China bioaccumulation of organochlorine pesticides in human and fish fats. Hum Ecol Risk Assess 12:402–407
45. van Geen A, Protus T, Cheng Z, Horneman A, Seddique AA, Hoque MA, Ahmed KM (2004) Testing groundwater for arsenic in Bangladesh before installing a well. Environ Sci Technol 38:6783–6789
46. Lafferty B (2008) Kinetics of Arsenic transformations in the soil environment. The 2008 Joint Annual Meeting
47. Smith PG (2007) Arsenic biotransformation in terrestrial organisms, A study of the transport and transformation of arsenic in plants fungi fur and feathers using conventional speciation analysis and X–ray absorption spectroscopy. PhD Thesis, Queen's University Canada
48. Smith AH, Lingas EO, Rahman M (2000) Contamination of drinking water by arsenic in Bangladesh a public health emergency. Bull World Health Organ 78:1093–1103

49. Opar A, Pfaff A, Seddique AA, Ahmed KM, Graziano JH, van Geen A (2007) Responses of 6500 households to arsenic mitigation in Araihazar Bangladesh. Health Place 13:164–172
50. Steinmaus C, Yuan Y, Bates MN, Smith AH (2003) Case control study of bladder cancer and drinking water arsenic in the Western United States. Am J Epidemiol 158:1193–1201
51. Tseng CH (2005) Blackfoot disease and arsenic: a never ending story. J Environ Sci Health C 23:55–74
52. Mandal BK, Suzuki KT (2002) Arsenic round the world: a review. Talanta 58:2–235
53. Buschmann J, Berg M, Stengel C, Winkel L, Sampson MK, Trang PTK, Viet PH (2008) Contamination of drinking water resources in the Mekong delta floodplains arsenic and other trace metals pose serious health risks to population. Environ Int 34:756–764
54. Chen YN, Chai LY, Shu YD (2008) Study of arsenic (V) adsorption on bone char from aqueous solution. J Hazard Mater 160:168–172
55. Johnston R, Heijnen H (2001) Safe water technology for arsenic removal. Technologies for arsenic removal from drinking water 1–22. http://www.bvsde.opsoms.org/bvsacd/arsenico/technologies/Han.pdf. Accessed July 2008
56. Lu X, Kruatrachue M, Pokethitiyook P, Homyok K (2004) Removal of cadmium and zinc by water hyacinth *Eichhornia crassipes*. ScienceAsia 30:93–103
57. Tyler G, Pahlsson AM, Bengtsson G, Baath E, Tranvik L (1989) Heavy metal ecology and terrestrial plants microorganisms and invertebrates a review. Wat Air Soil Pollut 47:189–215
58. Dang VBH, Doan HD, Dang–Vu T, Lohi A (2009) Equilibrium and kinetics of biosorption of cadmium(II) and copper(II) ions by wheat straw. Bioresour Technol 100:211–219
59. Adriano DC (2001) Trace elements in terrestrial environments biochemistry bioavailability and risks of metals. Springer Verlag, New York
60. Peng K, Li X, Luo C, Shen Z (2006) Vegetation composition and heavy metal uptake by wild plants at three contaminated sites in Xiangxi area China. J Environ Sci Health Part A 40:65–76
61. Baszynski T (1986) Interference of Cd^{2+} in functioning of the photosynthetic apparatus of higher plants. Acta Soc Bot Pol 99:291–304
62. Drasch GA (1993) Increase of cadmium body burden for this century. Sci Tot Environ 67:75–89
63. Demirbas E, Kobya M, Senturk E, Ozkan T (2004) Adsorption kinetics for the removal of chromium (VI) from aqueous solutions on the activated carbons prepared from agricultural wastes. Wat SA 30:533–539
64. Carrasquero Durán A, Flores I, Perozo C, Pernalete Z (2006) Immobilization of lead by a vermicompost and its effect on white bean (Vigna Sinenis var Apure) uptake. Int J Environ Sci Technol 3:203–212
65. Brune A, Urbach W, Dietz KJ (1994) Compartmentation and transport of zinc in barley primary leaves as basis mechanisms involved in zinc tolerance. Plant Cell Environ 17:153–162
66. Aksu Z (2005) Application of biosorption for the removal of organic pollutants a review. Process Biochem 40:997–1026
67. Kidwai M, Mohan R (2005) Green chemistry: an innovative technology. Foundations Chem 7:269–287
68. Saeed A, Iqbal M (2003) Bioremoval of cadmium from aqueous solution by black gram husk (*Cicer arientinum*). Wat Res 37:3472–3480
69. Singh KK, Rastogi R, Hasan SH (2005) Removal of cadmium from wastewater using agricultural waste rice polish. J Hazard Mater 121:51–58
70. Pérez–Marín AB, Meseguer Zapata V, Ortuño JF, Aguilar M, Sáez J, Lloréns M (2007) Removal of cadmium from aqueous solutions by adsorption onto orange waste. J Hazard Mater 139:122–131
71. Göksungur Y, Üren S, Güvenç U (2005) Biosorption of cadmium and lead ions by ethanol treated waste baker's yeast biomass. Bioresour Technol 99:103–109
72. Sarı A, Tuzen M (2008) Biosorption of cadmium(II) from aqueous solution by red algae (*Ceramium virgatum*) Equilibrium kinetic and thermodynamic studies. J Hazard Mater 157:448–454

73. Tunali S, Kiran I, Akar T (2005) Chromium(VI) biosorption characteristics of *Neurospora crassa* fungal biomass. Miner Eng 18:681–689
74. Melo JS, D'Souza SF (2004) Removal of chromium by mucilaginous seeds of *Ocimum basilicum*. Bioresour Technol 92:151–155
75. Elangovan R, Philip L, Chandraraj K (2008) Biosorption of hexavalent and trivalent chromium by palm flower (*Borassus aethiopum*). Chem Eng J 141:99–111
76. Pehlivan E, Altun T (2008) Biosorption of chromium(VI) ion from aqueous solutions using walnut hazelnut and almond shell. J Hazard Mater 155:378–384
77. Aravindhan R, Madhan B, Raghava Rao J, Unni Nair B (2004) Recovery and reuse of chromium from tannery wastewaters using *Turbinaria ornata* seaweed. J Chem Technol Biotechnol 79:1251–1258
78. Jain M, Garg VK, Kadirvelu K (2009) Chromium(VI) removal from aqueous system using *Helianthus annuus* (sunflower) stem waste. J Hazard Mater 162:365–372
79. Laurino JP (2008) Removal of Lead (II) Ions by poly 2 octadecyl butanedioic acid isothermal and kinetic studies. J Macromol Sci A Pur Appl Chem 45:612–619
80. Şölener M, Tunali S, Safa Özcan A, Özcan A, Gedikbey T (2008) Adsorption characteristics of lead (II) ions onto the clay poly (methoxyethyl) acrylamide (PMEA) composite from aqueous solutions. Desalination 223:308–322
81. Shen W, Chen S, Shi S, Li X, Zhang X, Hu W, Wang H (2009) Adsorption of Cu (II) and Pb (II) onto diethylenetriamine bacterial cellulose. Carbohydr Polym 75:110–114
82. Nadeem R, Nasir MH, Hanif MS (2009) Pb (II) sorption by acidically modified *Cicer arientinum* biomass. Chem Eng J 150:40–48
83. Margesin R, Schinner F (2001) Bioremediation (natural attenuation and biostimulation) of diesel oil contaminated soil in an Alpine glacier skiing area. Appl Environ Microbiol 67(7):3127–3133
84. Paul JF, McDonald ME (2005) Development of empirical geographically specific water quality criteria A conditional probability analysis approach. J Am Wat Resour Assoc 41(5) art. no. 04095:1211–1223
85. Raji C, Anirudhan TS (1998) Copper-impregnated sawdust carbon for the treatment of as (III) rich water. J Sci Ind Res 57(1):10–15
86. Gardea-Torresdey JL, Tiemann KJ, Armendariz V, Bess-Oberto L, Chianelli RR, Rios J, Parsons JG, Gamez G (2000) Characterization of Cr (VI) binding and reduction to Cr (III) by the agricultural byproducts of *Avena monida* (Oat) biomass. J Hazard Mater 80(1–3):175–188
87. Robinson T, McMullan G, Marchant R, Nigam P (2001) Remediation of dyes in textile effluent a critical review on current treatment technologies with a proposed alternative. Biores Technol 77:247–255
88. Willmott N, Guthrie J, Nelson G (1998) The biotechnology approach to colour removal from textile effluent. J Soc Dyers Colour 114:38–41
89. Borchert M, Libra JA (2001) Decolorization of reactive dyes by the white rot fungus *Trametes versicolor* in sequencing batch reactors. Biotechnol Bioeng 75(3):313–321
90. Beydilli MI, Pavlostathis SG, Tincher WC (1998) Decolorization and toxicity screening of selected reactive azo dyes under methanogenic conditions. Wat Sci Technol 38(4–5 -5 4):225–232
91. Zissi U, Lyberatos G (2001) Improvement in bioreactor productivities using free radicals HOCl induced overproduction of xanthan gum from *Xanthomonas campestris* and its mechanism. Biotechnol Bioeng 72(1):62–68
92. Pointing SB (2001) Feasibility of bioremediation by white-rot fungi. Appl Microbiol Biotechnol 57(1–2):20–33
93. Field JA, Barber LB II, Thurman EM, Moore BL, Lawrence DL, Peake DA (1992) Fate of alkylbenzenesulfonates and dialkyltetralinsulfonates in sewage contaminated groundwater. Environ Sci Technol 26(6):1140–1147
94. Novotny V, Witte JW (1997) Ascertaining aquatic ecological risks of urban stormwater discharges. Water Res 31(10):2573–2585

95. Bumpus John A, Aust Steven D (1985) Studies on the biodegradation of organopollutants by a white rot fungus. United States Environmental Protection Agency, Office of Research and Development, (Report) EPA, 404–410
96. Yadav JS, Reddy CA (1993) Degradation of benzene toluene ethylbenzene and xylenes (BTEX) by the lignin degrading basidiomycete *Phanerochaete chrysosporium*. Appl Environ Microbiol 59(3):756–762
97. Baker JA (1989) Case studies in organic contaminant hydrogeology. Environ Geol Wat Sci 14(1):17–33
98. Entry JA, Watrud LS, Reeves M (1999) Accumulation of 137Cs and 90Sr from contaminated soil by three grass species inoculated with mycorrhizal fungi. Environ Pollut 104(3):449–457
99. Raskin I, Smith RD, Salt DE (1997) Phytoremediation of metals: using plants to remove pollutants from the environment. Curr Opin Biotechnol 8(2):221–226
100. Meagher RB (2000) Phytoremediation of toxic elemental and organic pollutants. Current Opin Plant Biol 3(2):153–162
101. Lasat MM (2002) Phytoextraction of toxic metals: a review of biological mechanisms. J Environ Qual 31(1):109–120
102. McGrath RJ, Styles P, Thomas E, Neale S (2002) Integrated high-resolution geophysical investigations as potential tools for water resource investigations in karst terrain. Environ Geol 42(5):552–557
103. Garbisu C, Alkorta I (2001) Phytoextraction A cost-effective plant based technology for the removal of metals from the environment. Bioresour Technol 77(3):229–236
104. Ghosh M, Singh SP (2005) A review on phytoremediation of heavy metals and utilization of its byproducts. Appl Ecol Environ Res 3(1):1–18
105. Pilon-Smits E (2005) Phytoremediation. Annu Rev Plant Biol 56:15–39
106. Boyd CE (1970) Vascular aquatic plants for mineral nutrient removal from polluted waters. Econ Bot 24:95–103
107. Cornwell DA, Zoltek J Jr, Patrinely CD, Furman TS, Kim JI (1977) Nutrient removal by water hyacinths. J Wat Pollut Contr Fed 49:57–65
108. Stewart KK (1970) Nutrient removal potential of various aquatic plants. Hyacinth Contr J 8:34–35
109. Wooten JW, Dodd JD (1976) Growth of water hyacinth in treated sewage effluent. Econ Bot 30:29–37
110. Scheffield CW (1967) Water hyacinth for nutrient removal. Hyacinth Contr J 6:27–30
111. Yount JL (1964) Aquatic nutrient reduction and possible methods. Report of the 35th Anniversary Meeting, FL Anti–mosquito Association, pp 83–85
112. Wolverton BC, Mckown MM (1976) Water hyacinth for removal of phenols from polluted waters. Aquatic Bot 30:29–37
113. Seidal K (1976) Macrophytes and water purification. In: Tourbier J, Pierson RW (eds) Biological control for water pollution. Pennsylvania University Press, Pennsylvania, pp 109–121
114. Pip E, Stepaniuk J (1992) Cadmium, copper and lead in sediments. Arch fur Hydrobilogie 124:337–355
115. Holm LG, Plucknett DL, Pancho V, Herberger JP (1977) The world's worstweeds: distribution and biology.University Press of Hawaii, Honolulu, 609 p
116. Mitchell DS (1976) The growth and management of *Eichhornia crassipes* and *Salvinia* spp. in their native environment and in alien situations. In: Varshney CK, Rzoska J (eds) Aquatic weeds in Southeast Asia 396. Dr. W. Junk, The Hague
117. Salati E (1987) Edaphic–phytodepuration: a new approach to waste water treatment. In: Reddy KR, Smith WH (eds) Aquatic plants for water treatment and resource recovery. Magnolia, Orlando, pp 199–208
118. Nor YM (1990) The absorption of metal ions by *Eichhornia crassipes*. Chem Speciation Bioavailability 2:85–91
119. Tiwari S, Dixit S, Verma N (2007) An effective means of bio–filtration of heavy metal contaminated water bodies using aquatic weed *Eichhornia crassipes*. Environ Monit Assess 129:253–256

120. Pinto CLR, Caconia A, Souza MM (1987) Utilization of water hyacinth for removal and recovery of silver from industrial waste water. Wat Sci Technol 19(10):89–101
121. Wolverton BC (1989) Aquatic plant/microbial filters for treating septic tank effluent in wastewater treatment. In: Hammer DA (ed) Municipal industrial and agricultural waste. Lewis, Chelsea
122. Brix H (1993) Macrophytes–mediated oxygen transfer in wetlands: transport mechanism and rate. In: Moshiri GA (ed) Constructed wetlands for water quality improvement. Lewis, Ann Arbor/London
123. Johnston CA (1993) Mechanism of water wetland water quality interaction. In: Moshiri GA (ed) Constructed wetland for water quality improvement. Lewis, Ann Arbor, pp 293–299
124. Stowell R, Ludwig R, Colt J, Tchobanoglous T (1981) Concepts in aquatic treatment design. J Environ Eng 112:885–894
125. Rai PK (2009) Heavy metal phytoremediation from aquatic ecosystems with special reference to macrophytes. Crit Rev Environ Sci Technol 39(9):697–753
126. Bradley BR, Daigger GT, Rubin R, Tchobanoglous G (2002) Evaluation of onsite wastewater treatment technologies using sustainable development criteria. Clean Technol Environ Policy 4:87–99
127. Lapena L, Cerezo M, Garcia–Augustin P (1995) Possible reuse of treated municipal wastewater for *Citrus* spp. plant irrigation. Bull Environ Contam Toxicol 55:697–703
128. Viessman W Jr, Hammer MJ (1998) Water supply and pollution control, 6th edn. Addison Wesley Longman Inc, Menlo Park
129. Padmanabhan PVA, Sreekumar KP, Thiyagarajan TK, Satpute RU, Bhanumurthy K, Sengupta P, Dey GK, Warrier KGK (2006) Nano–crystalline titanium dioxide formed by reactive plasma synthesis. Vacuum 80:11–12
130. Gaya UI, Abdullah AH (2008) Heterogeneous photocatalytic degradation of organic contaminants over titanium dioxide: a review of fundamentals, progress and problems. J Photochem Photobiol C: Photochem Rev 9:1–12
131. Yang H, Cheng H (2007) Controlling nitrite level in drinking water by chlorination and chloramination. Sep Purif Technol 56:392–396
132. Lu J, Zhang T, Ma J, Chen Z (2009) Evaluation of disinfection by–products formation during chlorination and chloramination of dissolved natural organic matter fractions isolated from a filtered river water. J Hazard Mater 162:140–145
133. Coleman HM, Marquis CP, Scott JA, Chin SS, Amal R (2005) Bactericidal effects of titanium dioxide–based photocatalysts. Chem Eng J 113:55–63
134. Serpone N, Sauve G, Koch R, Tahiri H, Pichat P, Piccinini P, Pelizetti E, Hidaka H (1996) Standardization protocol of process efficiencies and activation parameters in heterogeneous photocatalysis: relative photonic efficiencies. J Photochem Photobiol A: Chem 94:191–203
135. Ensminger D (1973) Ultrasonics: the low— and high—intensity applications. Marcel Dekker, New York
136. Thompson LH, Doraiswamy LK (1999) Sonochemistry: science and engineering. Ind Eng Chem Res 38:1215–1249
137. Suslick KS (1990) Sonochemistry. Science 247:1439–1445
138. Riesz P, Berdahl D, Christman CL (1985) Free radical generation by ultrasound in aqueous and nonaqueous solutions. Environ Health Perspect 64:233–252
139. Liang J, Komarov S, Hayashi N, Kasai E (2007) Improvement in sonochemical degradation of 4—chlorophenol by combined use of Fenton—like reagents. Ultrason Sonochem 14:201–207
140. Zeng L, McKinley JW (2006) Degradation of pentachlorophenol in aqueous solution by audible–frequency sonolytic ozonation. J Hazard Mater 135:218–225
141. Lim MH, Kim SH, Kim YU, Khim J (2007) Sonolysis of chlorinated compounds in aqueous solution. Ultrason Sonochem 14:93–98
142. Guo Z, Gu C, Zheng Z, Feng R, Jiang F, Gao G, Zheng Y (2006) Sonodegradation of halomethane mixtures in chlorinated drinking water. Ultrason Sonochem 13:487–492
143. Matouq MA-D, Al-Anber ZA (2007) The application of high frequency ultrasound waves to remove ammonia from simulated industrial wastewater. Ultrason Sonochem 14(3):393–397

144. Mason TJ, Collings A, Sumel A (2004) Sonic and ultrasonic removal of chemical contaminants from soil in the laboratory and on a large scale. Ultrason Sonochem 11:205–210
145. Moriwaki H, Takagi Y, Tanaka M, Tsuruho K, Okitsu K, Maeda Y (2005) Sonochemical decomposition of perfluorooctane sulfonate and perfluorooctanoic acid. Environ Sci Technol 39:3388–3392
146. Vecitis CD, Wang Y, Cheng J, Park H, Mader BT, Hoffmann MR (2010) Sonochemical degradation of perfluorooctane sulfonate in aqueous film—forming foams. Environ Sci Technol 44:432–438
147. Francony A, Petrier C (1996) Sonochemical degradation of carbon tetrachloride in aqueous solution at two frequencies: 20 kHz and 500 kHz. Ultrason Sonochem 3:77–82
148. Lee M, Oh J (2010) Sonolysis of trichloroethylene and carbon tetrachloride in aqueous solution. Ultrason Sonochem 17:207–212
149. Suri RPS, Nayak M, Devaiah U, Helmig E (2007) Ultrasound assisted destruction of estrogen hormones in aqueous solution: effect of power density, power intensity and reactor configuration. J Hazard Mater 146:472–478
150. Isariebel QP, Carine JL, Ulises — Javier JH, Anne — Marie W, Henri D (2009) Sonolysis of levodopa and paracetamol in aqueous solutions. Ultrason Sonochem 16:610–616
151. Neppolian B, Jung H, Choi H, Lee JH, Kang J-W (2002) Sonolytic degradation of methyl tert-butyl ether: the role of coupled Fenton process and persulphate ion. Water Res 19:4699–4708
152. Tuziuti T, Yasui K, Iida Y et al (2004) Effect of particle addition on sonochemical reaction. Ultrasonics 42:597–601
153. Shimizu K, Matsuda Y, Nonomura T, Ikeda H, Tamura N, Kusakari S, Kimbara J, Toyoda H (2007) Dual protection of hydroponic tomatoes from rhizosphere pathogens *Ralstonia solanacearum* and *Fusarium oxysporum* f.sp. *radicis lycopersici* and airborne conidia of *Oidium neolycopersici* with an ozone-generative electrostatic spore precipitator. Plant Pathol 56:987–997
154. Abdullah AZ, Ling PY (2010) Heat treatment effects on the characteristics and sonocatalytic performance of TiO_2 in the degradation of organic dyes in aqueous solution. J Hazard Mater 173:159–167
155. Hill JM, Marchant TR (1996) Modelling microwave heating. Appl Math Model 20:3–15
156. Wu T-N (2008) Environmental perspectives of microwave applications as remedial alternatives: review. Pract Periodical Hazard Toxic Radioactive Waste Manag 12:102–115
157. Nüchter M, Ondruschka B, Bonrath W, Gum A (2004) Microwave assisted synthesis – a critical technology overview. Green Chem 6:128–141
158. Jones DA, Lelyveld TP, Mavrofidis SD (2002) Microwave heating applications in environmental engineering—a review. Resour Conserv Recycl 34:75–90
159. Cravotto G, Di Carlo S, Tumiatti V (2005) Degradation of persistent organic pollutants by Fenton's reagent facilitated by microwave or high–intensity ultrasound. Environ Technol 26:721–724
160. Theron J, Walker JA, Cloete TE (2008) Nanotechnology and water treatment: applications and emerging opportunities. Crit Rev Microbiol 34(1):43–69
161. Bottero JY, Rose J, Wiesner MR (2006) Nanotechnologies: tools for sustainability in a new wave of water treatment processes. Integr Environ Assess Manag 4(2):391–395
162. Terrones M (2003) Science and technology of the twenty-first century: synthesis, properties and applications of carbon nanotubes. Ann Rev Mater Res 33:419–501
163. Tribolet P, Kiwi–Minsker L (2005) Carbon nanofibers grown on metallic filters as novel catalytic materials. Catal Today 102:15–22
164. Sato S, Takahashi R, Sodesawa T, Nozaki F, Jin XZ, Suzuki S (2000) Mass–transfer limitation in mesopores of Ni–MgO catalyst in liquid–phase hydrogenation. J Catal 191(2):261–270
165. Garcia J, Gomes HT, Serp Ph, Kalck Ph, Figueiredo JL, Faria JL (2006) Carbon nanotube supported ruthenium catalysts for the treatment of high strength wastewater with aniline using wet air oxidation. Carbon 44:2384–2391
166. Daneshvar N, Aber S, Seyed Dorraji MS, Khataee AR, Rasoulifard MH (2007) Photocatalytic degradation of the insecticide diazinon in the presence of prepared nanocrystalline ZnO powders under irradiation of UV–C light. Sep Purif Technol 58:91–98

167. Shi WY, Gu JZ, Wu WJ (2011) Removal of dye from textile dyeing wastewater by using oxidized multiwalled carbon nanotubes. Adv Mater Res 193:343–344
168. Xiao H, Ai Z, Zhang L (2009) Nonaqueous sol − gel synthesized hierarchical CeO_2 nanocrystal microspheres as novel adsorbents for wastewater treatment. J Phys Chem C 113(38):16625–16630
169. Hu J, Chen G, Lo IMC (2006) Selective removal of heavy metals from industrial wastewater using maghemite nanoparticle: performance and mechanisms. J Environ Eng 132(7):709–715
170. Zhang Y, Xu S, Luo Y, Pan S, Ding H, Li G (2011) Synthesis of mesoporous carbon capsules encapsulated with magnetite nanoparticles and their application in wastewater treatment. J Mater Chem 21:3664–3671
171. Lau WJ, Ismail AF (2009) Polymeric nanofiltration membranes for textile dye wastewater treatment: preparation, performance evaluation, transport modelling, and fouling controls — a review. Desalination 245:321–348
172. Choi JH, Fukushi K, Yamamoto K (2007) A submerged nanofiltration membrane bioreactor for domestic wastewater treatment: the performance of cellulose acetate nanofiltration membranes for long–term operation. Sep Purif Technol 52:470–477
173. Mo JH, Lee YH, Kim J (2008) Treatment of dye aqueous solutions using nanofiltration polyamide composite membranes for the dye wastewater reuse. Dyes Pigments 76(2):429–434
174. Curcio E, Ji X, Quazi AM, Barghi S (2010) Hybrid nanofiltration–membrane crystallization system for the treatment of sulfate wastes. J Membrane Sci 360:493–498
175. Misra AK (2011) Impact of urbanization on the hydrology of Ganga Basin (India). Water Resour Manag 25:705–719

Chapter 2
Water Pathways Through the Ages: From Early Aqueducts to Next Generation of Wastewater Treatment Plants

Giusy Lofrano Ph.D., Jeanette Brown, and Giovanni De Feo

2.1 Introduction

As stated by philosopher Thales from Miletus (ca. 636–546 B.C.), "Hydor (Water) is the beginning of everything". Thales understood that water is life and without water no living thing will survive.

The availability of high-quality water as well as management of wastewater is a prerequisite for every urban development in every part of the world. People have always had to deal with the need of having enough water for their use and subsequently manage wastewater for the protection of public health and the environment [1].

Over the centuries, civilizations developed structures to bring drinking water to the population centers, especially through the use of aqueducts. There are many historical records on the purpose and design of aqueducts. Many of these aqueducts, though thousands of years old, are still standing. But, there is limited historical information on how wastewater was removed from population centers. This is surprising since the lack of sanitation affects human development to the same or even greater extent as the lack of clean water. Although there is typically a stigma to discussing waste treatment, in reality as sanitation improved so did public health, economic conditions, and the evolution of man [2].

G. Lofrano Ph.D. (✉)
Civil and Environmental Engineer, via Asiago, 1-84132 Salerno, Italy
e-mail: glofrano@unisa.it

J. Brown
Manhattan College, Riverdale, NY, USA
e-mail: jbrown421@optonline.net

G. De Feo
Department of Industrial Engineering, University of Salerno, 1-84084 Fisciano, Salerno, Italy
e-mail: g.defeo@unisa.it

S.K. Sharma and R. Sanghi (eds.), *Advances in Water Treatment and Pollution Prevention*, DOI 10.1007/978-94-007-4204-8_2,

For centuries, wastewater management was not given much, if any, consideration. The importance of proper sanitation for the protection of public health was not understood by cities until the nineteenth century [3–5]. However, some early cultures did understand that wastewater had to be managed and furthermore could be used beneficially. For example, the early civilizations of Syria and Palestine had water filtering techniques and channels which conveyed wastewater to farm fields for use as a fertilizer.

In most early cultures, wastewater was disposed of in the streets and near population centers creating serious impacts on public health and the environment. Numerous epidemics occurred throughout Europe until the nineteenth century [3, 6–8]. Sadly, when it came to waste management and sanitation, even countries those that suffered epidemics, tended to have short memories.

Throughout history, wastewater management has presented people and governments with far-reaching technical and political challenges. The story of water and wastewater management is at once a story of human ingenuity and human frailty [7, 9].

A number of keystone events defined the speed at which environmental management evolved through the ages. Some of these events were scientific, such as stream purification models, while others were socioeconomic such as two world wars [4, 10, 11]. However, according to the recent Human Development Report [7], the lesson from the past is that progress in wastewater management and sanitation was driven above all by political coalitions uniting industrialists, municipalities, and social reformers. This means that if on one side developing new technologies as well as appropriate strategies for wastewater management is required, on the other side there must be an urgent need to overcome the stigma of a polluted environment [2].

Although several historians and economists have described the evolution of wastewater management through the ages [9, 12–15], they, as it is normal, often lack an engineering perspective. Several studies have reconstructed traces of ancient dams, aqueducts, and pipes [16], which provided water for human consumption but archaeological research has largely neglected the difficulty of wastewater management [17]. Sewers and primitive treatment is omitted from archaeology and historical research and forgotten. This chapter intends to highlight those sanitation systems as well as wastewater management strategies that were developed in different periods and cultures, why they were developed and where we are today. As stated by Walker (1971, cit. in [18]): “Analysis of many problems is generally simplified by putting them into historical perspective”.

2.2 Water Pathways from Aqueducts to Sewer

2.2.1 Aqueducts

Today, the most popular image of aqueducts is associated with those designed by the Romans who elevated them throughout their entire length on lines of arches. However, the Romans were not the first to construct aqueducts. In Persia, India,

Egypt, and other areas of the Middle East, there were systems to convey water hundreds of years before the Romans began building them [19]. For example, around 691 B.C., the Assyrians built an 80 km long limestone aqueduct 10 m high and 30 m wide that brought fresh water to the city of Nineveh.

Aqueducts were also built by Greeks before the Romans since Minoan Era [20]. One of the most famous aqueducts in ancient Greece is the tunnel (underground aqueduct) of Eupalinos for the water supply of Samos [21].

2.2.1.1 Aqueducts During Minoan Era

Angelakis and coworkers [21–23] published studies carried out on aqueducts built during the Minoan era. Their studies date a systematic evolution of water management in Crete to early in the Bronze Age, i.e., the Early Minoan period (ca. 3500–2150 B.C.). Usually, aqueducts were used to collect water from springs located apart from settlements [22]. Water was transported through aqueducts by closed or opened terracotta pipes and/or opened or covered channels of various dimensions and sections. Remains of aqueducts can be seen in Gournia, Karfi, Knossos (Mavrokolympos), Malia, Mochlos, Minoa, and Tylissos. Minoan hydrologists and engineers were aware of some of the basic principles and practices of water sciences which are still used today. With their knowledge, they emphasized construction and operation of aqueducts as a means to supply water to the people [21].

Most knowledge of how Minoan cities were supplied with potable water is mainly acquired from the Palace of Knossos. The water distribution system at Knossos, as well as the mountainous terrain and available springs, made aqueducts possible [24, 25]. According to studies by Angelakis et al. [23] and Mays et al. [25], the Minoan inhabitants of Knossos depended partially on wells, but mostly on water provided by the Kairatos River to the east of the low hill of the palace, and on springs. Indications suggest that the water supply system of the Knossos palace initially relied on the spring of Mavrokolybos, a limestone spring located 450 m southwest of the palace [26]. Due to growth in population, it was necessary to use springs from longer distances to meet the increased demand. Thus, an aqueduct made of terracotta pipe provided a conduit to convey water from a perennial spring on the Gypsadhes hill to meet the needs of the community [24, 27] as well as the aqueduct of Tylissos and the aqueduct of Manlia. The remnants of the aqueduct of Tylissos, which was 1.4 km long and extended from the spring Saint Mamas located northwest of the palace, suggests that part of the aqueduct was constructed from closed pipes and part of it was a curved channel [23]. At the end of the conduits, a stone-made tank was used for water pretreatment, mainly to remove sediments and/or suspended solids, perhaps the first water treatment plant on record. The aqueduct of Manlia was 2.4 km long and probably fed by a spring located west of the hilly area of Profitis Elias "Holly Hillock." This aqueduct was also a combination of closed terracotta pipes and opened channels [23].

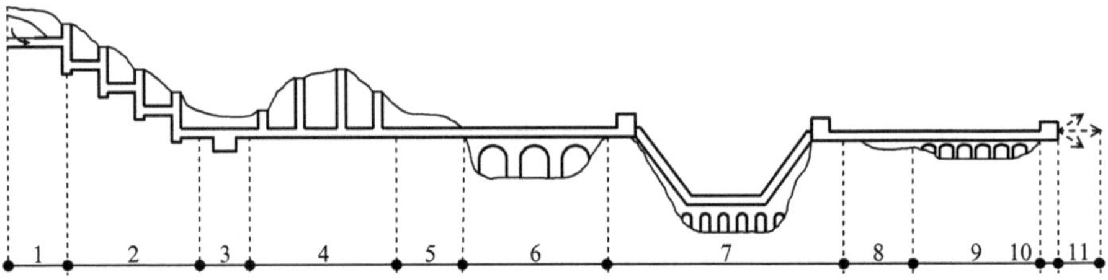

Fig. 2.1 Flow sheet and components of a Roman aqueduct: (*1*) source – *caput aquae*; (*2*) steep chutes (dropshafts); (*3*) settling tank; (*4*) tunnel and shafts; (*5*) covered trench; (*6*) aqueduct bridge; (*7*) inverted siphon; (*8*) *substruction*; (*9*) arcade; (*10*) distribution basin/*castellum aquae divisorium*; (*11*) water distribution system (Reprinted from De Feo and Napoli [15] with kind permission of © IWA Publishing)

2.2.1.2 Romans Aqueducts

The Romans were brilliant managers and engineers, and their systems rivaled modern technology [2]. Rome's water system is one of the marvels of the ancient world. The aqueducts represented an outward symbol of all that Rome stood for and all that Rome had to offer [28]. Regarding it, Pliny wrote in the *Naturalis Historia*, "the greatest wonder the world has ever seen" (XXXVI, 24, 123). Figure 2.1 is a diagram showing the components of a typical Roman aqueduct.

One of the largest Roman aqueducts is the Augustan Aqueduct Serino-Naples-Miseno (Serino aqueduct) in the Campania Region, Southern Italy, which was originally 96 km long [16]. However, about 49 km of branches were added making the total length of the Serino aqueduct complex around 145 km. After two millennia, the *Fontis Augustei* still continues to supply clear and freshwater for the inhabitants of Naples using the headwaters from Acquaro-Pelosi at 380 m above sea level and the Urciuoli at 330 m. De Feo and Napoli [16] investigated the historical development of "Serino Aqueduct," dating its construction between 33 and 12 B.C., when *Marcus Vipsanius Agrippa* was *curator aquarum* in Rome. Its principal use was to furnish the Roman fleet of Misenum with freshwater and also to supply water for the increasing demand of the important commercial harbor of *Puteoli* as well as drinking water for big cities such as *Cumae* and *Neapolis*.

The Aegean island Lesvos preserves one of the most impressive Roman aqueducts in Roman Greece dating to about the second or third century A.D. It was used as the water supply of Mytilene, the capital of the island, and for water supply and irrigation for the south eastern area of the island. The aqueduct was fed by the lake of Megali Limni (big lake), at the Olympus Mountain, and by other secondary springs, such as in the Agiassou area (i.e., Karini). The aqueduct passed through a varied landscape; thus, it includes sections built on the soil surface, as well as, tunnels, and bridges [26]. The Lesvos aqueduct was 22 km long, with a uniform slope of 0.0096 m/m, 0.65–1.10 m deep and 0.35–0.64 m wide (Karakostantinou 2006 cit. in [26]). It had a maximum capacity of about 127,000 m^3 per day. A 170 m section of the aqueduct still remains at the village of Moria. The section is 27 m high and

Fig. 2.2 (**a**) Remains of the Roman aqueduct in Moria, a Lesvian village at 6 km from Mytilene town; (**b**) Pillars and the arcs of the arches

consists of 17 arches, also called "Kamares" laying on their column (Fig. 2.2a). Each of the openings is comprised of three arches which are supported by a pillar. The tops of the pillars each have cymatium and abacus architectural details. The walls are built in the "emplekton" system, with two parallel rows of stone filled with rubble. The pillars and the arcs of the arches (Fig. 2.2b) are built of local marble which was readily available (Karakostantinou 2006 cit. in [26]).

One of the most complex and least studied aspects of Roman water supply is the urban distribution system. The study of the ruins of Pompeii gives a clearer understanding of a typical Roman urban water distribution and treatment system.

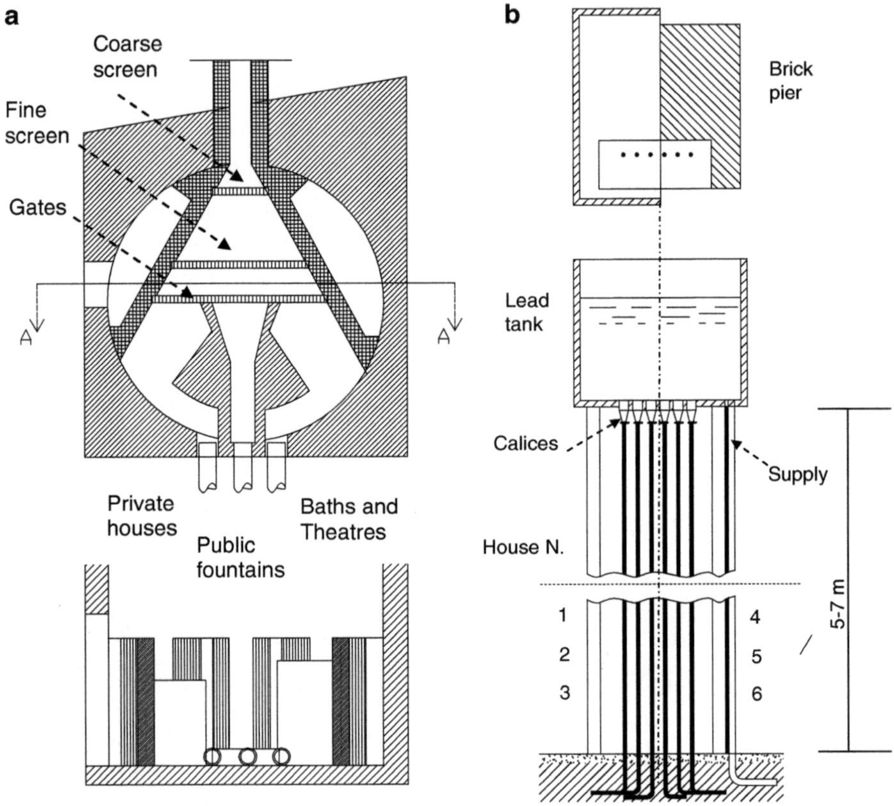

Fig. 2.3 (**a**) Plan of distribution arrangements inside *castellum divisorium* (or *castellum dividiculum*) at Porta Vesuvii: plan et elevation; (**b**) Pompeii water tower or *castellum secondarium* (Source: Refs. [25, 29])

A branch of Serino aqueduct ran from Sarno to Pompeii, terminating in the *castellum divisorium,* a large water tower at the highest point of the town near Porta Vesuvii.

The entire *castellum divisorium* was housed inside a large brick building and consisted of a shallow circular basin (Fig. 2.3a). After passing two coarse and fine screens to remove the worst impurities, the water then flowed through three parallel exit distribution channels; the entrance to each was regulated by wooden gates which would control water supply in case of drought [28]. The high water pressure was alleviated by smaller secondary water towers called *castellum secondarium* through a series of lead pipes (Fig. 2.3b). The water towers were lead tanks positioned on top of brick masonry pillars and were 6 m high.

2.2.2 *Ancient Sewers*

The water that was collected from springs, conveyed by aqueducts and used by humans was now contaminated with various impurities thus becoming wastewater. Wastewater caused odor, disease, attracted pests, and polluted the environment. In most places, wastewater just flowed into the streets or in ditches, but some cultures actually developed a system of sewers.

2.2.2.1 Roman Sewers

As stated above, much is known and has been written about Rome's water supply [4, 16, 28]. Much less has been said of the impact of wastewater management on Roman lifestyle. Although sewer and water pipes were not inventions of the Romans, since they were already present in other Eastern civilizations, they were certainly perfected by the Romans. The Romans resumed the engineering works of the Assyrians and turned their concepts into major infrastructure to serve all the citizens [2]. In addition to the famous aqueducts for supply of freshwater, ancient Rome had an impressive sewage system. The most famous as well the largest known ancient sewer is the Cloaca Maxima.

Cloaca Maxima

The *Cloaca Maxima* was built during the dynasty of Tarquin (VI century B.C.), nearly three centuries before the first aqueduct (Aqua Appia 312 B.C.). Initially constructed to drain the marsh on which Rome was later built, it originally stretched more than 100 m through the center of the *Forum Romanum*, between the *Basilica Aemilia* and *Julia.* It was 4.5 m wide and 3.3 m high, and was about 12 m below the present ground elevation [30]. It was built so solidly and with such foresight that it was used by the Romans for over 2,500 years and a section close to the "Torre dei Conti" is still working today [2].

The *Cloaca Maxima* spread throughout the city center. New shafts drained each of the imperial fora, the area around the Carcer, Temples of Saturn and Castor, and a large duct running alongside the Via Sacra fed into the main channel in front of the *Basilica Aemilia* [31].

West Colosseum Drainage Channel

As is often the case, drainage channels and sewer systems were found by accident at the archaeological excavation of the West Collector's Colosseum. The superintendent of Antiquities and the Fine Arts in Rome had ordered the cleaning of the floor below the arena of the Colosseum, as part of routine maintenance work.

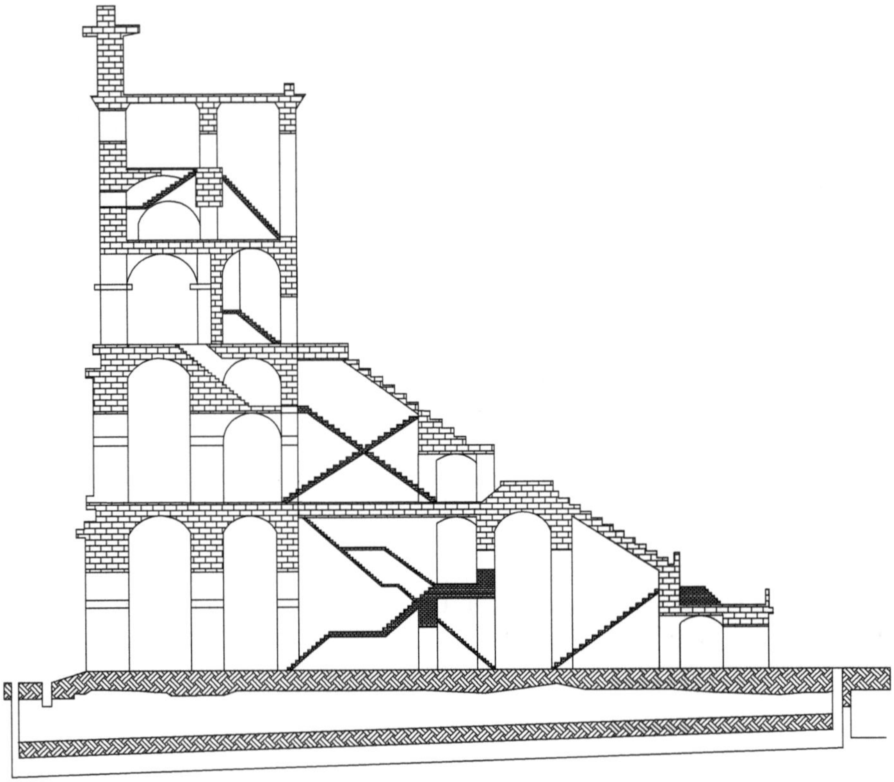

Fig. 2.4 West Collector in the architectonic system of the Coliseum

Superintendent Gianfilippo Carettoni invested funds to excavate the West Collector and found it completely blocked. Archaeologists determined the duct was 130 cm across and made of brick paving stones which dated to the time of Domitian, first emperor of the Flavian Dynasty (81–96 B.C.). The walls of the duct were originally built in oak planks wedged between the floor and the foundations, which were later covered with concrete. The position of the West Collector in the architectonic system of the Colosseum is shown in Fig. 2.4.

2.2.3 *After the Fall of Roman Empire*

The end of the Roman Empire led to the deterioration of the aqueducts and sanitation systems. Drainage and water supplies as well as the coastal road were no longer usable [2]. When the Roman Empire collapsed, the sanitary dark ages began and lasted for over 1,000 years (476–1800). People would toss waste into the streets, wastewater would run in open trenches along the walkways, and chamber pots would be emptied out of windows A few medieval cities, particularly in central and northern Italy including Milan, Arezzo, Bologna, and Florence had well intentioned programs using municipal statutes to control environmental conditions to improve city life and some cities in England also practiced good sanitation but these were the exceptions.

2.2.3.1 Black Plague

During the middle age, epidemics raged through the majority of European cities. Waste leached into the groundwater and wells. Rivers in London and Paris were open sewers. Waste provided food for rats and although the Black Plague in 1349 was not directly caused by poor sanitation, it had to be an indirect cause. Where there was good sanitation, such as at the Christchurch Monastery in Canterbury, England, no one contracted the Black Plague.

After the plague, scientists in England began to understand the connection between disease and sanitation practices and convinced the leaders of the country they must control the random disposal of waste products. In 1372, Edward III forbade throwing waste into the Thames River. In 1388, there was an act of Parliament which "forbade the throwing of waste into ditches, rivers, and water courses." During the fifteenth century, Henry VI established a Commission of Sewers. The word "sewer" comes from old English and means seaward from the fact that the drains in London were open ditches which pitched toward the Thames River and then to the sea.

2.2.3.2 Sewers of Paris

Around 1200, Phillipe Auguste had Parisian streets paved, incorporating a drain for wastewater in the middle. The first underground sewers were built in Paris in 1370 beneath Rue Montmartre and drained to a tributary of the Seine River. During the reign of Napoleon, 130 km of vaulted sewers were constructed and by 1805 there were over 182 miles of sewers. In 1850, the prefect for the Seine Baron Haussmann and the engineer Eugene Belgrand designed the present Parisian sewer and water supply networks. They built a double water supply network (one for drinking water and one for non-drinking water) and a sewer network which was 600 km long by 1878 [2].

2.3 Wastewater Treatment Technology Evolution

Today, wastewater treatment is performed through a series of processes which include preliminary treatment, primary treatment, secondary (biological treatment), and advanced treatment. Preliminary treatment removes rags, pieces of wood, grit, and other items which cannot be treated and will cause problems with further downstream units. Primary treatment is a physical separation of solid and liquids with heavier solids settling to the bottom of a tank and removed for further treatment. Secondary treatment is a biological treatment used to remove carbonaceous pollutants, and advanced treatment is used primarily to remove nutrients and trace organics.

In the nineteenth century, there were outbreaks of cholera which brought about the first centralized wastewater treatment technology which were early versions of primary treatment such as:

- 1860 – Septic Tank
- 1868 – Trickling sand filter
- 1893 – Septic tanks effluent treated with sand filters

2.3.1 Septic Tanks and Primary Treatment

The septic tank system is believed to have originated in France. In 1860, L.H. Mouras of Vesoul designed a cesspool in which the inlet and outlet pipes dipped below the water surface thus forming a water seal. Successively, Donald Cameron in 1895 patented septic tanks. The process gained great popularity and one observer commented, "Since the septic tank idea gained favor, every designer of sewage tanks has used the name septic for his tank, and apparently with good reason, for originally the word septic meant simply bacterial, just as the word anti-septic means anti-bacterial" (Melosi 2000 cit. [4]).

One problem connected with the settling methods for wastewater treatment right from the beginning was the disposal of the resulting sludge. The solution to this problem was the Imhoff Tank, patented in 1906, which consisted of a cylindrical settling tank and a digestion tank directly underneath. The Imhoff Tank, being the model for other so-called two-storied facilities, introduced anaerobic sludge digestion as the most commonly used method of sludge treatment in Germany. In the process of sludge treatment, digestion was followed by dewatering on drying beds and transportation for agricultural purposes [10]. In the United States, primary treatment was the most common form of wastewater treatment until 1972 when the passage of the Clean Water Act mandated secondary treatment.

2.3.2 Trickling Sand Filter

Up to 1900, virtually all sewage treatment, where it existed, was carried out by land treatment (irrigation). As the population continued to expand, it became more and more difficult to find sufficient areas of land on the fringes of the towns and cities [4]. The idea that there was a way to purify wastewater through the use of microorganisms gradually began to emerge around the end of the nineteenth century [2]. The first experiments carried out by Frankland in 1870 on filtration through soil, led to the concept of the tricking filter which is an attached growth (biofilm) process. In one of his experimental filters containing coarse porous gravel at Beddington Sewage Farm in Croydon, south of London, it was found that the wastewater loading rate of 0.045 m^3/m^3 of bed per day produced a well-nitrified (conversion of ammonia to nitrate) effluent and that the "filter" showed no signs of clogging after 4 months of operation (Stanbridge 1976 cit. in [4]). These early forms of attached growth secondary treatment were never patented but were commonly used throughout the United Kingdom. Around the turn of the century, trickling filters were introduced as the first artificial biological treatment method. Compared with irrigation fields, they were characterized by a much smaller need of space.

2.3.3 Activated Sludge Processes

Wastewater treatment has significantly evolved since early in the twentieth century. We have progressed from simply removing solids to producing high-quality effluent for direct water reuse. In 1900, Wigand patented a type of attached growth process, and in 1916, Poujoulat developed a second patent which was the predecessor to Rotating Biological Contactors (RBCs) [32].

In November 1912, Dr. Gilbert Fowler of the University of Manchester visited the United States in connection with the pollution of New York harbor [4]. Coming back to England, he described to his colleagues, Edward Ardern and William Lockett, some experiments that he had seen at the Lawrence Experimental Station of the Massachusetts State Board of Health, in which sewage was aerated in a bottle which had been internally coated with green algae. Ardern and Lockett published the results of their studies in 1914 on *Journal of the Society of Chemical Industry* [33]. Since that time, suspended growth (activated sludge) processes were developed in England and then brought to the United States. As secondary treatment became common, public health and the environment improved significantly. By the mid-1980s, almost every treatment plant in the United States used some form of biological treatment.

2.3.4 Advances in Treatment: Submerged Aerated Biofilters (BAFs)

Submerged aerated biofilters capable of removing solids and organic matter without requiring additional sedimentation did not appear until the end of the 1980s [34, 35]. The origins of BAFs can be tracked back to contact aerators, which have been used for more than 50 years [36]. Mendoza and Stephenson [37] dated the first report of these reactor types to 1913, where aerated tanks containing layers of slate supporting a biofilm were employed for sewage treatment. In the 1930s, work was carried out using the tank filter, otherwise known as the "Emscher Filter". These reactors used coarse slag instead of slate layers as the support medium, which led to an increase in dissolved oxygen content of the sewage and hence better treatment. The first submerged contact aeration process was developed in 1939 (Wilford and Conlon 1957 cit. in [37]). This introduced the concept of using two stages of biological treatment with rock and media. Griffith [38] reported that by 1943, submerged contact aeration units with sedimentation tanks were capable of achieving greater than 80% removal of BOD_5 at 74 plants treating domestic sewage in the United States. Modality and process conditions of submerged biological aerated filters are shown in Fig. 2.5.

2.3.5 Alternatives to Mechanical Treatment Processes

Capital investments and maintenance costs have been the most common constraints to wastewater treatment plant installation in small communities and in developing countries. This situation gave rise to widespread interest worldwide in using "natural systems" for wastewater treatment. The U.S. EPA established the Small Flows Research Program in 1973 to help support viable alternative wastewater treatment systems in areas where site conditions precluded the use of conventional systems. Among them, constructed wetlands represent a very promising technology. From the technical point of view, constructed wetlands consist of low deep basins, with long Hydraulic Retention Time (HRT), filled up with sand, gravel, and earth on a protective coating of waterproof sheath and planted with macrophytes [39, 40]. Constructed wetlands are usually classified according to the different types of macrophytes and hydraulic regime (Fig. 2.6).

First studies were carried out at pilot scale at the Max Planck Institute, Germany, in 1952. Today, constructed wetlands are found commonly in Great Britain, Canada, the United States, Mexico, India, South Africa, Brazil, Australia, Italy, and many European countries ranging in size from small single-family residential systems to major facilities treating up to 12 million gallons per day of wastewater. These systems are becoming more acceptable to local and regional regulators as the data on their performance continue to prove the cost-effective nature of their operation.

2.3.6 Advanced Wastewater Treatment

With greater understanding of the impact of wastewater on the environment and more sophisticated analytical methods, advanced treatment is becoming more common in developed countries, primarily for the removal of nitrogen, phosphorus, or both. In 1962, Lutdzack and Ettinger put forward the use of an anoxic zone to achieve biological denitrification in an activated sludge process, introducing a

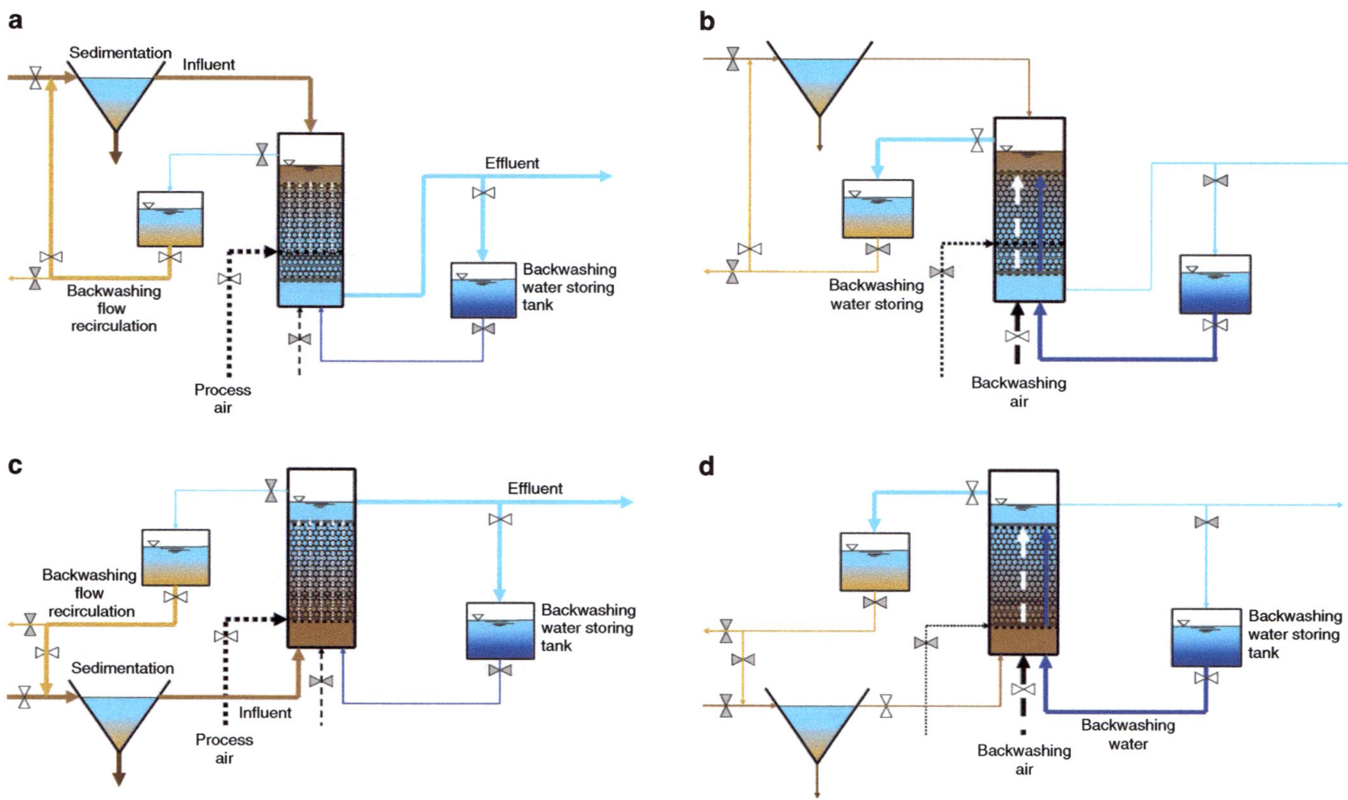

Fig. 2.5 Modalities and process conditions of submerged biological aerated filters: (**a**) down-flow modality, process condition; (**b**) down-flow modality, backwashing condition; (**c**) up-flow modality, process condition; (**d**) up-flow modality, backwashing condition (Source: Ref. [14])

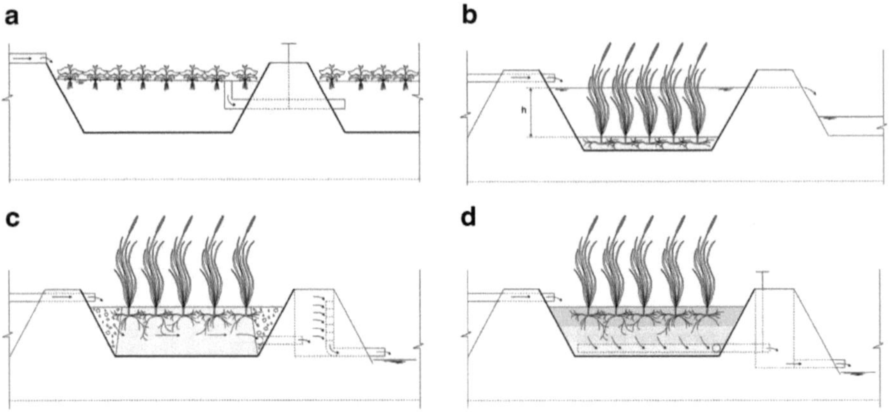

Fig. 2.6 (**a**) Aquaculture; (**b**) Free water surface (FWS); (**c**) Subsurface flow system-horizontal (SFS-H); (**d**) Subsurface flow system-vertical (SFS-V)

practice that is now commonly applied [41]. Studies carried out by Downing et al. [18] are now incorporated into design methods of biological nitrification. Ten years later, James Barnard developed and patented the biological removal of nitrogen and phosphorous in a single sludge system [8–10]. The MLE (modified Lutdzack and Ettinger) process and the five-stage Bardenpho process are widely used throughout the world for both nitrogen and phosphorus removal and can achieve effluent total nitrogen less than 4 mg/L and effluent total phosphorus of less than 0.1 mg/L.

2.3.7 *Disinfection Practices, Nineteenth Century and Twentieth Century*

In the nineteenth century, many scientists believed that odor was the cause of disease, so chemicals such as chlorine were used as deodorants [21]. In 1854, chloride of lime was used to deodorize London's sewage. In 1859, the Metropolitan Board of Works, London, showed that a dosage of 400 lbs/MG of Calcium Chloride could delay purification of raw sewage for as much as 4 days. By 1880, scientists began to understand pathogenic bacteria and their association with specific disease. For example, calcium chloride was used to treat feces from typhoid patients before disposal to sewers.

In 1893, in Hamburg, Germany, chlorine was first used on a plant scale for disinfection of wastewater and in 1906, ozone was used in France as a disinfection agent. In 1909, compressed, liquefied chlorine was commercially available and by 1914, equipment had been designed for metering and applying chlorine gas to wastewater. The first recorded use of ultraviolet light for disinfection was in France in 1916. By the early twenty-first century, ultraviolet disinfection has become the most common disinfection process in use.

2.3.8 Next Generation of Wastewater Treatment

2.3.8.1 Advanced Oxidation Processes

Once advanced treatment and the reduction of nitrogen and phosphorous as well as the prevention of eutrophication was employed at most treatment plants, the removal of biorecalcitrant and emerging contaminants became the next goal for wastewater treatment and reuse. Although biological treatment processes are, at present, the cheapest and the most environmental friendly, their potential for removal of nonbiodegradable COD and micro-pollutants is limited. Therefore there is an increasing need to focus on new treatment technologies. Advanced oxidation processes (AOPs) are promising methods for wastewater contaminated by biorecalcitrant compounds. AOPs are based on the generation of highly reactive hydroxyl radicals (HO•) which are powerful oxidants ($E = 2.8$ V) that are able to mineralize quickly and unselectively a broad range of organic compounds.

They are produced on-site most commonly by combinations of ultraviolet irradiation (UV), ozone, hydrogen peroxide, and (photo) catalysts (mainly transition metal salts) forming rather compact, homogenous oxidation systems. The heterogeneous type of AOPs particularly involve TiO_2-mediated semiconductor (photo) catalytic processes.

2.3.8.2 Decentralized Systems

Decentralized systems are also being developed and include urine source separation [27, 41], although this practice is rather ancient considering that Romans usually separate urine for laundry (*fullonica*), wool laundry (*officinal lanifricatoria*), dyeing (*officina infectoria*), and tannery (*officina coriariorum*) [1].

Decentralized systems have both advantages and disadvantages in cost and performance, but they have their place in the overall management of wastes [50].

2.4 Evolution of Pollutant Identification and Characterization

Since 1880, scientists began to understand pathogenic bacteria and their association with specific disease and to understand that waterborne pathogens had a significant effect on public health. Once bacterial pollution of water was understood, there also became an understanding that there were other pollutants in wastewater but unable to measure those pollutants.

A milestone in pollution control was the concept of biochemical oxygen demand (BOD). This test is used globally to determine the carbonaceous and nitrogenous demand on a system for oxygen and is the universal measure of pollution. The significance and use of this test has evolved over many years. According to Baird and Smith [18], the history of BOD can be subdivided in three main periods. The first is characterized by the discovery and description of oxygen utilization in wastewater treatment. Pasteur's pioneering microbiology and Winkler's development of dissolved oxygen measurement in water afforded early sanitary scientists an understanding and ability to measure the relationship between organic substances and dissolved oxygen in receiving water bodies. In 1912, the Eighth Report of the Royal Commission on Sewage Disposal of London established the first standards and tests for BOD analysis to be applied to sewage and sewage effluents. These standards were copied by many other countries [2]. The second period was characterized by the development and refinement of the measurements of dissolved oxygen and BOD and the third period began in 1972 and is characterized by the use of BOD_5 test for regulatory compliance and process control.

Since the development and refinement of the BOD test, other tests have been developed which allow for much more detailed characterization of wastewater pollutants. Atomic absorption, gas chromatography, mass spectroscopy, liquid chromatography, and other techniques allow determination of metals, organic compounds, and nutrients at very low concentrations [20]. There is currently technology available to examine DNA of various microorganisms used in the treatment process and will allow for better process optimization. With instrumentation, pollutants can be determined in very low concentrations (nanogram); therefore, their impact on receiving waters can better be defined and treatment technologies developed. This is especially important for those pollutants that are endocrine disruptors such as pharmaceutical and hormones.

2.5 Conclusion

In 2005, the British Medical Association and the American Medical Association declared modern wastewater treatment as having the greatest positive impact on public health and longevity. Without modern wastewater treatment technologies, life expectancy would be less than 40 years. By reviewing the history and evolution of waste treatment, it is easy to understand its importance. But as we develop technology to better analyze for pollutants, we must also continue to develop technology to manage those pollutants.

This history will continue to evolve as long as there are human beings on Earth. "And so this process must continue in a never-ending cycle" (Aristotle, *Metereologica* II, 3).

Acknowledgment The authors wish to thank Dr. Anna Carrafiello for the redaction of Fig. 2.3.

References

1. De Feo G, Laureano P, Drusiani R, Angelakis AN (2010) Water and wastewater management technologies through the centuries. Water Sci Technol: Water Supply 10(3):337–349
2. Lofrano G, Brown J (2010) Wastewater management through the ages: a history of mankind. Sci Total Environ 408:5254–5264
3. Brown J (2005) The early history of wastewater treatment and disinfection. In: World water congress 2005: impacts of global climate change – proceedings of the 2005 world water and environmental resources congress, Anchorage, AK, 15–19 May 2005
4. Cooper PF (2007) Historical aspects of wastewater treatment. In: Lens P, Zeeman G, Lettinga G (eds) Decentralised sanitation and reuse: concepts, systems and implementation. IWA Publishing, London
5. Vuorinen HS, Juuti PS, Katko TS (2007) History of water and health from ancient civilizations to modern times. Water Sci Technol 7:49–57
6. Aiello AE, Larson MS, Sedlak R (2008) Hidden heroes of the health revolution sanitation and personal hygiene. Am J Infect Control 36:128–151
7. HDR (Human Development Report) (2006) Beyond scarcity: power, poverty and the global water crisis. United Nations Development Programme, New York
8. Lucking B (1984) Evaluating the sanitary revolution: typhus and typhoid in London 1851–1900. In: Woods R, Woodward J (eds) Urban Disease and Mortality in Nineteenth Century England. Batsford, London/New York, pp 102–119
9. Sorcinelli P (1998) Storia sociale dell'acqua. Riti e Culture. Mondadori, Milano
10. Seeger H (1999) The history of German wastewater treatment. Eur Water Manag 2:51–56
11. Shifrin NS (2005) Pollution management in twentieth century. J Environ Eng Asce 131:676–691
12. Maneglier H (1994) Storia dell'acqua. SugarCo, Milano
13. Neri Serneri S (2007) The construction of the modern city and the management of water resources in Italy, 1880–1920. J Urban Hist 33:813–827
14. Sori E (2001) La città e i rifiuti - Ecologia urbana dal Medioevo al primo Novecento. Il Mulino, Saggi, Bologna, Italy
15. Tarr JA (1985) Historical perspectives on hazardous wastes in the United States. Waste Manag Res 3:95–113
16. De Feo G, Napoli RMA (2007) Historical development of the Augustan aqueduct in Southern Italy: twenty centuries of works from Serino to Naples. Water Sci Technol 7:131–138
17. Tolle-Kastenbein R (2005) Archeologia dell'Acqua (Water archaeology). Longanesi, Roma
18. Baird BR, Smith RK (2002) Third century of biochemical oxygen demand. Water Environment Federation, Alexandria
19. Breasted JH (1906) Ancient records of Egypt: historical documents from the earliest times to the Persian conquest, vol 5. University of Chicago Press, Chicago
20. Adam JP (1988) L'Arte di Costruire presso i Romani. Materiali e Tecniche (Roman building: materials and techniques). Longanesi, Milano
21. Angelakis AN, Koutsoyiannis D, Tchobanoglous G (2005) Urban wastewater and stormwater technologies in ancient Greece. Water Res 39:210–220
22. Angelakis AN, Spyridakis SV (1996) The status of water resources in Minoan times: a preliminary study. In: Angelakis AN, Issar AS (eds) Diachronic climatic impacts on water resources with emphasis on Mediterranean region. Springer, Heidelberg, pp 161–191
23. Angelakis AN, Savvakis YM, Charalampakis G (2007) Aqueducts during the Minoan era. Water Sci Technol 7:95–101
24. Mays LW (2007) Ancient urban water supply systems in arid and semi-arid regions. In: Proceedings of international symposium on new directions in urban water management. UNESCO, Paris, France, 12–14 Sept 2007. Korea Water Resources Association. http://www.kwra.or.kr. Accessed Feb 2010

25. Mays LW, Koutsoyiannis D, Angelakis AN (2007) A brief history of urban water supply in antiquity. Water Sci Technol: Water Supply 7(1):1–12
26. De Feo G, Mays LW, Angelakis AN (2011) Water and wastewater management technologies in the ancient Greek and Roman civilizations. In: Wilderer P (ed) Treatise on water science, vol 4. Academic, Oxford, pp 3–22
27. Graham JW (1987) The palaces of Crete. Princeton University Press, Princeton
28. Hodge AT (2002) Roman aqueducts & water supply, 2nd edn. Gerald Duckworth & Co. Ltd, London
29. Kretzschmer F (1966) La technique romaine (trans. Breur J, Ulrix F). La Renaissance du Livre, Brussels
30. Hopkins J (2007) The cloaca maxima and the monumental manipulation of water in archaic Rome. The Water Rome 4:1–15
31. Narducci P (1889) Sulla fognatura della città di Roma. Descrizione tecnica. Forzani, Roma
32. Patwardhan AW (2003) Rotating biological contactors: a review. Ind Eng Chem Res 42:2035–2051
33. Ardern E, Lockett WT (1914) Experiments in the oxidation of sewage without the aid of filters. J Soc Chem Ind 33:524
34. De Feo G (2004) Wastewater treatment with submerged biological aerated filters (Trattamento delle acque reflue mediante biofiltri aerati a letto sommerso). PhD thesis in Civil Engineering for the environment and territory, Cycle II, new series, University of Salerno, Faculty of Engineering, BNI 2005–6275 T, Bibl. National Central Florence Location: TDR.2004.5510, Inventory: CF20045510 (manual request), in Italian
35. Pujol R, Hamon M, Kandel X, Lemmel H (1994) Biofilters: flexible, reliable biological reactors. Water Sci Technol 11:33–38
36. Rusten B (1984) Wastewater treatment with aerated submerged biological filters. J Water Pollut Control F56:424–431
37. Mendoza-Espinosa L, Stephenson T (1999) A review of biological aerated filters (BAFs) for wastewater treatment. Environ Eng Sci 16(3):201–216
38. Griffith LB (1943) Contact aeration for sewage treatment. Eng News Rec 28:60–64
39. EPA Manual (2000) Constructed wetlands treatment of municipal wastewater. United States Environmental Protection Agency, Office of Research and Development, Cincinnati
40. Kadlec RH, Knight RL (1996) Treatment wetlands. Lewis Publishers, Boca Raton/New York
41. Lutdzack FT, Ettinger MB (1962) Controlling operation to minimize activated sludge effluent nitrogen. J Water Pollut Control F 34:124–138
42. Downing AL, Painter HA, Knowles G (1964) Nitrification in the activated sludge process. J Inst Sew Purif 2:130
43. Barnard JL (1973) Biological denitrification. Water Pollut Control 72:705–720
44. Barnard JL (1974) Cut P and N without chemicals. Water Waste Eng 11:41–44
45. Barnard JL (1975) Biological nutrients removal without the addition of chemicals. Water Res 9:485–490
46. Gayman M (2008) Glimpse into London's early sewers. Cleaner Magazine. www.swopnet.com/engr/londonsewers
47. Jönsson H, Stenström TA, Svensson J, Sundin A (1997) Source separated urine-nutrient and heavy metal content, water saving and faecal contamination. Original research article. Water Sci Technol 35(9):145–152
48. Rossi L, Lienert J, Larsen TA (2009) Real-life efficiency of urine source separation. Original research article. J Environ Manage 90(5):1909–1917
49. Wilderer PA, Schreff D (2000) Decentralized and centralized wastewater management: a challenge for technology developers. Water Sci Technol 41(1):1–8
50. Ettinger MB (1965) Developments in detection of trace organic contaminants. J Am Water Work Assoc 57:453–457

Chapter 3
The Removal of Illicit Drugs and Metabolites During Wastewater and Drinking Water Treatment

Alexander L.N. van Nuijs and Adrian Covaci

3.1 Introduction

An increasing concern has been raised in recent years about the occurrence of pharmaceutically active compounds in the aquatic environment. Since they have been detected in a wide variety of water samples (drinking water, groundwater, surface water, influent and effluent wastewater) in concentrations ranging from the ng/L to the μg/L level, these compounds have been recognized as emerging contaminants [1]. For some compounds, an association has been made with their environmental concentration and negative effects on aquatic organisms [2], but for most compounds, little is known about the long-term effects derived from simultaneous exposure of these compounds on the aquatic environment.

While the occurrence, removal, and fate of pharmaceuticals was investigated in several papers, the presence of illicit drugs and their metabolites in water samples has not been intensively investigated. However, these substances are widely used worldwide and as a consequence, it is likely that, comparable with pharmaceuticals, these compounds are present in the water cycle. In general, municipal WWTPs have not been designed to remove residues of pharmaceutically active compounds and the ability of such compounds to pass through WWTPs depends on various factors, including the chemical and biological persistence of the individual compound, its sorption behavior, its evaporation via water vapor, and the technology used for sewage purification.

The presence of illicit drugs and their metabolites has been evidenced in (influent and effluent) wastewater, surface water, and drinking water in concentrations

A.L.N. van Nuijs • A. Covaci (✉)
Toxicological Centre, University of Antwerp, Universiteitsplein 1, 2610 Antwerp, Belgium
e-mail: alexander.vannuijs@ua.ac.be; adrian.covaci@ua.ac.be

S.K. Sharma and R. Sanghi (eds.), *Advances in Water Treatment and Pollution Prevention*, DOI 10.1007/978-94-007-4204-8_3,

up to μg/L levels, with the highest concentrations observed for cocaine (COC) and its most important metabolite benzoylecgonine (BE) [3]. In analogy with pharmaceuticals, illicit drugs and their metabolites enter the aquatic cycle mainly through excretion after human consumption in wastewater. If an incomplete removal of these compounds during wastewater treatment and drinking water treatment is obtained, they may reenter surface water (via effluents) and drinking water (via surface or ground water), respectively.

Since illicit drugs (and their metabolites) are often highly pharmaceutically active, their presence in the aquatic environment can lead to risks comparable with pharmaceuticals. However, until now, this has not been investigated for illicit drugs. To decrease the risk of negative effects on the aquatic environment, research could also be oriented to optimize wastewater and drinking water treatment procedures for the removal of these substances. Consequently, the concentrations of these compounds in surface and drinking water will be low or negligible, leading to lower exposure levels.

In this chapter, an overview of the current knowledge of the removal of illicit drugs and their metabolites during wastewater and drinking water treatment is given together with a critical discussion regarding future research needs in this field.

3.2 Removal of Illicit Drugs and Metabolites During Wastewater Treatment

3.2.1 Cocaine and Metabolites

For this group of illicit drugs, the removal during wastewater treatment of the two compounds that are in the highest concentrations measured in surface water, COC and BE, was investigated by several research groups. Different wastewater treatment technologies, conventional activated sludge (CAS), trickling filters, and biological filtration have been investigated together with wastewater treatment plants (WWTPs) applying just primary treatment [4–10].

The majority of the removal experiments were done in WWTPs working with the CAS technique. The results of these studies showed that the application of CAS allowed an average removal percentage higher than 90% for COC and BE. Only in Croatia, decreased removal efficiencies (~50%) were observed [9], probably due to a short residence time in this WWTP. A high removal of cocainics (>93%) was also observed in one Spanish WWTP [8] (Fig. 3.1). Overall, we can conclude that the CAS technique works efficient in removing COC and BE from influent wastewater.

Metcalfe et al. compared the working of the CAS technology for the removal of COC and BE with a WWTP that applies only primary treatment [7]. Their results showed that in the case of only primary treatment, removal efficiencies were lower than 40% for COC and BE compared with the earlier mentioned values of >90% for

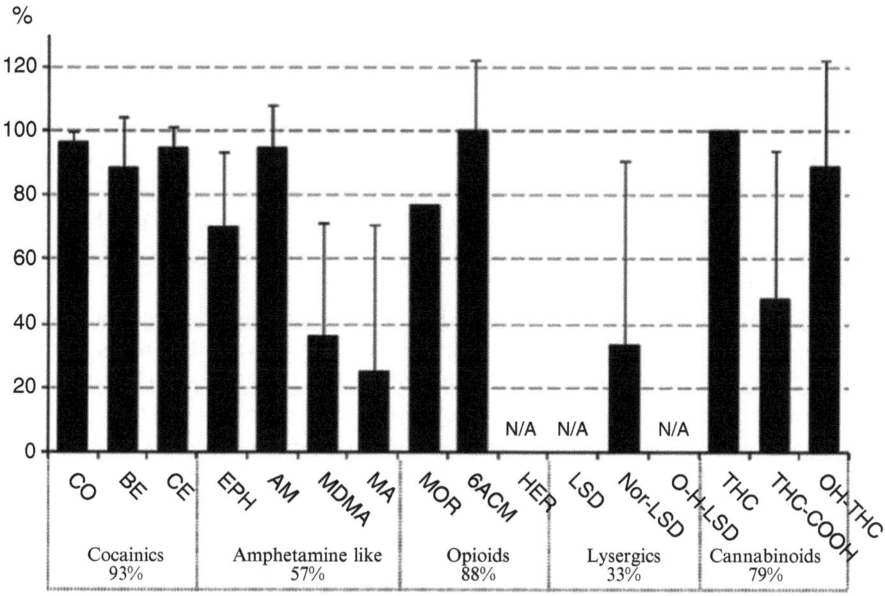

Fig. 3.1 Average removal of illicit drugs during CAS wastewater treatment (Reproduced from Postigo et al. with kind permission of © Elsevier [8])

the CAS technique. It has to be pointed out that the residence time in the WWTP working only with primary treatment is five times lower which can also contribute to lower removal efficiencies.

In two studies, the CAS technology was compared with a treatment based on biological filters [5, 8]. Both experiments concluded that the removal efficiencies in the WWTPs working with biological filters were lower than the CAS-based WWTPs, but no absolute values were given. Karolak et al. did not conclude if the lower removal of COC and BE with biological filters is due to the technology itself or due to the lower residence time in the WWTP that is accompanied with this kind of treatment [5].

Kasprzyk-Hordern et al. evaluated the use of trickling filters for the removal of COC and BE from wastewater. With observed removal efficiencies lower than 25%, and for BE even higher concentrations in effluent than in influent, this kind of wastewater treatment technology seems not efficient for removing both compounds from wastewater [6].

In conclusion, the most applied CAS technique is efficient in removing COC and BE from wastewater (>90% removal). Other techniques, such as biological filters and trickling filters, show removal efficiencies lower than 50% and as a result relatively high loads of COC and BE can be released in surface water through effluents.

3.2.2 Amphetamine-like Stimulants

In the amphetamine-like stimulant (ALS) group, removal experiments were executed for the following compounds: amphetamine (AMP), methamphetamine (METH), methylenedioxymethamphetamine (MDMA), methylenedioxyamphetamine (MDA), and methylenedioxyethylamphetamine (MDEA) [4–9, 11, 12]. Following wastewater treatment technologies were evaluated: CAS, biological filters, trickling filters, reverse osmosis, return activated sludge (RAS), and just primary treatment.

Highly variable removal rates for all ALS compounds were observed for WWTPs applying the CAS technology (Fig. 3.1). In general, removal efficiencies were the highest for AMP, in most cases higher than 90%. For MDMA, METH, and MDA, the average removal was lower (on average 50%). However, it has to be pointed out that high ranges for removal rates were reported and that in some cases even higher concentrations in effluent than in influent were observed. A possible explanation for this phenomenon is the transformation from one compound into another during wastewater treatment together with deconjugation processes.

Three studies concluded that the use of the biological filters or just a primary treatment significantly decreased the removal efficiency for the ALS [5, 7, 8]. Loganathan and coworkers briefly mention in their paper that the use of RAS is efficient in removing METH from wastewater [12]. However, no numbers are given and as a consequence it is difficult to compare RAS with other techniques. The use of trickling filters in the treatment of wastewater containing AMP seems as efficient as CAS, as was observed by Kasprzyk-Hordern and colleagues. For both technologies, removal efficiencies higher than 90% were observed [6].

Boleda et al. [11] evaluated the use of reverse osmosis for the removal of ALS from wastewater. They observed that for MDMA, AMP, and METH this technique works well with removal >80%. Moreover, they observed less variability in the removal compared with the CAS technology. For MDA and MDEA, the removal was lower (between 50% and 60%), but the variability was again low.

Except for biological filters, all techniques that were evaluated for the removal of ALS showed rather high removal rates. Important to point out is that for CAS, highly variable removal rates were observed, probably related to transformation process from one compound to another during this treatment. The application of reverse osmosis seems to overcome this problem and is the most efficient technique for the removal of ALS from wastewater.

Only a limited number of papers have reported on the occurrence of ALS in sewage sludge and biosolids resulting from WWTPs. Kaleta et al. [13] measured concentrations of AMP ranging between 5 and 300 ng/g dry weight in Austrian raw and treated sewage sludge. In US sludge, Jones-Lepp and Stevens [14] have detected and quantified METH at lower levels up to 4 ng/g dry weight.

3.2.3 Opiates and Related Compounds

Experiments show that morphine (MOR) and 6-monoacetylmorphine (6-MAM) are fairly well removed during CAS wastewater treatment, with an average removal rate of 80% for MOR and 90% for 6-MAM [8, 9, 15] (Fig. 3.1). For MOR, large variations in removal efficiencies were observed which can be related with the cleavage of morphine conjugates during the treatment process. Biological filters showed less efficient removal of the two compounds in comparison with CAS [8].

Methadone (MTD) and its major metabolite 2-ethylidene-1,5-dimethyl-3,3-diphenylpyrrolidine (EDDP) were not well removed during wastewater treatment based on CAS. Only 30% and 2% were removed for MTD and EDDP, respectively [9, 15]. These two compounds fairly withstand CAS wastewater treatment and as a result they can easily enter surface water.

For codeine (COD), a similar removal as for MTD was observed, with only 33% retained during CAS treatment [15]. Also for this compound, an important input in surface water can be expected. However, Boleda also evaluated the use of reverse osmosis for the removal of COD from wastewater and they concluded that this technique is highly efficient in removing this compound from wastewater [11].

3.2.4 Cannabinoids

In this group, the removal of tetrahydrocannabinol (THC), 11-hydroxytetrahydrocannabinol (OH-THC), and 11-nor-9-carboxytetrahydrocannabinol (THC–COOH) during wastewater treatment were investigated [8, 15] (Fig. 3.1). Similar results were found in both studies with an excellent removal of THC (>97%) and OH-THC (90%), but low removal rates (40% on average) for THC–COOH. A possible explanation for this is the cleavage of conjugates (glucuronides and/or sulfates) of this metabolite during wastewater treatment, similarly with what was earlier mentioned for the opiates.

3.2.5 Other Compounds

One study investigated the removal of ketamine during wastewater treatment based on the CAS technique and found removal rates higher than 80%, concluding that it is efficiently removed [4].

N-desmethyl lysergic acid diethylamide (nor-LSD) showed very low or no removal during CAS wastewater treatment, as was observed by Postigo et al. [8].

3.3 Removal of Illicit Drugs and Metabolites During Drinking Water Treatment

3.3.1 Cocaine and Metabolites

The behavior of COC and BE during different drinking water treatment steps was evaluated [16] (Fig. 3.2). In the first step consisting of prechlorination, coagulation/flocculation, and sand filtration, COC and BE were only removed at 13% and 9% from the source water (being surface water). The low observed values can be explained by the low reactivity of both analytes to chlorine. The following ozonation step had removal rates of 24% and 43% for COC and BE. These rather low values can be due to the absence of active sites (e.g., double of triple carbon bonds) on both compounds that are reactive to ozone. The next treatment with granulated activated carbon filters (GAC) resulted in high removal rates, most probably to efficient adsorption. For COC, a complete removal was observed, while for BE 72% was removed in this step. In the last postchlorination step, another 27% of the BE was removed. The total removal percentages were 100% for COC and 90% for BE, and as a result concentrations of BE up to 130 ng/L were found in the processed drinking water.

3.3.2 Amphetamine-like Stimulants

AMP, MDMA, METH, and MDA were followed through the treatment process of surface water to drinking water [16] (Fig. 3.2). AMP, METH, and MDA were completely removed in the first step of the treatment (prechlorination, coagulation/flocculation, sand filtration), because of their high reactivity to chlorine. MDMA was only removed in small amounts (23%) in this step. Future research is needed to investigate why this compound is behaving different compared to the other strongly related compounds. MDMA is removed in the next steps (ozonation, GAC filtration, and postchlorination) for 28%, 88%, and 100% respectively. As a result, no ALS compound could be detected in the processed drinking water.

3.3.3 Opiates and Related Compounds

The behavior of MOR, COD, MTD, and EDDP was investigated during drinking water treatment [15]. The drinking water treatment plant treats surface water in four steps: (1) Prechlorination, flocculation/coagulation, and sand filtration, (2) ozonation, (3) GAC filtration, and (4) postchlorination. In the first step, MOR and COD were efficiently removed with removal percentages of 96% and 88%,

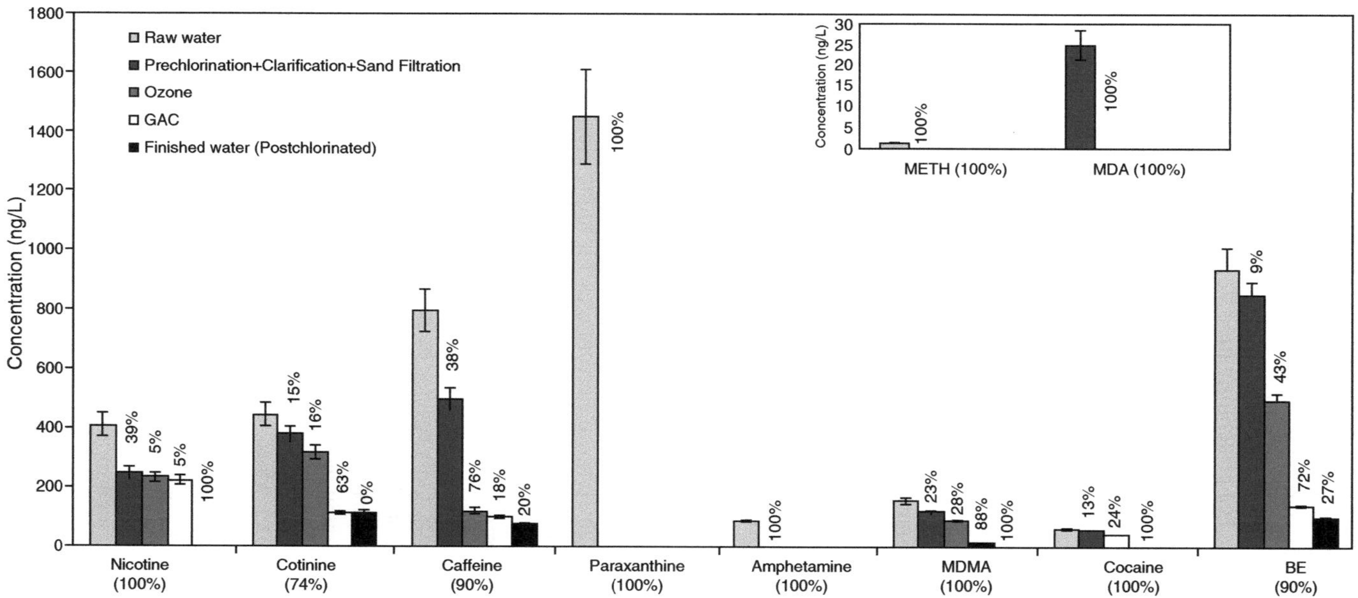

Fig. 3.2 The removal of illicit drugs during drinking water treatment. The elimination percentages of the compounds after each treatment are included on each *bar* and the total removal percentages are included in *parentheses*. Nicotine, cotinine, caffeine, and paraxanthine are not discussed in this chapter (Reproduced from Huerta-Fontela et al. [16] with kind permission of © The American Chemical Society (2008))

respectively. For MTD and EDDP, lower removal rates were observed (54% and 28%), resulting from a different reactivity for chlorine. The following ozonation step resulted in a complete removal of MOR. COD, MTD, and EDDP were only removed for 56%, 44%, and 37%. The GAC filtration then completely removed COD, while MTD and EDDP surprisingly resisted during this treatment step (59% and 52% removal). In the last postchlorination step, 54% of the MTD concentrations and 28% of the EDDP concentrations were removed. This elimination resulted in the absence of COD and MOR in the finished drinking water (100% removal), while MTD and EDDP could be identified in drinking water (91% and 88% removal).

3.3.4 Cannabinoids

THC and THC–COOH were already completely removed in the first drinking water treatment step (prechlorination, flocculation/coagulation, and sand filtration), and as a result these compounds were not detected in finished drinking water [15].

3.4 Conclusions and Future Research

This chapter documents that experiments have been carried out to investigate the removal of a wide range of illicit drugs during wastewater and drinking water treatment. For wastewater treatment, the application of the CAS technology revealed in most experiments high removal percentages for some compounds (COC, BE, MOR, 6-MAM, ketamine, THC, OH–THC), while for other compounds (THC–COOH, COD, MTD, EDDP, nor-LSD) low removal rates were observed. In the ALS group, highly variable removal was seen. Other wastewater treatment techniques than CAS were also investigated, including reverse osmosis, trickling filters, biological filters, and RAS. Only reverse osmosis showed a better performance than CAS for ALS and COD; other techniques showed to be less or equally efficient. The removal of illicit drugs during drinking water treatment was only investigated in one drinking water treatment plant and showed complete removal of all compounds except BE, MTD, and EDDP. However, the concentrations of these compounds found in finished drinking water were low, up to 130 ng/L for BE. Yet it is not known if such concentrations can affect human health during prolonged and chronic exposure through drinking water.

Future research in the removal of illicit drugs is however still needed. All studies executed until now were only based on simple experiments comparing influent and effluent concentrations. However, in order to better understand the processes occurring during wastewater treatment, biodegradation studies should be conducted under laboratory-controlled conditions simulating WWTPs and DWTPs. The following processes should be investigated more in detail: (1) The disposal of illicit drugs and metabolites onto sludge, (2) deconjugation reactions of conjugates to parent compound or metabolite which could lead to higher effluent concentrations, (3) transformation reactions of metabolites and parent compounds. To further decrease the input of illicit drugs and metabolites into surface water and drinking water, alternative and new treatment techniques should also be investigated, such as membrane bioreactors, ozonation and photocatalytic processes, constructed wetlands, advanced sorbents, and nanotechnology.

Acknowledgments Alexander van Nuijs and Dr. Adrian Covaci are grateful to the Flanders Scientific Funds for Research (FWO) for financial support. We also acknowledge the funding of a FWO project on pharmaceuticals and illicit drugs in wastewater (G. 0177.09 N).

References

1. Richardson SD (2009) Water analysis: emerging contaminants and current issues. Anal Chem 81:4645–4677
2. Khetan SK, Collins TJ (2007) Human pharmaceuticals in the aquatic environment: a challenge to green chemistry. Chem Rev 107:2319–2364
3. Zuccato E, Castiglioni S (2009) Illicit drugs in the environment. Philos Trans R Soc A 367: 3965–3978
4. Huerta-Fontela M, Galceran MT, Martin-Alonso J, Ventura F (2008) Occurrence of psychoactive stimulatory drugs in wastewaters in north-eastern Spain. Sci Total Environ 397:31–40
5. Karolak S, Nefau T, Bailly E, Solgadi A, Levi Y (2010) Estimation of illicit drugs consumption by wastewater analysis in Paris area (France). Forensic Sci Int 200:153–160
6. Kasprzyk-Hordern B, Dinsdale RM, Guwy AJ (2009) The removal of pharmaceuticals, personal care products, endocrine disruptors and illicit drugs during wastewater treatment and its impact on the quality of receiving waters. Water Res 43:363–380
7. Metcalfe C, Tindale K, Li H, Rodayan A, Yargeau V (2010) Illicit drugs in Canadian municipal wastewater and estimates of community drug use. Environ Pollut 158:3179–3185
8. Postigo C, Lopez de Alda MJ, Barcelo D (2010) Drugs of abuse and their metabolites in the Ebro River bassin: occurrence in sewage and surface water, sewage treatment plants removal efficiency, and collective drug usage estimation. Environ Int 36:75–84
9. Terzic S, Senta I, Ahel M (2010) Illicit drugs in wastewater of the city of Zagreb (Croatia) – estimation of drug abuse in a transition country. Environ Pollut 158:2686–2693
10. van Nuijs ALN, Pecceu B, Theunis L, Dubois N, Charlier C, Jorens PG, Bervoets L, Blust R, Neels H, Covaci A (2009) Spatial and temporal variations in the occurrence of cocaine and benzoylecgonine in waste- and surface water from Belgium and removal during wastewater treatment. Water Res 43:1341–1349
11. Boleda R, Majamaa K, Aerts P, Gomez V, Galceran T, Ventura F (2010) Removal of drugs of abuse from municipal wastewater using reverse osmosis membranes. Desalin Water Treat 21:122–130
12. Loganathan B, Phillips M, Mowery H, Jones-Lepp TL (2009) Contamination profiles and mass loadings of macrolide antibiotics and illicit drugs from a small urban wastewater treatment plant. Chemosphere 75:70–77
13. Kaleta A, Ferdig M, Buchberger W (2006) Semiquantitative determination of residues of amphetamine in sewage sludge samples. J Sep Sci 29:1662–1666
14. Jones-Lepp TL, Stevens R (2007) Pharmaceuticals and personal care products in biosolids/sewage sludge: the interface between analytical chemistry and regulation. Anal Bioanal Chem 387:1173–1183

15. Boleda R, Galceran T, Ventura F (2009) Monitoring of opiates and cannabinoids and their metabolites in wastewater, surface water and finished water in Catalonia, Spain. Water Res 43:1126–1136
16. Huerta-Fontela M, Galceran MT, Ventura F (2008) Stimulatory drugs of abuse in surface waters and their removal in a conventional drinking water treatment plant. Environ Sci Technol 42:6809–6916

Chapter 4
Removal of Dyes and Pigments from Industrial Effluents

Tjoon Tow Teng and Ling Wei Low

4.1 Introduction

4.1.1 Dyes and Pigments

Colour is introduced into dyeing materials using substances known as dyes and pigments [1]. Dyes can be defined as soluble coloured substances which can be applied mainly to textile materials from solution in water, whereas, pigments are insoluble compounds integrated by a dispersion process into products such as paints, printing inks, and plastics [1].

Dye molecules comprises of two main components: chromophores and auxochromes. Chromophores are known as group of atoms principally responsible for the colour of dye. The most important chromophores are the azo (–N=N–), carbonyl (C=O), methine (–CH=), and nitro ($-NO_2$) groups. Auxochromes are useful in providing an essential "enhancement" of the colour [1]. Common auxochrome groups (include hydroxyl (OH) and amino (NR_2) groups) normally increase the intensity of the colour and shift the absorption to longer wavelengths of light [1, 2].

According to Bogacki et al. [3], pigments can be classified into two categories: homogenous pigments and mixtures. Homogenous pigments contain similar type of pigments particles, e.g., colourful metal oxides and metallic powders. On the other hand, mixtures are obtained by chemical coating the stain of a suitable carrier such as the active, highly dispersed silica, precipitated calcium carbonate. Mixtures can also be obtained by mechanical mixing of intensely staining pigment with filler.

T.T. Teng (✉) • L.W. Low
School of Industrial Technology, Universiti Sains Malaysia, Penang 11800, Malaysia
e-mail: ttteng@usm.my; low_lingwei@yahoo.com

S.K. Sharma and R. Sanghi (eds.), *Advances in Water Treatment and Pollution Prevention*, DOI 10.1007/978-94-007-4204-8_4,

4.1.2 Application of Dyes and Pigments

Mankind has used dyes for thousands of years. About 180,000 years ago, Neanderthal man was believed to be the earliest mankind to use colourant. In 1856, the world's first commercially successful synthetic dye was discovered by William Henry Perkin and by the end of the nineteenth century, 10,000 new synthetic dyes had been developed and manufactured [1].

Synthetic dyes are solely recalcitrant organic compounds that are extensively used in various branches of textile industries [4–9], pulp and paper industries [10, 11], leather tanning industries [12–14], food industries [15–18], agricultural research [19, 20], hair colourings [21], light-harvesting arrays [22], etc. Synthetic dyes have been used to control the efficacy of sewage [23], groundwater tracing [24], and determination of surface area of activated sludge [25].

The chemical classes of dyes used more frequently at industrial scale are the azo, anthraquinone, sulphur, indigoid, triphenylmethyl, and phthalocyanine derivatives [26]. All dyes used in industries are designed to resist fading upon exposure to sweat, water, light, and chemicals including oxidizing agents, and microbial attack [27].

There are more than 100,000 commercial dyes with a rough estimated production of 7×10^5 to 1×10^6 tons per year [28–30]. The exact data on the quantity of dyes discharged to the environment are unknown. A loss of 1–2% in production and 1–10% loss in use are assumed [26]. From the statistics, extensive application of synthetic dyes can cause considerable environmental pollution and are dangerous to the health of living organisms.

The impact and toxicity of dyes containing wastewater have been extensively studied. The dye classes, dye description, and typical pollutants associated with various dyes are listed in Table 4.1.

From Table 4.1, types of pollutants associated with various dyes are colour, organic acids, and unfixed dyes. Most of the dyes are water soluble, except disperse dye.

Dyeing effluents are hardly decolourised by conventional wastewater treatment methods [27, 33]. In addition to their visual effect and their adverse impact in terms of chemical oxygen demand (COD), biochemical oxygen demand (BOD), and total organic carbon (TOC), many synthetic dyes may cause the formation of toxic carcinogenic metabolites (high concentration of nutrients, presence of chlorinated phenolic compounds, sulphur, and lignin derivatives).

High volumetric rate of industrial effluent discharge and increasing stringent legislation make the search for appropriate treatment technologies. This is an important task for a greener tomorrow [34].

Table 4.1 Typical characteristics of dyes used in textile dyeing process

Dye class	Description	Types of pollutants associated with various dyes	Reference
Acid	Anionic compounds with high water solubility	Colour; organic acids; unfixed dyes	[31]
Basic	Very bright dyes, water-soluble cationic dyes, applied in weakly acidic dyebaths	Colour; organic acids; unfixed dyes	[31]
Direct	Water-soluble, anionic compounds can be applied directly to cellulosics without mordants (a substance that will form a coordination complex with the dye. Useful in setting dyes on fabrics)	Colour; salt; unfixed dye; cationic fixing agents; surfactant; defoamer; leveling and retarding agents; finish; diluents	[31, 32]
Disperse	Water insoluble	Colour; organic acids; carriers; leveling agents; phosphates; defoamers; lubricants; dispersants; diluents	[32]
Mordant	Final colour largely dependent on the choice of mordant. An example of mordant is potassium dichromate	Mordants (particularly in heavy metals categories) will cause hazardous to health	[31]
Reactive	Water-soluble, anionic compounds, largest dye class	Colour; salt; alkali; unfixed dye; surfactants; defoamer; diluents; finish	[31]
Sulphur	Organic compounds containing sulphur or sodium sulphide	Colour; alkali; oxidizing agents; reducing agent; unfixed dye	[32]
Vat	Oldest dyes, more chemically complex, water insoluble	Colour; alkali; oxidizing agents; reducing agent	[31, 32]

4.2 Technologies Available for Colour Removal

Industrial wastewater treatment consists of four processes [35], namely, pretreatment, primary treatment, secondary treatment, and tertiary treatment. Pretreatment process includes equalization and neutralization where industrial wastewater is pretreated with equalization and neutralization before being discharged to municipal sewerage systems or to a central industrial sewerage system [36]. After pretreatment step, physical or chemical separation techniques (primary treatment) are applied to remove suspended solids in the wastewater [36], followed by a secondary treatment. Secondary treatment/biological treatment uses microorganisms to stabilize the waste components before going to tertiary treatment. The tertiary treatment (physical chemical treatment) processes include adsorption, ion-exchange, stripping, chemical oxidation, and membrane separations. The final steps for the wastewater treatment process are the sludge processing and disposal steps.

Generally, the methodologies adopted to treat dye wastewater can be classified into three categories: (1) physical (2) physicochemical or chemical, and

(3) biological. These methods can be used either singularly or together in various combinations. Many treatment processes combine two or more treatment technologies to provide a better or more efficient treatment.

4.2.1 Physical Methods

Physical treatment involves particle separation processes; no gross chemical or biological changes are involved. Examples of physical treatment processes are adsorption and ion exchange. These processes are particle size dependent [37].

4.2.1.1 Adsorption

Adsorption process has been widely used in industrial dye wastewater treatment, because adsorption is a well-known equilibrium separation process and is effective in wastewater treatment [38–40]. The term adsorption refers to a process where molecules of gas, liquid, or dissolved solids (adsorbate) adhere to a surface (adsorbent) [41]. There are two types of adsorption processes, namely, physisorption and chemisorption. Physisorption involves the attraction between the solid surface and the adsorbed molecules via weak van der Waals forces. It is a readily reversible process and includes both mono and multilayer coverage of adsorbate molecules on the adsorbent's surface [42]. Chemisorption involves higher strength chemical bonding which is difficult for the removal of chemisorbed species from the solid surface [36]. It is an irreversible reaction and includes monolayer coverage of adsorbate molecules on the adsorbent's surface [43].

Transportation of adsorbate to the porous media adsorbent usually occurs through four main steps [44, 45], namely, bulk solution transport, external diffusion, intraparticle diffusion, and adsorption. Bulk solution transport process involves transportation of the adsorbate from bulk solution to hydrodynamic boundary layer (liquid–solid interphase) surrounding the adsorbent [45]. After the transportation of adsorbate to the hydrodynamic boundary layer surrounding the adsorbent, the adsorbate must then pass through the layer to the adsorbent surface (external diffusion). This transportation process is caused by molecular diffusion. The distance the adsorbate travels is largely dependent on the velocity of the bulk solution. The thickness of the boundary layer affects the rate of transportation. The thicker the boundary layer, the lower the rate of transportation becomes [44, 45]. After the adsorbate has passed through the boundary layer to the active site, intraparticle diffusion takes place. There are two types of diffusion that affect the intraparticle transportation, namely, pore diffusion and surface diffusion. Pore diffusion takes place when molecules diffuse from the solution into the pores of the adsorbate, whereas surface diffusion takes place along the adsorbent surface [45]. Lastly, the adsorbate is attached onto the available active sites on the adsorbent surface. This step occurs very fast; therefore, one of the preceding diffusion steps controls the rate of mass transfer.

Few years ago, use of activated carbon as adsorbent in adsorption process was found to be effective in wastewater treatment, but the cost of activated carbon is too high. Therefore, there is a need to search for cheaper and effective adsorbents. Recently, a number of studies have focused on low-cost adsorbent that is capable in degrading and adsorbing dyes from wastewater. A material could be assumed to be low cost if it is abundant in nature, inexpensive, requires little processing, and is effective in the treatment process [46, 47]. Low-cost adsorbents could be obtained from any agriculture and industry by-products, natural materials, and biosorbents that contain high surface area and porosity [47]. Adsorption of dyes onto low-cost adsorbents is largely dependent on dye properties such as molecular structure and type, number and position of substituent in the dye molecule [48], source of raw materials used, preparation and treatment conditions such as pyrolysis temperature and activation time, surface chemistry, surface charge, pore structure, and surface area of the adsorbent [49].

Non-conventional Low-Cost Adsorbents

Non-conventional low-cost adsorbents can be divided into three different categories, i.e., agriculture and industry waste, natural materials, and bioadsorbents.

Agriculture and Industry Waste Materials

Agriculture and industry waste materials are assumed to be low-cost adsorbents because they are abundant and require less processing [49]. Bagasse, a by-product of sugar milling process, is the crushed remnant of sugarcane stalks left after the extraction of juice. Bagasse consists of three components namely, pith, fibre, and rind mixed in different proportions [50]. Low et al. [51, 52] studied the adsorption of a basic dye (methylene blue (MB)) on various acid-modified bagasses. A novel bagasse pretreatment method was used that could effectively remove organic compounds from the bagasse that contribute to chemical oxygen demand (COD) in the dye aqueous solution. Pretreatment with thorough washing of adsorbent using boiling distilled water was performed instead of conventional washing using distilled water at room temperature. The highest percentage of colour removal was 99.45% and COD reduction was 99.36% using sulphuric acid-modified bagasse at pH 9.0. The results showed the removal and reduction efficiency of the colour and COD, respectively, were almost the same. This is because the dye species that contributes to the COD has been completely removed. It can be concluded that the adsorption efficiency of low-cost adsorbent is largely dependent on the adsorbent pretreatment steps. This adsorption process followed pseudo-second order, and Langmuir as well as Freundlich isotherms, and was endothermic in nature.

Ofomaja and Ho [53] used palm kernel fibre as adsorbent for the removal of anionic dye (4-bromoaniline-1,8-dihydronaphthalene-3,6-disodiumsulphate) from aqueous solutions. The results obtained were well described by Langmuir isotherm. The equilibrium monolayer capacity of palm kernel fibre for the adsorption of this

Table 4.2 Adsorption capacities for some agriculture and industry waste materials

Dye	Adsorbent	Adsorption capacity (mg/g)	Working isotherm	Reference
Basic dye				
MB	Teak wood bark	914.59	Langmuir isotherm	[54]
MB	Papaya seeds	555.55	Langmuir isotherm; pseudo-second-order kinetic model	[55]
Basic Red-22	Sugar industry mud	519.00	Langmuir isotherm	[56]
MB	Grass waste	457.64	Langmuir isotherm; pseudo-second-order kinetic model	[57]
MB	Pomelo peel	344.83	Langmuir isotherm; pseudo-second-order kinetic model	[58]
MB	Jackfruit peel	285.71	Langmuir isotherm; pseudo-second-order kinetic model	[59]
MB	Banana stalk waste	243.90	Langmuir isotherm; pseudo-second-order kinetic model	[60]
MB	Garlic peel	82.64	Freundlich isotherm; pseudo-second-order kinetic model	[61]
MB	Rubber seed shell	82.64	Freundlich isotherm; pseudo-second-order kinetic model	[62]
Reactive dye				
Reactive Blue 171	Fly ash	1.860	Freundlich isotherm; pseudo-second-order kinetic model	[63]
Reactive Red	Coconut tree flower carbon (CFC); Jute fibre carbon (JFC)	CFC = 181.9 JFC = 200	Langmuir isotherm; pseudo-second-order kinetic model	[64]
Acid dye				
Acid Red 1	Coal fly ash	92.59–103.09	Langmuir isotherm; pseudo-second-order kinetic model	[65]
Acid Black 1; Acid Blue 193	Fly ash	Acid Blue 193 = 10.937 Acid Black 1 = 10.331	Freundlich isotherm; pseudo-second-order kinetic model	[63]

anionic dye was 0.0705 mmol/g at 297 K and palm kernel dosage of 5 g/dm^3. The thermodynamic analysis showed that the system was exothermic in nature. Up to 66.44% of the dye adsorbed could be desorbed by using distilled water.

The adsorption capacities for some agriculture and industry waste materials are listed in Table 4.2.

Table 4.3 Adsorption capacities for some natural materials

Dye	Adsorbent	Adsorption capacity (mg/g)	Working isotherm	Reference
Basic dye				
MB	Montmorillonite clay	289.12	Langmuir isotherm; pseudo-second-order kinetic model	[68]
MB	Bentonite	151.00–175.00	Langmuir and Redlich-Peterson isotherms	[69]
MB	Fibrous clay minerals	85.00	Langmuir isotherm; pseudo-first-order kinetic model	[70]
MB	Palygorskite	50.80	Pseudo-second-order kinetic model	[71]
MB	Glass wool	2.24	Langmuir and Freundlich isotherms	[72]
Reactive dye				
RY-145; RB-B	Sorel's cement	RY-145 = 107.67 RB-B = 103.14	Langmuir isotherm	[73]
Direct dye				
Congo red	Australian kaolin	More than 85% colour removal	Langmuir isotherm; pseudo-first-order kinetic model	[74]

Natural Materials

Recently, there has been a growing interest in utilizing natural materials such as clay, zeolites, and other siliceous materials as adsorbents. Alpat et al. [66] studied the adsorption of basic dye, Toluidine Blue O from aqueous solution using Turkish zeolite (clinoptilolite) as adsorbent. Kinetic studies showed that adsorption of Toluidine Blue O on clinoptilolite was well fitted to pseudo-second-order kinetic model. The maximum adsorption capacity of clinoptilolite was 2.1×10^{-4} mol/g at solution pH 11.0. The results showed the adsorption process fitted to both the Langmuir and Freundlich models. Other than Turkish zeolite, researchers have also used other natural materials as adsorbents such as calcined clay materials and lime. They were used as adsorbents to remove diazo direct dye (congo red) from aqueous solutions [67]. The results showed that the removal efficiency of congo red dye using calcined clay materials and lime was more than 94%. The dye removal was governed by combined physico-chemical reactions of adsorption, ion-exchange, and precipitation. The adsorption process was well described by pseudo-second-order kinetic model and Freundlich isotherm mode. Table 4.3 shows the adsorption capacities of a few natural materials.

Biomass

The use of biological materials as adsorbents in wastewater treatment is known as bioadsorption or biosorption. Use of bioadsorbent in adsorption process can reduce dye concentration to parts-per-billion (ppb) levels. Bioadsorption is a novel, cheap, and effective process [49].

Recently, the use of dead or living biomass, fungi, and other microbial cultures as bioadsorbents is increasing because bioadsorbents are available in large quantities and low cost. Yu et al. [75] studied the polymethacrylic acid and polyamic acid-modified biomass of baker's yeast to remove MB from aqueous solution. The results showed the effect of pH and ionic strength on adsorption capacity of the modified biomass was not significant. The adsorption process follows Langmuir isotherm with maximum MB uptake capacities of 869.6 mg/g of polymethacrylic acid-modified biomass of baker's yeast and 680.3 mg/g of polyamic acid-modified biomass of baker's yeast. Table 4.4 shows adsorption capacities for some bioadsorbents.

Agriculture and industry waste materials (Table 4.2), natural materials (Table 4.3), and biomass (Table 4.4) have shown to be good alternatives to expensive commercial activated carbon. Most of the adsorption processes follow pseudo-second-order kinetic and Langmuir isotherm. In some cases, the adsorption process follows both Langmuir isotherm as well as Freundlich isotherm, where the adsorbents contain heterogeneous moieties which are uniformly distributed on the surface of the adsorbents [52]. Adsorbate concentration, pH, contact time, adsorbent dosage, shaking speed, and temperature play an important role in the adsorption process. With the use of these low-cost adsorbents, wastewater treatment process can become relatively economical. However, the usage of these low-cost adsorbents is still in the developing stage; more work is needed to investigate on the mechanisms of the adsorption process. Ion exchange may play an important role in the uptake of dyes. Comparison on the cost of using activated carbon and various non-conventional, low-cost adsorbents are strongly recommended in order to promote the use of low-cost adsorbents in industrial scale. Undoubtedly, low-cost adsorbents offer a lot of promising benefits for commercial use in future.

4.2.1.2 Ion Exchange

Ion exchange is a process where wastewater is passed over the ion exchange resin where a constituent in the wastewater exchanges an ion in the resin until the available exchange sites are saturated. Employing this technique, both cationic and anionic dyes can be removed from the effluent successfully [86]. The advantage of ion-exchange resin catalysts over other commonly used solid acid catalysts is their relatively high concentration of active sites, and it is this factor that makes them such effective catalysts for facile reaction [87]. The main drawback of ion-exchange method is that ion exchange cannot accommodate a wide range of dyes [86].

Table 4.4 Adsorption capacities for some bioadsorbents

Dye	Adsorbent	Adsorption capacity (mg/g)	Working isotherm	Reference
Basic dye				
MB	*Caulerpa lentillifera*	417.00	Langmuir isotherm; pseudo-second-order kinetic model	[76]
Basic Red 18	Activated sludge biomass	285.71	Langmuir and Freundlich isotherm; pseudo-second-order kinetic model	[77]
MB	Alga *Sargassum muticum* seaweed	279.20	Langmuir and Freundlich isotherm; pseudo-first-order kinetic model	[78]
MB	Dead macro fungi	232.73	Langmuir isotherm	[79]
MB	Green alga *Ulva lactuca*	40.20	Langmuir and Freundlich isotherm; pseudo-second-order kinetic model	[80]
MB	The brown alga *Cystoseira barbatula* Kutzing	38.61	Langmuir isotherm; pseudo-second-order kinetic model	[81]
MB	*Posidonia oceanica* (L) fibres	5.56	Langmuir isotherm	[82]
Reactive dye				
Reactive Black 8; Brown 9; Green 19; Blue 38	*Rhizopus nigricans*	90–96% adsorption of the selected dyes	Langmuir and Freundlich isotherm	[83]
Reactive Black 5	*Klebsiella* sp. UAP-b5	N/A	Pseudo-second-order kinetic model	[84]
Acid dye				
Acid Blue 40	*Thuja orientalis*	97.06	Langmuir isotherm; pseudo-second-order kinetic model	[85]

4.2.2 *Physicochemical and Chemical Methods*

Physicochemical methods and chemical methods are hard to differentiate. Both methods involve chemical reactions to improve water quality. Coagulation, membrane filtration, and advance oxidation process (AOP) such as ozonation,

H_2O_2/pyridine/Cu system, Fenton's process, photocatalysis, and ultrasonic irradiation are some examples of physicochemical or chemical treatment processes. Generally, physico-chemical and chemical treatment processes are economically feasible but are sometimes costly due to the cost of chemicals. Large quantities of concentrated sludge produced and pH dependence are the main drawbacks of these processes [88, 89].

4.2.2.1 Advance Oxidation Processes (AOP)

Colour dye effluents are more resistant to mild oxidation conditions. There is a need to use more powerful oxidizing agents in order to achieve higher removal efficiency [90]. Advance oxidation processes (AOP) are technologies based on intermediacy of hydroxyl and other radicals to oxidize toxic and non-biodegradable compounds to various by-products [91].

H_2O_2/Pyridine/Cu(II) System

H_2O_2/pyridine/Cu(II) system is one of the AOP that uses a chelating ligand and a transition metal forming the transition metal complex to produce superoxide anions ($^*O_2$). These anions function in the degradation of wide ranges of polycyclic aromatic hydrocarbons [92]. Pyridine coordinates with copper to form cupric complex [93]:

$$\mathrm{Cu(II)} \xrightarrow{\text{pyridine}} \mathrm{Cu(II)complex} \tag{4.1}$$

Cupric complex with H_2O_2 is reduced to cuprous complex [93]:

$$\mathrm{Cu(II)complex} + \mathrm{H_2O_2} \rightarrow \mathrm{Cu(II)(O_2H)^-} + \mathrm{H^+} \tag{4.2}$$

$$\mathrm{Cu(II)(O_2H)^-} \rightarrow \mathrm{Cu(I)^*O_2H} \tag{4.3}$$

Cuprous complex then reacts with excess H_2O_2 degrading them to generate reactive radicals such as superoxide anions and hydroxyl radicals [93].

$$\mathrm{Cu(I)^*O_2H} + \mathrm{H_2O_2} \rightarrow \mathrm{Cu(I)} + {}^*\mathrm{O_2} + {}^*\mathrm{OH} + \mathrm{H_2O} \tag{4.4}$$

Lim et al. [94] studied the treatment of disperse dye (Terasil Red R) wastewater using H_2O_2/pyridine/Cu(II) system. Ninety-nine percent colour removal at the optimal concentration of 5.5 mM H_2O_2, 38 mM pyridine, and 1.68 mM Cu(II) was obtained. The COD reduction of the dye solution was up to 90%. This treatment system shows good potential in decolourising and reducing COD of the dye effluents. The main drawback is the sludge production (170 mg/L). Higher amount of sludge produced would generally mean higher costs for disposal.

Ozonation

Ozonation is an effective oxidation process which has been widely used to decolourise synthetic dyes. Ozone can selectively oxidize molecules with unsaturated bonds (–C=C– or –N=N–) and aromatic structures [95]. Oxidation using ozone leads to degradation of chlorinated hydrocarbons, phenols, pesticides, and aromatic hydrocarbons [96]. Ozone can react in two pathways: direct and indirect. Direct pathway involves the ozone molecule itself acting as electron acceptor [90]. Tehrani-Bagha et al. [97] studied the decolourisation of Reactive Blue 19 using ozone. The most important decomposition products are sulphate, nitrate, formate, and acetate. Complete decolourisation of RB 19 was achieved. The lower COD reduction (55.00%) compared with 100% colour removal was attributed to incomplete oxidation of organic materials [97, 98]. The effect of ozonation on the toxicity of wastewater effluents has also been determined using nematode *Caenorhabditis elegans*. The data showed that the toxicity highly depended on the type of dye to be decomposed [99].

Ozonation can therefore effectively decolourise dye containing wastewater via oxidative cleavage of the conjugated system of the dye molecule. This treatment system does not form any residual or sludge [100] and it leaves the effluent with no colour and low COD. Another advantage is that ozone can be used as an oxidant in its gaseous state; this does not increase the volume of wastewater and sludge [96]. The disadvantage of ozonation process is its short half-life, making it an expensive process [96].

Fenton's Process

Fenton's process uses hydrogen peroxide and ferrous ions as Fenton's reagent for the oxidation of toxicants present in wastewater. The Fenton system uses ferrous ions to react with hydrogen peroxide to produce hydroxyl radicals (OH^*) with powerful oxidizing abilities to degrade certain toxic contaminants [101]. Fenton's process performs best with high initial dye concentrations and at low initial pH [102]. Advantages of this process include high ability to remove colour, reduce COD and toxicity of dye containing wastewater. Since the mechanism involves flocculation, impurities are transferred from the wastewater to the sludge which is ecologically questionable. It has conventionally been incinerated to produce power, but such disposal is far from environment friendly [90].

The Fenton's process is dependent on the final floc formation and its setting quality [103]. Cationic dyes do not coagulate at all [103]; acid, direct, vat, mordant, and reactive dyes usually coagulate, but the resulting floc is of poor quality and does not settle well [2]. To overcome this problem, Fenton sludge recycling system was developed, where Fe (III) sludge deposition is eliminated [90]. A comparative study of the oxidation of disperse dyes by an electrochemical process, ozone, hypochlorite, and Fenton's process showed that Fenton's process was the best among the oxidation processes in terms of high colour removal and COD reduction [104].

Ultrasonic Irradiation

Ultrasonic irradiation has proven to be an effective method for degrading organic effluents into less toxic compounds [105]. High-power ultrasound produces strong cavitation in the aqueous solution causing shock wave and reactive free radicals by the violent collapse of the cavitation bubble [106]. Singla et al. [107] studied the sonolytic degradation of acid dye (acid orange 24) at various initial concentrations in water. Pang et al. [108] and Wang et al. [109] studied the sonocatalytic degradation of organic dyes in the aqueous solutions. The degradation of the dye occurred through hydroxyl radical attack at the bubble/solution interface.

Photocatalysis

Photocatalysis using TiO_2 as the photocatalyst is an emerging technology. This process could be operated at ambient conditions and the catalyst itself is cheap, non-toxic, and commercially available at various crystalline forms [110]. Lin and Lee [111] coated TiO_2 powder onto the surface of a dendritic silver (Ag) carrier to synthesize TiO_2/Ag for decolourising Procion Red MX-5B. The photocatalytic degradation of organic dyes in the presence of nanostructured TiO_2 has been reviewed by Khataee and Kasiri [112].

4.2.2.2 Liquid–Liquid Extraction

Liquid–liquid extraction (LLE) is a process that involves the transport of a solute from an aqueous phase to an organic phase [113]. LLE is based on the principle that a solute in the immiscible solvents can distribute itself in a certain ratio between two immiscible solvents, one of which is usually aqueous phase (water) and the other an organic phase (organic solvent). Extraction process largely depends on mass transfer rate [114]. Muthuraman et al. [115] studied the extraction and recovery of MB from industrial wastewater using LLE. Benzoic acid served as the extractant. Ninety-nine percent of the dye was extracted from the aqueous solution and the extracted dye in the organic phase was back extracted into sulphuric acid solutions. Muthuraman and Teng [116] also studied the recovery of an acid dye, methyl red, from an aqueous solution by LLE where xylene was used as an extractant (carrier). The extraction of methyl red from aqueous solution was dependent on initial dye concentration. Under optimum conditions, dye extraction efficiency varied from 97% to 90% in 5 min at the aqueous to organic phase volume ratio (A/O) from 1 to 10. NaOH was found to be a suitable medium for the stripping process. It is sometimes not easy to find a suitable solvent with specific characteristics such as low-cost, non-toxic, non-flammable, chemically stable, and inert toward other components of the system.

One of the other prominent LLE techniques is supported liquid membrane (SLM). SLM requires less energy and is simpler to operate compared with LLE, which involves multi-shaking and solvent recovery steps [115]. SLM is a separation

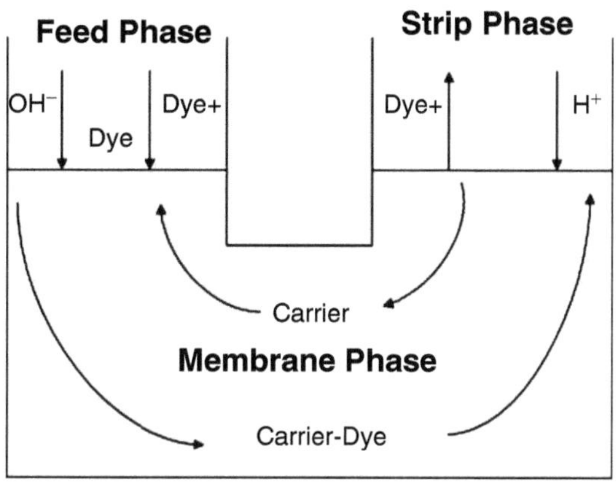

Fig. 4.1 Schematic description of transport of dye through liquid membrane

process that uses a thin layer of organic solution absorbed in the pores of a polymeric support [117]. This membrane acts as an intermediary, separating the feed phase containing dye species from the strip phase. The dye was then transported to and recovered from the strip solution. Muthuraman and Teng [117] studied the removal and recovery of Rhodamine B (basic dye) from an aqueous solution using SLM. Vegetable oil was used as the liquid membrane. The results showed that vegetable oil can effectively recover Rhodamine B dye from aqueous solution under optimized conditions.

The transportation of dyes across the SLM system can be explained in three suggested mechanisms [117]:

1. In an alkaline medium, the cationic dye remains as unionized dye molecule (neutral).
2. Diffusion of the unionized dye into the organic phase takes place due to the like nature of dye and the organic phase (hydrophobic).
3. At the membrane-strip solution interface under acidic condition, unionized dye molecule is converted into ionized molecule (i.e., stripping of the dye takes place).

Figure 4.1 shows the schematic description of transport of dye through liquid membrane.

Other than SLM, emulsion liquid membrane process (ELM) is known to be one of the effective methods for separation when the material being extracted is present in very low concentration [118]. Three advantages of ELM are: (1) relatively low-energy consumption when compared to other separation processes [119], (2) high efficiency because larger surface area is available for mass transfer to take place [120], and (3) high selectivity when carrier agents are used in the membrane phase that bind with target compounds [120]. ELM consists of a stripping phase encapsulated by an organic phase (membrane), which in turn contains the extractant and surfactant in organic diluents to stabilize the emulsion droplet. Therefore, ELM process involves simultaneous extraction and stripping in one step [121].

4.2.2.3 Coagulation-Flocculation and Chemical Precipitation Method

Coagulation-flocculation process can be applied as a pretreatment, a posttreatment, or even as a main treatment [31]. Basically, coagulation is a chemical addition and mixing process. Coagulation achieves charge neutralization or particles destabilization through addition of different charge of ion which allows the particles to agglomerate and subsequently to be removed [37]. Flocculation is always used in conjunction with coagulation. Flocculation bridges the stabilized particles together and forms bigger aggregates (flocs) in size. This process is normally accomplished by slow mixing because slow mixing allows particles to collide and adhere with each other until it is heavy enough to settle [122].

The most commonly used coagulant in wastewater treatment is ferric chloride ($FeCl_3$), ferric sulphate ($Fe_2(SO_4)_3$), aluminium sulphate ($Al(SO_4)_3$), ferrous sulphate ($FeSO_4$) and calcium hydroxide ($Ca(OH)_2$), polyaluminium chloride (PACl), lime, and alum [122, 123]. Flocculants that are used for wastewater treatment can be classified into three categories [124]:

1. Inorganic flocculants
2. Organic synthetic flocculants
3. Natural flocculants

Short detention time and low capital cost make chemical coagulation a widely used technique in wastewater treatment [90]. A comparison of the colour removal efficiency and settling properties among three coagulants, namely, aluminium sulphate (alum), polyaluminium chloride (PACl), and magnesium chloride ($MgCl_2$) had been reported by Tan et al. [125]. Flocs formed by $MgCl_2$ treatment were found to give shorter settling time than the alum and PACl treatment. The presence of magnesium in the aqueous solution enhances the removal of pollutants. The effective pH range for the treatment of textile dye wastewater using magnesium chloride as coagulant is between 10.5 and 11.0. The highest percentage of colour removal and COD reduction are 97.90% and 88.40%, respectively. Liao and Randtke [126], Folkman and Wachs [127], and Black and Christman [128] reported that with the presence of Mg^{2+} ion, good coagulation could be achieved. This treatment process is pH dependent.

Recently, increasing demand for environmental-friendly technologies has led to the search for natural, available, and cheap materials to be used as coagulants in wastewater treatment. The use of inorganic salts such as aluminium chloride, aluminium sulphate, magnesium chloride, lime, and ferric chloride have received attention for their potential health impacts [129]. Using these materials as coagulants may result in several environmental consequences: (1) an increase in metal concentration in water which may cause severe impact to human health; (2) production of toxic sludge; and (3) dispersion of arylamide oligomers which may also be a health hazard. Hence, alternative coagulants and flocculants have been considered for environmental applications [130]. Examples of natural coagulants which can be applied in dye wastewater treatment are listed in Table 4.5.

Table 4.5 Alternative coagulants to be used in dye wastewater treatment by coagulation-flocculation method

Natural coagulant	Dye	Efficiency	Reference
Chitosan	Acid Blue 92	99.00% colour removal	[131]
Sludge	Acid Red 119	96.53% colour removal	[132]
Moringa oleifera seed	Alizarin Violet 3R	95.00% colour removal	[133]
Steel industrial wastewater	Disperse Blue 79	99.00% colour removal and 94.00% COD reduction	[134]

Table 4.5 clearly indicates that the performance of natural coagulants in dye wastewater treatment is good; more than 90% of colour removal and COD reduction of dyes were achieved using some natural coagulants. Their non-toxicity and their biodegradability make them effective and environmental-friendly coagulants for the removal of dye containing wastewater.

Coagulation-flocculation process is effective, though in some cases it produces large amount of sludge. More research is needed to search for highly available, low-cost, non-toxic, and efficient coagulants.

4.2.3 *Biological Methods*

Biological treatment methods use microorganisms, mostly bacteria to decompose organic matters in wastewater. This treatment method is relatively inexpensive, having low running costs and the end products of complete mineralization are not toxic and stable [36]. Compared to chemical and physical treatments, biological treatments require less energy and chemicals. Generally, biological treatment methods can be divided into aerobic (involving oxygen) and anaerobic (without oxygen) methods. The former involves the usage of free or dissolved oxygen by microorganisms (aerobes). The microorganisms convert the organic wastes to biomass and carbon dioxide. The latter involves the degradation process in the absence of oxygen where complex organic wastes are degraded into carbon dioxide, methane, and water through three basic steps namely hydrolysis, acidogenesis (including acetogenesis), and methanogenesis [135].

Aerobic treatment methods are commonly used in the treatment of organic wastewaters for achieving high degree of treatment efficiency and are suitable for the treatment of low strength wastewater (biodegradable COD concentrations less than 1,000 mg/L), while anaerobic treatment is used in waste treatment based on the concept of resource recovery and utilization. Anaerobic treatment methods are suitable for the treatment of high strength wastewaters (biodegradable COD concentrations over 4,000 mg/L) [136]. A comparison of aerobic and anaerobic treatment systems is shown in Table 4.6.

Table 4.6 shows that both aerobic and anaerobic treatments are able to remove organic matters in water bodies. However, sludge production in aerobic treatment

Table 4.6 Comparison of some characteristics of aerobic and anaerobic treatments

Feature	Aerobic	Anaerobic	Reference
Organic removal	High efficiency	High efficiency	[135]
Quality of effluent	Good	Modest to poor	[137]
Sludge production	High	Low	[138]
Nutrient requirement	High	Low	[135]
Alkalinity requirement	Low	High for certain industrial waste	[137]
Energy requirement	High	Low to moderate	[135]
Temperature dependent	Low	High	[137]
Start-up time	2–4 weeks	2–4 months	[135]
Odour	Less opportunity for odours	Potential odour problems	[138]
Bioenergy and nutrient recovery	No	Yes	[137]
Organic loading rate	Modest	High	[135]

is higher than anaerobic treatment. The combinations of different anaerobic and aerobic processes have been applied to treat a broad range of industrial wastewater including textile industry wastewater. For wastewater containing high COD between 25,000 and 100,000 mg/L, the usage of anaerobic-aerobic processes might be suitable where the processes result in high organic matter removal efficiency, smaller amount of aerobic sludge produced, and no pH adjustment is needed [139].

4.2.3.1 Aerobic Treatment

Bacterial Treatment

The use of bacteria to degrade azo dyes started in 1970s with reports of *Bacillus subtilis*, followed by *Aeromonas hydrophila*, and *Bacillus cereus* [90]. A detailed list of bacteria used by different researchers for decolourisation of industrial effluent containing azo dye is given in Table 4.7.

Aerobic treatment using bacteria are shown to be effective in decolourising more than 80% of industrial effluents containing azo dyes.

Fungal Treatment

The main aim of using fungi in wastewater treatment is to purify the effluent by consumption of organic substances, thus reducing wastewater COD and BOD. The other aim is to obtain some valuable product, such as fungal biomass for protein-rich animal feed or some specific fungal metabolite [146]. A few researchers have studied the capability of fungal strains in decolourising azo and triphenyl methane

Table 4.7 Bacteria employed for the decolourisation of industry effluents containing azo dyes

Dye	Organisms used	Efficiency/Conditions	Reference
Reactive dye			
Hydrolysed Remazol Brilliant Violet 5R	Sequencing batch reactor (SBR) inoculated with sludge collected in a full-scale, continuous activated sludge plant	90% of colour removal 24 h cycle with a sludge retention time of 15 days and an aerated reaction phase of 10 h	[140]
Remazol Black B	*Paenibacillus azoreducens* sp. nov	98% Remazol Black B removed 24 h at a dye concentration of 100 mgdm^{-3} and a temperature of 37°C	[141]
Reactive azo dyes	*Pseudomonas luteola*	59–99% of colour removal 28°C static incubator at a dye concentration of 100 mg/L	[142]
Reactive Black 5	Newly isolated thermophilic microorganisms	>80% colour removal in less than 12 h	[143]
Remazol Blue and Reactive Black B	Immobilized thermophilic cyanobacterial strain *Phormidium* sp.	Remazol Blue: the highest dye removal was 50.3% Reactive Black B: the highest dye removal was 60%	[144]
Acid dye			
Acid Black 172	*Pseudomonas* sp. *DYI*	Optimal condition for decolourisation: pH 6.23, 30°C, 8.0 mM of Fe3^{+}, and 10.0 g/L of NaH_2PO_4	[145]

dyes [147–149]. Many genera either in living or dead form of fungi have been employed for the removal of dye in the aqueous solutions as shown in Table 4.8.

Three possible mechanisms are involved in decolourisation of dye aqueous solution using fungi [161]: biosorption, biodegradation, and bioaccumulation. Biosorption does not involve metabolic energy in the binding of solutes to the biomass [162]. Biodegradation is an energy-dependent process. This process involves breakdown of dyes into various by-products through the action of various enzymes. Bioaccumulation can be defined as accumulation of pollutants by actively growing cells by metabolism [163]. Fungi have advantages over bacteria because fungi are able to degrade more complex and a variety of substrates. The advantages of fungal treatment are cost-effectiveness, environmental-friendly, toxic compound removal, and sludge reduction [164].

Table 4.8 Fungi employed for the decolourisation of industrial effluent

Dye	Organism	Mechanism	Reference
Reactive dye			
Remazol Blue, Remazol Black B, Remazol Red RB	*Saccharomyces cerevisiae*	Bioaccumulation	[150]
Everzol Turquoise Blue G	*Coriolus versicolor*	Biodegradation	[151]
Reactive Brilliant Red X-3B	*Penicillium* sp. QQ	Biosorption	[152]
Reactive Red 2	Modified fungal biomass	Biosorption	[153]
Gemazol Turquoise Blue-G	*Rhizopus arrhizus*	Biosorption	[154]
Gray Lanaset G	*Trametes versicolor*	Biodegradation	[155]
Synozol Red HF6N, Synazol Yellow HF2GR	*Aspergillus niger*	Biosorption	[156]
Drimarene Red, Drimarene Blue	*Aspergillus foetidus*	Biosorption	[157]
Basic dye			
Astrazon Red	*Funalia trogii*	Biosorption and Biodegradation	[158]
Malachite Green	*Fomes scerodermeus*	Biodegradation	[159]
Disperse dye			
Disperse Orange 3, Disperse Yellow 3	*Pleurotus ostreatus*	Biodegradation	[160]

4.2.3.2 Anaerobic Treatment

The use of anaerobic process in dye wastewater treatment has been investigated since the early 1970s. This treatment process could be a cheap alternative compared to aerobic systems as expensive aeration is omitted and problems with bulking sludge are avoided [36]. This treatment system helps in decolourisation of dyes, by rendering them amenable to further aerobic treatment and degradation [165]. Typically, anaerobic breakdown yields methane and hydrogen sulphide [166].

Experiments indicated that the decolourisation of reactive water-soluble azo dyes was achieved under anaerobic conditions using glucose as carbon source [166]. The synthetic dye, tartrazine was found to be readily decolourised in an anaerobic baffled reactor [167, 168]. Wijetunga et al. [169] reported the effect of organic load on decolourisation of textile wastewater containing acid dyes in upflow anaerobic sludge blanket reactor (UASB). Over 85% of colour removal was observed for Acid Red 131, Acid Yellow 79, and Acid Blue 204. Acid Red 131 and Acid Yellow 79 were decolourised through biodegradation while Acid Blue 204 was decolourised due to adsorption onto anaerobic granules. Decolourisation of Congo Red (CR) and real textile wastewater was measured in one-and two-stage anaerobic treatment system. Two-stage anaerobic treatment system is more stable compared to one-stage anaerobic treatment. The results also showed that HRT reduction did not affect the decolourisation of dye. Thus, electron transfer was not a concern in the study [4].

4.2.4 Thermal Methods

4.2.4.1 Wet Air Oxidation

Wet air oxidation uses high temperature (at 300°C) and high pressure (>10 MPa) for the treatment of textile industry effluents that contain high chemical oxygen demand concentration of 10,000–100,000 mg/L [170–173]. Wet air oxidation converts all organic contaminants to carbon dioxide and water. This process can also remove oxidizable inorganic components such as cyanides and ammonia. For industrial effluent with COD concentration less than 5,000 mg/L, this treatment process is not recommended, because this process is not cost-effective compared to other chemical or biological processes.

4.2.4.2 Thermolysis

Thermolysis (catalytic/uncatalytic) is a chemical process where heat is used to decompose a substance to another substance which is stable and usable. This process is reported to have improved the biogasification of biomass due to breakdown of lower molecular weight compounds [174]. Thermolysis process can be used followed by coagulation-flocculation-filtration, catalytic wet air oxidation, or aerobic biological oxidation [170, 171, 175–177].

4.2.5 Combined Treatment Methods

In order to treat dye wastewater that contains high chemical oxygen demand values and high organic compounds such as phenols, dyes, tensioactives with good economics and high degree of energy efficiency, a few combined physical, chemical, and biological processes have been studied and listed in Table 4.9.

4.3 Conclusions

Although a number of successful physical, chemical, and biological processes have been applied, cost-effective removal of colour from effluents remains a challenge in industrial dye wastewater treatment. Wide ranges of pH, salt concentrations, and chemical structure of dye often add to the complication of dye wastewater treatment. Most of the industrial wastewaters containing anionic and cationic dyes are treated using adsorption process. Inexpensive, locally available, and effective materials can be used as adsorbents to replace expensive commercial-activated carbon for the removal of dyes from aqueous solution. Low-cost adsorbents offer a lot of promising

Table 4.9 Combination of various treatment methods

Combination method	Process	Reference
Physical/chemical or chemical/physical		
Coagulation/adsorption	Coagulation was used as a main treatment process followed by adsorption for the removal of reactive dyes (C.I. Reactive Red 45 and C.I. Reactive Green 8) from synthetic wastewater 99.9% of colour removal and 90% of COD reduction were achieved for both dyes	[178]
Coagulation-flocculation/ adsorption	Removal of Black 5 and Orange 16 was investigated using combined coagulation-flocculation/ adsorption method The removal efficiencies were 90% for Black 5 and 84% for Orange 16	[179]
Coagulation-flocculation/ nanofiltration	$FeCl_3$ was used as coagulant agent for the removal of Remazol Black 133 B, Procion Crimson H-EXL, Procion Navy H-EXL, Procion Yellow H-EXL, Procion Blue H-EXL 98% of colour removal for all dyes was achieved for all dyes except Remazol Black 133 B that achieved a 90%	[180]
Fenton's process/ nanofiltration	Three combinations of Fenton's process and nanofiltration are studied using a synthetic solution of eosin dye: Scheme 1 is Fenton's process followed by nanofiltration; Scheme 2 is nanofiltration followed by Fenton's process; Scheme 3 is two steps of nanofiltration The results showed: Scheme 1 is found suitable for eosin removal for dye concentration up to 70 mg/L; Scheme 2 is not suitable for dye removal; Scheme 3 is the most suitable among all the three schemes. It reduces eosin concentration to less than 1 mg/L	[181]
Chemical/biological or biological/chemical		
Photocataytic/anaerobic	Photocatalytic process based on immobilized titanium dioxide was used as pre/post-treatment method to biological anaerobic treatment for the treatment of azo, anthraquinone, and phthalocyanine textile dyes Photocatalysis was able to decolourise more than 90% of dye from the solutions The end products of this treatment process were non-toxic toward methanogenic bacteria	[182]

(continued)

Table 4.9 (continued)

Combination method	Process	Reference
Ozonation/biological degradation with biofilm	Treatment of dye wastewater using ozonation produces toxic and carcinogenic by-products Therefore, combined treatment of ozonation and subsequent biological degradation with a biofilm for colour removal and COD reduction was investigated. 96% of colour removal for Remazol Black B dye was obtained There is an increase in toxicity level after ozonation but after submitting the ozonized synthetic wastewater to biological treatment with a biofilm the toxicity level decreases	[183]
Thermal/chemical		
Thermolysis/coagulation	Thermolysis alone resulted in 92.85% of colour removal and 77.9% of COD reduction at pH 12 Coagulation of fresh composite waste using aluminium potassium sulphate resulted in 95.4% and 88.62% of colour removal and COD reduction, respectively, at pH 8 Coagulation of the supernatant obtained after treatment by catalytic thermolysis resulted in close to 100% colour removal and 97.3% of COD reduction	[174]

benefits for commercial use in future. Chemical treatment methods such as advance oxidation processes (AOP), and coagulation-flocculation are effective in treating a wide range of dye aqueous solutions, but sludge disposal still remains a problem. Ozonation, photocatalytic, and sonocatalytic degradation methods are increasingly used in wastewater treatment because they do not increase the volume of wastewater and sludge, but production of toxicity by-products prevents its wide acceptance. Use of combination methods for the removal of dyes containing wastewaters has received increasing attention.

References

1. Christie RM (2001) Colour chemistry. The Royal Society of Chemistry, Cambridge, UK, pp 1–200
2. Raghavacharya C (1997) Color removal from industrial effluents – a comparative review of available technologies. Chem Eng World 32:53–54
3. Bogacki MB, Iwona M, Andrzej K (2004) Application of experimental design for optimization of physicochemical properties of the inorganic pigment, iron (III) silicate. Dyes Pigments 61:149–164

4. Firmino PIM, Da Silva MER, Cervantes FJ, Dos Santos AB (2010) Colour removal of dyes from synthetic and real textile wastewaters in one- and two-stage anaerobic systems. Bioresour Technol 101:7773–7779
5. Gupta GS, Shukla SP, Prasad G, Singh VN (1992) China clay as an adsorbent for dye house wastewaters. Environ Technol 13:925–936
6. Moisés TP, Patricia BH, Barrera-Diaz CE, Gabriela RM, Reyna NR (2010) Treatment of industrial effluents by a continuous system: eletrocoagulation-activated sludge. Bioresour Technol 101:7761–7766
7. Shukla SP, Gupta GS (1992) Toxic effects of omega chrome red ME and its treatment by adsorption. Ecotoxicol Environ Staf 24:155–163
8. Sokolowska-Gajda J, Freeman HS, Reife A (1996) Synthetic dyes based on environmental considerations: 2. Iron complexed formazan dyes. Dyes Pigments 30:1–20
9. Srinivasan A, Viraraghavan T (2010) Decolorization of dye wastewaters by biosorbents: a review. J Environ Manag 91:1915–1929
10. Jain CK, Kumar A, Hayssam Izazy M (2009) Color removal from paper mill effluent through adsorption technology. Environ Monit Assess 149:343–348
11. Kalyani KSP, Balasubramanian N, Srinivasakannan C (2009) Decolorization and COD reduction of paper industrial effluent using electro-coagulation. Chem Eng J 151:97–104
12. Kabadasil I, Tunay O, Orhon D (1999) Wastewater control and management in a leather tanning district. Water Sci Technol 40:261–267
13. Kanth SV, Venba R, Madhan B, Chandrababu NK, Sadulla S (2008) Studies on the influence of bacterial collagenase in leather dyeing. Dyes Pigments 76:338–347
14. Paschoal FMM, Anderson MA, Zanoni MVB (2009) Simultaneous removal of chromium and leather dye from simulated tannery. J Hazard Mater 166:531–537
15. Bhat RV, Mathur P (1998) Changing scenario of food colours in India. Curr Sci 74:198–202
16. Doehlert DC, Simsek S, Wise ML (2009) The green oat story: possible mechanisms of green color formation in oat products during cooking. J Food Sci 74:226–231
17. Meimaridou A, Haasnoot W, Noteboom L, Mintzas D, Pulkrabova J, Hajslova J, Nielen MWF (2010) Color encoded microbeads-based flow cytometric immunoassay for polycyclic aromatic hydrocarbons in food. Anal Chim Acta 672:9–14
18. Slampova A, Smela D, Vondrackova A, Jancarova I, Kuban V (2001) Determination of synthetic colorants in foodstuffs. Chem List 95:163–168
19. Cook SMF, Linden DR (1997) Use of rhodamine WT to facilitate dilution and analysis of atrazine samples in short-term transport studies. J Environ Qual 26:1438–1441
20. Kross BC, Nicholson HF, Ogilvie LK (1996) Methods development study for measuring pesticide exposure to golf course workers using video imaging techniques. Appl Occup Environ Hyg 11:1346–1351
21. Scarpi C, Ninci F, Centini M, Anselmi C (1998) High-performance liquid chromatography determination of direct and temporary dyes in natural hair colourings. J Chromatogr A 796:319–325
22. Wagner RW, Lindsey JS (1996) Boron-dipyrromethane dyes for incorporation in synthetic multi-pigment light-harvesting arrays. Pure Appl Chem 68:1373–1380
23. Morgan-Sagastume JM, Jimenez B, Noyola A (1997) Tracerstudies in a laboratory and pilot scale UASB reactor. Environ Technol 18:817–826
24. Field MS, Wilhelm RG, Quinlan JF, Aley TJ (1995) An assessment of the potential adverse properties of fluorescent tracer dyes used for groundwater tracing. Environ Monit Assess 38:75–97
25. Sorensen BL, Wakeman RJ (1996) Filtration characterization and specific surface area measurement of activated sludge by rhodamine B adsorption. Water Res 30:115–121
26. Forgacs E, Cserhati T, Oros G (2004) Removal of synthetic dyes from wastewaters: a review. Environ Int 30:953–971
27. Wesenberg D, Kyriakides I, Agathos SN (2003) White-rot fungi and their enzymes for the treatment of industrial dye effluents. Biotechnol Adv 22:161–187

28. Christie RM (2007) Environmental aspects of textile dyeing. Woodhead, Boca Raton/Cambridge
29. Hunger K (2003) Industrial dyes: chemistry, properties, applications. Wiley-VCH, Weinheim/Cambridge
30. Husain Q (2006) Potential applications of the oxidoreductive enzymes in the decolorization and detoxification of textile and other synthetic dyes from polluted water: a review. Crit Rev Biotechnol 26:201–221
31. Dos Santos AB, Cervantes FJ, Van Lier JB (2007) Review paper on current technologies for decolourisation of textile wastewaters: perspectives for anaerobic biotechnology. Bioresour Technol 98:2369–2385
32. Kirk O (1993) Encyclopedia of chemical technology, vol 8. Wiley, New York
33. Shaul GM, Holdsworth TJ, Dempsey CR, Dostal KA (1991) Fate of water soluble azo dyes in the activated sludge process. Chemosphere 22:107–119
34. O'Neill C, Hawkes FR, Hawkes DL, Lourenco ND, Pinheiro HM, Delee W (1999) Colour in textile effluents—sources, measurement, discharge consents and simulation: a review. J Chem Technol Biotech 74:1009–1018
35. Perry RH, Green DW, Maloney JO (1997) Perry's chemical engineers' handbook, 7th edn. McGraw-Hill, New York
36. Gupta VK, Suhas (2009) Application of low-cost adsorbents for dye removal – a review. J Environ Manag 90:2313–2342
37. Fitzpatrick CSB, Gregory J (2003) Coagulation and filtration. In: Mara D, Horan NJ (eds) The handbook of water and wastewater microbiology. Academic, London
38. Bansal RC, Goyal M (2005) Activated carbon adsorption. Taylor & Francis Group, Boca Raton
39. Danis TG, Albanis TA, Petrakis DE, Pomonis PJ (1998) Removal of chlorinated phenols from aqueous solutions by adsorption on alumina pillared clays and mesoporous alumina aluminum phosphates. Water Res 32:295–302
40. Freeman HM (1989) Standard handbook of hazardous waste treatment and disposal, 2nd edn. McGraw-Hill, New York
41. Cheremisinoff NP (2002) Handbook of water and wastewater treatment technologies. Butterworth-Heinemann, Boston
42. Alley ER (2000) Water quality control handbook. McGraw-Hill Inc, New York
43. Weber WJ (1972) Physiochemical processes for water quality control. Wiley, New York, pp 199–259
44. Snoeyink VL, Summers RS (1999) Adsorption of organic compounds. In: Letterman RD (ed) Water quality and treatment: a handbook of community water supplies, 5th edn. American Water Works Association/McGraw-Hill Inc, New York
45. Wang XS, Zhou Y, Jiang Y, Sun C (2008) The removal of basic dyes from aqueous solutions using agricultural by-products. J Hazard Mater 157:374–385
46. Bailey SE, Olin TJ, Bricka M, Adrian DD (1999) A review of potentially low-cost sorbents for heavy metals. Water Res 33:2469–2479
47. Crini G (2006) Non-conventional low-cost adsorbents for dye removal: a review. Bioresour Technol 97:1061–1085
48. Reife A, Freeman H (1996) Environmental chemistry of dyes and pigments. Wiley, New York
49. Rafatullah M, Sulaiman O, Hashim R, Ahmad A (2010) Adsorption of methylene blue on low-cost adsorbents: a review. J Hazard Mater 177:70–80
50. Rasul MG, Rudolph V, Carsky M (1999) Physical properties of bagasse. Fuel 78:905–910
51. Low LW, Teng TT, Alkarkhi AFM, Ahmad A, Morad N (2011) Optimization of the adsorption conditions for the decolorization and COD reduction of methylene blue aqueous solution using low-cost adsorbent. Water Air Soil Pollut 214:185–195
52. Low LW, Teng TT, Ahmad A, Morad N, Wong YS (2011) A novel pretreatment method of lignocellulosic material as adsorbent and kinetic study of dye waste adsorption. Water Air Soil Pollut 218:293–306

53. Ofomaja AE, Ho YS (2007) Equilibrium sorption of anionic dye from aqueous solution by palm kernel fibre as sorbent. Dyes Pigments 74:60–66
54. McKay G, Porter JF, Prasad GR (1999) The removal of dye colours from aqueous solutions by adsorption on low-cost materials. Water Air Soil Pollut 114:423–438
55. Hameed BH (2009) Evaluation of papaya seeds as a novel non-conventional low-cost adsorbent for removal of methylene blue. J Hazard Mater 162:939–944
56. Magdy YH, Daifullah AAM (1998) Adsorption of a basic dye from aqueous solutions onto sugar-industry-mud. Waste Manag 18:219–226
57. Hameed BH (2009) Grass waste: a novel sorbent for the removal of basic dye form aqueous solution. J Hazard Mater 166:233–238
58. Hameed BH, Mahmoud DK, Ahmad AL (2008) Sorption of basic dye from aqueous solution by pomelo (*Citrus grandis*) peel in a batch system. Colloids Surf A 316:78–84
59. Hameed BH (2009) Removal of cationic dye from aqueous solution using jackfruit peel as non-conventional low-cost adsorbent. J Hazard Mater 162:344–350
60. Hameed BH, Mahmoud DK, Ahmad AL (2008) Sorption equilibrium and kinetics of basic dye from aqueous solution using banana stalk waste. J Hazard Mater 158:499–506
61. Hameed BH, Ahmad AA (2009) Batch adsorption of methylene blue from aqueous solution by garlic peel, an agricultural waste biomass. J Hazard Mater 164:870–875
62. Oladoja NA, Asia IO, Aboluwoye CO, Oladimeji YB, Ashogbon AO (2008) Studies on the sorption of basic dye by rubber (*Hevea brasiliensis*) seed shell. Turk J Eng Environ Sci 32: 143–152
63. Sun DS, Zhang XD, Wu YD, Liu X (2010) Adsorption of anionic dyes from aqueous solution on fly ash. J Hazard Mater 181:335–342
64. Senthilkumaar S, Kalaamani P, Porkodi K, Varadarajan PR, Subburaam CV (2006) Adsorption of dissolved reactive red dye from aqueous phase onto activated carbon prepared from agricultural waste. Bioresour Technol 97:1618–1625
65. Hsu TC (2008) Adsorption of an acid dye onto coal fly ash. Fuel 87:3040–3045
66. Alpat SK, Özbayrak Ö, Alpat S, Akçay H (2008) The adsorption kinetics and removal of cationic dye, Toluidine Blue O, from aqueous solution with Turkish zeolite. J Hazard Mater 151:213–220
67. Vimonses V, Jin B, Chow CWK (2010) Insight into removal kinetic and mechanisms of anionic dye by calcined clay materials and lime. J Hazard Mater 177:420–427
68. Almeida CAP, Debacher NA, Downs AJ, Cottet L, Mello CAD (2009) Removal of methylene blue from colored effluents by adsorption on momtmorillonite clay. J Colloid Interface Sci 332:46–53
69. Hong S, Wen C, He J, Gan F, Ho YS (2009) Adsorption thermodynamics of methylene blue onto bentonite. J Hazard Mater 167:630–633
70. Hajjaji M, Alami A, El-Bouadili A (2006) Removal of methylene blue from aqueous solution by fibrous clay minerals. J Hazard Mater 135:188–192
71. Al-Futaisi A, Jamrah A, Al-Hanai R (2007) Aspects of cationic dye molecule adsorption to palygorskite. Desalination 214:327–342
72. Chakrabarti S, Dutta BK (2005) Note on the adsorption and diffusion of methylene blue in glass fibers. J Colloid Interface Sci 286:807–811
73. Hassan SSM, Awwadb NS, Aboterikaa AHA (2009) Removal of synthetic reactive dyes from textile wastewater by Sorel's cement. J Hazard Mater 162:994–999
74. Vimonses V, Lei SM, Jin B, Chow CWK, Saint C (2009) Adsorption of Congo red by three Australian kaolins. Appl Clay Sci 43:465–472
75. Yu JX, Li BH, Sun XM, Yuan J, Chi RA (2009) Polymer modified biomass of bakers yeast for enhancement adsorption of methylene blue, rhodamine B and basic magenta. J Hazard Mater 168:1147–1154
76. Marungrueng K, Pavasant P (2007) High performance biosorbent (*Caulerpa lentillifera*) for basic dye removal. Bioresour Technol 98:1567–1572
77. Gulnaz O, Kaya A, Matyar F, Arikan B (2004) Sorption of basic dyes from aqueous solution by activated sludge. J Hazard Mater 108:183–188

78. Rubin E, Rodriguez P, Herrero R, Cremades J, Barbara I, Sastre de Vicente ME (2005) Removal of methylene blue from aqueous solutions using as biosorbent *Sargassum muticum*: an invasive macroalga in Europe. J Chem Technol Biotechnol 80:291–298
79. Maurya NS, Mittal AK, Cornel P, Rother E (2006) Biosorption of dyes using dead macro fungi: effect of dye structure, ionic strength and pH. Bioresour Technol 97:512–521
80. El-Sikaily A, Khaled A, El-Nemr A, Abdelwahab O (2006) Removal of methylene blue from aqueous solution by marine green alga *Ulva lactuca*. Chem Ecol 22:149–157
81. Caparkaya D, Cavas L (2008) Biosorption of methylene blue by a brown alga *Cystoseira barbatula* Kutzing. Acta Chim Solv 55:547–553
82. Ncibi MC, Hamissa AMD, Fathallah A, Kortas MH, Baklouti T, Mahjoub B, Seffen M (2009) Biosorptive uptake of methylene blue using Mediterranean green alga *Enteromorpha* spp. J Hazard Mater 170:1050–1055
83. Kumari K, Abraham TE (2007) Biosorption on anionic textile dyes by nonviable biomass of fungi and yeast. Bioresour Technol 98:1704–1710
84. Elizalde-González MP, Fuentes-Ramírez LE, Guevara-Villa MRG (2009) Degradation of immobilized azo dyes by *Klebsiella sp*. UAP-b5 isolated from maize bioadsorbent. J Hazard Mater 161:769–774
85. Akar T, Ozcan AS, Tunali S, Ozcan A (2008) Biosorption of a textile dye (Acid Blue 40) by cone biomass of *Thuja orientalis*: estimation of equilibrium, thermodynamic and kinetic parameters. Bioresour Technol 99:3057–3065
86. Mishra G, Tripathy M (1993) A critical review of the treatments for decolouration of textile effluent. Colourage 40:35–38
87. Ali HG, Gehad RE, Ahmed Z (2007) Kinetics of the oxidative decolorization of Reactive Blue-19 by acidic bromate in homogeneous and heterogeneous media. Dyes Pigments 73:90–97
88. Kace JS, Linford HB (1975) Reduced cost flocculation of a textile dyeing wastewater. J Water Pollut Control Fed 47:1971
89. Lee JW, Choi SP, Thiruvenkatachari R, Shim WG, Moon H (2006) Evaluation of the performance of adsorption and coagulation processes for the maximum removal of reactive dyes. Dyes Pigments 69:196–203
90. Anjaneyulu Y, Sreedhara Chary N, Samuel Suman Raj D (2005) Decolourization of industrial effluents-available methods and emerging technologies – a review. Rev Environ Sci Bio/Technol 4:245–273
91. Klavarioti M, Mantzavinos D, Kassinos D (2009) Removal of residual pharmaceuticals from aqueous systems by advanced oxidation processes. Environ Int 35:402–417
92. Watanabe T, Koller K, Messner K (1998) Copper-dependent depolymerization of lignin in the presence of fungal metabolite, pyridine. J Biotechnol 62:221–230
93. Pecci L, Montefoschi G, Cavallini D (1997) Some new details of the copper-hydrogen peroxide interaction. Biochem Biophy Res Comm 235:264–267
94. Lim CL, Morad N, Teng TT, Ismail N (2009) Treatment of Terasil Red R dye wastewater using H_2O_2/pyridine /Cu(II) system. J Hazard Mater 168:383–389
95. Adams CD, Gorg S (2002) Effect of pH and gas-phase ozone concentration on the decolourization of common textile dyes. J Environ Eng 128:293–298
96. Xu Y, Lebrun RE (1999) Treatment of textile dye plant effluent by nanofiltration. Membr Sep Sci Technol 34:2501–2519
97. Tehrani-Bagha AR, Mahmoodi NM, Menger FM (2010) Degradation of persistent organic dye from colored textile wastewater by ozonation. Desalination 260:34–38
98. Sevimli MF, Kinaci C (2002) Decolorization of textile wastewater by ozonation and Fenton's process. Water Sci Technol 45:279–286
99. Hitchcock DR, Law SE, Wu J, Williams PL (1998) Determining toxicity trends in the ozonation of synthetic dye wastewaters using the nematode *Caenorhabditis elegans*. Arch Environ Contam Toxicol 34:259–264
100. Ince NH, Gonene DT (1997) Treatability of a textile azo dye by UV/H_2O_2. Environ Technol 18:179–185

101. Behnajady MA, Modirshahla N, Ghanbary F (2007) A kinetic model for the decolorization of C.I. Acid Yellow 23 by Fenton process. J Hazard Mater 148:98–102
102. Marco SL, Albino AD, Ana S, Carla A, José AP (2007) Degradation of a textile reactive azo dye by a combined chemical-biological process: Fenton's reagent-yeast. Water Res 41: 1103–1109
103. Robinson T, McMullan G, Marchant R, Nigam P (2001) Remediation of dyes in textile effluent: a critical review on current treatment technologies with a proposed alternative. Bioresour Technol 77:247–255
104. Balchioglu IA, Arslan I, Sacan MT (2001) Homogeneous and heterogeneous advanced oxidation of two commercial reactive dyes. Environ Technol 22:813–822
105. Shrestha RA, Pham TD, Sillanpää M (2009) Effect of ultrasonication on removal of persistent organic pollutants (POPs) from different types of soils. J Hazard Mater 170:871–875
106. Naddeo V, Belgiorno V, Napoli RMA (2007) Behaviour of natural organic matter during ultrasonic irradiation. Desalination 210:175–182
107. Singla R, Grieser F, Ashokkumar M (2009) Sonochemical degradation of martius yellow dye in aqueous solution. Ultrason Sonochem 16:28–34
108. Pang YL, Abdullah AZ, Bhatia S (2010) Effect of annealing temperature on the characteristics, sonocatalytic activity and reusability of nanotubes TiO_2 in the degradation of Rhodamine B. Appl Catal B: Environ 100:393–402
109. Wang J, Lv YH, Zhang L, Liu B, Jiang RZ, Han GX, Xu R, Zhang XD (2010) Sonocatalytic degradation of organic dyes and comparison of catalytic activities of CeO_2/TiO_2, SnO_2/TiO_2 and ZrO_2/TiO_2 composites under ultrasonic irradiation. Ultrason Sonochem 17:642–648
110. Doll TE, Frimmel FH (2004) Kinetic study of photocatalytic degradation of carbamazepine, clofibric acid, iomeprol and iopromide assisted by different TiO_2 materials determination of intermediates and reaction pathways. Water Res 38:955–964
111. Lin YC, Lee HS (2010) Effects of TiO_2 coating dosage and operational parameters on a TiO_2/Ag photocatalysis system for decolorizing Procion red MX-5B. J Hazard Mater 179: 462–470
112. Khataee AR, Kasiri MB (2010) Photocatalytic degradation of organic dyes in the presence of nanostructured titanium dioxide: influence of the chemical structure of dyes. J Mol Catal A: Chem 328:8–26
113. Cox M, Rydberg J (2004) Introduction to solvent extraction. In: Rydberg J, Cox M, Musikas C, Choppin GR (eds) Solvent extraction principles and practice. Marcel Dekker, New York, pp 1–10
114. Woo LD, Hi HW, Yup HK (2000) Removal of an organic dye from water using a predispersed solvent extraction. Sep Sci Technol 35:1951–1962
115. Muthuraman G, Teng TT, Leh CP (2009) Extraction and recovery of methylene blue from industrial wastewater using benzoic acid as an extractant. J Hazard Mater 163:363–369
116. Muthuraman G, Teng TT (2009) Extraction of methyl red from industrial wastewater using xylene as an extractant. Prog Nat Sci 19:1215–1220
117. Muthuraman G, Teng TT (2009) Use of vegetable oil in supported liquid membrane for the transport of Rhodamine B. Desalination 249:1062–1066
118. Fouad EA, Bart HJ (2008) Emulsion liquid membrane extraction of zinc by a hollow-fiber contactor. J Membr Sci 307:156–168
119. Naim MM, Monir AA (2003) Desalination using supported liquid membranes. Desalination 153:361–369
120. Frankenfield JW, Li NN (1987) Recent advance in liquid membrane technology, Handbook of separation process technology. Wiley, New York, pp 840–861
121. Prashant SK, Vijaykumar VM (2002) Application of liquid emulsion membrane (LEM) process for enrichment of molybdenum from aqueous solutions. J Membr Sci 201:123–135
122. Semerjian L, Ayoub GM (2003) High-pH- Magnesium coagulation-flocculation in wastewater treatment. Adv Environ Res 7:389–403
123. Annadurai G, Sung SS, Lee DJ (2004) Simultaneous removal of turbidity and humic acid from high turbidity stormwater. Adv Environ Res 8:713–725

124. Xia S, Zhang Z, Wang X, Yan A, Chen L, Zhao Z, Leonard D, Jaffrezic-Renault N (2007) Production and characterization of a bioflocculant by Proteus mirabilis TJ-1. Bioresour Technol 99:6520–6527
125. Tan BH, Teng TT, Mohd Omar AK (2000) Removal of dyes and industrial dye wastes by magnesium chloride. Water Res 34:597–601
126. Liao MY, Randtke SJ (1986) Predicting the removal of soluble organic contaminants by lime softening. Water Res 20:27–35
127. Folkman Y, Wachs AM (1973) Removal of algae from stabilization pond effluents by lime treatment. Water Res 7:419–428
128. Black AP, Christman RF (1961) Electrophoretic studies of sludge particles produced in lime-soda softening. J Am Water Works Assoc 53:737
129. Roussy J, Van Vooren M, Dempsey BA, Guibal E (2005) Influence of chitosan characteristics on the coagulation and the flocculation of bentonite suspensions. Water Res 39:3247–3258
130. Bolto B, Gregory J (2007) Organic polyelectrolytes in water treatment. Water Res 41: 2301–2324
131. Szygula A, Guibal E, Palacín MA, Ruiz M, Sastre AM (2009) Removal of an anionic dye (Acid Blue 92) by coagulation-flocculation using chitosan. J Environ Manag 90:2979–2986
132. Sadri Moghaddam S, Alavi Maghaddam MR, Arami M (2010) Coagulation-flocculation process for dye removal using sludge from water treatment plant: optimization through response surface methodology. J Hazard Mater 175:651–657
133. Beltrán-Heredia J, Sánchez-Martín J, Delgado-Regalado A, Jurado-Bustos C (2009) Removal of Alizarin Violet 3R (anthraquinonic dye) from aqueous solutions by natural coagulants. J Hazard Mater 170:43–50
134. Anouzla A, Abrouki Y, Souabi S, Safi M, Rhbal H (2009) Colour and COD removal of disperse dye solution by a novel coagulant: application of statistical design for the optimization and regression analysis. J Hazard Mater 166:1302–1306
135. Chan YJ, Chong MF, Law CL, Hassel DG (2009) A review on anaerobic-aerobic treatment of industrial and municipal wastewater. Chem Eng J 155:1–18
136. Seghezzo L, Zeeman G, Van Lier JB, Hamelers HVM, Lettinga G (1998) A review: the anaerobic treatment of sewage in UASB and EGSB reactors. Bioresour Technol 65:175–190
137. Yeoh BG (1995) Anaerobic treatment of industrial wastewaters in Malaysia. In: Post conference seminar on industrial wastewater management in Malaysia, Kuala Lumpur, Malaysia
138. Leslie Grady CP Jr, Daigger GT, Lim HC (1999) Biological wastewater treatment, 2nd edn., revised and expanded. CRC Press, Boca Raton
139. Cervantes FJ, Pavlostathis SG, Van Haandel AC (2006) Advanced biological treatment processes for industrial wastewaters: principles and applications. IWA publishing, London
140. Lourenco ND, Novais JM, Pinheiro HM (2000) Reactive textile dye colour removal in a sequencing batch reactor. Water Sci Technol 42:321–328
141. Meehan C, Bjourson AJ, McMullan G (2001) *Paenibacillus azoreducens sp. nov.*, a synthetic azo dye decolorizing bacterium from industrial wastewater. Int J Syst Evol Microbiol 51: 1681–1685
142. Hu TL (2001) Kinetics of azoreductase and assessment of toxicity of metabolic products from azo dyes by *Pseudomonas luteola*. Water Sci Technol 43:261–269
143. Deive FJ, Domínguez A, Barrio T, Moscoso F, Morán P, Longo MA, Sanromán MA (2010) Decolorization of dye Reactive Black 5 by newly isolated thermophilic microorganisms from geothermal sites in Galicia (Spain). J Hazard Mater 182:735–742
144. Ertuğrul S, Bakir M, Dönmez G (2008) Treatment of dye-rich wastewater by an immobilized thermophilic cyanobacterial strain: *Phormidium* sp. Ecol Eng 32:244–248
145. Du LN, Yang YY, Li G, Wang S, Jia XM, Zhao YH (2010) Optimization of heavy metal-containing dye Acid Black 172 decolorization by Pseudomonas sp. DYI using statistical designs. Int Biodeterior Biodegrad 64:566–573
146. Pant D, Adholeya A (2007) Biological approaches for treatment of distillery wastewater: a review. Bioresour Technol 98:2321–2334

147. Bumpus JA, Brock BJ (1988) Biodegradation of crystal violet by the white rot fungus phanerochaete chrysosporium. Appl Environ Microbiol 54:1143–1150
148. Sani R, Banerjee U (1999) Decolorization of acid green 20, a textile dye, by the white rot fungus *Phanerochaete chrysosporium*. Adv Environ Res 2:485–490
149. Vasdev K, Kuhad RC, Saxena RK (1995) Decolorization of triphenylmethane dyes by a bird's nest fungus *Cyathus bulleri*. Curr Microbiol 30:269–272
150. Aksu Z (2003) Reactive dye bioaccumulation by *Saccharomyces cerevisiae*. Process Biochem 38:1437–1444
151. Kapdan IK, Kargia F, McMullanb G, Marchant R (2000) Effect of environmental conditions on biological decolourization of textile dyestuff by C. versicolor. Enzym Microb Technol 26:381–387
152. Gou M, Qu YY, Zhou JT, Ma F, Tan L (2009) Azo dye decolorization by a new fungal isolate, *Penicillium sp.* QQ and fungal-bacterial cocultures. J Hazard Mater 170:314–319
153. Akar T, Divriklioglu M (2010) Biosorption applications of modified fungal biomass for decolorization of Reactive Red 2 contaminated solutions: batch and dynamic flow mode studies. Bioresour Technol 101:7271–7277
154. Aksu Z, Cagatay SS (2006) Investigation of biosorption of Gemazol Turquise Blue-G reactive dye by dried *Phizopus arrhizus* in batch and continuous systems. Sep Purif Technol 48:24–35
155. Blanquez P, Caminal G, Sarra M, Vicent T (2007) The effect of HRT on the decolourization of the Grey Lanaset G textile dye by *Trametes versicolor*. Chem Eng J 126:163–169
156. Khalaf MA (2008) Biosorption of reactive dye from textile wastewater by non-viable biomass of *Aspergillus niger* and *Spirogyra sp.*. Bioresour Technol 99:6631–6634
157. Sumathi S, Manju BS (2000) Uptake of reactive textile dyes by *Aspergillus foetidus*. Enzym Microb Technol 27:347–355
158. Yesilada O, Sing S, Asma D (2002) Decolourization of textile dye Astrazon Red FBL by *Funalia trogii* pellets. Bioresour Technol 81:155–157
159. Papinutti VI, Forchiassin F (2004) Modification of malachite green by *Fomes sclerodermeus* and reduction of toxicity to *Phanerochaete chrysosporium*. FEMS Microbiol Lett 231:205–209
160. Zhao X, Hardin IR (2007) HPLC and spectrophotometric analysis of biodegradation of azo dyes by *Pleurotus ostreatus*. Dyes Pigments 73:322–323
161. Kaushik P, Malik A (2009) Fungal dye decolourization: recent advances and future potential. Environ Int 35:127–141
162. Tobin JM, White C, Gadd GM (1994) Metal accumulation by fungi: applications in environment biotechnology. J Ind Microbiol 13:126–130
163. Aksu Z, Donmez G (2005) Combined effects of molasses sucrose and reactive dye on the growth and dye bioaccumulation properties of *Candida tropicalis*. Process Biochem 40: 2443–2454
164. More TT, Yan S, Tyagi RD, Surampalli RY (2010) Potential use of filamentous fungi for wastewater sludge treatment. Bioresour Technol 101:7691–7700
165. Krull R, Hempel DC (1994) Biodegradation of naptholenosulphenic acid containing sewage in a two stage treatment plant. Bioprocess Eng 10:229–234
166. Carliell CM, Barclay SJ, Buckely CA (1996) Treatment of exhausted reactive dye bath effluent using anaerobic digestion: laboratory and full scale trails. Water SA 22:225–233
167. Bell J, Plumb JJ, Buckley CA, Stuckey DC (2000) Treatment and decolorization of dyes in an anaerobic baffled reactor. J Environ Eng 126:1026–1032
168. Plumb JJ, Bell J, Stuckey DC (2001) Microbial populations associated with treatment of an industrial dye effluent in an anaerobic baffled reactor. Appl Environ Microbiol 67:3226–3235
169. Wijetunga S, Li XF, Chen J (2010) Effect of organic load on decolourization of textile wastewater containing acid dyes in upflow anaerobic sludge blanket reactor. J Hazard Mater 177:792–798
170. Lei L, Chen G, Hu X, Yue PL (2000) Homogeneous catalytic wet-air oxidation for the treatment of textile wastewater. Water Environ Res 72:147–151

171. Lei L, Hu X, Chen G, Porter JF, Yue PL (2000) Wet air oxidation of desizing wastewater from textile industry. Ind Eng Chem Res 39:2896–2901
172. Lei L, Hu X, Chu HP, Yue PL (1997) Catalytic wet air oxidation of dyeing and desizing wastewater. Water Sci Technol 65:311–319
173. Lei L, Hu X, Yue PL (1998) Improved wet oxidation for the treatment of dyeing wastewater concentrate from membrane separation process. Water Res 32:2753–2759
174. Kumar P, Prasad B, Mishra IM, Chand S (2008) Treatment of composite wastewater of a cotton textile mill by thermolysis and coagulation. J Hazard Mater 151:770–779
175. Bayramoglu M, Kobya M, Can OT, Sozbir M (2004) Operating cost analysis of electrocoagulation of textile dye wastewater. Sep Purif Technol 37:117–125
176. Lei L, Wang D (2000) Wet oxidation of PVA-containing desizing wastewater. Chin J Chem Eng 8:52–56
177. Lin SH, Lin CM (1993) Treatment of textile waste effluent by ozonation and chemical coagulation. Water Res 27:1743–1748
178. Papić S, Koprivanac N, Loncárić Bozć A, Metes A (2004) Removal of some reactive dyes from synthetic wastewater by combined Al(III) coagulation/carbon adsorption process. Dyes Pigments 62:291–298
179. Furlan FR, Da Silva LGDM, Morgado AF, De Souza AAU, De Souza SMAGU (2010) Removal of reactive dyes from aqueous solutions using combined coagulation/flocculation and adsorption on activated carbon. Resour Conserv Recycl 54:283–290
180. Riera-Torres M, Gutiérrez-Bouzán C, Crespi M (2010) Combination of coagulation-flocculation and nanofiltration techniques for dye removal and water reuse in textile effluents. Desalination 252:53–59
181. Banerjee P, DasGupta S, De S (2007) Removal of dye from aqueous solution using a combination of advance oxidation process and nanofiltration. J Hazard Mater 140:95–103
182. Harrelkas F, Paulo A, Alves MM, El-Khadir L, Zahraa O, Pons MN, Van Der Zee FP (2008) Photocatalytic and combined anaerobic-photocatalytic treatment of textile dyes. Chemosphere 72:1816–1822
183. De Souza SMDAGU, Bonilla KAS, De Souza AAU (2010) Removal of COD and color from hydrolyzed textile azo dye by combined ozonation and biological treatment. J Hazard Mater 179:35–42

Chapter 5
Heavy Metal Removal Through Biosorptive Pathways

Jinsheng Sun, Yulan Ji, Fang Cai, and Jing Li

5.1 Introduction

Rapid industrialization and urbanization have resulted in elevated levels of toxic heavy metals. This has been entangled, almost everywhere, in most industrial activities involving leakage and redistribution of heavy metals, such as mining and metalliferous smelting, metallurgy, iron and steel, electroplating, electrolysis, electro-osmosis, leatherworking, photography etc. Subsequently, these courses result in gradual exhaustion of metal land mineral resources. Especially, wastewaters produced from these industrial activities not only bring about serious environmental effect, but also threaten human health and ecosystem. These heavy metals, for the convenience of analyses, reportedly fall into three families, including toxic metals (such as Hg, Cr, Pb, Zn, Cu, Ni, Cd, As, Co, Sn, etc.), precious metals (such as Pd, Pt, Ag, Au, Ru etc.), and radionuclides (such as U, Th, Ra, Am, etc.) [1].

Methods for removing metal ions from aqueous solution mainly consist of physical, chemical, and biological technologies. Traditional physiochemical methods include chemical precipitation, oxidation or reduction, filtration, ion exchange, electrochemical treatment, reverse osmosis, membrane technology, and evaporation recovery. Most of these are ineffective or excessively expensive when the metal concentrations are less than 100 mg/L [2, 3]. For example, usual physical pathways like ion exchange, activated carbon, and membrane adsorption belong to the latter costly group, especially when treating large amounts of wastewater. In recent years, in a way of financial convenience, applying biotechnology in controlling and

---

J. Sun (✉) • Y. Ji • F. Cai
School of Chemical Engineering and Technology, Tianjin University, Tianjin 300072, People's Republic of China
e-mail: jssun2006@vip.163.com

J. Li
Engineering Department, China Tianjin Chemical Engineering Design Institute, No. 2 Gu Shan Road, Tianjin 300193, People's Republic of China

S.K. Sharma and R. Sanghi (eds.), *Advances in Water Treatment and Pollution Prevention*, DOI 10.1007/978-94-007-4204-8_5,

removing metal pollution has been paid much attention, and gradually becomes hot topic, for potential application, in the field of metal pollution control. Further alternative process is biosorption, which utilizes various certain natural materials of biological origin, including bacteria, fungi, yeast, algae, etc. These biosorbents possess instinctive metal-sequestering features to decrease the concentration of heavy metal ions in solution from ppm to ppb level. It can effectively capture dissolved metal ions and carry them out of dilute complex solutions with high efficiency and celerity; sound features as an ideal candidate for treating high volume and low concentration complex wastewaters [1].

This metabolism-independent mechanism is called biosorption [4]. According to Volesky and Holan [5], who presented an extensive review on biosorption, the strong biosorbent behavior of certain types of microbial biomass toward metallic ions is a function of the chemical makeup of microbial cells. In fact, the biomass in practice is nonliving and all cells are metabolically inactive. Compared with conventional treatment methods, biosorption has the following advantages [6–9]:

- High efficiency and selectivity for heavy metals in low concentrations
- Energy-saving
- Broad operational range of pH and temperature
- Easy reclamation of heavy metal
- Easy recycling of the biosorbent

Heavy metal removal by biosorption has been investigated during the last several decades [10]. The capability of some living microorganisms to accumulate metallic elements has first been observed from toxicological point of view [10]. However, further researches revealed that inactive/dead microbial biomass can passively bind metal ions via various physicochemical mechanisms. This discovery then led biosorption to the frontier of researches and an active field for the removal of metal ions. A large quantity of materials has been investigated as biosorbents for extensive removal of metals or organics. The tested biosorbents can be basically classified into the following categories: Bacteria (e.g., *Bacillus subtillis*), fungi (e.g., *Rhizopus arrhizus*), yeast (e.g., *Saccharomyces cerevisiae*), algae, industrial wastes (e.g., *S. cerevisiae* waste biomass from fermentation and food industry), agricultural wastes (e.g., corn core), and other polysaccharide materials, etc. [11]. The role of some groups of microorganisms has been given enough concern, such as bacteria, fungal, yeast, algae, etc.

Some potential biomaterials with high metal-binding capacity have been identified, such as *Sargassum natans*, *Bacillus subtillis*, *Rhizopus arrhizus*, *Saccharomyces cerevisae*, and waste microbial biomass from the fermentation and food industries [5, 8, 9, 12–17]. Among these materials or biosorbents, some types bind and collect the majority of heavy metals with no specific priority, while others can even be specific for certain types of metals [5].

Researches on biosorption focus on the biosorbents, the biosorption mechanism, and large-scale experiments. Although many biological materials can bind heavy metals, only those with sufficiently high metal-binding capacity and selectivity for heavy metals are suitable for use in a full-scale biosorption process. Mechanisms responsible for biosorption are understood to a limited extent, and commercial application of microbial biomass as a biosorbent suffered from problems associated with the physical characteristics of this material [18]. Attempts to apply biosorption process into practice have proved to be futile [19]. So, the biosorption process is basically at the stage of laboratory-scale study. Also, great efforts have to be made to improve biosorption process, including immobilization of biomaterials, improvement of regeneration and reuse, optimization of biosorption process, etc. In sum, theoretical and practical development about biosorption should be allowed to forge ahead.

The aim of this chapter is to present the state-of-the-art biosorbent investigation and to compare results found in the literature. The effects of environmental factors on biosorption capacities, biosorption mechanisms, advantages of biotechnology, regeneration/reuse of biosorbents, as well as proposals about improving biotechnology are presented and discussed. What is more, current hot topics, trends, and outlook of this field are also touched upon.

5.2 Brief Review of Physicochemical Removal Technology

Before the introduction, this brief review offers a control to the biological pathways that will be discussed in the next paragraph.

Conventional physicochemical methods, such as activated carbon, precipitation, reverse osmosis, membrane separation, chemical oxidation, and ionic exchange and so on, have been commonly used. However, most of those are less effective, excessively expensive, and eco-unfriendly when initial metal concentrations are in the range of 10–100 mg/L [20]. For instance, chemical precipitation and electrochemical treatment are not very effective, especially when metal ion concentration in aqueous solution is among 1–100 mg/L, and also produce large quantity of sludge required to treat with still more great difficulty [21].

The advantages and limitations in application of these traditional methods are given below in detail as follows:

5.2.1 Activated Carbon

It is a crude form of graphite with a random or amorphous structure, which is highly porous, exhibiting abroad range of pore sizes, from visible cracks, crevices, and slits of molecular dimensions [22]. And with its specific high surface area, microporous character, and chemical nature of their surface, activated carbons have made them potential adsorbents for the removal of heavy metals from industrial wastewater [23, 24]. In spite of its advantages, it also shows many shortcomings such as high cost to prepare, reactivation resulting in a loss of the carbon, performance type dependent of carbon, being nonselective, and so on. Although modified activated carbon [25] can improve the selectivity, its preparation cost roars up subsequently.

5.2.2 Chemical Precipitation

This pathway is widely used for heavy metal removal from inorganic effluent [26, 27]. Typically, the metal precipitates from the solution in the form of hydroxide [28]. Since lime or calcium hydroxide is the most commonly employed precipitant agent, this method gains availability at most corners of the world. Lime precipitation, another option, can be trusted to effectively treat inorganic effluent with a metal concentration of higher than 1,000 mg/L with a simple process. For instance, it helped to remove heavy metals such as Zn(II), Cd(II), and Mn(II) cations with initial metal concentrations of 450, 150, and 1,085 mg/L, respectively, in a batch continuous system [29]. In spite of its advantages, chemical precipitation requires a large amount of chemicals to reduce metals to an acceptable level for discharge [30]. And generally it can hardly be used to handle low concentration of metal wastewater, which is below 100 mg/L [21]. Other drawbacks are its excessive sludge production that requires further treatment, the increasing cost of sludge disposal, slow metal precipitation, and the long-term environmental impacts of sludge disposal [31–33].

5.2.3 Chemical Oxidation and Reduction

Adding oxidizing or reducing agent makes toxic substances in wastewater being oxidized or reduced to nontoxic or low toxic substances. It has been reported that heavy metals such as Mn^{2+}, Cu^{2+}, Pb^{2+}, Cd^{2+}, Cr^{3+}, and Hg^{2+} could be effectively removed using potassium ferrate(VI) as oxidation and co-precipitation [34, 35]. However, currently chemical oxidation and reduction was generally used as a pretreatment for wastewater before more powerful procedures.

5.2.4 Membrane Separation

Membrane separation methods like electrodialysis (ED), nanofiltration (NF), ultrafiltration (UF), and reverse osmosis (RO) [36] have received considerable attention for the treatment of inorganic effluent, since they are capable of removing not only suspended solid and organic compounds, but also inorganic contaminants such as heavy metals [37–39]. Although it can be applied in occasions of low metal ion concentrations, for example, Tzanetakis's [40] removal efficiency of Co(II) and Ni(II) were 90% and 69%, at initial metal concentrations of 0.84 and 11.72 mg/L, respectively. The main disadvantages of these processes are inadequate selectivity, high energy consumption, and immature technologies, which limit the popularization of this technology.

5.2.5 *Ion Exchange*

This is a method that uses ion-exchange resin or agent to exchange ions in dilute solution to achieve the purpose of extraction or removal of certain ions in solution. And it is widely used for the recovery and removal of metals from process and waste streams in chemical process industries [41–43].

In general, ion exchange is reportedly effective to treat inorganic effluent within a relatively wide range of metal concentrations from 10 mg/L [44, 45] to 10–100 mg/L [46–48], or even up to higher than 100 mg/L [49, 50]. Furthermore, unlike chemical precipitation, ion exchange does not present any sludge disposal problems [51]. Despite these advantages, it also has some limitations in treating wastewater laden with heavy metals. Suitable ion exchange resins are not available for all heavy metals, and the capital and operational cost is high [52].

5.3 Biological Removal Technology

In recent decades, more and more attention have been paid to biotechnological removal of heavy metal and application of natural biological origin, including bacteria, fungi, yeast, algae, etc. These methods have the advantage over low operating cost, minimization of the pollutant concentration, and high efficiency in detoxifying the dilute effluents.

5.3.1 *Heavy Metals Removal by Bacteria*

5.3.1.1 Introduction

Novel methods of metal removal and recovery based on biological materials have been considered and widely used. Bacteria are the most abundant and versatile of microorganisms and constitute a significant fraction of the entire living terrestrial biomass of $\sim 10^{18}$ g [53]. In addition, because of their small size, ubiquity, ability to grow under controlled conditions, and resilience to a wide range of environmental situations, bacteria were often used to remove heavy metal ions [54]. Certain types of microbial biomass can retain relatively high quantities of metals on their cell wall due to its microstructure. Bacteria species such as *Bacillus*, *Pseudomonas*, *Streptomyces*, *Escherichia*, and *Micrococcus* have been tested for uptake metals or organics. Moreover, many factors such as biomass type and concentration, metal concentration, pH, and biomass–metal contact time affect removal ratios of heavy metals.

Table 5.1 summarizes some of the important results of metal biosorption using bacterial biomasses, according to published references [2, 11]. Metal uptake capacity does not necessarily reach the maximum values in the field application.

Table 5.1 Bacterial biomass used for metal removal

Metal ions	Bacteria species	Biosorption capacity (mg/g)	References
Pb	*Bacillus* sp.	92.3	[55]
	Bacillus firmus	467	[56]
	Corynebacterium glutamicum	567.7	[57]
	Enterobacter sp.	50.9	[58]
	Pseudomonas aeruginosa	79.5	[59]
	Pseudomonas aeruginosa	0.7	[60]
	Pseudomonas putida	270.4	[61]
	Pseudomonas putida	56.2	[62]
	Streptomyces rimosus	135.0	[63]
Zn	*Streptomyces rimosus*	30	[64]
	Bacillus firmus	418	[56]
	Aphanothece halophytica	133.0	[65]
	Pseudomonas putida	6.9	[62]
	Pseudomonas putida	17.7	[66]
	Streptomyces rimosus	30.0	[64]
	Streptomyces rimosus	80.0	[64]
	Streptoverticillium cinnamoneum	21.3	[67]
	Thiobacillus ferrooxidans	82.6	[68]
	Thiobacillus ferrooxidans	172.4	[69]
Cu	*Bacillus firmus*	381	[56]
	Bacillus sp.	16.3	[55]
	Bacillus subtilis	20.8	[70]
	Enterobacter sp.	32.5	[58]
	Micrococcus luteus	33.5	[70]
	Pseudomonas aeruginosa	23.1	[59]
	Pseudomonas cepacia	65.3	[71]
	Pseudomonas putida	6.6	[62]
	Pseudomonas putida	96.9	[61]
	Pseudomonas putida	15.8	[66]
	Pseudomonas stutzeri	22.9	[70]
	Sphaerotilus natans	60	[72]
	Sphaerotilus natans	5.4	[72]
	Streptomyces coelicolor	66.7	[73]
	Thiobacillus ferrooxidansa	39.8	[69]
Cd	*Aeromonas caviae*	155.3	[74]
	Enterobacter sp.	46.2	[58]
	Pseudomonas aeruginosa	42.4	[59]
	Pseudomonas putida	8.0	[62]
	Pseudomonas sp.	278.0	[75]
	Staphylococcus xylosus	250.0	[75]
	Streptomyces pimprina	30.4	[76]
	Streptomyces rimosus	64.9	[77]

(continued)

Table 5.1 (continued)

Metal ions	Bacteria species	Biosorption capacity (mg/g)	References
Fe(III)	*Streptomyces rimosus*	122.0	[78]
Cr(IV)	*Bacillus coagulans*	39.9	[79]
	Bacillus megaterium	30.7	[79]
	Zoogloea ramigera	2	[80]
	Aeromonas caviae	284.4	[74]
	Bacillus coagulans	39.9	[79]
	Bacillus licheniformis	69.4	[81]
	Bacillus megaterium	30.7	[79]
	Bacillus thuringiensis	83.3	[82]
	Pseudomonas sp.	95.0	[75]
	Staphylococcus xylosus	143.0	[75]
Ni	*Bacillus thuringiensis*	45.9	[83]
	Streptomyces rimosus	32.6	[84]
Pd	*Desulfovibrio desulfuricans*	128.2	[85]
	Desulfovibrio fructosivorans	119.8	[85]
	Desulfovibrio vulgaris	106.3	[85]
Pt	*Desulfovibrio desulfuricans*	62.5	[85]
	Desulfovibrio fructosivorans	32.3	[85]
	Desulfovibrio vulgaris	40.1	[85]
U	*Arthrobacter nicotianae*	68.8	[86]
	Bacillus licheniformis	45.9	[86]
	Bacillus megaterium	37.8	[86]
	Bacillus subtilis	52.4	[86]
	Corynebacterium equi	21.4	[86]
	Corynebacterium glutamicum	5.9	[86]
	Micrococcus luteus	38.8	[86]
	Nocardia erythropolis	51.2	[86]
	Zoogloea ramigera	49.7	[86]
Th	*Arthrobacter nicotianae*	75.9	[86]
	Bacillus licheniformis	66.1	[86]
	Bacillus megaterium	74.0	[86]
	Bacillus subtilis	71.9	[86]
	Corynebacterium equi	46.9	[86]
	Corynebacterium glutamicum	36.2	[86]
	Micrococcus luteus	77.0	[86]
	Zoogloea ramigera	67.8	[86]

Some uptake values were experimentally observed, and some were predicted by the Langmuir model. Table 5.1 also provides the basic information to evaluate the possibility of using bacterial biomass for the removal of metal ions.

Large quantities of metals can be accumulated by a variety of processes dependent or independent on metabolism. Both living and dead biomass as well as cellular products such as polysaccharides can be used for this purpose. The basic principle of biosorption lies in biomass which are metabolically inactive [87].

Investigations from Ziagova et al. [75] proved that nonliving cells are advantageous over living counterparts for operating without an auxiliary cultivation system, immunity to toxic wastes around, and low storage requirement. A study [88] demonstrated that living and dead cells of *Bacillus sphaericus* OT4b31 showed a biosorption of 25% and 44.5% of Cr, respectively, while *B. sphaericus* IV(4)10 showed a biosorption of 32% and 45%, respectively. Practically biosorption experiments usually use heat-killed cells or bacterial biomass through heat treatment to break cell wall, exposing more functional groups, thus achieving maximum binding capacity [79, 89]. Biosorption is responsible for metal concentration by nonliving biomass owing to the absence of metabolic activity necessary for intracellular metal accumulation [90, 91].

5.3.1.2 Effect of Environmental Factors on Biosorption

Many researchers found that some environmental factors, such as pH, biomass concentration, metal concentration, and biomass-metal contact time, affected biosorption process, result in different removal effects. Further, the kinetics of adsorption has been examined and isotherms models, Langmuir and Freundlich, have been overwhelmingly used to analyze the equilibrium data.

Effect of pH

For biosorption of heavy metal ions, pH is one of the most important environmental factors. This value of solution strongly influences not only the site dissociation of the biomass surface, but also the solution chemistry of the heavy metals: hydrolysis, complexation by organic and/or inorganic ligands, redox reactions, precipitation, speciation, and biosorption availability of the heavy metals [92]. Generally, at low pH values, several cytomembrane functional groups such as amine, phosphonate, sulfonate, carboxyl, and hydroxyl groups are probably associated with the hydrogen ions, making the overall surface charge on the microorganisms positive [93]. On the contrary, rising pH increases the negative charge at the surface of the cells until all relevant functional groups are deprotonated, which favors electrochemical attraction and adsorption of cations. However, when pH climbs to a certain value, for example, as high as 7.0, precipitation of the metal can occur by the formation of metal hydroxides [94]. The difference in optimum pH values may be attributed to the different components of the cell wall of these species. Reportedly, this optimum is organism dependent, due to different adsorptive sites of microorganisms [95]. For example, Ziagova et al. [75] observed that optimum pH values were estimated to be 1.0 for *Staphylococcus xylosus* with 36% biosorption yield and between 4.0 and 5.0 for *Pseudomonas* sp. with 42% biosorption yield in the case of chromium which exists in the solution as anionic group. They revealed that *S. xylosus* is a Gram positive bacterium with its surface consisting of peptidoglycans, teichoic, and teichuronic acids. These substances provide most of the anionic groups of the cell and are protonated at low pH and facilitate the interaction of chromium. Also, the study of Hancock [96] indicated that the optimum pH value for chromium biosorption was estimated to be 4.0. However, Gialamouidis [87] demonstrated that this value for Mn(II) biosorption was 6.0 for both of *Pseudomonas* sp. and *S. xylosus*, which suggested that the adsorption of metals onto the biomass could be ruled by ionic attraction.

Effect of Biomass Concentration

Most studies reported that high biomass concentrations result in low metal sorption due to electrostatic interactions between cells, which protect binding sites from metal occupation [97]. However, increase of the biomass concentration results in reducing the equilibrium concentration of metal solution owing to the fact that there are more available sites for interaction on biomass. The study of Gialamouidis et al. [87] showed that the biosorption capacity of metal ions is inversely proportional to biomass concentration, when the initial concentration of metal ions is kept constant with *Pseudomonas* sp. concentrations over 4.0 g/L. Ziagova et al. [75] observed that chromium removal with *Pseudomonas* sp. does not change significantly at higher than 1.0 g/L biomass concentrations. However, in the case of chromium, there is a significant increase in the metal uptake around 40% by altering biomass concentration of *S. xylosus* from 1.0 to 8.0 g/L. This may be attributed to the higher number of available binding sites, which interact with chromium ions of the solution, as it has been reported before [98, 99]. At higher than 8.0 g/L biomass concentration, there is no further positive effect on chromium removal, which probably results from limitations in metal ion mobility.

Effect of Contact Time

The kinetics of metal uptake, assumed to be a passive physical adsorption at the cell surface, is very rapid and occurs in a very short time after the microorganisms have come into contact with metal ions [100]. The adsorption speed of uranium by *Citrobacter freudii* was very fast in the first 20 min and then slowed down gradually and reached equilibrium after 50 min [101]. In the study of Ziagova et al. [75], the amount of adsorbed Cd(II) reached the highest value, 50%, within the first 30 min with *Pseudomonas* sp. and 55% after 1.0 h with *S. xylosus*., respectively. Gialamouidis et al. [87] indicated that the biosorption capacities of *Pseudomonas* sp. and *S. xylosus* for Mn^{2+} increased with contact time rapidly and thereafter it proceeded at a lower rate and finally attained equilibrium within 10 min. This result is important because equilibrium time is one of the parameters for economical wastewater treatment plant application [24]. This behavior suggests that in the initial stage adsorption takes place rapidly on the external surface of the adsorbent followed by a slower internal diffusion process, which may be the rate-determining step.

5.3.1.3 Biosorption Models

Kinetics Model

In order to examine the mechanism of biosorption process such as mass transfer and chemical reaction, a suitable kinetic model is needed to describe the data. The linear pseudo-first-order equation is given as follows [102]:

$$\log(q_{eq} - q_t) = \log q_{eq} - \frac{k_1}{2.303}t \tag{5.1}$$

where q_t and q_{eq} are the amounts of metal ions adsorbed at time t and equilibrium (mg/g), respectively, and k_1 is the rate constant of pseudo-first-order adsorption process (min^{-1}).

The linear pseudo-second-order equation is given [102]:

$$\frac{\mathrm{t}}{q_t} = \frac{1}{k_2 q_{eq}^2} + \frac{1}{q_{eq}}t \tag{5.2}$$

where k_2 is the equilibrium rate constant of pseudo-second-order biosorption (g/mg min).

These results of Gialamouidis et al. [87] strongly suggest that the biosorption of Mn(II) onto *Pseudomonas* sp. and *S. xylosus* cells is most appropriately represented by a pseudo-second-order rate process. In the study of Xie et al. [101], the adsorption kinetics of uranium by *C. freudii* showed that the pseudo-second-order model fitted the absorption curve much better than the pseudo-first-order model, which indicated the adsorption rate was proportional to the number of unoccupied sites. Song et al. [103] demonstrated that the equilibrium of Au(III) biosorption was achieved within 5 min and the adsorption yield reached 100%, whereas that of Cu(II) was only 66.61% after 30 min in the binary system of Au(III) and Cu(II) by magnetotactic bacteria. They suggested that the first-order equation of Lagergren in most cases does not fit well for the whole range of contact time [104]. The pseudo-second-order equation can be used in this case assuming that the measured concentrations are equal to cell surface concentrations, and it is more likely to predict behavior over the whole range of adsorption and is in agreement with an adsorption mechanism being the rate controlling step. Consequently, they demonstrated that the experimental data were fitted well into the pseudo-second-order kinetic model in the single and binary component situations.

Isotherm Models

Langmuir and Freundlich isotherm equations were frequently used to describe the equilibrium state for single-ion adsorption experiments. The theoretical basis of Langmuir equation relies on the assumption. That is, there is a finite number of

Table 5.2 Adsorption isotherm parameters in the different conditions

Metal ions	Bacteria species	Langmuir constants Q_{max} (mg/g)	$b \times 10^{-3}$ (L/mg)	R^2	Freundlich constants K_f (L/g)	$1/n$	R^2	References
Cd	*Paeudomonas* sp.	278	9	0.901	9.08	0.531	0.998	[75]
	S. xylosus	250	8	0.912	7.00	0.547	0.996	[75]
	Ochrobactrum anthropi	37.3	37	0.971	44.64	2.45	0.961	[107]
Cr	*Pseudomonas* sp.	95	7	0.949	1.42	0.676	0.997	[75]
	S. xylosus	143	5	0.967	1.33	0.758	0.974	[75]
	Ochrobactrum anthropi	86.2	6	0.932	2.45	1.60	0.981	[107]
Au	*Magnetotactic bacteria*	37.7	978	0.9939	27.56	0.0637	0.9901	[103]
Cu	*Magnetotactic bacteria*	36.8	160.5	0.9985	16.09	0.091	0.9428	[103]
	Ochrobactrum anthropi	32.6	52	0.991	3.00	1.72	0.816	[107]
Mn	*Pseudomonas* sp.	109	13	0.99	4.30	0.536	0.93	[87]
	S. xylosus	59	26	0.97	3.22	0.533	0.99	[87]

binding sites which are homogeneously distributed over the adsorbent surface of the cells, having the same affinity for adsorption of a single molecular layer and there is no interaction between adsorbed molecules. The mathematical description of the equation is considered.

$$Q = \frac{Q_{\max} b C_i}{1 + b C_i} \tag{5.3}$$

where C_i is the metal residual concentration in solution (mg/L); $Q_{\max}$ is the maximum specific uptake corresponding to sites saturation (mg/g), and b is the biomass-metal binding affinity (mg/L) [105].

The Freundlich equation is the empirical relationship whereby it is assumed that the adsorption energy of a metal binding to a site on an adsorbent depends on whether or not the adjacent sites are already occupied. This empirical equation is presented.

$$Q = K_f \cdot (C_i)^{1/n} \tag{5.4}$$

where K_f and n are constants indicating adsorption capacity and adsorption intensity, respectively [106]. The results of some studies about Langmuir and Freundlich isotherm models are presented in Table 5.2.

5.3.1.4 Selective Adsorption in the Case of Multicomponent System

Competitive biosorption takes place in the multicomponent system. Under most conditions, aqueous sources of precious metals contain not only one but also several kinds of metal ions. Although many published reports are available on the single-metal biosorption, limited attention has been paid to the biosorption of multimetal

Table 5.3 The Freundlich parameters and the correlation coefficients

	Au(III) ions			Cu(II) ions		
$C_{0,other\ metal}$/mg/L	K_F	n	R^2	K_F	n	R^2
0	27.56	15.6985	0.9901	16.09	11.0254	0.9428
80	44.33	22.8833	0.9265	1.27	5.1125	0.9693
320	38.64	10.2669	0.9214	0.04	1.4656	0.9857

Note: $C_{0,\text{ other metal}}$ represents the other metal concentration in solution. When $C_{0,\text{ other metal}} = 0$, it is single ion solution; when not, it is binary system

Table 5.4 The Langmuir parameters and the correlation coefficients

	Au(III) ions			Cu(II) ions		
$C_{0,other\ metal}$/mg L^{-1}	Q^0 (mg/g)	b (L/mg)	R^2	Q^0 (mg/g)	b (L/mg)	R^2
0	37.7	0.9780	0.9939	36.8	0.1605	0.9985
80	48.8	4.1000	1.0000	22.4	0.0046	0.9260
320	47.6	5.2500	1.0000	27.7	0.0007	0.9026

Note: The meaning of $C_{0,\text{ other metal}}$ is the same as Table 5.3

ion system [108]. Summarization of these present literatures concerning biosorption of this more complicated system revealed different biosorptive features from those of single counterparts [103]. To make this reasonable, competition among the different metal ions for the cellular surface binding sites is believed to certainly occur and depend to some extent on the ionic characteristics for certain microbe. The examination of the effects of binary metal ions in various combinations, discussed in this section, is more representative of the actual environmental problems as it has been also reported [109], since real wastewaters are polluted with more than one toxic metals.

Song et al. [103] observed that the maximum adsorption capacity for Au(III) by magnetotactic bacteria in Au-Cu binary system was nearly doubled as that in the single system. The results demonstrated that Cu(II) has a "catalyzing" effect on Au(III) sorption during competitive biosorption process, which also indicates that the MTB cell walls have high selectivity toward Au(III) ions in binary-solute biosorption process.

Besides, Langmuir model and Freundlich model have been also introduced to describe both single and binary system biosorption. The Freundlich and Langmuir constants evaluated from linear regression analysis are listed in Tables 5.3 and 5.4, respectively. Within the metal concentration range of 80–500 mg/L, both models were suitable for describing the adsorption isotherms of Au(III) or Cu(II) ions; that of Au(III) was more fitted by Langmuir model and that of Cu(II) by Freundlich in binary system.

In the study of Wang et al. [108], simultaneous biosorptions of Ag(I) and Cu(II) ions by *Magnetospirillum gryphiswaldense* (MSR-1) were evaluated using several mediums that contained 2 mmol/L each of the two metal ions. In each medium, the molar ratios of Ag(I) and Cu(II) were different, which were 5:1, 4:1, 3:1, 2:1, 1:1,

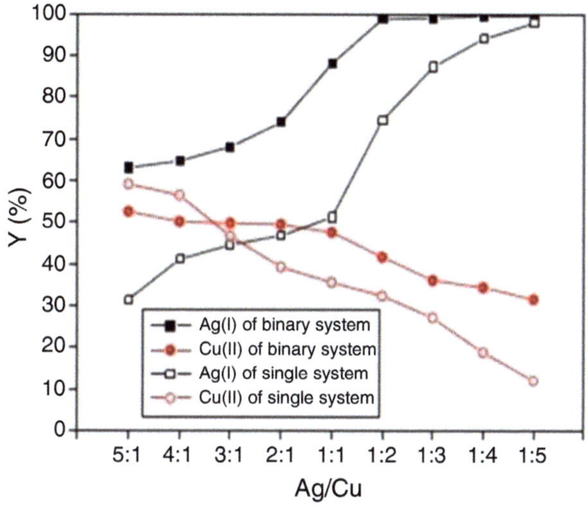

Fig. 5.1 The results of Ag(I)–Cu(II) binary system adsorption experiments (Reprinted from Ref. [108]. With kind permission of © Elsevier BV (2011))

1:2, 1:3, 1:4, 1:5. The results were compared with that of single systems, which included the same molar concentration of Ag(I) and Cu(II).

As shown in Fig. 5.1, the biosorption capacities of Ag(I) on MSR-1 in the presence of Cu(II) were higher than that of noncompetitive conditions. Interestingly, when the molar ratios of Ag(I) and Cu(II) were less than 4:1, the biosorption capacities of Cu(II) on MSR-1 were higher than that of noncompetitive cases, whereas those were lower under the molar ratios of Ag(I) to Cu(II) of both 5:1 and 4:1. They deduced that the most possible factor which affected the phenomenon of selective biosorption was the ions themselves.

In addition, they demonstrated that the adsorption isotherm of Ag(I) can also be perfectly expressed by the Langmuir model just as that in the single ion solution. Nevertheless, both models are mathematically unsuitable for describing Cu(II) adsorption isotherms in binary system, with the regression coefficients (R^2) less than 0.95. So, a new mathematically consistent Eq. (5) was selected for accurate description of Cu(II).

$$q_e = a(1 + be^{-kC_e}) \tag{5.5}$$

The constants in this case evaluated from regression analysis were $a = 3.50917$, $b = 7.38993$, $k = 0.11474$, with a high regression coefficient (R^2) 0.99343. This model, not fit for the single metal system in this study, is under further checking among vast analogous competitive adsorptive systems to find universality before it is introduced to mathematically describe the prohibited one(s). Undoubtedly, it has an important guiding significance for the future research.

The results indicated that biosorption in binary mixtures is a rather complicated adsorption mechanism affected by a lot of factors, including surface charge, functional groups, properties of the solution (e.g., pH), and the adsorbates (e.g., concentration, ionic strength, and weight). In addition, there are many interactions not only between the metal ions but also between the metal ions and the surface of the biomass [75]. If highly selective biosorbents for a certain metal ion were created, great progress would be made in the removal of heavy metal ions.

5.3.1.5 Regeneration of Biosorbents

The entire biosorption process for metal removal include sorption followed by desorption, i.e., to concentrate the solute. Biotechnological exploitation of biosorption technology for removal of heavy metal(s) depends on the efficiency of the regeneration of biosorbent after metal desorption. It is important to reclaim the biosorbents especially, since the biomass preparation is costly. Therefore, nondestructive recovery by mild and cheap desorbing agents is desirable for regeneration of biomass and reuse in multiple cycles. Appropriate eluants are necessary to attain the above-mentioned objective, which strongly depends on the type of biosorbent and the mechanism of biosorption [21]. Also, the eluant must meet the following requirements: (1) Non-damaging to the biomass, (2) less costly, (3) environmental friendly, and (4) effective [11]. Acidic and alkaline condition were used for desorption. The eluants such as $CaCl_2$ with HCl, HCl with EDTA, NaOH were reported [11]. According to the study of Chen et al. [110], approximately 74.8% of the adsorbed Cu was desorbed from nonliving *Pseudomonas putida* CZ1 by 1.0 M CH_3COOK, and more than 80% Zn. Gialamouidis et al. [87] observed that more than 88% of Mn(II) adsorbed by *Pseudomonas* sp. was released into the HNO_3 solution at the first cycle. After three cycles, biosorption ability of *Pseudomonas* sp. reduced by 77.4%, which may be due to the fact that HNO_3 solution modifies the cell wall, while the desorption ability remains to high levels.

The efficiency of desorption is often expressed by the S/L ratio, i.e. solid to liquid ratio. The solid represents the solid sorbent (in mg dry wt) and the liquid represents the amount of eluant applied (in mL). High values of S/L are desirable for complete elution and to make the process more economical. Sometimes, metal-selective elution is desirable and it is dependent on metal sequestration mechanism. Dilute mineral acids, EDTA, carbonates and bicarbonates, NH_4OH, $KHCO_3$, and KCN have been used to remove metal(s) from the loaded biomass [111].

To date, less attention has been paid to investigate the regeneration ability of the biosorbent, more relevant work is necessary for future field biosorption application.

5.3.1.6 Biosorption Mechanism Discussion

One of them is bioaccumulation, based on the incorporation of metals inside the living biomass. Another process is biosorption, in which metallic ions remain at the cellular surface by different mechanisms [11]. Elucidation of mechanisms active in metal biosorption is essential for successful exploitation of the phenomenon and for regeneration of biosorbent materials in multiple reuse cycles. The complex nature of biosorbent materials makes this task particularly challenging [5]. These results show that in binary mixtures, biosorption is a rather complicated adsorption mechanism affected by a lot of factors, including surface charge, functional groups, properties of the solution (e.g., pH), and the adsorbates (e.g., concentration, ionic strength, and weight). In addition, there are many interactions, not only between the metal ions, but also between the metal ions and the surface of the biomass. Velásquez and Dussan [88] suggested that in dead cells metals could have been adhered to surface molecules such as the S-layer which is a porous one that can have saturation velocity. Metallic ions must find a union target and pass through other cellular components before this happens. What is more, in living cells cellular density increases as time elapses; therefore, there are more available binding sites for metals. Additionally, if the metal gets inside the cell, it first needs to join surface molecules and then gets in through different mechanisms. So, S-layer proteins might execute a trapping role of metallic ions in both living and dead cells, being a potential alternative for bioremediation processes of heavy metals in field. Algae, fungi, yeast, and bacteria can remove heavy metals from aqueous solutions by binding the cationic metals onto negatively charged functional groups distributed on their cell walls, such as carboxyl and phosphoryl groups [112, 113]. According to Volesky and Holan [5], who presented an extensive review on biosorption, the strong biosorbent behavior of certain types of microbial biomass toward metallic ions is a function of the chemical makeup of microbial cells. Beveridge [114] reported that bacteria are excellent biosorbents due to their high surface-to-volume ratios and high content of potentially active chemisorptions sites such as teichoic acid in their cell walls. The study of Gialamouidis et al. [87] is very interesting. They calculated thermodynamic parameters such as enthalpy change (ΔHo), entropy change (ΔSo), and free energy change of the sorption (ΔGo), demonstrating that the heats of biosorption of Mn(II) on *Pseudomonas* sp. and *S. xylosus* cells were found to be of the same order as the heat of chemisorptions whereas the heat of biosorption on *Blakeslea trispora* cells is lower and in the range of physical adsorption. Besides, equilibrium between the cell surface and the metal ions is usually rapidly attained and easily reversible, because the energy requirements are limited. Studies of copper or zinc interactions with microorganisms showed that Cu or Zn may be associated with the functional groups on the cellular surface [115–118], intercellular accumulation, and storage via active cationic transport systems [114, 119], or precipitation of Cu sulfides at the biofilm surface [120]. In terms of biosorption, polysaccharides, proteins, and lipids on bacterial cell walls offer many functional groups such as carboxylate, hydroxyl, phosphate, amine, and sulphate groups which can bind metal ions. This natural affinity of biological compounds for metallic elements could contribute to the purification of metal-contaminated wastewater [18]. Vullo et al. [18] thought that metal biosorption depends upon the available surface negative charges. Metal ions biosorption could affect these surface negative charges and thus bacterial electric properties. Electrophoretic mobility measurements are generally done to understand bacterial cell electric properties [121–123], determining Zeta potential values.

Moreover, in their study, for concentrations greater than 0.05 mM, no changes are observed in Z potential, which could be explained by the saturation of the cell surface sites available for Cd. Xie et al. [101] observed that the adsorption capacity of uranium by *C. freudii* increased after the strain was pretreated with 0.5 mol/L NaOH. Proteins, lipids, and some dissolved polysaccharides on the surface of the bacteria could be removed by NaOH. The macromolecular configuration of the cell wall was destroyed and became relaxed, which could improve the diffusion capability and facilitate the entry of metal ions into the cell. In addition, acetyl in testa could be removed by alkali, and amidocyanogen could be exposed to form Louis alkali, which increased the active sites and the adsorption capacity eventually. By sequential pretreatment with NaOH and methanol-thick hydrochloric acid, the carboxyl on the *C. freudii* walls was locked and the adsorption capacity of uranium decreased by 45.66% compared with raw bacterium strain, which confirmed that carboxyl, as one of the active sites, played an important role in adsorption.

It is necessary to continue searching for the most promising biosorbents from an extremely large pool of readily available and inexpensive biomaterials [8, 9]. The mechanism involved in metal biosorption is far from well understood up to date. It is our mission to investigate the microbe-metal interactions and obtain the mechanism of metal uptake by biosorbents.

Bacteria may either possess the capacity for biosorption of many elements or, alternatively, depending on the species, may be element specific. It is likely that, in the future, microorganisms will be tailored for a specific element or a group of elements, using recombinant DNA technology which is based on genetic modification using endorestrictive nucleases [53].

5.3.2 *Heavy Metals Removal by Fungi*

5.3.2.1 Introduction

Powered by another mechanism, fungi also reveal the feature of metal capturers. There are three groups of fungi that have major practical importance: molds, yeasts, and mushrooms. Filamentous fungi and yeasts have been investigated for biosorption of heavy metal ions in many instances.

It is known for a long time that metallic ions are important to fungal metabolism. The presence of heavy metals had a significant impact on not only the metabolic activities of fungal cultures, but also the commercial fermentation processes. The results from these studies led to a notion of using fungi for the removal of toxic metals from wastewater and recovery of precious metals from process waters [124]. Fungal cell walls are mainly 80–90% polysaccharide, with proteins, lipids, polyphosphates, and inorganic ions, making up the wall-cementing matrix [21]. It is because of these components that both living and dead fungal cells have an extraordinary ability for taking up toxic and precious metals.

The biosorption of heavy metal by filamentous fungi and yeast have been reported and reviewed in many articles. The fungal organisms are widely used in a variety of large-scale industrial fermentation processes. For example, strains of *Aspergillus* are used in the production of ferrichrome, kojic acid, gallic acid, itaconic acid, citric acid, and enzymes like amylases, glucose isomerase, pectinase, lipases, and glucanases while *Saccharomyces cerevisiae* in the food and beverage industries. The use of fungi biomass as an adsorbent for heavy-metal pollution control not only can generate economic benefits by recycling industries wastes, but also reduce the burden of disposal costs associated with the waste biomass produced. Alternatively, the biomass can also be grown using unsophisticated fermentation techniques and inexpensive growth media [125].

This section will review and summarize the removal of heavy metals by yeast (*Saccharomyces* spp.) and filamentous fungi (such as *Penicillium* sp., *Aspergillus* sp., *Mucor* sp., *Rhizopus* sp.) from aqueous solutions.

5.3.2.2 Fungal Biosorbents

Research has proved that yeast and filamentous fungi can remove toxic metals, recover precious metals, and clean radionuclides from aqueous solutions to various extents. Yeasts of genera *Saccharomyces*, *Candida*, different species of *Penicillium*, under some circumstances, as well as *Aspergillus* are efficient biosorbents for heavy metal ions. Table 5.5 summarizes some of the important results of metal biosorption using fungal biomass.

Except for these four kinds of fungal biomass widely covered by the published reports, there are also a lot of other fungi employed for the same purpose.

The cell wall of *R. arrhizus* involves a high content of chitin. The ability of chitin to complex metal ions has been confirmed [148]. Viable *R. arrhizus* could remove Cu(II) with the maximum specific uptake capacity of 10.76 mg/g at 75 mg/L, the initial Cu(II) concentration [148]. Of the six species of inactivated fungal mycelia, *R. arrhizus*, *Mucor racemosus*, *Mycotypha africana*, *Aspergillus nidulans*, *A. niger*, and *Schizosaccharomyces pombe*, *R. arrhizus* exhibited the highest capacity ($q_{max} = 213$ μmol/g). Further experiments with different cellular fractions of *R. arrhizus* showed that Zn was predominantly bound to cell-wall chitin and chitosan ($q_{max} = 312$ μmol/g). *R. arrhizus* were reported to adsorb Th(IV) [152], Pb(II) [153], Cd(II), Ni(II), and Cr(III) [154]. Brady and Tobin [155] investigated the freeze-dried *R. arrhizus* biomass for its potential to absorb the hard metal ion Sr^{2+} and the borderline metal ions Mn^{2+}, Zn^{2+}, Cd^{2+}, Cu^{2+}, and Pb^{2+} from aqueous solutions.

Biosorption of metal ions such as Li^{+}, Ag^{+}, Pb^{2+}, Cd^{2+}, Ni^{2+}, Zn2+, Cu^{2+}, Sr^{2+}, Fe^{2+}, Fe^{3+}, and Al^{3+} by *Rhizopus nigricans* biomass was studied, with the maximum biosorption capacity for the individual metal ions ranging from 160 to 460 μmol/g [156].

Table 5.5 Fungal biomasses for removal of heavy metals from aqueous solution

Biosorbents	Metal ions	Biosorption capacity(mg/g)	References
Saccharomyces cerevisiae	Pb	270.3	[126–129]
	Ni	46.3	[127, 129]
	Cr	32.6	[127, 128]
	Cd	31.75	[128–130]
	Cu	6.35	[128, 129]
	Zn	35.31	[128, 129]
	Hg	64.19	[129]
	Co	24.75	[129]
	Fe	16.8	[131]
	Ag	59	[131]
	Pd	40.6	[132]
	Pt	44.0	[133]
	U	180	[131]
	Th	63	[134]
Candida albicans, Candida utilis	Pb	833.33	[135]
	Cu	23	[136, 139]
	Cr	7	[137, 139]
	Zn	28	[139]
	Cd	19	[139]
	Pb	39	[139]
Candida tropicalis	Zr	179	[138]
Penicillium chrysogenum	Pb	55	[140]
	Cd	56	[141]
	Ni	260	[142]
	Cu	92	[140]
Penicillium simplicissimum	Cd	52.5	[143]
	Zn	65.5	[143]
	Pb	76.9	[143]
	Cu	112.3	[144]
Penicillium purpurogenum	As	35.6	[145]
	Hg	70.4	[145]
	Cd	110.4	[145]
	Pb	252.8	[145]
Penicillium canescens	As	26.4	[146]
	Hg	54.8	[146]
	Cd	102.7	[146]
	Pb	213.2	[146]
Penicillium digitatum	Cd	3.5	[15]
	Pb	5.5	[15]
Penicillium griseofulvum	Cu	1.51	[147]
Penicillium italicum	Cu	2	[2]
	Zn	0.2	[2]
Penicillium notatum	Cu	80	[125]
	Zn	23	[125]
	Cd	5.0	[125]

(continued)

Table 5.5 (continued)

Biosorbents	Metal ions	Biosorption capacity(mg/g)	References
Penicillium janthinellum	U	52.7	[125]
Aspergillus niger	Cu	9.53	[148]
	Cd	3.43	[149]
	Pb	7.24	[149]
	Ni	0.96	[149]
Aspergillus terreus	Cu	180	[150]
	Fe	164.5	[151]
	Cr	96.5	[151]
	Ni	19.6	[151]
	Cu	224	[151]

The live and dead white rot fungus *Trametes versicolor* entrapped in Ca-alginate beads were able to adsorb Cd(II). The maximum experimental biosorption capacity was 102.3 ± 3.2 and 120.6 ± 3.8 mg/g, respectively [157].

A white rot fungus species *Lentinus sajorcaju* biomass was entrapped into alginate gel via a liquid curing method in the presence of Ca(II) ions. The maximum experimental biosorption capacity for entrapped live and dead fungal mycelia of *L. sajorcaju* were found to be 104.8 ± 2.7 mg/g and 123.5 ± 4.3 mg/g, individually [158].

5.3.2.3 Influencing Factors of Removal of Heavy Metals by Fungi

Pretreatment

The study of Yunsong Zhang et al. [159] showed that ethanol/caustic pretreatment could increase the biosorption capabilities of Cu^{2+} on *Saccharomyces cerevisiae*. In addition, SEM and Zeta potential of the three samples account for that caustic and ethanol-pretreatment resulted in the change of baker's yeast surface structure and charge which is relative to adsorption. These results demonstrate that the increase of biosorption capacity for Cu^{2+} by ethanol and caustic-baker's yeast was attributed to the increase and exposure of carboxyl and amino groups on the surface of biomass sample.

Biosorption of chromium ions was found to vary with the type of pretreatment. All pretreatment methods (acid (0.1N H_2SO_4); alkali (0.01N NaOH); acetone (50%, v/v); formaldehyde (10%, v/v); cetyl trimethyl ammonium bromide (CTAB) (5%, w/v); polyethylemine (PEI) (1%, w/v); and 3-(2-amino ethyl amino) propyl trimethoxy silane (APTS) (3%, v/v)) increased the biosorption of chromium compared to the autoclaved *A. niger* biomass [160]. Park et al. [161] also reported approximately 30% removal of chromium using *A. niger* biomass. The acetone pretreated biomass showed a biosorption capacity of 1.8 mg/g, which was slightly higher compared to autoclaved biomass.

pH

Solution's initial pH is a critical parameter for adsorption experiments [162]. This parameter strongly influences the solution chemistry of the metals, the activity of functional groups (carboxylate, phosphate and amino groups) on the cell wall, as well as the competition of metallic ions for the binding site [1]. At low pH (<1.0), the biosorption capacity for metal ions is very low, because large quantity of hydrogen ions competes with metal ions at sorption sites. As the pH increases, more negatively charged cell surface become available thus facilitating great metal uptake [163]. However, metal precipitates at high pH values (>7.0) thus inhibiting the contact of metal with the most fungal biomass.

Ting Fan et al. [143] investigated the effects of the initial pH on biosorption of Cd(II), Zn(II), and Pb(II) ions in aqueous solution by *Penicillium simplicissimum*. Maximum biosorption capacities were obtained at pH 4.0, 6.0, and 5.0 for Cd(II), Zn(II), and Pb(II), respectively.

The biosorption of nickel(II) ions by deactivated protonated yeast was observed with respect to the initial pH by V. Padmavathy et al. [164]. The adsorption capacity was pH dependent with a maximum value of 11.4 mg/g at a pH of 6.75. The cadmium(II) ion adsorption capacity of Baker's yeast increased with increasing pH, and was maximum at a pH of 6.5 [165].

Contact Time

It is reported that biosorption of chromium consisted of two phases [161]: a primary rapid phase and a secondary slow phase. The rapid phase lasted for about 15 min and accounted for a major portion in the total metal biosorption. Chromium removal of 72% was observed in this rapid phase. A maximum chromium removal of 76% was achieved after 2 h. From 2 h to the end of the kinetic study (12 h), biosorption was slow and chromium removal varied by $\pm 2\%$. This rapid initial uptake is consistent with reports on the biosorption of Cr(VI) by other researchers [166–170].

This is because the kinetics of metal adsorption is usually rapid during the early period of contact between the adsorbent and adsorbate possibly by reason of electrostatic attraction. The active adsorption sites of the adsorbent become involved in metal complexation as soon as the adsorbent is introduced into the system. It has been suggested that the slow second phase of metal ions uptake may be due to the reduced availability of active sites [171].

Initial Metal Ion Concentration

As a rule, increasing the initial metal concentration results in an increase in the biosorption capacity because the initial metal concentration provides a driving force to overcome mass transfer resistances between the biosorbent and biosorption

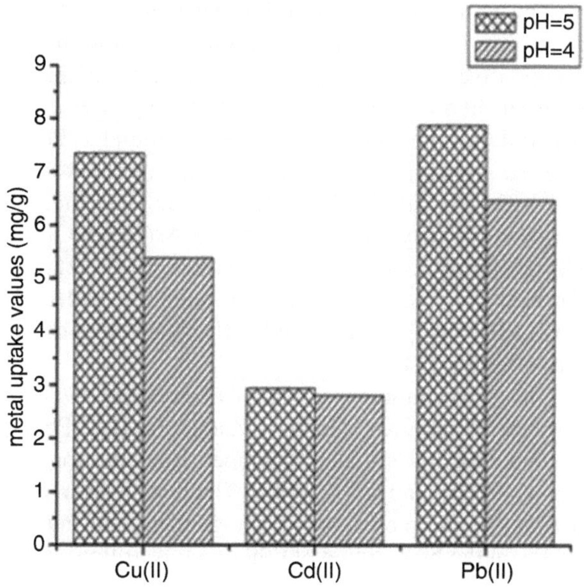

Fig. 5.2 Competitive metal uptake values (Reprinted from Ref. [103]. With kind permission of © Elsevier (2007))

medium so that higher sorption capacities were obtained at higher initial concentrations. The adsorption yields are determined at different initial metal ion concentrations. Increasing the metal ion concentration generally caused a decrease in the biosorption yield. In the case of lower concentrations, the ratio of initial number of metal ions to the available sorption sites was low and higher biosorption yields were obtained. At higher concentrations, the available sites of biosorption became fewer and the saturation of the sorption sites was observed. So biosorption yields decreased.

The effect of initial metal ion concentration on the biosorption capacity of *Aspergillus flavus* was studied by Tamer Akar and Sibel Tunali [172]. The biosorption of Pb(II) and Cu(II) ions on *A. flavus* increased with increasing initial concentration of metal ions, becoming saturated at 200 and 150 mg/L for Pb(II) and Cu(II) ions, respectively.

5.3.2.4 Selectivity and Competitive Biosorption by Fungi

Competitive biosorption of Cd^{2+} and Pb^{2+} together with Cu^{2+} ions with ethanol treated yeast cells were conducted for solution containing 25 mg/L of each metal ions at pH 4 and 5 [130]. In their study [130], the highest biosorption capacity of Cu^{2+} ions was determined with caustic treated waste baker's yeast cells; but for comparison, ethanol treated yeast biomass was also used for biosorption of Cu^{2+} ions in competitive biosorption studies. The competitive metal uptake values are shown in Fig. 5.2.

The competitive biosorption capacities of the yeast biomass for most metal ions were lower than non-competitive conditions. The decreased metal uptake in competitive conditions was thought to be a response to increased competition between like charged species for binding sites of the ethanol treated yeast cells. The order of the sorption capacity was found as Pb > Cu > Cd for both pH values.

P. canescens at competitive or noncompetitive cases exhibited the same preferential order: Pb(II) > Cd(II) > Hg(II) > As(III). Under noncompetitive conditions, the sorption capacity was 26.4 mg/g for As(III), 54.8 mg/g for Hg(II), 102.7 mg/g for Cd(II), and 213.2 mg/g for Pb(II) respectively. The competitive adsorption capacity for the heavy metal ions were 2.0 mg/g for As(III), 5.8 mg/g for Hg(II), 11.7 mg/g for Cd(II), and 32.1 mg/g for Pb(II), when the initial concentration of the metal ions was 50 mg/L [146].

The biosorption capacities of *Penicillium simplicissimum* in the binary and trinary metal mixture of Cd(II) + Zn(II), Zn(II) + Pb(II), and Cd(II) + Zn(II) + Pb(II) system were lower than noncompetitive conditions that of unitary system of Cd(II), Zn(II), Pb(II), and Cd(II) [143]. The results clearly demonstrated that the combined action of multiple ions was antagonistic and that Cd(II) exerted the inhibitoriest effect on the biosorption of other metals, followed by Pb(II) and Zn(II).

A similar phenomenon had been observed in the binary adsorption Pb(II) and Cu(II) biosorption onto *Aspergillus flavus*, where it was shown that Pb(II) and Cu(II) strongly competed with each other; the biosorption capacities of the binary metal mixture were lower than that of noncompetitive conditions; and the uptake capacities of Pb(II) ions were smaller influence than Cu(II) ions [172].

All these published experiments illustrated prohibited competitive biosorptive capacities, in comparison with the non-competitive cases, for the biomass to uptake metal ions. The most likely reason for this antagonistic effect is the competition for adsorption sites on the cell surfaces and/or the screening effect by the competing metal ions.

5.3.2.5 Mechanism of Metal Uptake

DAI Shu-juan et al. [173] provided that cadmium can be adsorbed from electroplating wastewater by waste *Saccharomyces cerevisiae* and accumulated mainly in form of chemical chelating. Electrostatic attraction, hydrogen bonding, and van der Waals force all function in adsorption process. And $-NH_2-$, $-C{=}O-$, $-C{=}O-NH-$, $-CH_3$, $-OH$ are the main adsorption groups.

Fourier transform, infrared, and X-ray photoelectron spectroscopies were used to discover that carboxyl, amide, and hydroxyl groups on the biomass surface were involved in the sorption of copper and cadmium by *Penicillium chrysogenum* and ion exchange and complexation dominated the sorption process [174].

The understanding of the mechanism by which microorganisms accumulate metals is crucial to the development of microbial processes for concentration, removal, and recovery of metals from aqueous solution. Metabolism-independent metal binding to the cell walls and external surfaces is the only mechanism present in the case of non-living biomass. Metabolism-independent uptake essentially involves adsorption process such as ionic, chemical, and physical adsorption. A variety of ligands located on the fungal walls are known to be involved in metal chelation. These include carboxyl, amine, hydroxyl, phosphate, and sulfhydryl groups. Metal ions could be absorbed by complexing with negatively charged reaction sites on the cell surface [175, 176]. The relative importance of each functional group is often difficult to resolve [177]. Microbial cell wall is rich in polysaccharide and glycoproteins such as glucans, chitin, mannans, and phospho-mannans. These polymers form abundant source of metal binding ligands. Cell walls of fungi present a multi-laminate architecture where up to 90% of their dry mass consists of amino or non-amino polysaccharides [178]. In general, the fungal cell wall can be regarded as a two-phase system consisting of the chitin skeleton framework embedded in an amorphous polysaccharide matrix [178]. Up to 30% of *Aspergillus niger* biomass is comprised of an association of chitin and glucan [179]. Chitin and chitosan components of the cell wall are suggested to be important for metal uptake [180, 181].

5.3.2.6 Modeling of Biosorption: Isotherm and Kinetic Models

Analysis of equilibrium data is important for developing an equation that can be used for design purposes. There are several isotherm and kinetic models to describe the sorption phenomena.

Isotherm Modeling

The biosorption isotherms by biomasses are usually described by the well-known adsorption isotherm models of Freundlich and Langmuir (see Sect. 5.3.1.3).

The parameters of Langmuir and Freundlich isotherm model are presented in Table 5.6. Broadly speaking, most of the data were fit better to Langmuir isotherm model than Freundlich isotherm model.

Except for these two models, Redlich–Peterson and Temkin adsorption isotherms were also used to describe the sorption phenomena. Arzu Y. Dursun [183] used Langmuir, Freundlich, and Redlich–Peterson adsorption models to describe the sorption phenomena of Cu(II) and Pb(II) to dried *A. niger*. In view of the results, the isotherms appeared to follow the Langmuir model more closely than the other models at all the temperatures studied. However, the Freundlich and Redlich–Peterson adsorption models also seemed to agree well with the experimental data considering that percentage error values were lower than 11.5. Jing-song Wang [184] used Langmuir, Freundlich, and Temkin isotherms models to describe the sorption phenomena of U(VI) to *A. fumigatus*. Based on the correlation coefficient R^2, the Langmuir isotherm model was the best model to describe the experimental data, but the negative value of qmax indicated that this model is not suitable for

Table 5.6 Adsorption isotherm parameters

Metal ions	Bacteria species	Langmuir constants Q_{max} (mg/g)	$b \times 10^{-3}$ (L/mg)	R^2	Freundlich constants K_F	n	R^2	References
Pb	*Candida albicans*	828.5	60.0	0.995	51.309	4.716	0.935	[135]
	Penicillium simplicissimum	144.9	590	0.999	33.0	3.27	0.9016	[144]
Cr	*Aspergillus niger*	3.98	1.1	0.95	2.57	6.6	0.95	[149]
Ni	*Aspergillus niger*	2.2	140	0.98	1.47	1.79	0.98	[149]
Cd	*Aspergillus niger*	18.08	20	0.997	6.26	4.4	0.671	[182]
Cu	*Saccharomyces cerevisiae*	10.27	15.13	0.9968	0.4048	1.762	0.9644	[159]
	Penicillium simplicissimum	106.4	193	0.9978	19.3	3.02	0.9102	[144]
Zn	*Aspergillus niger*	26.1	120	0.998	6.95	3.6	0.770	[182]

this case. Compared with the other two models, it is clear that both the Freundlich and Temkin isotherm models fitted with the experimental data well, and the former model was a better fit than the latter.

Kinetic Modeling

The biosorption kinetics by biomasses is usually described by first-order rate equation and the second-order rate equation (see Sect. 5.3.1.3).

In most cases, the first-order equation of Lagergren does not fit well over the entire adsorption period and is generally applicable for the initial 30–50 min of the sorption process [185]. Different to the pseudo first-order model, the pseudo-second-order model predicts the behavior over the whole duration of adsorption and is in agreement with the feature of the adsorption mechanism in the rate-controlling step.

Kuber C. Bhainsa et al. [186] investigated the kinetic of biosorption for Th by *Aspergillus fumigatus*. Data obtained from the kinetic of uptake when modeled with pseudo-second-order equation showed excellent fitting and the kinetic rate constant and R^2 values are >0.99. This suggests that the kinetics of Th uptake followed Lagergren's pseudo-second-order equation. This was also further confirmed by calculating the q_e based on pseudo-second-order reaction that showed very good fit between the experimental data and the predicted curve obtained by modeling the uptake value.

Removal of cadmium and zinc ions from aqueous solution by living *Aspergillus niger* was investigated by LIU Yun-guo et al. [182]. The parameters of pseudo-first and pseudo-second order rate kinetics are presented. The theoretical Q_{eq} values estimated from the first-order kinetic give significantly different values compared with the experimental values, and the correlation coefficients are also found to be slightly lower. The correlation coefficients for the second-order kinetic model and the theoretical values of Q, also agree well with the experimental ones.

Almost all of the kinetic data of sorption by fungi biomass kinetic were more fitted to pseudo-second-order equation.

5.3.3 *Heavy Metals Removal by Algae*

5.3.3.1 Introduction

Algae, primarily marine microalgae, are of special interest in search for and the development of new biosorbents materials due to their high sorption capacity and their ready availability in practically unlimited quantities in the seas and oceans [187, 188]. Actually a vast array of biological materials, especially bacteria, yeasts, fungi, and algae have received increasing attention for heavy metal removal and recovery due to their good performance, low cost, and large available quantities. In general, the heavy metal uptake capacities varied significantly for different types of biomass studied. For divalent heavy metal ions, the reported values for bacterium biomass typically ranged from 0.05 to 0.2 mmol/g, fungi and yeast 0.2–0.5 mmol/g, fresh water algae 0.5–1.0 mmol/g, and marine algae 1.0–1.5 mmol/g [189].

From the published literature, among the three groups of algae (red, green, brown algae) brown algae received the most attention. Higher uptake capacity has been found for brown algae than for red and green algae [190]. Table 5.7 summarized the results achieved with brown algae, green algae, and red algae. We know that brown algae have since proven to be the most effective and promising substrates among other biosorbents.

The development of new biosorbent materials is a research area full of challenge, as algae biomass is an accessible, economically competitive, and renewable source. Therefore, its potential value in the treatment of heavy metal wastewater is obvious.

5.3.3.2 Influencing Factors on Biosorption

Effect of pH

This parameter plays an important role for the biosorption process of heavy metal ions from aqueous solution, according to a variety of earlier studies [192–194]. Whereas the calculation from the solubility product equilibrium constant (K_{sp}) [195] demonstrated that the suitable pH ranges for the various metal ions were slightly different, i.e., experiments for Cu^{2+} sorption was performed for the pH range of 1–6, Cd^{2+} at pH of 1–8, Zn^{2+} at pH of 1–7, and Pb^{2+} at pH of 1–7.5. For the algae biosorption, pH is also a key factor. Algal biomasses contain high content of carboxyl groups from mannuronic and guluronic acids on the cell wall polysaccharides, which suggests that the biosorption process could be affected by changes in the solution pH [196]. Some examples are cited of algae biosorption on different metal ions.

Table 5.7 Uptake of metals by algal biomass

Metal ions	Algae	q_{max} (mmol/g)
Cd	*Ascophyllum nodosum* (B)	0.338 – 1.913
	Chaetomorha linum (G)	0.48
	Chlorella vulgaris (G)	0.30
	Codium fragile (G)	0.0827
	Corallina officinalis (R)	0.2642
	Fucus vesiculosus (B)	0.649
	Gracilaria edulis (R)	0.24
	Gracilaria Salicornia (R)	0.16
	Padina sp.(B)	0.53
	Padina tetrastomatica	0.53
	Porphira columbina (R)	0.4048
	Sargassum sp. (B)	1.40
	Sargassum baccularia (B)	0.74
	Sargassum natans (B)	1.174
	Sargassum siliquosum (M)	0.73
Ni	*Ascophyllum nodosum*	1.346 – 2.316
	Chlorella miniata	0.237
	Chlorella vulgaris	0.205 – 1.017
	Chondrus crispus (R)	0.443
	Codium taylori (G)	0.099
	Fucus vesiculosus	0.392
	Galaxaura marginata (R)	0.187
	Padina gymnospora (B)	0.170
	Sargassum fluitans (B)	0.409
	Sargassum natans	0.409
	Sargassum vulgare (M)	0.085
	Scenedesmus obliquus	0.5145
Pb	*Ascophyllum nodosum*	1.313 – 2.307
	Chlorella vulgaris	0.47
	Cladophora glomerata (G)	0.355
	Chondrus crispus	0.941
	Codium taylori	1.815
	Fucus vesiculosus	1.105 – 2.896
	Galaxaura marginata	0.121
	Gracilaria corticata (R)	0.2017 – 0.2605
	Padina gymnospora	0.314
	Padina tetrastomatica (B)	1.049
	Polysiphonia violacea (R)	0.4923
	Sargassum fluitans	1.594
	Sargassum hystrix (B)	1.3755
	Sargassum natans	1.1487 – 1.221
	Sargassum vulgare	1.100
	Ulva lactuca (G)	0.61
	Undaria pinnatifida (B)	1.945

(continued)

Table 5.7 (continued)

Metal ions	Algae	q_{max} (mmol/g)
Cu	*Chlorella miniata* (G)	0.366
	Chlorella vulgaris	0.254 – 0.758
	Sargassum sp.	1.08
	Scenedesmus obliquus (G)	0.524
Zn	*Chlorella vulgaris*	0.37
Cr(VI)	*Chlorella vulgaris*	0.534 – 1.525
	Sargassum sp.	1.30 – 1.3257
	Scenedesmus obliquus	1.131
Fe(III)	*Chlorella vulgaris*	0.439

Source [191]
(*B*) Brown alga, (*G*) Green alga (*R*) Red alga

Experimental biosorption on green algae *Spirogyra* species studied by V. K. Gupta [197, 198] showed that the percent adsorption of Cr(VI) increases with pH from pH 1.0 to 2.0 and thereafter decreases with further increase in pH. And the maximum adsorption at all the concentrations takes place at pH 2.0. As to lead sorption, when the pH of the lead solution (100 and 200 mg/L) increased from 2.99 to 7.04, the adsorption capacity of lead was changed, i.e., it first increased from pH 2.99 to pH 5.0 and then dramatically decreased up to pH 7.04. The results stated strong pH dependence of biosorption. This is consistent with the results obtained for the other adsorbent systems [199, 200].

Additionally, cadmium (II) uptake by *C. vulgaris* studied by Z. Aksu [185] is also a function of solution pH. The results showed that the biosorption of cadmium (II) increased with pH up to 4.0 and then declined with further increase in pH. And the maximum equilibrium uptake value was found to be 62.3 mg/g at pH 4.0.

Summarizing, almost all the relative publications revealed similar information that a good control of pH is of great significance for metal biosorption by algae.

Effect of Temperature

The binding of most metals to microorganisms by biosorption observed to enhance as temperature rising [201–203]. Also, increasing temperature is known to stir up the diffusion rate of adsorbate molecules within pores as a result of decreasing solution viscosity and will also modify the equilibrium capacity of the adsorbent for a particular adsorbate [204]. Thus temperature is also an important factor affecting the adsorption.

Research of cadmium (II) and nickel (II) biosorption with *C. vulgaris* carried out by Z. Aksu [185, 208] manifested that the equilibrium uptake of cadmium (II) and nickel (II) ions to the dried green alga was significantly affected by the temperature. The uptake of cadmium (II) decreased with an increasing temperature, while nickel (II) increased. And their optimum adsorption temperature was 20°C and 45°C, respectively. That is, cadmium (II) adsorption is normally exothermic; thus, the extent of adsorption generally increases with decreasing temperature. This result is typical for biosorption of metals involving no energy-mediated reactions, where metal removal from solution is due to purely physical/chemical interactions between the biomass and metal in solution. Increasing the temperature reduced the biosorption capacity of biomass. Nevertheless, nickel (II) adsorption was endothermic; thus, the extent of adsorption increased with increasing temperature. Besides, the sorption of nickel (II) ions by dried *C. vulgaris* may involve not only physical but also chemical sorption. So at higher temperatures, an increase in active sites occurs due to bond rupture.

There are many studies that show the information on the effect of temperature in recent years, which are associated with the adsorption mechanism [196, 198, 206].

Effect of Contact Time

The contact time has also been evaluated as one of the most important factors affecting the biosorption efficiency. The alga *Chlorella vulgaris* was reported to exhibit the feature of rapid uptake of nickel and then reach the equilibrium in 30–60 min [205]. Also, the study of Sari A. [196] showed the biosorption efficiency of Cd (II) ions by *C. virgatum*. The biosorption efficiency increases with rising contact time up to 60 min, after which it is almost constant. Another biosorption of lead from aqueous solutions by green algae (*Spirogyra* sp.) [198] demonstrated that maximum adsorption took place within first 100 min. In fact, the equilibrium time needed for the different metal–biomass systems reportedly range from 1 to 3 h, that is, about 90% of the total metal ion sorption was achieved within 60 min [207].

Overall, reported experiments on contact time reveal, without exception, that the sorption took place in two stages: A very rapid surface adsorption followed by a long period of equilibrium. Adsorption got slowed down in the later stage because initially a large number of vacant surface sites may be available for adsorption and after a certain point of time, the remaining vacant surface sites may be difficult to occupy due to forces between the solute molecules of the solid and bulk phase [208–210]. The diminishing removal with increasing time may also attribute to intraparticle diffusion process dominating over adsorption [211].

Effect of Biomass Dosage

The amount of biosorbent used for the treatment of heavy metals is also an important parameter which influences the biosorption capacity. The biosorption of Cu(II) in the work of Karthikeyan [212] showed an increased uptake of Cu(II) with the alga quantity which can be accounted for the higher dose of adsorbent in the solution and the greater availability of exchangeable sites for the ions. But the maximum uptake was attained at about 100 mg. And he analyzed that this trend was caused by the formation of biosorbent aggregates at higher biomass concentration, which in turn could reduce the effective surface area available for the biosorption. In fact, many studies [196, 198, 213] showed the similar phenomenon that with increasing biomass dosage, the absorption efficiency experienced an inverted-U curve arousal.

5.3.3.3 Adsorption Mechanism

Metal ion binding during biosorption processes has been found to involve complex mechanisms, such as ion-exchange, complexation, electrostatic attraction, and microprecipitation [5]. In fact, for many years, it had been thought that in the process of biosorption, metal ions bind to highly developed surface of a biosorbent via the mechanism of physical adsorption [214]. Whereas the biosorption of Cr^{3+}, Cd^{2+}, and Cu^{2+} ions by blue–green algae *Spirulina* sp. studied by Chojnacka [215] showed that the mechanism of biosorption is rather chemisorptions than physical adsorption (ion-exchange), which was further confirmed by the low surface area associated with physical adsorption and the presence of cations that appeared in the solution after biosorption. The maximum contribution of physical adsorption in the overall biosorption process was evaluated to be 3.7%. Study of the mechanisms of biosorption by Raize [216] manifested that the main cadmium cation sequestration mechanism by the algal biomass was apparently chelation, while that of nickel was mainly ion exchange. Lead cations exhibit higher affinity to the algal biomass, and their binding mechanism includes a combination of ion exchange, chelation, and reduction reactions, accompanied by metallic lead precipitation on the cell wall matrix. The SEM method was used by Han et al. [217] to probe that the surface complexation was the exact mechanism in Cr(III) biosorption by *Chlorella miniata*. In fact, among various proposed mechanisms, ion exchange was thought to be the most important process for the algal biomass [214]. Ion exchange mechanism was thought to be reversible, and more than 90% of the biosorbed metal could be recovered through acid washing [9].

5.3.3.4 Biosorption Models

Isotherm Modeling

The equilibrium data, commonly known as adsorption isotherms, are the basic requirements for the design of the adsorption systems. Several isotherm equations, such as Langmuir, Freundlich, Redlich–Peterson, D-R isotherms, and so on, have been used for the equilibrium modeling of biosorption systems. Langmuir and Freundlich isotherms (see Sect. 5.3.1.3) were widely used to quantitatively describe metal sorption by algae [196, 198, 205, 206, 218–220].

Additionally, some authors explain that Langmuir isotherm corresponds to a dominant ion-exchange mechanism while the Freundlich isotherm shows adsorption–complexation reactions taking place in the adsorption process

Table 5.8 Langmuir and Freundlich parameters for the sorption of the test metals by the selected algae

		Langmuir constant			Freundlich constant		
Algae	Metal ion	q_{max} (mg/g)	b (L/mg)	R^2	K_F ($mg^{1-n}/g\ l^n$)	n	R^2
Spirogyra neglecta	Cu	40.83	0.034	0.96	5.44	0.370	0.89
	Cd	27.95	0.047	0.97	5.01	0.320	0.89
	Zn	31.51	0.034	0.99	4.23	0.368	0.94
	Ni	26.30	0.042	0.99	4.27	0.366	0.92
	Pb^{2+}	90.19	0.015	0.97	4.75	0.513	0.92
Pithophora oedogonia	Cu	23.08	0.069	0.98	5.72	0.265	0.92
	Cd	13.07	0.038	0.98	2.02	0.344	0.93
	Zn	8.98	0.103	0.99	2.98	0.214	0.94
	Ni	11.81	0.039	0.98	1.84	0.342	0.95
	Pb^{2+}	71.13	0.020	0.96	5.28	0.461	0.91
Cladophora calliceima	Cu	14.08	0.088	0.98	4.26	0.229	0.92
	Cd	9.14	0.098	0.97	2.95	0.218	0.92
	Zn	8.51	0.101	0.98	2.80	0.216	0.95
	Ni	7.67	0.074	0.99	2.00	0.257	0.94
	Pb^{2+}	40.50	0.022	0.94	5.29	0.414	0.88
Hydrodictyon reticulatum	Cu	8.72	0.103	0.96	2.81	0.221	0.98
	Cd	7.20	0.071	0.97	1.83	0.262	0.98
	Zn	3.70	0.115	0.92	1.26	0.212	0.99
	Ni	13.86	0.021	0.99	1.14	0.444	0.96
	Pb^{2+}	24.00	0.054	0.97	4.87	0.299	0.90
Aulosira fertilissima	Cu	21.77	0.060	0.99	4.83	0.285	0.92
	Cd	14.57	0.087	0.98	4.28	0.236	0.92
	Zn	19.15	0.049	0.99	3.63	0.310	0.93
	Ni	4.16	0.235	0.83	2.05	0.145	0.99
	Pb^{2+}	31.12	0.042	0.97	4.99	0.338	0.89

[221, 222]. Furthermore, most of the equilibrium data were fit better to Langmuir isotherm model than Freundlich isotherm model. The parameters of Langmuir and Freundlich isotherm model are presented in Table 5.8 [223].

Kinetic Models

In order to investigate the mechanism of biosorption and potential rate-controlling step such as mass transport and chemical reaction processes, kinetic models are used to test the experimental data. Numerous models (e.g., Elovich, diffusion, pseudo-first-order, and pseudo-second-order equations) have been suggested to analyze the kinetics of sorption process. Kinetic models based on the capacity of the adsorbent such as the first-order equation and second-order expression were commonly used (see Sect. 5.3.1.3).

In the work of O.M. Freitas [224], experimental data suggested that the pseudo-first-order model successfully describes the kinetics of the biosorption of Zn on *L. hyperborea* and Cd on *B. bifurcate*. The results of Aravindhan Rathinam [28] also suggest that kinetics of cadmium biosorption by H. valentiae biomass followed pseudo-first-order kinetic model very well.

The pseudo-second-order kinetic model has been applied successfully to metal biosorption by several algae [185, 198, 200, 206, 225–228]. Additionally, these parameters in equations can change depending on experimental conditions as it was found by P. Lodeiro [225].

The Weber and Morris sorption kinetic model [229], a kind of diffusion model, was employed to investigate the sorption mechanism in the study of Prasert Pavasant [195].

To sum up, kinetic model is an important part of the study of heavy metal biosorption, from which we can get the information of the control steps and mechanism of the process.

5.3.4 Plant Leaves and Their Extraction

5.3.4.1 Introduction

There is a considerable potential for adopting a natural, abundant, and economical metal adsorption system, and plant leaves and their extraction, which are cheap and easily available in a great supply, could be used as an adsorbent for the removal of heavy metals from aqueous solution. In fact, tree leaves and their extraction are frequently employed in heavy metal removal [230–236]. Studies show that adsorption capability also relies on leaves used. Examples of efficient types of plant leaves for removal of metal ions are reed [237] for cadmium, poplar for lead and copper [238, 239], cinchona for copper [239], pine for nickel [240], and cypress for aluminum [241]. Moreover, R. Salim [232] conducted cadmium removal experiments from aqueous solutions by 20 species of plant leaves and their combinations. The results showed that most efficient types of plant leaves for cadmium removal were those of styrax, plum, pomegranate, and walnut.

5.3.4.2 Effects of Environmental Factors on Biosorption

Effect of pH

The pH of the sorbate solution is considered one of the most important environmental factors affecting the biosorption process. This factor is capable of influencing not only the binding site dissociation state, but also the solution chemistry of the target metal [242]. Successful biosorption of base metal cations usually takes place in the range of pH 3–7, and is extremely pH dependent. Özer A. and Özer D. [243]

reported an optimal pH value for lead and nickel uptake of 5.0, while Vianna LNL [244] found that maximal copper, cadmium, and zinc uptakes happened at pH 4.5, and decreased significantly when the pH was dropped to 3.5 or 2.5.

H. Benaissa [231] studied the removal of copper by dried sunflower leaves finding the maximum copper sorption occurs at around initial pH 5–6. A. Sharma and K.G. Bhattacharyya [230] used leaf powder as biosorbent to reclaim Cd (II) obtaining continuous adsorption increases from 8.8% at pH 4.0 to 70.0% at pH 7.0, and 93.6% at pH 9.5. Similarly, R. Salim [232] found that the optimum experimental condition reached at pH 4.1. Rice bran was also used for the removal of Cr (VI) from wastewater [245] and the maximum removal yield was 99.4% also at a low pH of 2.0. Even grape stalks wastes came in handy to absorb copper and nickel ions and I. Villaescusa [246] found a pH-dependent profile for the course. Maximum sorption for both metals was found to occur at around pH 5.5–6.0. Y. Bulut and Z. Baysal [247] used wheat bran as a sorbate on biosorption of Pb (II) and found that the Pb (II) adsorption by wheat bran increases with pH. It can be observed that the removal of Cu (II) and Pb (II) exhibits similar trend, i.e., it increases with increasing pH, climbing to maximum at a certain pH range. Bin Yu and Y. Zhang [248] reported that the maximum removal of Pb (II) by sawdust sorption occurs around pH 5.0, while that of Cu (II) 7.0. Furthermore, the greatest increases in the sorption rate of metal ions on sawdust were observed in a range of pH from 2 to 8 for copper and 2–5 for lead.

Effect of Temperature

As one of the factors affecting the biosorption efficiency, temperature was studied in many papers [249–253]. S. Qaiser [254] studied the biosorption of lead from aqueous solution by *Ficus religiosa* leaves and found that the temperature change in the range of 20–40 C affected the biosorption capacity and the maximum removal was observed at 25°C. In M.S. Al-Masri's [255] study, the influence of temperature on U, Pb, and Cd removal capacities using different parts of poplar trees is different. The maximum U and Cd removal capacity using leaves occurred at a temperature of 25°C while the maximum U, Pb, and Cd removal capacities using branches occurred at a temperature of 35°C.

5.3.4.3 Mechanism Analysis

The mechanism of metal biosorption varies with the metal species and type of biosorbent [217]. To bind and accumulate these pollutants, different previously mentioned mechanisms reportedly also empower the course, solo or ensemble, such as physical adsorption, complexation, ion exchange, and surface microprecipitation [203, 256–258]. For example, S. Qaiser [254] reported the release of Ca, Mg, and Na ions during lead biosorption revealing ion exchange as the suitable major removal mechanism. Javad Zolgharnein [259] in his study of Cr (VI) adsorption onto *Elaeagnus* tree leaves found chemisorption as the predominant mechanism of this special biosorption.

5.3.5 Summary

In fact, in recent years, applying biotechnology in controlling and removing metal pollution has been paid much attention, and gradually becomes hot topic in the field of metal pollution control because of its potential application. And biosorption is an alternative process, which provides a variety of sorption materials, including bacteria, fungi, yeast, algae, plant leaves, etc. stated above. These biosorbents with relatively high metal-binding capacity and selectivity possess metal-sequestering property and are suitable for the extraction of metal ions from large volumes of water. Therefore it is an ideal candidate for the treatment of high-volume and low concentration complex wastewaters.

5.4 Combination and Improvement of Different Pathways

Commercial application of microbial biomass as a biosorbent may suffer from the problem of how to remove or condense the metal ions-loaded microorganisms from aqueous solution to facilitate reclaiming in the next step. Generally, millipore filtration and evaporation methods have been carried out, but the high cost of membrane and high energy consumption impede them from wide application. Song et al. [103] studied competitive biosorption of Au(III) and Cu(II) ions by magnetotactic bacteria, which may be significant for developing promising biosorbents. Besides, Wang et al. [108] observed that Cu (II) promoted the adsorption of Ag (I) in the competitive biosorption of Ag (I) and Cu (II) on *Magnetospirillum gryphiswaldense* (MSR-1), which was confirmed magnetotactic bacterium [260], was reported to be sensitive to magnetic field and might be one of the promising options for solving this problem by means of external field. These bacteria were observed to be able to move along the local magnetic field lines, which is a very useful feature to facilitate them to be easily separated from the solutions. This unique finding indicates the high possibility to recover heavy metal ions from wastewater using the method of "MTB biosorption and magnetic separation", which was simple, effective, and environmentally friendly. However, it is still faced with another problem. For example, Song et al. [261] applied a magnetotactic bacterium, *Stenotrophomonas* sp., to remove Au(III) from contaminated wastewater. The analyses from FTIR and XRD confirmed that the reduction of Au(III) to Au(0) by the reductants on the MTB biomass occurred, and the deposition of nanocrystal Au(0) particles, ranging from 24.7 to 31.4 nm, could be estimated on the biomass surface. Although easy to remove the MTB-nano Au(0) group by the external magnetic field, it may be a great challenge to separate MTB from nano Au(0), because both nano Au(0) and nanometer-scale magnetosomes own magnetism.

Also, according to combination of abiotic and biotic components, Chen et al. [110] used the bacteria-mineral composite as a geochemically reactive solid and quantified its trace metal ion scavenging ability. In their study, the first quantitative comparison of the metal-binding capacities of *P. putida* CZ1–goethite composite to its individual components was evaluated. As a consequence, the nonliving cells–goethite composite retained approximately 82% more Zn than that predicted by its solo counterpart.

What is more, some attention was focused on immobilization technology for heavy metal removal biomaterial [262–265]. Entrapment, cellular aggregation, and surface fixation are the most common methods to support this technology [266–270]. The choice of an adequate matrix for cell immobilization affects the process performance, since the metal biosorption efficiency can be affected by these heterogeneous systems [271, 272]. Microorganisms loaded natural and synthetic adsorbents have also been used for separation and preconcentration of heavy metals at trace levels. *S. cerevisiae* loaded on sepiolite was utilized as biosorbent for copper(II), zinc(II), and cadmium(II) in natural waters [264]. Bag et al. [91] proposed a biosorptive enrichment procedure for Cr (III) and Cr (VI) ions by *Saccharomyces cerevisiae*-loaded sepiolite. *S. cerevisiae* and *Chlorella vulgaris* loaded on silica gel has been used for the separation–preconcentration of Pt^{2+} and Pd^{2+} [273]. Vullo et al. [18] made another progress by successfully immobilizing *P. veronii* 2E on inert surfaces such as Teflon membranes, silicone rubber, and polyurethane foams. In addition, *P. veronii* 2E was found to be able to grow on all three surfaces and develop a film over the matrix surfaces, form aggregates, and adhere to glass during batch cultures. Removal rate of Pb(II) reaches the summit 97.7% at pH 4, at 200 mg/L of initial Pb(II) concentration with 10 g/L of *P. sanguineus* beads, prepared by dropping a mixture of 1.5% (w/v) sodium alginate solution and *P. sanguineus* mycelial mat into a 2% (w/v) $CaCl_2$ solution [274].

Furthermore, the idea about combination of abiotic and biotic components coupled with the immobilization theory and magnetic separation technology. Magnetic particles have been applied as a new sorbent to adsorb metal ions, in which the difficulty of separation was resolved in an external magnetic field [275]. Based on these researches, the biofunctional magnetic bead was synthesized and utilized in the wastewater treatment [276]. Li et al. [276] removed heavy metal in wastewater by bio-functional magnetic beads constituted by the powder of *Rhizopus cohnii* and Fe_3O_4 particles coated with alginate and polyvinyl alcohol (PVA). Then magnetic separation technology could make the separation of solid and liquid phase easier. The combined technique of biosorption and magnetic separation holds the advantages of flexibility, eco-friendly characteristics, and economic in operational cost. Thus, the current investigation into this technique is very important and inspiring.

5.5 Comparative Conclusions of the Methods

Every before-mentioned method for heavy metal removal, especially biosorption has actually its advantages and disadvantages for recovering heavy metals. This means in some cases, two or more methods may be chose to treat the target solution to work their way through drastically. Please refer to Table 5.9 for detail.

5.6 Recovery and Reuse of Heavy Metal

Heavy metal ions from the contaminated water could be reduced into metal nanoparticles by physical, chemical, and biological technology [277, 278]. Most of these physical and chemical methods need extreme conditions like temperature, pressure etc. Chemical reducing agents are believed to be associated with environmental toxicity or biological hazards, whereas comparatively safer reluctant like citrate, ascerbate, simple sugars like glucose, fructose etc. are not efficient in productivity. It has been, therefore, of increasing interest to develop efficient green biological reduction [278]. What is more, current chemical methods of metal nanoparticle synthesis have shown limited success and is expected that the use of a biological approach may overcome many of these obstacles. The exploitation of microorganisms for the biosynthesis of metal nanoparticles is an area of research that has received increasing interest over the last decade. The use of living microbes as a tool for nanoparticle biosynthesis has been researched extensively [279]. The size and shape-dependent physicochemical and optoelectronic properties of metal nanoparticles have important applications in catalysis, biosensing, recording media, and optics and so on [280]. For example, gold nanoparticles (Au NPs) have potentially exciting applications in hyperthermia of tumors, optical coatings, and scanning tunneling microscopes as conductive tips [281]. Especially, biological method available for low metal concentration, less than 100 mg/L, could exert the advantage of simple course, easy operation, low cost, little pollution, and high recovery rate of heavy metal etc. [18]. If all of the removal technologies could be applied in to commercial production, not only effluents are purified but also the toxic heavy metal ions are effectively reused to make a green and perfect recycle between the nature and our industrial society. In sum, researches on heavy metal removal give a wide vision and a promising future to development of all the fields.

5.7 Current Hot Topics, Trends, and Outlook of This Field

Nowadays, more researche has been focused on the search for alternative and innovative wastewater treatment techniques. The interesting one was the utilization of biological materials such as algae, fungi, and bacteria for the metal removal and

Table 5.9 Advantages and disadvantages of the methods

Methods	Advantages	Disadvantages
Activated carbon	High surface area	High cost to prepare activated carbon Reactivation results in a loss of the carbon Performance dependent on the type of carbon used and non-selective
Chemical precipitation	Effectively treat inorganic effluent with a metal concentration of higher than 1,000 mg/L with a simple process	The demand of a large amount of chemicals Generally it cannot be used to handle low concentration of metal wastewater, which is below 100 mg/L The long-term environmental impacts of sludge disposal
Chemical redox	Be applicable for several metal ions	Used as a pretreatment for wastewater treatment
Membrane separation	Can be application in occasions where the concentration of metal ions is low	Inadequate selectivity High energy consumption Immature technologies
Ion exchange	Effective to treat inorganic effluent with a wide metal concentration of less than 10 mg/L to higher than 100 mg/L Does not present any sludge disposal problems	Suitable ion exchange resins are not available for all heavy metals The capital and operational cost is high
Biosorption	Growth-independent, non-living biomass is not subject to toxicity limitation of cells Biomass can be procured from the existing fermentation industries, which is essentially a waste after fermentation. The process is not governed by the physiological constraint of living microbial cells Because non-living biomass behave as an ion exchanger, the process is very rapid and takes place between few minutes to few hours. Metal loading on biomass is often very high, leading to very efficient metal uptake Because cells are non-living, processing conditions are not restricted to those conducive for the growth of cells. In other words, a wider range of operating conditions such as pH, temperature and metal concentration is possible. No aseptic conditions are required for this process Metal can be desorbed readily and then recovered if the value and amount of metal recovered are significant and if the biomass is plentiful, metal-loaded biomass can be incinerated, thereby eliminating further treatment	Early saturation can be problem i.e. when metal interactive sites are occupied, metal desorption is necessary prior to further use, irrespective of the metal value The potential for biological process improvement (e.g. through genetic engineering of cells) is limited because cells are not metabolizing. Because production of the adsorptive agent occurs during pre-growth, there is no biological control over characteristic of biosorbent. This will be particularly true if waste biomass from a fermentation unit is being utilized There is no potential for biologically altering the metal valency state. For example less soluble forms or even for degradation of organometallic complexes

recovery among these techniques [21]. In other studies, biomasses have exhibited economic, rapid adsorption, and eco-friendly characteristics [2]. However, it is still necessary to continue the steps towards the world of these promising biosorbents from an extremely large pool of readily available and inexpensive biomaterials [8, 9]. The mechanism involved in metal biosorption is far from well understood up to date. It is the mission of this field to investigate the microbe-metal interactions and obtain the mechanism of metal uptake by biosorbents and then grasp this tool to change the world.

Also in the field, it is a hot and rising topic to synthesize metallic nanoparticles using biological technology by simultaneously reducing heavy metal ions level in the wastewater. The development of these green techniques for the controlled synthesis of metallic nanoparticles of well-defined size and shape is a big challenge and numerous chemical methods, aimed at controlling the physical properties of the particles, are still in the development stage and problems are often experienced with stability of the nanoparticle preparations, control of the crystal growth, and aggregation of the particles [115, 117, 118]. Although, these researches are still in their infancy of theoretic and experimental phase, the coming fruits are sure to be of excitingly wide perspective and practicability in the near future.

Acknowledgments The authors greatly acknowledge the financial support from the National Natural Science Foundation, P.R. China (Project No. 21076155). We herein express our thankfulness to the Analysis Center of Tianjin University for Product Purity testing.

References

1. Wang JL, Chen C (2006) Biosorption of heavy metals by *Saccharomyces cerevisiae*: a review. Biotechnol Adv 24:427–451. doi:10.1016/j.biotechadv.2006.03.001
2. Ahluwalia SS, Goyal D (2007) Microbial and plant derived biomass for removal of heavy metals from wastewater. Bioresour Technol 98:2243–2257. doi:10.1016/j.biortech.2005.12.006
3. Patterson JW, Minear R, Gasca E, Petropoulou C (1998) Industrial discharges of metals to water. In: Allen HE, Garrison AW, Luther GW (eds) Metals in surface waters, 3rd edn. Ann Arbor Press, Michigan
4. Alexander M (1999) Biodegradation and bioremediation, 2nd edn. Academic, San Diego
5. Volesky B, Holan ZR (1995) Biosorption of heavy metals. Biotechnol Progress 11:235–250. doi:10.1021/bp00033a001
6. Chen YS, Sun QJ, Chen J, Zhang YY (1997) Research on technology of biosorption of heavy metals. Adv Environ Sci 5:34–43. doi:10.1021/bp00033a001
7. Chen M, Gan YR (1999) Biosorption of heavy metal. Chem Ind Eng 16:19–25. doi:cnki:ISSN:1004–9533.0.1999–01–004
8. Kratochvil D, Volesky B (1998) Biosorption of Cu from ferruginous wastewater by algae biomass. Water Res 32:2760–2768. doi:10.1016/S0043-1354(98), 00015-3
9. Kratochvil D, Volesky B (1998) Advances in the biosorption of heavy metals. Trends Biotechnol 16:291–300. doi:10.1016/S0167-7799(98), 01218-9
10. Volesky B (1990) Biosorption and biosorbents. In: Volesky B (ed) Biosorption of heavy metals. CRC Press, Boca Raton
11. Vijayaraghavan K, Yun YS (2008) Bacterial biosorbents and biosorption. Biotechnol Adv 26:266–291. doi:10.1016/j.biotechadv.2008.02.002

12. Gavrilesca M (2004) Removal of heavy metals from the environmental by biosorption. Eng Life Sci 4:219–232. doi:10.1002/elsc.200420026
13. Davis TA, Volesky B, Mucci A (2003) A review of the biochemistry of heavy metal biosorption by brown algae. Water Res 37:4311–4330. doi:10.1016/S0043-1354(03), 00293-8
14. Volesky B (2001) Detoxification of metal-bearing effluents: biosorption for the next century. Hydrometallurgy 59:203–216. doi:10.1016/S0304-386X(00), 00160-2
15. Veglio F, Beolchini F (1997) Removal of metals by biosorption: a review. Hydrometallurgy 44:301–316. doi:10.1016/S0304-386X(96)00059-X
16. Kapoor A, Viraraghavan T (1995) Fungi biosorptiondan alternative treatment option for heavy metal bearing wastewaters: a review. Bioresour Technol 53:195–206. doi:10.1016/0960-8524(95)00072-M
17. White C, Wilkinson SC, Gadd GM (1995) The role of microorganisms in biosorption of toxic metals and radionuclides. Int Biodeter Biodegr 35:17–40. doi:10.1016/0964-8305(95), 00036-5
18. Vullo DL, Ceretti HM, Daniel MA, Ramırez SAM, Zalts A (2008) Cadmium, zinc and copper biosorption mediated by *Pseudomonas veronii* 2E. Bioresour Technol 99:5574–5581. doi:10.1016/j.biortech.2007.10.060
19. Tsezos M (2001) Biosorption of metals. The experience accumulated and the outlook for technology development. RefDoc. http://cat.inist.fr/?aModele=afficheN&cpsidt=898441. Accessed 2001
20. Aksu Z (1998) Biosorption of heavy metals by micro algae in batch and continuous systems. In: Wong YS, Tam NFY (eds) Wastewater treatment with algae. Springer, Berlin
21. Wang JL, Chen C (2009) Biosorbents for heavy metals removal and their future. Biotechnol Adv 27:195–226. doi:10.1016/j.biotechadv.2008.11.002
22. Hamerlinck Y, Mertens DH (1994) In: Vansant EF (ed) Activated carbon principles in separation technology. Elsevier, New York
23. Dobrowolski R, Stefaniak E (2000) Study of chromium (VI) adsorption from aqueous solution on to activated carbon. Adsorpt Sci Technol 18:97–106. doi:10.1260/0263617001493314
24. Kadirvelu K, Namasivayam C (2003) Activated carbon from coconut coirpith as metal adsorbent: adsorption of Cd (II) from aqueous solution. Adv Environ Res 7:471–478. doi:10.1016/S1093-0191(02), 00018-7
25. Gomez-Serrano V, Macias-Garcia A, Espinosa-Mansilla A, Valenzuela-Calahorro A (1998) Adsorption of mercury, cadmium and lead from aqueous solution on heat-treated and sulphurized activated carbon. Water Res 32:1–4. doi:10.1016/S0043-1354(97), 00203-0
26. Benefield LD, Morgan JM (1999) Chemical precipitation. In: Letterman RD (ed) Water quality and treatment. McGraw-Hill, New York
27. US Environmental Protection Agency (EPA) (2000) Chemical precipitation. US EPA, Washington, DC, EPA832-F-00–018
28. Tunay O (2003) Developments in the application of chemical technologies to wastewater treatment. Water Sci Technol 48:43–52
29. Charerntanyarak L (1999) Heavy metals removal by chemical coagulation and precipitation. Water Sci Technol 39:135–138. doi:10.1016/S0273-1223(99), 00304-2
30. Juttner K, Galla U, Schmieder H (2000) Electrochemical approaches to environmental problems in the process industry. Electrochim Acta 45:2575–2594. doi:10.1016/S0013-4686(00)00339-X
31. Yang XJ, Fane AG, Mac Naughton S (2001) Removal and recovery of heavy metals from wastewater by supported liquid membranes. Water Sci Technol 43:341–348
32. Bose P, Bose MA, Kumar S (2002) Critical evaluation of treatment strategies involving adsorption and chelation for wastewater containing copper, zinc, and cyanide. Adv Environ Res 7:179–195. doi:10.1016/S1093-0191(01), 00125-3
33. Wingenfelder U, Hansen C, Furrer G, Schulin R (2005) Removal of heavy metals from mine water by natural zeolites. Environ Sci Technol 39:4606–4613. doi:10.1021/es048482s

34. Jain A, Sharma VK, Mbuya OS (2009) Removal of arsenic by Fe (VI), Fe (VI)/Fe (III), and Fe (VI)/Al (III) salts: effect of pH and anions. J Hazard Mater 169:339–344. doi:10.1016/j.jhazmat.2009.03.101
35. Jiang JQ (2007) Research progress in the use of ferrate (VI) for the environmental remediation. J Hazard Mater 146:617–623. doi:10.1016/j.jhazmat.2007.04.075
36. Kurniawan TA, Chan GYS, Lo WH, Babel S (2006) Physico-chemical treatment techniques for wastewater laden with heavy metals. Chem Eng J 118:83–98. doi:10.1016/j.cej.2006.01.015
37. Ahn KH, Song KG, Cha HY, Yeom IT (1999) Removal of ions in nickel electroplating rinse water using low-pressure nanofiltration. Desalination 122:77–84. doi:10.1016/ S0011-9164(99), 00029-6
38. Juang RS, Shiau RC (2000) Metal removal from aqueous solutions using chitosan-enhanced membrane filtration. J Membr Sci 165:159–167. doi:10.1016/S0376-7388(99), 00235-5
39. Ujang Z, Anderson GK (1996) Application of low-pressure reverse osmosis membrane for Zn^{2+} and Cu^{2+} removal from wastewater. Water Sci Technol 34:247–253. doi:10.1016/S0273-1223(96), 00811-6
40. Tzanetakis N, Taama WM, Scott K, Jachuck RJJ, Slade RS, Varcoe J (2003) Comparative performance of ion exchange membrane for electrodialysis of nickel and cobalt. Sep Purif Technol 30:113–127. doi:10.1016/S1383-5866(02), 00139-9
41. Fernández Y, Maraón E, Castrillón L, Vázquez I (2005) Removal of Cd and Zn from inorganic industrial waste leachate by ion exchange. J Hazard Mater 126:169–175. doi:10.1016/j.jhazmat.2005.06.016
42. Lee IH, Kuan YC, Chern JM (2006) Factorial experimental design for recovering heavy metals from sludge with ion-exchange resin. J Hazard Mater 138:549–559. doi:10.1016/j.jhazmat.2006.05.090
43. Juang RS, Kuo HC, Liu FY (2006) Ion exchange recovery of Ni (II) from simulated electroplating waste solutions containing anionic ligands. J Hazard Mater 128:53–59. doi:10.1016/j.jhazmat.2005.07.027
44. Sapari N, Idris A, Hisham N (1996) Total removal of heavy metal from mixed plating rinse wastewater. Desalination 106:419–422. doi:10.1016/S0011-9164(96), 00139-7
45. Gode F, Pehlivan E (2003) A comparative study of two chelating ion exchange resins for the removal of chromium (III) from aqueous solution. J Hazard Mater 100:231–243. doi:10.1016/S0304-3894(03), 00110-9
46. lvarez-Ayuso EA, Garcia-Sanchez A, Querol X (2003) Purification of metal electroplating wastewaters using zeolites. Water Res 37:4855–4862. doi:10.1016/j.watres.2003.08.009
47. Papadopoulos A, Fatta D, Parperis K, Mentzis A, Harambous KJ, Loizidou M (2004) Nickel uptake from a wastewater stream produced in a metal finishing industry by combination of ion-exchange and precipitation methods. Sep Purif Technol 39:181–188. doi:10.1016/j.seppur.2003.10.010
48. Keane MA (1998) The removal of copper and nickel from aqueous solution using Y zeolite ion exchangers. Colloid Surf A 138:11–20. doi:10.1016/S0927-7757(97), 00078-2
49. Ali AA, Bishtawi RE (1997) Removal of lead and nickel ions using zeolite tuff. J Chem Technol Biotechnol 69:27–34. doi:10.1002/(SICI)1097-4660(199705
50. Lin SH, Kiang CD (2003) Chromic acid recovery from waste acid solution by an ion exchange process: equilibrium and column ion exchange modeling. Chem Eng J 92:193–199. doi:10.1016/S1385-8947(02), 00140-7
51. Dobrevsky I, Todorova-Dimova M, Panayotova T (1996) Electroplating rinse wastewater treatment by ion exchange. Desalination 108:277–280. doi:10.1016/S0011-9164(97), 00036-2
52. Ahmed S, Chughtai S, Keane MA (1998) The removal of cadmium and lead from aqueous solution by ion exchange with Na–Y zeolite. Sep Purif Technol 13:57–64. doi:10.1016/S1383-5866(97), 00063-4
53. Mann H (1990) Removal and recovery of heavy metals by biosorption. In: Volesky B (ed) Biosorption of heavy metals. CRC press, Boca Raton

54. Urrutia MM (1997) General bacterial sorption processes. In: Wase J, Forster C (eds) Biosorbents for metal ions. CRC Press, London
55. Tunali S, Cabuk A, Akar T (2006) Removal of lead and copper ions from aqueous solutions by bacterial strain isolated from soil. Chem Eng J 115:203–211. doi:10.1016/j.cej.2005.09.023
56. Salehizadeh H, Shojaosadati SA (2003) Removal of metal ions from aqueous solution by polysaccharide produced from *Bacillus firmus*. Water Res 37:4231–4235. doi:10.1016/S0043-1354(03), 00418-4
57. Choi SB, Yun YS (2004) Lead biosorption by waste biomass of *Corynebacterium glutamicum* generated from lysine fermentation process. Biotechnol Lett 26:331–336. doi:10.1023/B:BILE.0000015453.20708.fc
58. Lu WB, Shi JJ, Wang CH, Chang JS (2006) Biosorption of lead, copper and cadmium by an indigenous isolate *Enterobacter* sp J1 possessing high heavy-metal resistance. J Hazard Mater 134:80–86. doi:10.1016/j.jhazmat.2005.10.036
59. Chang JS, Law R, Chang CC (1997) Biosorption of lead, copper and cadmium by biomass of *Pseudomonas aeruginosa* PU21. Water Res 31:1651–1658. doi:10.1016/S0043-1354(97), 00008-0
60. Lin CC, Lai YT (2006) Adsorption and recovery of lead(II) from aqueous solutions by immobilized *Pseudomonas aeruginosa* PU21 beads. J Hazard Mater 137:99–105. doi:10.1016/j.jhazmat.2006.02.071
61. Uslu G, Tanyol M (2006) Equilibrium and thermodynamic parameters of single and binary mixture biosorption of lead(II) and copper(II) ions onto *Pseudomonas putida*: effect of temperature. J Hazard Mater 135:87–93. doi:10.1016/j.jhazmat.2005.11.029
62. Pardo R, Herguedas M, Barrado E, Vega M (2003) Biosorption of cadmium, copper, lead and zinc by inactive biomass of *Pseudomonas putida*. Anal Bioanal Chem 376:26–32. doi:10.1007/s00216-003-1843-z
63. Selatnia A, Boukazoula A, Kechid N, Bakhti MZ, Chergui A, Kerchich Y (2004) Biosorption of lead (II) from aqueous solution by a bacterial dead *Streptomyces rimosus* biomass. Biochem Eng J 19:127–135. doi:10.1016/j.bej.2003.12.007
64. Mameri N, Boudries N, Addour L, Belhocine D, Lounici H, Grib H (1999) Batch zinc biosorption by a bacterial nonliving *Streptomyces rimosus* biomass. Water Res 33:1347–1354. doi:10.1016/S0043-1354(98), 00349-2
65. Incharoensakdi A, Kitjaharn P (2002) Zinc biosorption from aqueous solution by a halotolerant cyanobacterium *Aphanothece halophytica*. Curr Microbiol 45:261–264. doi:10.1007/s00284-002-3747-0
66. Chen XC, Wang YP, Lin Q, Shi JY, Wu WX, Chen YX (2005) Biosorption of copper(II) and zinc(II) from aqueous solution by *Pseudomonas putida* CZ1. Colloids Surf B Biointerfaces 46:101–107. doi:10.1016/j.colsurfb.2005.10.003
67. Puranik PR, Paknikar KM (1997) Biosorption of lead and zinc from solutions using *Streptoverticillium cinnamoneum* waste biomass. J Biotechnol 55:113–124. doi:10.1016/S0168-1656(97), 00067-9
68. Celaya RJ, Noriega JA, Yeomans JH, Ortega LJ, Ruiz-Manriquez A (2000) A biosorption of Zn(II) by *Thiobacillus ferrooxidans*. Bioprocess Eng 22:539–542. doi:10.1007/s004499900106
69. Liu HL, Chen BY, Lan YW, Cheng YC (2004) Biosorption of Zn(II) and Cu(II) by the indigenous *Thiobacillus thiooxidans*. Chem Eng J 97:195–201. doi:10.1016/S1385-8947(03), 00210-9
70. Nakajima A, Yasuda M, Yokoyama H, Ohya-Nishiguchi H, Kamada H (2001) Copper biosorption by chemically treated *Micrococcus luteus* cells. World J Microbiol Biotechnol 17:343–347. doi:10.1023/A:1016638230043
71. Savvaidis I, Hughes MN, Poole RK (2003) Copper biosorption by *Pseudomonas cepacia* and other strains. World J Microbiol Biotechnol 19:117–121. doi:10.1023/A:1023284723636
72. Beolchini F, Pagnanelli R, Toro L, Veglio F (2006) Ionic strength effect on copper biosorption by *Sphaerotilus natans*: equilibrium study and dynamic modeling in membrane reactor. Water Res 40:144–152. doi:10.1016/j.watres.2005.10.031

73. Ozturk A, Artan T, Ayar A (2004) Biosorption of nickel(II) and copper(II) ions from aqueous solution by *Streptomyces coelicolor* A3(2). Colloids Surf B Biointerfaces 34:105–111. doi:10.1016/j.colsurfb.2003.11.008
74. Loukidou MX, Karapantsios TD, Zouboulis AI, Matis KA (2004) Diffusion kinetic study of cadmiurn (II) biosorption by *Aeromonas caviae*. J Chem Technol Biotechnol 79:711–719. doi:10.1002/jctb.1043
75. Ziagova M, Dimitriadis G, Aslanidou D, Papaioannou X, Tzannetaki EL, Liakopoulou-Kyriakides M (2007) Comparative study of Cd(II) and Cr(VI) biosorption on *Staphylococcus xylosus* and *Pseudomonas* sp. in single and binary mixtures. Bioresour Technol 98:2859–2865. doi:10.1016/j.biortech.2006.09.043
76. Puranik PR, Chabukswar NS, Paknikar KM (1995) Cadmium biosorption by *Streptomyce pimprina* waste biomass. Appl Microbiol Biotechnol 43:1118–1121. doi:10.1007/BF00166935
77. Selatnia A, Bakhti MZ, Madani A, Kertous L, Mansouri Y (2004) Biosorption of Cd^{2+} from aqueous solution by a NaOH-treated bacterial dead *Streptomyces rimosus* biomass. Hydrometallurgy 75:11–24. doi:10.1016/j.hydromet.2004.06.005
78. Selatnia A, Boukazoula A, Kechid N, Bakhti MZ, Chergui A (2004) Biosorption of Fe^{3+} from aqueous solution by a bacterial dead *Streptomyces rimosus* biomass. Process Biochem 39:1643–1651. doi:10.1016/S0032-9592(03), 00305-4
79. Srinath T, Verma T, Ramteke PW, Garg SK (2002) Chromium (VI) biosorption and bioaccumulation by chromate resistant bacteria. Chemosphere 48:427–435. doi:10.1016/S0045-6535(02), 00089-9
80. Nourbakhsh M, Sag Y, Ozer D, Aksu Z, Kutsal T, Caglar A (1994) A comparative study of various biosorbents for removal of chromium(VI) ions from industrial wastewaters. Process Biochem 29:1–5. doi:10.1016/0032-9592(94), 80052-9
81. Zhou M, Liu YG, Zeng GM, Li X, Xu WH, Fan T (2007) Kinetic and equilibrium studies of Cr(VI) biosorption by dead *Bacillus licheniformis* biomass. World J Microbiol Biotechnol 23:43–48. doi:10.1007/s11274-006-9191-8
82. Sahin Y, Ozturk A (2005) Biosorption of chromium(VI) ions from aqueous solution by the bacterium *Bacillus thuringiensis*. Process Biochem 40:1895–1901. doi:10.1016/j.procbio.2004.07.002
83. Ozturk A (2007) Removal of nickel from aqueous solution by the bacterium *Bacillus thuringiensis*. J Hazard Mater 147:518–523. doi:10.1016/j.jhazmat.2007.01.047
84. Selatnia A, Madani A, Bakhti MZ, Kertous L, Mansouri Y, Yous R (2004) Biosorption of Ni^{2+} from aqueous solution by a NaOH-treated bacterial dead *Streptomyces rimosus* biomass. Miner Eng 17:903–911. doi:10.1016/j.mineng.2004.04.002
85. de Vargas I, Macaskie LE, Guibal E (2004) Biosorption of palladium and platinum by sulfatereducing bacteria. J Chem Technol Biotechnol 79:49–56. doi:10.1002/jctb.928
86. Nakajima A, Tsuruta T (2004) Competitive biosorption of thorium and uranium by *Micrococcus luteus*. J Radioanal Nucl Chem 260:13–18. doi:10.1023/B:JRNC.0000027055.16768.1e
87. Gialamouidis D, Mitrakas M, Liakopoulou-Kyriakides M (2010) Equilibrium, thermodynamic and kinetic studies on biosorption of Mn(II) from aqueous solution by *Pseudomonas* sp., *Staphylococcus xylosus* and *Blakeslea trispora* cells. J Hazard Mater 182:672–680. doi:10.1016/j.jhazmat.2010.06.084
88. Velasquez L, Dussan J (2009) Biosorption and bioaccumulation of heavy metals on dead and living biomass of *Bacillus sphaericus*. J Hazard Mater 167:713–716. doi:10.1016/j.jhazmat.2009.01.044
89. Fehrmann C, Pohl P (1993) Cadmium adsorption by the non-living biomass of microalgae grown in axenic mass culture. J Appl Phycol 5:555–562. doi:10.1007/BF02184634
90. Godlewska-Żyłkiewicz B (2004) Preconcentration and separation procedures for the spectrochemical determination of platinum and palladium. Microchim Acta 147:189–210. doi:10.1007/s00604-004-0234-2
91. Bag H, Turker AR, Lale M, Tunceli A (2000) Separation and speciation of Cr (III) and Cr (VI) with *Saccharomyces cerevisiae* immobilized on sepiolite and determination of both species in water by FAAS. Talanta 51:895–902. doi:10.1016/S0039-9140(99), 00354-9

92. Esposito A, Pagnanelli F, Veglio F (2002) pH-related equilibria models for biosorption in singlemetal systems. Chem Eng Sci 57:307–313. doi:10.1016/S0009-2509(01), 00399-2
93. Duran C, Bulut VN, Gundogdu A, Soylak M, Belduz AO, Beris FS (2009) Biosorption of heavy metals by *Anoxybacillus gonensis* immobilized on Diaion HP-2MG. Sep Sci Technol 44:335–358. doi:10.1080/01496390802437131
94. Klimmek S, Stan HJ, Wilke A, Bunke G, Buchholz R (2001) Comparative analysis of the biosorption of cadmium, lead, nickel and zinc by algae. Environ Sci Technol 35:4283–4288. doi:10.1021/es010063x
95. Aksu Z, Akpinar D (2001) Competitive biosorption of phenol and Cr(VI) from binary mixtures onto dried anaerobic activated sludge. Biochem Eng J 7:183–193. doi:10.1016/S1369-703X(00), 00126-1
96. Hancock IC (1986) The use of Gram-positive bacteria for the removal of metals from aqueous solutions. In: Thompson R (ed) Trace metal removal from aqueous solution. Royal Chemistry Society, London
97. Sekhar KC, Subramanian S, Modak JM, Natarajan KA (1998) Removal of metal ions using an industrial biomass with reference to environmental control. Int J Miner Process 53:107–120. doi:10.1016/S0301-7516(97), 00061-6
98. Basci N, Kocadagistan E, Kocadagistan B (2004) Biosorption of copper(II) from aqueous solutions by wheat shells. Desalination 164:135–140. doi:10.1016/S0011-9164(04), 00172-9
99. Valdman E, Erijman L, Pessoa FLP, Leite SGF (2001) Continuous biosorption of Cu and Zn by immobilised waste biomass *Sargassum* sp. Process Biochem 36:869–873. doi:10.1016/S0032-9592(00), 00288-0
100. Singh S, Rai BN, Rai LC (2001) Ni(II) and Cr(VI) sorption kinetics by *Microcystis* in single and multimetallic systems. Process Biochem 36:1205–1213. doi:10.1016/S0032-9592(01), 00160-1
101. Xie SB, Yang J, Chen C, Zhang XJ, Wang QL, Zhang C (2008) Study on biosorption kinetics and thermodynamics of uranium by *Citrobacter freudii*. J Environ Radioact 99:126–133. doi:10.1016/j.jenvrad.2007.07.003
102. Ozacar M, Sengil IA (2003) Adsorption of reactive dyes on calcined alunite from aqueous solutions. J Hazard Mater 98:211–224. doi:10.1016/S0304-3894(02), 00358-8
103. Song HP, Li XG, Sun JS, Yin XH, Wang YH, Wu ZH (2007) Biosorption equilibrium and kinetics of Au(II1) and Cu(I1) on magnetotactic bacteria. Chin J Chem Eng 15:847–854. doi:10.1016/S1004-9541(08), 60013-0
104. Ho YS, McKay G (1999) Pseudo-second order model for sorption processes. Process Biochem 34:451–465. doi:10.1016/S0032-9592(98), 00112-5
105. Donmez G, Aksu Z (2002) Removal of chromium(VI) from saline wastewaters by *Dunaliella* species. Process Biochem 38:751–762. doi:10.1016/S0032-9592(02), 00204-2
106. Bayramoglou G, Celik G, Yalcin E, Yilmaz M, Arica MY (2005) Modification of surface properties of *Lentinus sajor-caju* mycelia by physical and chemical methods: evaluation of their Cr6+ removal efficiencies from aqueous medium. J Hazard Mater 119:219–229. doi:10.1016/j.jhazmat.2004.12.022
107. Ozdemir G, Ozturk T, Ceyhan N, Isler R, Cosar T (2003) Heavy metal biosorption by biomass of *Ochrobactrum anthropi* producing exopolysaccharide in activated sludge. Bioresour Technol 90:71–74. doi:10.1016/S0960-8524(03), 00088-9
108. Wang YH, Gao H, Sun JS, Li J, Su YX, Ji YL, Gong CM (2011) Selective reinforced competitive biosorption of Ag (I) and Cu (II) on *Magnetospirillum gryphiswaldense*. Desalination 270:258–263. doi:10.1016/j.desal.2010.11.053
109. Sag Y, Akcael B, Kutsal T (2003) Application of multicomponent adsorption models to the biosorption of Cr(VI), Cu(II), and Cd(II) ions on *Rhizopus arrhizus* from ternary metal mixtures. Chem Eng Commun 190:797–812. doi:10.1080/00986440302119
110. Chen XC, Chen LT, Shi JY, Wu WX, Chen YX (2008) Immobilization of heavy metals by *Pseudomonas putida* CZ1/goethite composites from solution. Colloids Surf B Biointerfaces 61:170–175. doi:10.1016/j.colsurfb.2007.08.002
111. Vieira R, Volesky B (2000) Biosorption: a solution to pollution. Int Microbiol 3:17–24

112. Ehrlich HL (1997) Microbes and metals. Appl Microbiol Biotechnol 48:687–692. doi:10.1007/s002530051116
113. Gadd GM (1990) Biosorption. Chem Ind 13: 421–426. RefDoc. http://cat.inist.fr/?aModele=afficheN&cpsidt=19367157. Accessed 1990
114. Beveridge TJ (1989) Role of cellular design in bacterial metal accumulation and mineralization. Annu Rev Microbiol 43:147–171. doi:10.1146/annurev.mi.43.100189.001051
115. BeveridgeTJ HMN, Lee H, Leung KT, Poole RK, Savvaidis I, Silver S, Trevors JT (1997) Metal–microbe interactions: contemporary approaches. Adv Microb Physiol 38:177–243. doi:10.1016/S0065-2911(08), 60158-7
116. Kretschmer KC, Gardea-Torresdey J, Chianelli R, Webb R (2002) Determination of copper binding in *Anabaena flos-aquae* purified cell walls and whole cells by X-ray absorption spectroscopy. Microchem J 71:295–304. doi:10.1016/S0026-265X(02)00022-X
117. YeeN BLG, Phoenix VR, Ferris FG (2004) Characterization of metal-cyanobacteria sorption reactions: a combined macroscopic and infrared spectroscopic investigation. Environ Sci Technol 38:775–782. doi:10.1021/es0346680
118. Toner B, Manceau A, Marcus MA, Millet DB, Sposito G (2005) Zinc sorption by a bacterial biofilm. Environ Sci Technol 39:8288–8294. doi:10.1021/es050528+
119. Chen XC, Shi JY, Chen YX, Xu XH, Xu SY, Wang YP (2006) Tolerance and biosorption of copper and zinc by *Pseudomonas putida* CZ1 isolated from metal-polluted soil. Can J Microbiol 52:308–316
120. White C, Gadd GM (2000) Copper accumulation by sulfate-reducing bacterial biofilms. FEMS Microbiol Lett 183:313–318. doi:10.1111/j.1574-6968.2000.tb08977.x
121. Borrok DM, Fein JB (2005) The impact of ionic strength on the adsorption of protons, Pb, Cd and Sr onto surfaces of Gram negative bacteria: testing non-electrostatic diffuse and triple monolayer models. J Colloid Interface Sci 286:110–126. doi:10.1016/j.jcis.2005.01.015
122. Hayashi H, Seiki H, Tsuneda S, Hirata A, Sasaki H (2003) Influence of growth phase on bacterial cell electrokinetic characteristics examined by soft particle electrophoresis theory. J Colloid Interface Sci 264:565–568. doi:10.1016/S0021-9797(03), 00418-1
123. Hetzer A, Daughney CJ, Morgan HW (2006) Cadmium ion biosorption by the thermophilic bacteria *Geobacillus stearothermophilus* and *G. thermocatenulatus*. Appl Environ Microbiol 72:4020–4027. doi:10.1128/AEM.00295–06
124. Kapoor A, Viraraghavan T (1997) Fungi as biosorbents. In: Wase J, Forster C (eds) Biosorbents for metal ions. CRC Press, London
125. Kapoor A, Viraraghavan T (1995) Fungal biosorption—an alternative treatment option for heavy metal bearing wastewaters: a review. Bioresour Technol 53:195–206. doi:10.1016/0960-8524(95)00072-M
126. Cabuk A, Akar T, Tunali S, Gedikli S (2007) Biosorption of Pb(II) by industrial strain of *Saccharomyces cerevisiae* immobilized on the biomatrix of cone biomass of *Pinus nigra*: equilibrium and mechanism analysis. Chem Eng J 131:293–300. doi:10.1016/j.cej.2006.12.011
127. Özer A, Özer D (2003) Comparative study of the biosorption of Pb (II), Ni (II) and Cr (VI) ions onto *S. cerevisiae*: determination of biosorption heats. J Hazard Mater 100:219–229. doi:10.1016/S0304–3894(03)00109–2
128. Mapolelo M, Torto N (2004) Trace enrichment of metal ions in aquatic environments by *Saccharomyces cerevisiae*. Talanta 64:39–47. doi:10.1016/j.talanta.2003.10.058
129. Al-Saraj M, Abdel-Latif MS, El-Nahal I, Baraka R (1999) Bioaccumulation of some hazardous metals by sol-gel entrapped microorganisms. J Non-Cryst Solids 248:137–140. doi:10.1016/S0022-3093(99), 00306-3
130. Goksungur Y, Uren S, Guvenc U (2005) Biosorption of cadmium and lead ions by ethanol treated waste baker's yeast biomass. Bioresour Technol 96:103–109. doi:10.1016/j.biortech.2003.04.002
131. Bustard M, McHale AP (1998) Biosorption of heavy metals by distillery-derived biomass. Bioprocess Eng 19:351–353. doi:10.1007/s004490050531
132. Chen C, Wang JL (2006) Review on biosorption of heavy metal by *Saccharomyces cerevisiae*. China Biotechnol 26:69–76. doi:cnki:ISSN:1671-8135.0.2006-01-014

133. Dhankhara R, Hoodaa A (2011) Fungal biosorption – an alternative to meet the challenges of heavy metal pollution in aqueous solutions. Environ Technol 32:467–491. doi:10.1080/09593330.2011.572922
134. Tsezos M (1997) Biosorption of lanthanides, actinides and related materials in biosorbents for metal ions. In: Wase J, Forster C (eds) Biosorbents for metal ions. CRC Press, London
135. Baysal Z, Çinar E, Bulut Y, Alkan H, Dogru M (2009) Equilibrium and thermodynamic studies on biosorption of Pb(II) onto *Candida albicans* biomass. J Hazard Mater 161:62–67. doi:10.1016/j.jhazmat.2008.02.122
136. Zu YG, Zhao XH, Hu MS, Ren Y, Xiao P, Zhu L, Cao YJ, Zhang Y (2006) Biosorption effects of copper ions on *Candida utilis* under negative pressure cavitation. J Environ Sci 18:1254–1259. doi:10.1016/S1001-0742(06), 60071-5
137. Yin H, He BY, Lu XY, Peng H, Ye JS, Yang F (2008) Improvement of chromium biosorption by UV–HNO_2 cooperative mutagenesis in *Candida utilis*. Water Res 42:3981–3989. doi:10.1016/j.watres.2008.07.005
138. Akhtar K, Akhtar MW, Khalid AM (2008) Removal and recovery of zirconium from its aqueous solution by *Candida tropicalis*. J Hazard Mater 156:108–117. doi:10.1016/j.jhazmat.2007.12.002
139. Muter O, Lubinya I, Millers D, Grigorjeva L, Ventinya E, Rapoport A (2002) Cr(VI) sorption by intact and dehydrated *Candida utilis* cells in the presence of other metals. Process Biochem 38:123–131. doi:10.1016/S0032-9592(02), 00065-1
140. Deng SB, Ting YP (2005) Characterization of PEI-modified biomass and biosorption of Cu(II), Pb(II) and Ni(II). Water Res 39:2167–2177. doi:10.1016/j.watres.2005.03.033
141. Holan ZR, Volesky B (1995) Accumulation of cadmium, lead, and nickel by fungal and wood biosorbents. Appl Biochem Biotechnol 53:133–146. doi:10.1007/BF02788603
142. Tan TW, Hu B, Su HJ (2004) Adsorption of Ni^{2+} on amine-modified mycelium of *Penicillium chrysogenum*. Enzyme Microb Technol 35:508–513. doi:10.1016/j.enzmictec.2004.08.035
143. Fan T, Liu YG, Feng BY, Zeng GM, Yang CP, Zhou M, Zhou HZ, Tan ZF, Wang X (2008) Biosorption of cadmium (II), zinc (II) and lead (II) by *Penicillium simplicissimum*: isotherms, kinetics and thermodynamics. J Hazard Mater 160:655–661. doi:10.1016/j.jhazmat.2008.03.038
144. Li XM, Liao DX, Xu XQ, Yang Q, Zeng GM, Zheng W, Guo L (2008) Kinetic studies for the biosorption of lead and copper ions by *Penicillium simplicissimum* immobilized within loofa sponge. J Hazard Mater 159:610–615. doi:10.1016/j.jhazmat.2008.02.068
145. Say R, Yılmaz N, Denizli A (2003) Biosorption of cadmium, lead, mercury, and arsenic ions by the fungus *Penicillium purpurogenum*. Sep Sci Technol 38:2039–2053. doi:10.1081/SS-120020133
146. Say R, Yilmaz N, Denizli A (2003) Removal of heavy metal ions using the fungus *Penicillium canescens*. Adsorpt Sci Technol 21:643–650. doi:10.1260/026361703772776420
147. Shah MP, Vora SB, Dave SR (1999) Evaluation of potential use of immobilized *Penicillium griseofulvum* in bioremoval of copper. Process Metall 9:227–235. doi:10.1016/S1572-4409(99), 80112-6
148. Dursun AY, Uslu G, Tepea O, Cuci Y, Ekiz HI (2003) A comparative investigation on the bioaccumulation of heavy metal ions by growing *Rhizopus arrhizus* and *Aspergillus niger*. Biochem Eng J 15:87–92. doi:10.1016/S1369-703X(02), 00187-0
149. Kapoor A, Viraraghavan T, Cullimore DR (1999) Removal of heavy metals using the fungus *Aspergillus niger*. Bioresour Technol 70:95–104. doi:10.1016/S0960-8524(98), 00192-8
150. Gulati R, Saxena RK, Gupta R (2002) Fermentation waste of *Aspergillus terreus*: a potential copper biosorbent. World J Microbiol Biotechnol 18:397–401. doi:10.1023/A:1015540921432
151. Dias MA, Lacerda ICA, Pimentel PF, de Castro HF, Rosa CA (2002) Removal of heavy metals by an *Aspergillus terreus* strain immobilized in a polyurethane matrix. Lett Appl Microbiol 34:46–50. doi:10.1046/j.1472-765x.2002.01040.x
152. Gadd GM, White C (1992) Removal of thorium from simulated acid process streams by fungal biomass—potential for thorium desorption and reuse of biomass and desorbent. J Chem Technol Biotechnol 55:39–44. doi:10.1002/jctb.280550107

153. Naja G, Mustin C, Berthelin J, Volesky B (2005) Lead biosorption study with *Rhizopus arrhizus* using a metal-based titration technique. J Colloid Interface Sci 292:537–543. doi:10.1016/j.jcis.2005.05.098
154. Bhattacharyya S, Pal TK, Basumajumdar A, Banik AK (2002) Biosorption of heavy metals by *Rhizopus arrhizus* and *Aspergillus niger*. J Indian Chem Soc 79:747–750
155. Brady JM, Tobin JM (1995) Binding of hard and soft metal ions to *Rhizopus arrhizus* biomass. Enzyme Microb Technol 17:791–796. doi:10.1016/0141-0229(95)00142-R
156. Kogej A, Pavko A (2001) Comparison of *Rhizopus nigricans* in a pelleted growth form with some other types of waste microbial biomass as biosorbents for metal ions. World J Microbiol Biotechnol 17:677–685. doi:10.1023/A:1012901224684
157. Arica MY, Kacar Y, Genc O (2001) Entrapment of white-rot fungus *Trametes versicolor* in Caalginate beads: preparation and biosorption kinetic analysis for cadmium removal from an aqueous solution. Bioresour Technol 80:121–129. doi:10.1016/S0960-8524(01), 00084-0
158. Bayramoglu G, Denizli A, Bektas S, Arica MY (2002) Entrapment of *Lentinus sajorcaju* into Caalginate gel beads for removal of Cd (II) ions from aqueous solution: preparation and biosorption kinetics analysis. Microchem J 72:63–76. doi:10.1016/S0026-265X(01), 00151-5
159. Zhang YS, Liu WG, Xu M, Zheng F, Zhao MJ (2010) Study of the mechanisms of Cu^{2+} biosorption by ethanol/caustic-pretreated baker's yeast biomass. J Hazard Mater 178:1085–1093. doi:10.1016/j.jhazmat.2010.02.051
160. Mungasavalli DP, Viraraghavan T, Jin YC (2007) Biosorption of chromium from aqueous solutions by pretreated *Aspergillus niger*: batch and column studies. Colloids Surf A Physicochem Eng Asp 301:214–223. doi:10.1016/j.colsurfa.2006.12.060
161. Park D, Yun YS, Park JM (2005) Use of dead fungal biomass for the detoxification of hexavalent chromium: screening and kinetics. Process Biochem 40:59–2565. doi:10.1016/j.procbio.2004.12.002
162. Saygideger S, Gulnaz O, Istifli ES, Yucel N (2005) Adsorption of Cd (II), Cu (II) and Ni (II) ions by *Lemna minor* L.: effect of physicochemical environment. J Hazard Mater 126:96–104. doi:10.1016/j.jhazmat.2005.06.012
163. Ahuja P, Gupta R, Saxena RK (1999) Zn^{2+} biosorption by *Oscillatroria anguistissima*. Process Biochem 34:77–85. doi:10.1016/S0032-9592(98), 00072-7
164. Padmavathy V, Vasudevan P, Dhingra SC (2003) Biosorption of nickel(II) ions on baker's yeast. Process Biochem 38:1389–1395. doi:10.1016/S0032-9592(02), 00168-1
165. Vasudevan P, Padmavathy V, Dhingra SC (2003) Kinetics of biosorption of cadmium on baker's yeast. Bioresour Technol 89:281–287. doi:10.1016/S0960-8524(03), 00067-1
166. Tewari N, Vasudevan P, Guha BK (2005) Study of biosorption of Cr(VI) by *Mucor hiemalis*. Biochem Eng J 23:185–192. doi:10.1016/j.bej.2005.01.011
167. Tunali S, Kiran I, Akar T (2005) Chromium(VI) biosorption characteristics of *Neurospora crassa* fungal biomass. Miner Eng 18:681–689. doi:10.1016/j.mineng.2004.11.002
168. Prakasham RS, Merrie JS, Sheela R, Saswathi N, Ramakrishna SV (1999) Biosorption of chromium VI by free and immobilized *Rhizopus arrhizus*. Environ Pollut 104:421–427. doi:10.1016/S0269-7491(98), 00174-2
169. Bai RS, Abraham TE (2001) Biosorption of Cr (VI) from aqueous solution by *Rhizopus nigrican*. Bioresour Technol 79:73–81. doi:10.1016/S0960-8524(00), 00107-3
170. Sag Y, Aktay Y (2002) Kinetic studies on sorption of Cr(VI) and Cu(II) ions by chitin, chitosan and *Rhizopus arrhizus*. Biochem Eng J 12:143–153. doi:10.1016/S1369-703X(02), 00068-2
171. Sag Y, Aktay Y (2000) Mass transfer and equilibrium studies for the sorption of chromium ions onto chitin. Process Biochem 36:157–173. doi:10.1016/S0032-9592(00), 00200-4
172. Akar T, Tunali S (2006) Biosorption characteristics of *Aspergillus flavus* biomass for removal of Pb(II) and Cu(II) ions from an aqueous solution. Bioresour Technol 97:1780–1787. doi:10.1016/j.biortech.2005.09.009
173. Dai SH, Wei DZ, Zhou DQ, Jia CY, Wang YG, Liu WG (2008) Removing cadmium from electroplating wastewater by waste *Saccharomyces cerevisiae*. Trans Nonferr Metal Soc China 18:1008–10013. doi:10.1016/S1003-6326(08), 60173-9

174. Deng SB, Ting YP (2005) Fungal biomass with grafted poly(acrylic acid) for enhancement of Cu(II) and Cd(II) biosorption. Langmuir 21:5940–5948. doi:10.1021/la047349a
175. Beveridge TJ, Murray RGE (1980) Sites of metal deposition in the cell walls of *Bacillus subtilis*. J Bacteriol 141:876–887
176. Gupta R, Ahuja P, Khan S, Saxena RK, Mohapatra M (2000) Microbial biosorbents: meetings challenges of heavy metals pollution in aqueous solution. Curr Sci 78:967–973. doi:10.1517/13543784.13.4.373
177. Strandberg GW, Shumate SE, Parrott JR (1981) Microbial cells as biosorbents for heavy metals: accumulation of uranium by *Saccharomyces cerevisiae* and *Pseudomonas aeruginosa*. Appl Environ Microbiol 41:237–245, PMCID:PMC243671
178. Farkas V (1980) Biosynthesis of cell wall of fungi. Microbiol Rev 44:117–141, PMCID: PMC281469
179. Muzzarelli RA, Tanfari F (1982) The chelating ability of chitinous material from *Aspergillus niger*, *Streptomyces*, *Mucor rouxii*, *Phycomyces blakeseanus* and *Choamephora curcurnitium*. In: Mirano S, Tokura S (eds) Chitin and Chitiosan. Japanese Society of Chitin and Chitiosan, Sapporo
180. Tsezos M, Volesky B (1981) Biosorption of uranium and thorium. Biotechnol Bioeng 23:583–604. doi:10.1002/bit.260230309
181. Fourest E, Roux JC (1992) Heavy metal biosorption by fungal mycelial by-products: mechanism and influence of pH. Appl Microbiol Biotechnol 37:399–403. doi:10.1007/BF00211001
182. Liu YG, Fan T, Zeng GM, Li X (2006) Removal of cadmium and zinc ions from aqueous solution by living *Aspergillus niger*. Trans Nonferr Metal Soc China 16:681–686. doi:10.1016/S1003-6326(06), 60121-0
183. Dursun AY (2006) A comparative study on determination of the equilibrium, kinetic and thermodynamic parameters of biosorption of copper(II) and lead(II) ions onto pretreated *Aspergillus niger*. Biochem Eng J 28:187–195. doi:10.1016/j.bej.2005.11.003
184. Wang JS, Hu XJ, Xie SB, Bao ZL (2010) Biosorption of uranium (VI) by immobilized *Aspergillus fumigatus* beads. J Environ Radioact 101:504–508, 10.1016/j.jenvrad.2010.03.002
185. Aksu Z (2001) Equilibrium and kinetic modelling of cadmium (II) biosorption by *C. Vulgaris* in a batch system: effect of temperature. Sep Purif Technol 21:285–294. doi:10.1016/S1383-5866(00), 00212-4
186. Bhainsa KC, D'Souza SF (2009) Thorium biosorption by *Aspergillus fumigatus*, a filamentous fungal biomass. J Hazard Mater 165:670–676. doi:10.1016/j.jhazmat.2008.10.033
187. Kuyicak N, Volesky B (1990) Biosorption by fungal biomass. In: Volesky B (ed) Biosorption of heavy metals. CRC press, Boca Raton
188. Rincon J, Gonzalez F, Ballester A, Blazquez ML, Munoz JA (2005) Biosorption of heavy metals by chemically-activated alga *Fucus vesiculosus*. J Chem Technol Biotechnol 80:1403–1407. doi:10.1002/jctb.1342
189. Yu Q, Matheickal JT, Yin P (1999) Heavy metal uptake capacities of common marine macro algal biomass. Water Res 33:1534–1537. doi:10.1016/S0043-1354(98), 00363-7
190. Brinza L, Dring MJ, Gavrilescu M (2007) Marine micro- and macro-algal species as biosorbents for heavy metals. Environ Eng Manag J 6:237–251
191. Romera E, Gonzalez F, Ballester A, Blazquez ML, Munoz JA (2006) Biosorption with algae: a statistical review. Crit Rev Biotechnol 26:223–235. doi:10.1080/07388550600972153
192. Matheickal JT, Iyengar L, Venkobachar C (1991) Sorption and desorption of Cu (II) by *Ganoderma lucidum*. Water Qual Res J Can 26:187–200
193. Fourest E, Canal C, Roux JC (1994) Improvement of heavy metal biosorption by mycelial dead biomasses (*Rhizopus arrhizus*, *Muchor miehei*, and *Pencillium chrysogenum*): pH control and cationic activation. FEMS Microbiol Rev 14:325–332. doi:10.1016/0168-6445(94), 90050-7
194. Matheickal JT, Yu Q (1996) Biosorption of lead from aqueous solutions by marine alga *Ecklonia radiate*. Water Sci Technol 34:1–7. doi:10.1016/S0273-1223(96), 00780-9
195. Pavasant P, Apiratikul R, Sungkhum V, Suthiparinyanont P, Wattanachira S, Marhaba TF (2006) Biosorption of Cu^{2+}, Cd^{2+}, Pb^{2+}, and Zn^{2+} using dried marine green macroalga *Caulerpa lentillifera*. Bioresour Technol 97:2321–2329. doi:10.1016/j.biortech.2005.10.032

196. Sari A, Tuzen M (2008) Biosorption of cadmium (II) from aqueous solution by red algae (*Ceramium virgatum*): equilibrium, kinetic and thermodynamic studies. J Hazard Mater 157:448–454. doi:10.1016/j.jhazmat.2008.01.008
197. Gupta VK, Shrivastava AK, Jain N (2001) Biosorption of chromium (VI) from aqueous solutions by green algae *Spirogyra* species. Water Res 35:4079–4085. doi:10.1016/ S0043-1354(01), 00138-5
198. Gupta VK, Rastogi A (2008) Biosorption of lead from aqueous solutions by green algae *Spirogyra* species: kinetics and equilibrium studies. J Hazard Mater 152:407–414. doi:10.1016/j.jhazmat.2007.07.028
199. Tuzun I, Bayramoglu G, Alcin YE, Basaran G, Celik G, Arica MY (2005) Equilibrium and kinetic studies on biosorption of Hg(II), Cd(II) and Pb(II)ions onto microalgae *Chlamydomonas reinhardtii*. J Environ Manage 77:85–92. doi:10.1016/j.jenvman.2005.01.028
200. Deng LP, Su YY, Su H, Wang XX, Zhu XB (2007) Sorption and desorption of lead(II) from wastewater by green algae *Cladophora fascicularis*. J Hazard Mater 143:220–225. doi:10.1016/j.jhazmat.2006.09.009
201. Ajmal M, Rao RAK, Ahmad R, Ahmad J (2000) Adsorption studies on *Citrus reticulata* (fruit peel of orange): removal and recovery of Ni (II) from electroplating wastewater. J Hazard Mater B 79:117–131. doi:10.1016/S0304-3894(00)00234-X
202. Aksu Z, Kutsal T (1990) A comparative study for biosorption characteristics of heavy metal ions with *C. vulgaris*. Environ Technol 11:979–987. doi:10.1080/09593339009384950
203. Donmez GC, Aksu Z, Ozturk A, Kutsal T (1999) A comparative study on heavy metal biosorption characteristics of some algae. Process Biochem 34:885–892. doi:10.1016/S0032-9592(99), 00005-9
204. Khezami L, Capart R (2005) Removal of chromium (VI) from aqueous solution by activated carbons: kinetic and equilibrium studies. J Hazard Mater 123:223–231. doi:10.1016/j.jhazmat.2005.04.012
205. Aksu Z (2002) Determination of the equilibrium, kinetic and thermodynamic parameters of the batch biosorption of nickel (II) ions onto *Chlorella vulgaris*. Process Biochem 38:89–99. doi:10.1016/S0032-9592(02), 00051-1
206. Pahlavanzadeh H, Keshtkar AR, Safdari J, Abadi Z (2010) Biosorption of nickel (II) from aqueous solution by brown algae: equilibrium, dynamic and thermodynamic studies. J Hazard Mater 175:304–310. doi:10.1016/j.jhazmat.2009.10.004
207. Sheng PX, Ting YP, Chen JP, Hong L (2004) Sorption of lead, copper, cadmium, zinc and nickel by marine algal biomass: characterization of biosorptive capacity and investigation of mechanisms. J Colloid Interface Sci 275:131–141. doi:10.1016/j.jcis.2004.01.036
208. Bishnoi NR, Pant A, Garima (2004) Biosorption of copper from aqueous solution using algal biomass. J Sci Ind Res 63:813–816, ISSN 0022–4456
209. Chand S, Agarwal VK, Kumar P (1994) Removal of hexavalent Cr from wastewater by adsorption. Indian J Environ Health 36:151–158, ISSN 0367–827X
210. Saravanane R, Sundararajan T, Sivamurthyreddy S (2002) Efficiency of chemically modified low cost adsorbents for the removal of heavy metals from wastewater: a comparative study. Indian J Environ Health 44:78–87
211. Deo N, Ali M (1992) Optimization of a new low cost adsorbent in removal of Cr(VI) from wastewater. Indian J Environ Protect 12:828–834
212. Karthikeyan S, Balasubramanian R, Iyer CSP (2007) Evaluation of the marine algae *Ulva fasciata* and *Sargassum* sp. for the biosorption of Cu(II) from aqueous solutions. Bioresour Technol 98:452–455. doi:10.1016/j.biortech.2006.01.010
213. Aravindhan R, Maharshi B, Sreeram KJ (2010) Biosorption of cadmium metal ion from simulated wastewaters using *Hypnea valentiae* biomass: a kinetic and thermodynamic study. Bioresour Technol 101:1466–1470. doi:10.1016/j.biortech.2009.08.008
214. Crist DR, Crist RH, Martin JR, Watson JR (1994) Ion exchange systems in proton–metal reaction with algal cell walls. FEMS Microbiol Rev 14:309–314. doi:10.1111/j.1574-6976.1994.tb00104.x

215. Chojnacka K, Chojnacki A, Gorecka H (2005) Biosorption of Cr^{3+}, $Cd^{2+}$and Cu^{2+} ions by blue–green algae *Spirulina* sp.: kinetics, equilibrium and the mechanism of the process. Chemosphere 59:75–84. doi:10.1016/j.chemosphere.2004.10.005
216. Raize O, Argaman Y, Yannail S (2004) Mechanisms of biosorption of different heavy metals by brown marine macroalgae. Biotechnol Bioeng 87:451–458. doi:10.1002/bit.20136
217. Han X, Wong YS, Tam NFY (2006) Surface complexation mechanism and modeling in Cr (III) biosorption by a microalgal isolate, *Chlorella miniata*. J Colloid Interface Sci 303:365–371. doi:10.1016/j.jcis.2006.08.028
218. Kalyani S, Rao PS (2004) Removal of nickel (II) from aqueous solutions using marine macroalgae as the sorbing biomass. Chemosphere 57:1225–1229. doi:10.1016/j.chemosphere.2004.08.057
219. Ajjabiv LC, Chouba L (2009) Biosorption of Cu2+ and Zn2+ from aqueous solutions by dried marine green macroalga *Chaetomorpha linum*. J Environ Manage 90:3485–3489. doi:10.1016/j.jenvman.2009.06.001
220. Sahmurova A, Türkmenler H (2010) Biosorption kinetics and isotherm studies of Cd(II) by dried *Enteromorpha compress* macroalgae cells from aqueous solutions. Clean Soil Air Water 38:936–941. doi:10.1002/clen.201000108
221. Murugesan GS, Sathishirkumar M, Suaminathan K (2006) Arsenic removal from groundwater by pretreated waste tea fungal biomass. Bioresour Technol 97:483–487. doi:10.1016/j.biortech.2005.03.008
222. Fiol N, Villascusa I, Martinez M, Mirralles N, Poch J, Seralos J (2006) Sorpton of Pb (II), Ni (II), Cu (II) and Cd (II) from aqueous solutions by olive stone waste. Sep Purif Technol 50:132–140. doi:10.1016/j.seppur.2005.11.016
223. Singh A, Mehta SK, Gaur JP (2007) Removal of heavy metals from aqueous solution by common freshwater filamentous algae. World J Microbiol Biotechnol 23:1115–1120. doi:10.1007/s11274-006-9341-z
224. Freitas OM, Martins RJE, Delerue-Matos CM, Boaventura RAR (2008) Removal of Cd (II), Zn (II) and Pb (II) from aqueous solutions by brown marine macro algae: kinetic modeling. J Hazard Mater 153:493–501. doi:10.1016/j.jhazmat.2007.08.081
225. Lodeiro P, Cordero B, Grille Z, Herrero R, Sastre de Vicente ME (2004) Physicochemical studies of cadmium (II) biosorption by the invasive alga in Europe, *Sargassum muticum*. Biotechnol Bioeng 88:237–247. doi:10.1002/bit.20229
226. Cruz CCV, da Costa ACA, Henriques CA, Luna AS (2004) Kinetic modeling and equilibrium studies during cadmium biosorption by dead *Sargassum* sp. biomass. Bioresour Technol 91:249–257. doi:10.1016/S0960–8524(03)00194–9
227. Martins BL, Cruz CCV, Luna AS, Henriques CA (2006) Sorption and desorption of Pb2+ ions by dead *Sargassum* sp. biomass. Biochem Eng J 27:310–314. doi:10.1016/j.bej.2005.08.007
228. Kumar YP, King P, Prasad VSRK (2006) Removal of copper from aqueous solution using *Ulva fasciata* sp., a marine green algae. J Hazard Mater 137:367–373. doi:10.1016/j.jhazmat.2006.02.010
229. Weber WJ, Morris JC (1962) Advance in water pollution research: removal of biological resistant pollutions from wastewater by adsorption. In: Proceedings of the international conference on water pollution symposium. Pergamon Press, Oxford
230. Sharma A, Bhattacharyya KG (2005) *Azadirachta indica* (neem) leaf power as a biosorbent for removal of Cd (II) from aqueous medium. J Hazard Mater B 125:102–112. doi:10.1016/j.jhazmat.2005.05.012
231. Benaissa H, Elouchdi MA (2007) Removal of copper ions from aqueous solutions by dried sunflower leaves. Chem Eng Process 46:614–622. doi:10.1016/j.cep. 2006.08.006
232. Salim R, Al-Subu M (2008) Efficiency of removal of cadmium from aqueous solutions by plant leaves and the effects of interaction of combinations of leaves on their removal efficiency. J Environ Manage 87:521–532. doi:10.1016/j.jenvman.2007.01.028
233. Al Rmalli SW, Dahmani AA, Abuein MM, Gleza AA (2008) Biosorption of mercury from aqueous solutions by powdered leaves of castor tree (*Ricinus communis* L.). J Hazard Mater 152:955–959. doi:10.1016/j.jhazmat.2007.07.111

234. Sangi M, Shahmoradi A, Zolgharnein J, Azimi GH, Ghorbandoost M (2008) Removal and recovery of heavy metals from aqueous solution using *Ulmus carpinifolia* and *Fraxinus excelsior* tree leaves. J Hazard Mater 155:513–522. doi:10.1016/j.jhazmat.2007.11.110
235. Zolgharnein J, Shamoradi A, Sangi MR (2008) Optimization of Pb (II) biosorption by Robinia tree leaves using statistical design of experiments. Talanta 76:528–532. doi:10.1016/j.talanta.2008.03.039
236. Zolgharnein J, Shahmoradi A (2010) Characterization of sorption isotherms, kinetic models, and multivariate approach for optimization of Hg (II) adsorption onto *Fraxinus* tree leaves. J Chem Eng Data 55:5040–5049. doi:10.1021/je1006218
237. Sayrafi O, Salim R, Sayrafi SA (1996) Removal of cadmium from polluted water using decaying leaves: effects of type of leaves and of concentration of cadmium. J Environ Sci Health A 31:2503–2513. doi:10.1080/10934529609376506
238. Salim R, Al-Subu M, Qashoa S (1994) Removal of lead from polluted water using decaying plant leaves. J Environ Sci Health A 29:2087–2114. doi:10.1080/10934529409376166
239. Al-Subu M, Salim R, Abu-Shqair I, Swaileh K (2001) Removal of dissolved copper from polluted water using plant leaves: effects of acidity and plant species. Rev Int Contam Ambient 17:91–96
240. Salim R (1988) Removal of nickel (II) from polluted water using decaying leaves-effects of pH and type of leaves. J Environ Sci Health A 23:183–197, 321–334. doi: 10.1080/10934528809375403
241. Salim R, Robinson JW (1985) Removal of dissolved aluminum released by acid rain using decaying leaves. J Environ Sci Health A 20:701–734. doi:10.1080/10934528509375253
242. Fiol N, Villaescusa I, Martínez M, Miralles N, Poch J, Serarols J (2006) Sorption of Pb (II), Ni (II), Cu (II) and Cd (II) from aqueous solution by olive stone waste. Sep Purif Technol 50:132–140. doi:10.1016/j.seppur.2005.11.016
243. Özer A, Özer D (2003) Comparative study of the biosorption of Pb (II), Ni (II) and Cr (VI) ion onto *S. cerevisiae*: determination of biosorption heats. J Hazard Mater B 100:219–228. doi:10.1016/S0304–3894(03)00109–2
244. Vianna LNL, Andrade MC, Nicoli JR (2000) Screening of waste biomass from *Saccharomyces cerevisiae*, *Aspergillus oryzae* and *Bacillus lentus* fermentations for removal of Cu, Zn and Cd by biosorption. World J Microbiol Biotechnol 16:437–440. doi:10.1023/A:1008953922144
245. Singh KK, Rastogi R, Hasan SH (2005) Removal of Cr (VI) from wastewater using rice bran. J Colloid Interface Sci 290:61–68. doi:10.1016/j.jcis.2005.04.011
246. Villaescusa I, Fiol N, Martinez M, Miralles N, Poch J, Serarols J (2004) Removal of copper and nickel ions from aqueous solutions by grape stalks wastes. Water Res 38:992–1002. doi:10.1016/j.watres.2003.10.040
247. Bulut Y, Baysal Z (2006) Removal of Pb (II) from wastewater using wheat bran. J Environ Manage 78:107–113. doi:10.1016/j.jenvman.2005.03.010
248. Yu B, Zhang Y, Shukla A, Shukla SS, Dorris KL (2001) The removal of heavy metals from aqueous solutions by sawdust adsorption: removal of lead and comparison of its adsorption with copper. J Hazard Mater 84:83–94. doi:10.1016/S0304-3894(01), 00198-4
249. Ho YS, Huang CT, Huang HW (2002) Equilibrium sorption isotherm for metal ions on tree fern. Process Biochem 37:1421–1430. doi:10.1016/S0032-9592(02), 00036-5
250. Wang R, Liao X, Shi B (2005) Adsorption behaviors of Pt(II) and Pd(II) on collagen fibre immobilized bayberry tannin. Ind Eng Chem Res 44:4221–4226. doi:10.1021/ie049069w
251. Ahluwalia SS, Goyal D (2005) Removal of heavy metals from waste tea leaves from aqueous solution. Eng Life Sci 5:158–162. doi:10.1002/elsc.200420066
252. Singh KK, Talat M, Hasan SH (2006) Removal of lead from aqueous solutions by agricultural waste maize bran. Bioresour Technol 97:2124–2130. doi:10.1016/j.biortech.2005.09.016
253. Reddya D, Harinatha Y, Seshaiah K (2010) Biosorption of Pb (II) from aqueous solutions using chemically modified *Moringa oleifera* tree leaves. Chem Eng J 162:626–634. doi:10.1016/j.cej.2010.06.010

254. Qaiser S, Saleemi AR, Umar M (2009) Biosorption of lead from aqueous solution by *Ficus religiosa* leaves: batch and column study. J Hazard Mater 166:998–1005. doi:10.1016/j.jhazmat.2008.12.003
255. Al-Masri MS, Amin Y (2010) Biosorption of cadmium, lead, and uranium by powder of poplar leaves and branches. Appl Biochem Biotechnol 160:976–987. doi:10.1007/s12010-009-8568-1
256. Aksu Z, Donmez G (2001) Comparison of copper (II) biosorptive properties of live and treated *Candida* sp. J Environ Sci Health A 36:367–381. doi:10.1081/ESE-100102928
257. Yang J, Volesky B (1999) Sorption of copper on algae and fungi. Environ Sci Technol 33:751–757
258. Aksu Z, Kutsal T, Gun S, Haciosmanoglu N, Gholaminejad M (1991) Investigation of biosorption of Cu(II), Ni(II) and Cr(VI) ions to activated sludge bacteria. Environ Technol 12:915–921. doi:10.1080/09593339109385086
259. Zolgharnein J, Shahmoradi A (2010) Adsorption of Cr (VI) onto elaeagnus tree leaves: statistical optimization, equilibrium modeling, and kinetic studies. J Chem Eng Data 55:3428–3437. doi:10.1021/je100157y
260. Tian JS, Li Y, Li JL (2008) High-yield growth and magnetosome formation by *Magnetospirillum gryphiswaldense* MSR-1 in an oxygen-controlled fermentor supplied solely with air. Appl Microbiol Biotechnol 9:389–397. doi:10.1007/s00253-008-1453-y
261. Song HP, Li XG, Sun JS, Xu XM, Han X (2008) Application of a magnetotactic bacterium, *Stenotrophomonas* sp. to the removal of Au(III) from contaminated wastewater with a magnetic separator. Chemosphere 72:616–621. doi:10.1016/j.chemosphere.2008.02.064
262. Krishna MVB, Chandrasekaran K, Rao SV, Karunasagar D, Arunachalam J (2005) Speciation of Cr(III) and Cr(VI) in waters using immobilized moss and determination by ICP-MS and FAAS. Talanta 65:135–142. doi:10.1016/j.talanta.2004.05.051
263. Godlewska-Zyłkiewicz B, Kozłowska M (2005) Solid phase extraction using immobilized yeast *Saccharomyces cerevisiae* for determination of palladium in road dust. Anal Chim Acta 539:61–67. doi:10.1016/j.aca.2005.02.051
264. Bag H, Lale M, Turker AR (1999) Determination of Cu, Zn and Cd in water by FAAS after preconcentration by baker's yeast (*Saccharomyces cerevisiae*) immobilized on sepiolite. Fresenius J Anal Chem 363:224–230. doi:10.1007/s002160051178
265. Menegario AA, Smichowski P, Polla G (2005) On-line preconcentration and speciation analysis of Cr(III) and Cr(VI) using baker's yeast cells immobilised on controlled pore glass. Anal Chim Acta 546:244–250. doi:10.1016/j.aca.2005.05.030
266. Akhtar N, Iqbal J, Iqbal M (2004) Removal and recovery of nickel (II) from aqueous solution by loofa sponge-immobilized biomass of *Chlorella sorokiana*: characterization studies. J Hazard Mat B 108:85–94. doi:10.1016/j.jhazmat.2004.01.002
267. Bender J, Phillips P (2004) Microbial mats for multiple applications in aquaculture and bioremediation. Bioresour Technol 94:229–238. doi:10.1016/j.biortech.2003.12.016
268. Chang WC, Hsu GS, Chiang SM, Su MC (2006) Heavy metal removal from aqueous solution by wasted biomass from a combined AS-biofilm process. Bioresour Technol 97:1503–1508. doi:10.1016/j.biortech.2005.06.011
269. Gadd GM (2000) Bioremedial potential of microbial mechanisms of metal mobilization and immobilization. Curr Opin Biotechnol 11:271–279. doi:10.1016/S0958-1669(00), 00095-1
270. Liu Y, Xu H, Yang S, Tay J (2003) A general model for biosorption of Cd, Cu, and Zn by aerobic granules. J Biotechnol 102:233–239. doi:10.1016/S0168-1656(03), 00030-0
271. Han R, Zhang J, Zou W, Xiao H, Shi J, Liu H (2006) Biosorption of copper(II) and lead(II) from aqueous solution by chaff in a fixed-bed column. J Hazard Mater 133:262–268. doi:10.1016/j.jhazmat.2005.10.019
272. Vijayaraghavan K, Jegan J, Palanivelu K, Velan M (2005) Biosorption of copper, cobalt and nickel by marine green alga *Ulva reticulate* in a packed column. Chemosphere 60:419–426. doi:10.1016/j.chemosphere.2004.12.016

273. Godlewska-Zylkiewicz B (2003) Biosorption of platinum and palladium for their separation preconcentration prior to graphite furnace atomic absorption spectrometric determination. Spectrochim Acta B Atom Spectrosc 58:1531–1540. doi:10.1016/S0584-8547(03), 00076-4
274. Azila YY, Mashitah MD, Bhatia S (2008) Process optimization studies of lead (Pb(II)) biosorption onto immobilized cells of *Pycnoporus sanguineus* using response surface methodology. Bioresour Technol 99:8549–8552. doi:10.1016/j.biortech.2008.03.056
275. Hu J, Chen GH, Irene MCL (2005) Removal and recovery of Cr(VI) from wastewater by maghemite nanoparticles. Water Res 39:4528–4536. doi:10.1016/j.watres.2005.05.051
276. Li HD, Li Z, Liu T, Xiao X, Peng ZH, Deng L (2008) A novel technology for biosorption and recovery hexavalent chromium in wastewater by bio-functional magnetic beads. Bioresour Technol 99:6271–6279. doi:10.1016/j.biortech.2007.12.002
277. Sau TK, Murphy CJ (2004) Room temperature, high-yield synthesis of multiple shapes of gold nanoparticles in aqueous solution. Am Chem Soc 126:8648–8649. doi:10.1021/ja047846d
278. Binupriya AR, Sathishkumar M, Vijayaraghavan K, Yun SI (2010) Bioreduction of trivalent aurum to nano-crystalline gold particles by active and inactive cells and cell-free extract of *Aspergillus oryzae* var. viridis. J Hazard Mater 177:539–545. doi:10.1016/j.jhazmat.2009.12.066
279. Riddin T, Gericke M, Whiteley CG (2010) Biological synthesis of platinum nanoparticles: effect of initial metal concentration. Enzyme Microb Technol 46:501–505. doi:10.1016/j.enzmictec.2010.02.006
280. Shankar SS, Rai A, Ankamwar B, Singh A, Ahmad A (2004) Biological synthesis of triangular gold nanoprisms. Nat Mater 3:482–488. doi:10.1038/nmat1152
281. Shankar SS, Rai A, Ahmad A, Sastry M (2005) Controlling the optical properties of lemongrass extract synthesized gold nanotriangles and potential application in infrared-absorbing optical coatings. Chem Mater 17:566–572. doi:10.1021/cm048292g

Chapter 6
Mesoporous-Assembled Nanocrystal Photocatalysts for Degradation of Azo Dyes

Thammanoon Sreethawong

6.1 Introduction

Currently, wastewater problems can be generally found all over the world. Various kinds of wastewaters have been generated from industrial processes. One of the major industrial water pollutions is wastewaters from textiles, fibers, fabrics, plastics, papers, leathers, and related industries, which consist of high unfixed organic dye concentrations of approximately 20%. Big concern has been paid to these organic dyes because of their prevalent application, noxious aromatic intermediate formation, and biorecalcitrance for conventional aerobic wastewater-treating processes [1–5]. A more complicated environmental problem related to the wastewaters is that the organic dyes are opposed to bacterial degradation, and accordingly they can inevitably be converted to carcinogenic compounds. Particularly, a plentiful type of synthetically colored organic compounds is azo dye [6–11], such as methyl orange, congo red, and acid black, which is characterized by the presence of azo group (–N=N–) commonly linked between aromatic rings.

Various technologies have been developed for azo dye degradation from the wastewaters, including physical methods (adsorption), chemical methods (chlorination and ozonation), biodegradation, reverse osmosis, combined coagulation and flocculation, electrochemical oxidation, as well as photocatalysis [12–16]. The existing physical/chemical methods have some disadvantages of being economically unfeasible (more energy and chemicals required), being unable to degrade the recalcitrant azo dyes and their aromatic intermediates completely, generating a large amount of undesired sludge that unavoidably causes secondary pollution problems, and greatly increasing the overall treatment cost due to complicated procedures. However, photocatalysis is an environmentally friendly process that employs solar

T. Sreethawong (✉)
Baan Klangmuang Luzern, Soi On-nut 46, Sukhumvit 77 Road,
Suanluang, Bangkok 10250, Thailand
e-mail: tsreethawong@hotmail.com

S.K. Sharma and R. Sanghi (eds.), *Advances in Water Treatment and Pollution Prevention*, DOI 10.1007/978-94-007-4204-8_6,

irradiation energy to perform catalytic reactions under ambient conditions [15]. Thus, photocatalysis technology has been extensively investigated for azo dye degradation from the wastewaters.

From literature, it is clearly seen that among the new oxidation methods called advanced oxidation processes (AOPs), heterogeneous photocatalysis has been increasingly considered to be a promising destructive technology due to its several advantages. Firstly, it can degrade the azo dye pollutants by destroying their complex molecules to less toxic substances in the presence of UV and near-UV light irradiation, which ultimately leads to the total degradation of the azo dye pollutants. Secondly, environmentally benign materials can be used as a semiconductor photocatalyst, especially titanium dioxide (titania, TiO_2). Thirdly, this process can be operated at room temperature and atmospheric pressure. Finally, the photocatalytic process has been attracting a considerable attention because of its comparatively low cost due to the use of solar light as the source of irradiation.

6.2 Background of Azo Dyes

6.2.1 General Remarks

The "azo dye" term is normally used to name synthetically colored organic compounds that are characterized by the presence of the azo group (–N=N–), which cannot be found in natural dyes. This divalent group can be linked to sp^2 hybridized carbon atom on one side and to an aromatic or heterocyclic nucleus on the other. It may be linked to an unsaturated molecule of the carboxylic, heterocyclic, or aliphatic type. On commercial scale, the azo dyes are the largest and most widely used class of synthetic organic dyes. More than 10,000 color index (CI) generic names are applied to differentiate commercial dyes, whereas approximately 4,500 are currently used, and more than half of these dyes belong to the azo class [17]. The azo dyes are extensively used by the industries to provide colors to natural and synthetic textiles, fibers, fabrics, plastics, papers, and leathers, and they can also be used to produce organic pigments and in the dyeing of leather and plastic.

6.2.2 Classification of Azo Dyes

The most comprehensive information including the constitution, properties, preparation methods, manufactures, and other coloring data of the azo dyes is the Color Index publication, which is jointly collected by the Society of Dyers and Colorists and the American Association of Textile Chemists and Colorists [18, 19]. In the Color Index, a dual classification system is used to categorize the azo dyes with respect to usage area and chemical constitution. As a result of their versatile applications, the azo dyes comprise the largest chemical class in terms of the number, economic price, and quantities produced.

Table 6.1 Color index of different azo dyes [20]

Type of azo dye	Range of color index
Monoazo	11,000–19,999
Diazo	20,000–29,999
Trisazo	30,000–34,999
Polyazo	35,000–36,999

Most of the dye manufacturers use both letters and numerals together in their product's commercial name in order to identify the dye color. The examples of letters used are as follows: B is blue; G is green; R is red; and Y is yellow; whereas, numerals, such as 2G (or GG), 3G, 4G, etc. indicate the shade of the color. In some cases, suffixed letters are also used to describe other properties, such as solubility, light fastness, and brightness. Moreover, the azo dyes can be subdivided with respect to the number of existing azo groups into mono-, dis-, tris-, tetrakis-, etc. [20]. Generally, the significance of both mono- and diazo dyes are quite identical, trisazo dyes are somewhat less significant, while tetrakisazo dyes, except for some of them, are far less significant. As a consequence, the azo dyes containing more than three azo groups are usually incorporated under the heading of polyazo dyes. Table 6.1 summarizes the color index of different azo dyes.

6.3 Removal of Azo Dyes from Wastewater

A number of textiles, fibers, fabrics, plastics, papers, leathers, and other related industries use the azo dyes in order to color their products and also consume a large quantity of water during their production processes. Consequently, they generate a significantly high amount of azo dye-contaminated wastewaters. Generally, water quality is realized by public perception according to its color; therefore, the color is the first issue to be recognized in wastewaters. The existence of very small amounts of azo dyes in water (even less than 1 ppm for some azo dyes) is visible and undesirable [12, 21]. According to their extremely good solubility in water, they are regularly found in various quantities in industrial wastewaters. Because restrictions on the organic content in industrial wastewater effluents become increasingly stringent, the contaminated azo dyes are required to be removed from the wastewaters before being discharged. This is mainly because, as aforementioned, most of the azo dyes are toxic and even carcinogenic, and they can cause a serious danger when exposed to aquatic living organisms. Nevertheless, the azo dye-containing wastewaters are rather difficult to be treated since the azo dyes are recalcitrant organic compounds and resistant to aerobic wastewater-treating processes, as well as stable to light, heat, and oxidizing agents [22–24]. Hence, during the past decades, several physical, chemical, and biological methods have been developed to remove the azo dyes from wastewaters [12], as briefly described below.

6.3.1 Physical Treatment Methods

Various physical methods have been widely employed to treat the azo dye-containing wastewaters, especially membrane- and filtration-based processes (such as reverse osmosis, nanofiltration, and electrodialysis) and adsorption processes. Even though they are very important in terms of applications, some disadvantages for each process have been found. The major drawbacks of the membrane processes is their restricted lifetime of usage since fouling can gradually occur on the membrane surface, and the replacement cost is comparatively high. Moreover, the adsorption is one of the most popular methods to have been scientifically studied and practically applied for the removal of azo dye pollutants from wastewaters since the well-optimized operation in combination with the most efficient unit design of the adsorption processes can result in acceptably high azo dye removal efficiency [25]. Particularly, the adsorption processes seem to provide a relatively promising option for the treatment of azo dye-contaminated wastewaters in the cases that the solid adsorbents are inexpensive, and any pretreatment steps are not necessarily required prior to their use.

6.3.2 Chemical Treatment Methods

Chemical methods are also efficient for the removal of azo dyes contaminated in the wastewaters. The examples of important chemical methods are coagulation or flocculation combined with flotation and filtration, electroflotation, precipitation-flocculation, electrokinetic coagulation, conventional oxidation methods by oxidizing agents (such as ozone), and irradiation or electrochemical processes. These chemical methods are typically costly as compared to the physical methods, and although the azo dyes can be effectively removed, the formation and accumulation of sludges cause a secondary disposal problem. The excessive chemical use also definitely leads to a secondary pollution problem. Recently, other interesting techniques, known as advanced oxidation processes (AOPs), which are based on the generation of very powerful oxidizing agents, such as hydroxyl radicals, have been successfully applied for the azo dye degradation [26]. Even if these methods are efficient for the treatment of azo dye-contaminated wastewaters, they are still very costly and practically infeasible because the huge consumptions of chemicals and electricity are the main problems. However, in replacing the electrical energy, the other important sources of abundantly available and renewable energy, especially solar energy, can be employed in order to drive the AOP via photocatalysis, as explained in detail in the next section.

6.3.3 Biological Treatment Methods

Biological methods are so far the most practically attractive when compared to the aforementioned physical and chemical methods. Fungal decolorization, microbial degradation, adsorption by living or dead microbial biomass, and bioremediation systems are the important examples of biological methods that are generally used to treat the wastewaters. It is because several kinds of microorganisms, including bacteria, yeasts, algae, and fungi, are able to accumulate and degrade azo dye pollutants contaminated in the wastewaters. However, these methods still have some drawbacks. Biological treatment requires a large area of land and is restricted by toxicity limitation of some chemicals that may be generated and released. The less flexibility in process design and operation is also another main drawback [27]. Furthermore, even though many organic azo dye molecules can be partly degraded by biological methods, there are still several recalcitrant azo dyes that withstand the biodegradability due to their complex chemical structure [22].

All of the aforementioned methods have both advantages and drawbacks. Currently, no single process is capable of sufficiently treating the wastewaters, primarily because of the complicated characteristics of the contaminated azo dye compounds. In practical use, a combination of different methods is definitely required in order to achieve the desired effluent quality [28, 29]. Nevertheless, this chapter mainly focuses on photocatalysis, which is an effective method for degrading the azo dyes present in the wastewaters.

6.4 Photocatalysis and Photocatalytic Degradation of Azo Dyes

6.4.1 Semiconductor

A semiconductor is a substance which possesses an electrical conductivity between those of an insulator and a conductor. Basically, semiconductor substances have electronic band structure governed by their crystalline properties. A semiconductor, which can be employed as a photocatalyst, is typically an oxide of metals, such as TiO_2, $SrTiO_3$, ZnO, WO_3, and SnO_2 [30, 31]. For the electrons confined in a semiconductor, their actual energy distribution is determined by the Fermi level and temperature of the electrons. At absolute zero temperature, all of the electrons have energy below the Fermi energy, but at nonzero temperature, the energy levels are randomized, and some electrons have energy above the Fermi level. Among the electronic bands occupied with the electrons, the band having the highest energy level is universally defined as the valence band (VB), while the one outside of this is defined as the conduction band (CB). The energy difference between the valence band and the conduction band is defined as the band gap [31]. The overall schematic of band gap energy is shown in Fig. 6.1.

The band gap can be imaginably considered as a partition, where the electrons must jump over from the valence band to the conduction band for being free. The energy quantity required to jump over the partition is widely known as the band gap energy (E_g, normally in the unit of eV). Only the electrons capable of jumping over the

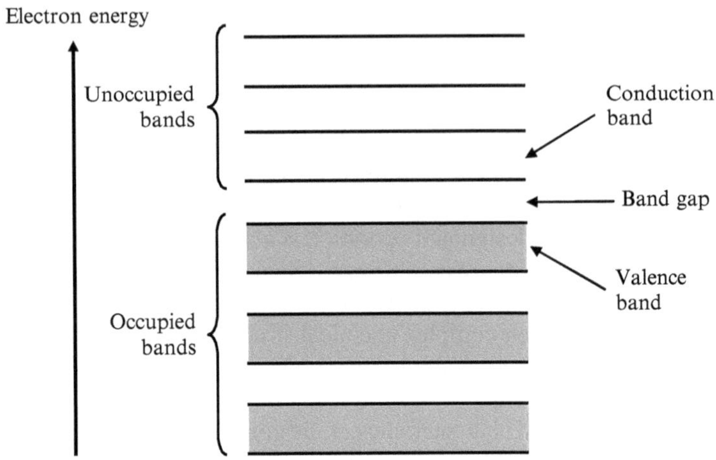

Fig. 6.1 The schematic of band gap energy [31]

Table 6.2 The band gap positions of some semiconductor photocatalysts [31, 32]

Semiconductor	Valence band level (eV)	Conduction band level (eV)	Band gap energy (eV)	Band gap wavelength (nm)
TiO_2	+3.1	−0.1	3.2	387
$SrTiO_3$	+3.1	−0.1	3.2	387
ZnO	+3.0	−0.2	3.2	387
WO_3	+3.0	+0.2	2.8	443
SnO_2	+4.1	+0.3	3.8	326

partition to enter the conduction band can freely move around. When a semiconductor is irradiated by light having suitable wavelengths with energy equal to or greater than its band gap energy, the valence band electrons can be excited to the conduction band. In the meantime, the positive holes are created and left in the valence band, and the number of holes is equal to that of electrons that have been excited to the conduction band. The valence band level, conduction band level, band gap energy, and band gap wavelength of some semiconductors are shown in Table 6.2.

6.4.2 Photocatalysts

Substances with semiconductor characteristics can be effectively employed as photocatalysts in photocatalytic systems because they are able to generate electrons and holes (both known as charge carriers) by absorbing appropriate light energies. The efficiency in separating the photoinduced charge carriers to prevent their recombination is a very crucial factor in determining the photocatalytic activity of a semiconductor photocatalyst. Titanium dioxide (titania, TiO_2) and strontium titanate

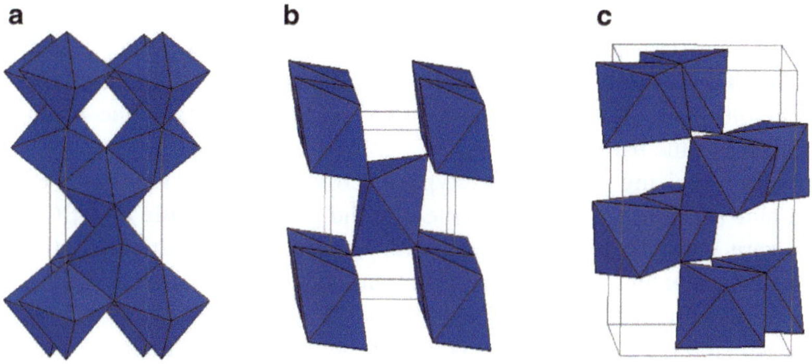

Fig. 6.2 Crystalline structures of (**a**) anatase, (**b**) rutile, and (**c**) brookite [36]

($SrTiO_3$) are the main examples of semiconductor photocatalysts in this chapter, and their properties, property modification, synthesis to achieve mesoporous-assembled structure, and application for photocatalytic azo dye degradation are explained.

TiO_2 is an important transition metal oxide, and it has received much increasing attention due to its chemical stability, photostability, nontoxicity, inexpensiveness, and other advantageous properties. Specifically, TiO_2 has been extensively employed in solar energy conversion, including photocatalysis applications [33–35]. The main three crystalline structures of TiO_2 found in nature are anatase (tetragonal), rutile (tetragonal), and brookite (orthorhombic). The structures of anatase, rutile, and brookite can be differentiated in terms of $TiO_2{}^{6-}$octahedral building blocks. The three crystalline structures differ by the distortion of each octahedral building block and by the connection pattern of chains of the octahedral building blocks. The anatase TiO_2 is formed from the connection of the vertices of octahedral building blocks; the rutile TiO_2 is formed from the connection of the edges of octahedral building blocks; and, the brookite TiO_2 is formed from the connection of both the vertices and edges of octahedral building blocks [36], as shown in Fig. 6.2.

In photocatalysis applications, both crystalline structures of anatase and rutile are usually used, while the anatase is reported to exhibit a superior photocatalytic activity as compared to the rutile for most of the photocatalytic reactions. The superior photocatalytic activity of the anatase phase can be explained by its slightly higher Fermi level and higher degree of surface hydroxylation (i.e., higher number of surface hydroxyl groups). In some cases of specific reactions, both crystalline phases are reported to possess relatively the same photocatalytic activity [37]. Moreover, the mixed crystalline phases (containing both anatase and rutile phases in structure) [38] or the rutile phase only [39] are also reported to exhibit a higher photocatalytic activity as compared to the anatase phase. The observed difference of the results may be reasonably contributed to the effects of various coexisting factors, such as specific surface area, pore size distribution, crystallite size, particle size distribution, particle agglomeration, and synthesis procedure, as well as the way to express the photocatalytic activity. Among various commercially available TiO_2 photocatalysts, the Degussa P-25 TiO_2, which is composed of a mixture of approximately 80% anatase and 20% rutile phases, is reported to be more active than both pure crystalline phases for many reactions. The enhancement in its photocatalytic activity may originate from its capability to efficiently separate the electron/hole pairs and prevent their undesired recombination due to the mixed phase nature of the particles.

Another promising photocatalyst type is perovskite-related material. Perovskite-related materials are simply represented by the general formula of ABO_3 (A represents rare earth or alkali element with or without its partial substitution by alkaline earth element, and B represents transition element, such as Ti, Ta, Ni, Fe, etc., with or without its partial substitution) [40, 41]. More specifically, strontium titanate ($SrTiO_3$) is a well-known cubic-perovskite oxide, which has received an increasing attention due to its excellent dielectric, photoelectric, optical, catalytic, and photocatalytic properties. $SrTiO_3$ is also considered to be applicable for various photocatalytic reactions. However, pure $SrTiO_3$ without a cocatalyst loading on its surface normally shows very low photocatalytic activity. Therefore, the modification of $SrTiO_3$ in several ways is necessary to obtain active photocatalysts for the photocatalytic degradation of azo dyes contaminated in the wastewaters. Particularly, the incorporation/loading of some effective transition and noble metals, such as Cr, Zr, Pt, and Au, into/onto $SrTiO_3$ has been proved to be a very efficient modification technique. For example, $SrTiO_3$ doped with Cr^{3+} led to an introduction of isolated energy levels within its band gap, so irradiated light can be absorbed at two levels, i.e., the band gap and sub-band gap, where the latter resulted in light absorption in the visible region [42]. The synthesis of Zr-doped $SrTiO_3$ by a sol-gel method and its photocatalytic degradation of methylene blue dye were also reported. It was shown that the 8% Zr content doped on $SrTiO_3$ provided the highest photocatalytic degradation rate due to free zirconium ions that induced $SrTiO_3$ to more efficiently absorb light [43].

6.4.3 Nanocrystalline Photocatalysts

Nanocrystalline photocatalysts are extremely small semiconductor particles, which are as small as few nanometers in particle size. During the past decades, the investigation of nanosized semiconductor particles and their photocatalytic applications has been one of the most attractive research areas in nanotechnology. The significance of these nanosized semiconductor particles stems from their unique physical, chemical, and photocatalytic properties. Interestingly, much research has revealed that some properties of nanosized semiconductor particles are different from those of the bulk materials. Nanosized particles possess properties falling into the transition region between the molecular and bulk phases. For the bulk materials, the electrons excited from the valence band to the conduction band by light absorption have comparatively high density of states, where they can move around with different kinetics energies. However, for the nanosized particles, the particle size is the same as or smaller than the size of the first excited state; therefore, the electrons and holes generated upon light irradiation cannot fit into such particle. Hence, as the size of the semiconductor particles is reduced below a critical value, the confinement of the generated charge carriers induces them to behave quantum characteristics. In the other words, this means that the bands split into separate electronic states in the valence and conduction bands, and the nanosized particles gradually behave in a similar manner to a huge atom. Nanosized semiconductor particles that possess size-dependent optical and electronic properties are widely known as quantized particles or quantum dots [44].

There are many advantages for the application of nanosized semiconductor particles. A main important advantage is the increase in band gap energy of semiconductor particles with decreasing particle size. As aforementioned, when the size of a semiconductor particle becomes lower than the critical diameter, the photogenerated charge carriers start to behave quantum characteristics, and the charge confinement produces separate electronic states. Consequently, there is a shift of the band edge and thus an increase in the band gap. This implies that the redox potentials of the valence band holes and the conduction band electrons can be adjusted by varying the size of the semiconductor particles. The increased driving force accordingly results in increasing the charge transfer rate at the particle surface. Hence, the use of nanosized semiconductor particles increases photocatalytic activity for many reaction systems, in which interfacial charge transfer is the rate-limiting step. Besides, the nanosized semiconductor particles can facilitate redox reactions to be able to readily occur, which cannot be easily enhanced in the bulk materials. The other benefit of nanosized particles is that the fraction of atoms located at the nanoparticle surface is very large, causing the nanosized particles to have high surface area-to-volume ratio, which in turn helps enhance their photocatalytic activity. Moreover, the nanosized particles require light with a shorter wavelength for band gap excitation, so a smaller portion of irradiated light is sufficiently useful for photocatalytic reactions [45].

6.4.4 Photocatalytic Degradation of Azo Dyes

There are two main mechanisms for azo dye degradation by photocatalysis using a semiconductor photocatalyst: photocatalytic oxidation under UV light irradiation and photosensitized oxidation under visible light irradiation [15].

6.4.4.1 Photocatalytic Oxidation

Basically, the conduction band electrons (e^-) and valence band holes (h^+) are generated when photocatalyst powders suspended in an aqueous solution are irradiated with light having energy equal to or higher than its band gap energy (UV light with $\lambda < 420$ nm in general cases), as mentioned before. The photogenerated electrons can both reduce the dissolved organic dye molecules and react with electron acceptors, such as O_2 adsorbed on the photocatalyst surface or dissolved in water, leading to its reduction to superoxide radical anion, $O_2^{\bullet -}$. The photogenerated holes can both oxidize the organic dye molecules to form oxidized species and react with OH^- or H_2O, leading to their transformation to hydroxyl radical, $OH^\bullet$. The resulting $O_2^{\bullet -}$ and $OH^\bullet$ radicals are reported to play a significant role in the photocatalytic degradation of azo dye molecules. The possible reactions occurring at the semiconductor photocatalyst surface to result in the azo dye degradation by photocatalytic oxidation can be expressed as follows [46–50]:

$$\text{Photocatalyst} + h\nu\,(\text{UV}) \rightarrow \text{Photocatalyst}\left(e_{CB}^- + h_{VB}^+\right)$$

$$\text{Photocatalyst}\left(h_{VB}^+\right) + H_2O \rightarrow \text{Photocatalyst} + H^+ + OH^\bullet$$

$$\text{Photocatalyst}\left(h_{VB}^+\right) + OH^- \rightarrow \text{Photocatalyst} + OH^\bullet$$

$$\text{Photocatalyst}\,(e_{CB}^-) + O_2 \rightarrow \text{Photocatalyst} + O_2^{\bullet -}$$

$$O_2^{\bullet -} + H^+ \rightarrow HO_2^\bullet$$

$$\text{Dye} + OH^\bullet,\ HO_2^\bullet,\ O_2^{\bullet -} \rightarrow \text{Degradation products}$$

$$\text{Dye} + h_{VB}^+ \rightarrow \text{Oxidation products}$$

$$\text{Dye} + e_{CB}^- \rightarrow \text{Reduction products}$$

The $O_2^{\bullet -}$ and $OH^\bullet$ radicals as well as $HO_2^\bullet$ radical subsequently formed, which possess an extremely strong oxidizing power, can oxidize most of azo dye molecules to less harmful end products, such as H_2O and CO_2. By this means, a lot of azo dyes that are intrinsically reactive toward hydroxyl radical can be degraded via photocatalysis, which is greatly affected by the valence band edge position of a semiconductor photocatalyst. The reduction pathway (i.e., direct reaction of azo dye with conduction band electrons) in the photocatalysis is also involved in the azo dye degradation, but in a less extent than the oxidation pathway (i.e., direct reaction of azo dye with valence band holes) [46, 51].

6.4.4.2 Photosensitized Oxidation

The detailed mechanism of photosensitized oxidation (also known as photoassisted degradation) by visible light irradiation ($\lambda > 420$ nm) totally differs from the above-mentioned mechanism under UV light radiation of photocatalytic oxidation. In this case, the mechanism starts with the excitation of the azo dye molecules adsorbed on the photocatalyst surface by visible light to their corresponding singlet or triplet states (^{1}Dye* or ^{3}Dye*), then followed by electron transfer from the excited dye molecules into the conduction band of the photocatalyst particles, whereas the dye molecules become unstable and are readily converted to the cationic dye radicals ($\text{Dye}^{\bullet +}$) that further undergoes their degradation. The possible reactions occurring at the semiconductor photocatalyst surface to result in the azo dye degradation by photosensitized oxidation can be expressed as follows [50, 52–55]:

$$\text{Dye} + h\nu\,(\text{visible}) \rightarrow {}^1\text{Dye}^* \text{ or } {}^3\text{Dye}^*$$

$${}^1\text{Dye}^* \text{ or } {}^3\text{Dye}^* + \text{Photocatalyst} \rightarrow \text{Dye}^{\bullet +} + \text{Photocatalyst}\,(e_{CB}{}^-)$$

$$\text{Photocatalyst}\,(e_{CB}{}^-) + O_2 \rightarrow O_2{}^{\bullet -} + \text{Photocatalyst}$$

$$\text{Dye}^{\bullet +} \rightarrow \text{Degradation products}$$

The cationic dye radicals can also both react with hydroxyl ions to cause their oxidation and react with $O_2{}^{\bullet -}$, $HO_2{}^{\bullet}$, or $HO^{\bullet}$ species to generate various intermediates that eventually lead to their degradation, as shown in the following:

$$\text{Dye}^{\bullet +} + OH^- \rightarrow \text{Dye} + HO^{\bullet}$$

$$\text{Dye} + 2HO^{\bullet} \rightarrow H_2O + \text{Oxidation products}$$

$$O_2{}^{\bullet -} + H^+ \rightarrow HO_2{}^{\bullet}$$

$$HO_2{}^{\bullet} + H^+ + \text{Photocatalyst}\,(e_{CB}{}^-) \rightarrow H_2O_2 + \text{Photocatalyst}$$

$$H_2O_2 + \text{Photocatalyst}\,(e_{CB}{}^-) \rightarrow HO^{\bullet} + HO^- + \text{Photocatalyst}$$

$$\text{Dye}^{\bullet +} + O_2{}^{\bullet -} \rightarrow \text{Degradation products}$$

$$\text{Dye}^{\bullet +} + HO_2{}^{\bullet}\,(\text{or } HO^{\bullet}) \rightarrow \text{Degradation products}$$

In experimental works, when either real or simulated solar light is employed, both photocatalytic and photosensitized mechanisms can simultaneously take place during the azo dye degradation, and both the photocatalyst and the light irradiation are required for the reaction. In the photocatalytic oxidation, the photocatalyst is irradiated and excited by UV light to initiate charge carrier formation. On the other hand, in the photosensitized oxidation, the dye molecules themselves rather than the photocatalyst are excited by visible light, followed by electron transfer into the photocatalyst conduction band. Nevertheless, it is still difficult to conclude that the photocatalytic oxidation mechanism is more capable of degrading azo dye molecules than the photosensitized oxidation mechanism since the photosensitized mechanism can help enhance the overall degradation efficiency and make the dye degradation more practical when using solar light [55].

6.5 Synthesis Procedures of Mesoporous-Assembled Nanocrystal Photocatalysts

6.5.1 Porous Materials

The International Union of Pure and Applied Chemistry (IUPAC) has classified different terminologies about porous materials with their definitions [56, 57] as

Table 6.3 Definitions about porous materials [56, 57]

Terminology	Definition
Porous material	Material with cavities or channels, which are deeper than their internal width
Micropore	Pore of internal width lower than 2 nm
Mesopore	Pore of internal width between 2 and 50 nm
Macropore	Pore of internal width higher than 50 nm
Pore size	Pore width (diameter of cylindrical pore or distance between opposite walls of slit)
Pore volume	Volume of pores determined by any given method
Surface area	Magnitude of total surface area determined by any given method under stated conditions

summarized in Table 6.3. As widely accepted, the pore size is identified as the pore width, i.e., the distance between two opposite walls of pore. The pore size has a clear and specific meaning when the geometrical shape of pores is well defined. However, in most cases, the effective pore size is commonly represented by the size of pores with the smallest dimension.

According to the IUPAC classification, porous materials can be generally divided into three major categories based on their principal pore size, as follows:

- Microporous materials (pore size lower than 2 nm) consist of amorphous silica and inorganic gel to crystalline materials, such as zeolites, aluminophosphates, gallophosphates, and related materials.
- Mesoporous materials (pore size between 2 and 50 nm) consist of the M41S family (e.g., MCM-41, MCM-48, MCM-50, etc.), other non-silica materials, such as double hydroxides, metal (e.g., titanium, zirconium, etc.) oxides and phosphates, and clays.
- Macroporous materials (pore size higher than 50 nm) consist of glass-related materials, aerogels, and xerogels.

Currently, microporous and mesoporous materials are regularly described as nanoporous materials. Even if mesoporous materials are proper for catalysis applications (including photocatalysis applications in this chapter), the pores of microporous materials are too small and may be easily blocked by metal particles during photocatalyst preparation with high metal loadings. The main characteristics and some important properties of microporous and mesoporous materials are given in Table 6.4. The significance of microporous materials is because of their broad industrial applications within catalytic areas, especially in petrochemistry and chemical synthesis [58]; however, the industrial applications of microporous materials are occasionally restricted by their relatively small pore sizes, as aforementioned. Much interest has been therefore devoted to mesoporous materials with the main attention to widen their catalysis applications to comparatively larger molecules. A few decades ago (1990s), the successful discovery of M41S materials has brought the focus of several researchers around the world to the synthesis of various mesoporous materials, which can be considered as multifunctional advanced materials [59, 60].

Table 6.4 Characteristics and some important properties of microporous and mesoporous materials

Material type	Microporous materials	Mesoporous materials
Pore size	<2 nm	2–50 nm
Main components for the synthesis	Silicon source, aluminum source, templating agent, solvent, and mineralizing agent	Silicon (or other metal) source, self-assembled templating agent, solvent, and catalyst (acid or base)
Advantages	Large surface area and adsorption capacity, the possibility of controlling the adsorption properties by tuning the hydrophobicity or hydrophilicity of the materials, the channel opening and cavities are in the range of many molecules of interest (0.5–1.2 nm), and the unique pore structure makes the materials able to recognize molecules	Controlled porosity, large accessible internal surface area and pore volume, a capacity that often leads to unusual magnetic, electronic, and optical properties, the ability to allow reactions with bulky substrates, grafting of large catalytically active species to the walls of the mesopores is achievable
Disadvantages	Not able to efficiently process molecules that are larger than their pore diameters (maximum 1–1.2 nm)	Poor thermal and hydrothermal stability

6.5.2 Sol-Gel Process

Sol-gel process has been extensively studied as an effective means to synthesize nanosized mesoporous materials. Because the mesoporous metal oxides, which are normally applied for catalysis/photocatalysis fields, can be thermally deactivated due to particle sintering and crystallite growth during their exposure to high-temperature treatment, and their catalytic/photocatalytic efficiency is strongly dependent on their specific surface area, the sol-gel process provides some unique advantages via a low-temperature process. It imparts a lower level of contamination of the synthesized materials and also yields a better stoichiometric and particle size control. In addition, this process does not require complicated equipment, and it provides a straightforward and efficient route for achieving nanosized particles, which are desirable for nano-photocatalysts with various functionalized properties [61–78]. Its molecular chemistry involves the formation of metal-oxo-polymer network via hydrolysis and polycondensation of metal precursors, such as metal alkoxides (i.e., the most important metal precursors for the sol-gel process), as follows [79]:

$$M - OR + H_2O \rightarrow M - OH + ROH$$

$$M - OH + M - OR \rightarrow M - O - M + R - OH$$

$$M - OH + M - OH \rightarrow M - O - M + H_2O$$

where M = Ti, Zr, etc., and R = alkyl group. Each individual stage during the sol-gel process, including the formation of colloidal sol particles, formation of gel network, drying of wet gel, and calcination of dried gel, can result in different degrees of crystallite growth and particle agglomeration. Therefore, the sol-gel process needs to be suitably controlled in order to attain highly active and reproducible mesoporous metal oxide powders. The relative hydrolysis and polycondensation rates greatly affect the structural and physicochemical properties of the resulting materials. Basically, the original precipitates derived from the sol-gel process possess amorphous structure in nature; thus, they require further thermal treatment to induce crystallization. The calcination step unavoidably leads to crystallite growth and particle agglomeration, and may also stimulate phase transformation. Thus, a surfactant is generally applied to prevent the particle agglomeration. The sol-gel process proceeded in the presence of a surfactant can enable a good control and dispersity of mesoporous metal oxide particle size.

There are several factors that affect the sol-gel process and the resulting mesoporous materials, particularly the reactivity (hydrolysis and polycondensation rates) of metal alkoxides, pH of the sol-gel medium, water-to-alkoxide ratio, sol-gel temperature, and nature of solvent and additive (e.g., acid catalyst, base catalyst, and templating surfactant). For example, the water-to-alkoxide ratio strongly influences the sol-gel chemistry and the structural properties of the hydrolyzed gel. A higher water-to-alkoxide ratio in the sol-gel medium guarantees a more complete hydrolysis of metal alkoxides, which consequently results in reducing the crystallite size of the calcined material. The utilization of acid or base catalyst is an alternative way to manipulate the hydrolysis and polycondensation rates of metal alkoxides–in the other words, the sol-gel reaction rates. It was reported that a smaller particle size and a narrower pore size distribution with a smaller mean pore diameter of a mesoporous metal oxide material were obtained via the sol-gel process with HCl addition in a suitable quantity [61]. The size of alkoxide group in metal alkoxides also exerts a significant effect in the particle size control. The metal alkoxides possessing more bulky alkoxide groups offer a slower hydrolysis rate due to their steric hindrance, which is very beneficial for the preparation of very fine colloidal sol particles [80].

6.5.3 Surfactant-Assisted Templating Sol-Gel Process

Surfactant-assisted templating sol-gel process is an effective approach for synthesizing mesoporous metal oxide materials based on the utilization of cooperatively co-assembled organic surfactant molecules in the form of micelles or liquid crystal phases as templates to support the growth of metal oxide materials induced by the sol-gel process [63]. After the complete gel formation, the templating surfactant molecules can be subsequently eliminated from the gel framework, e.g., by thermal treatment and chemical extraction. For this approach, the micellar structure of surfactant molecules and charge compatibility between surfactant head group and metal alkoxide species at the micellar interface are the critical parameters to

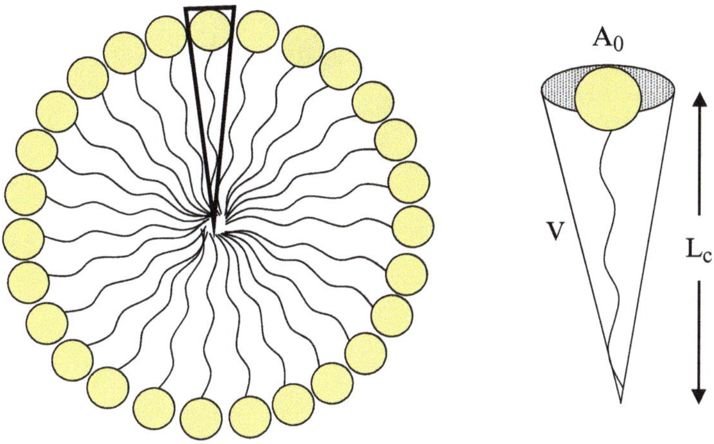

Fig. 6.3 Surfactant micellar structure [82]

control the framework structure and pore size of the resulting mesoporous metal oxide materials. Therefore, the synthesis of mesoporous metal oxide materials with desired shape and pore size can be succeeded by properly manipulating the processing conditions, e.g., types of metal alkoxide precursor and templating surfactant, pH of sol-gel medium, calcination condition, etc. Moreover, the critical packing parameter (P) is an important characteristic of surfactant micellar structure that can govern the formation of mesoporous structure of the obtained metal oxide materials [81, 82]. As seen from the surfactant micellar structure shown in Fig. 6.3, the P value is defined as:

$$P = \frac{V}{A_0 L_c}$$

where V is the topographical volume of hydrophobic portion of surfactant molecule, A_0 is the effective area of surface head group, and L_c is the critical length of hydrophobic tail of surfactant molecule.

The relationship between the P value and self-assembled surfactant shape is as follows: $0 \leq P \leq 1/3$ for sphere, $1/3 \leq P \leq 1/2$ for cylinder, and $1/2 \leq P \leq 1$ for bilayer. However, the P value is also strongly affected by some system conditions, e.g., pH, ionic strength, and temperature.

For the metal oxide particles synthesized by the surfactant-assisted templating sol-gel process, their mesoporous structure is formed by loose aggregation of crystalline metal oxide nanoparticles. The pore structures of metal oxide nanocrystals connect randomly and lack an apparent long-range order in the pore arrangement among the crystalline metal oxide nanoparticles [83]. It is apparent that the mesopores among the aggregated metal oxide nanoparticles exhibits the type IV IUPAC nitrogen adsorption-desorption isotherms with a clear hysteresis loop, which is the main feature of mesoporous-structured materials (as can be seen in the next

Fig. 6.4 A proposed possible model of the mesopore formation for the metal oxide nanocrystals synthesized by the surfactant-assisted templating sol-gel process

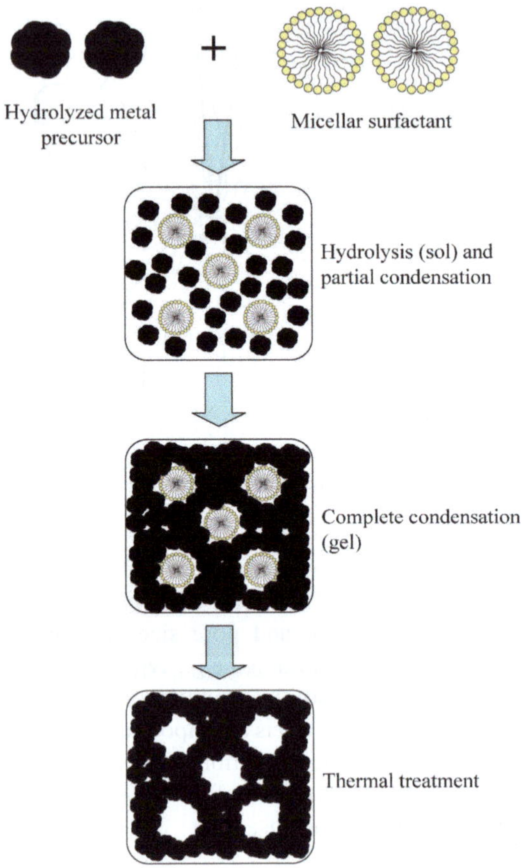

section). A proposed possible model of the mesopore formation for the metal oxide nanocrystals synthesized by the surfactant-assisted templating sol-gel process is shown in Fig. 6.4. As proposed in the model, the hydrolysis and condensation reactions of the metal precursor (either without or with chemical modification) take place when it is added to the micellar surfactant aqueous solution, and the organic-inorganic (surfactant-metal oxide) composite materials are formed. The interaction between surfactant micelles and the aggregated inorganic moieties occurs through a mechanism involving self-assembly with simultaneous condensation of inorganic moieties around the micellar surfactant head groups [84–86]. After the complete gel formation via the condensation, the micellar surfactant molecules highly dispersed and embedded inside the inorganic gel structure are thermally removed, rendering the inorganic moieties to coalesce and crystallize to form assembled particles with interconnected mesopores. The resulting products are therefore called mesoporous-assembled metal oxide nanocrystals.

6.6 Application of Mesoporous-Assembled Nanocrystal Photocatalysts for Azo Dye Degradation

The surfactant-assisted templating sol-gel process can be efficiently applied for synthesis of mesoporous-assembled nanocrystal photocatalysts. In this chapter, the synthesis and application for azo dye degradation of mesoporous-assembled TiO_2, $SrTiO_3$, and $SrTi_xZr_{1-x}O_3$ nanocrystal photocatalysts are mainly focused.

6.6.1 Mesoporous-Assembled TiO_2 Nanocrystal Photocatalyst

The mesoporous-assembled TiO_2 nanocrystal photocatalyst can be synthesized via the surfactant-assisted templating sol-gel process by using tetraisopropyl orthotitanate (TIPT) as the Ti precursor modified with acetylacetone (ACA) as the modifying agent and laurylamine hydrochloride (LAHC) as the structure-templating surfactant [63]. Firstly, the TIPT is homogeneously mixed with the ACA with the TIPT-to-ACA molar ratio of 1:1, leading to a modification of the co-ordination number of the Ti atom from 4 to 5, as shown in the equation below. This modification results in the change of mixture solution color from colorless to yellow. The resulting ACA-modified TIPT is much less active to the moisture in air than the TIPT itself, and can be applied as the starting mixture for the sol-gel synthesis.

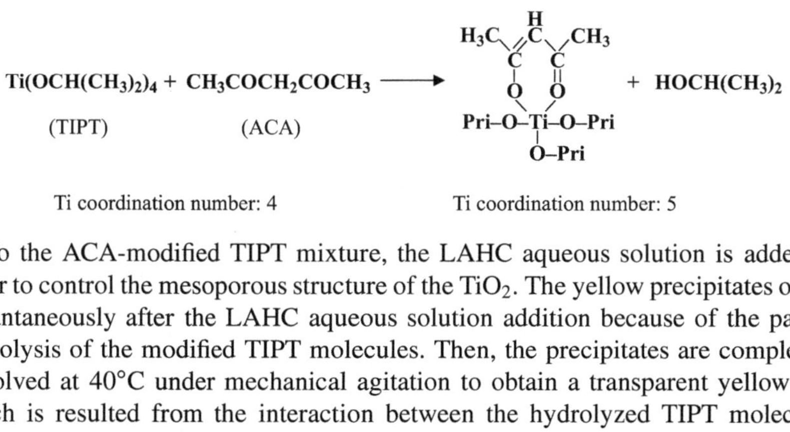

To the ACA-modified TIPT mixture, the LAHC aqueous solution is added in order to control the mesoporous structure of the TiO_2. The yellow precipitates occur instantaneously after the LAHC aqueous solution addition because of the partial hydrolysis of the modified TIPT molecules. Then, the precipitates are completely dissolved at 40°C under mechanical agitation to obtain a transparent yellow sol, which is resulted from the interaction between the hydrolyzed TIPT molecules and the hydrophilic head groups of micellar LAHC molecules, as well as from the suspension of these interacted species in the solution as microscopic colloidal particles. After that, a gelation is completed by thermal curing of the sol at 80°C due to the condensation of the modified TIPT molecules attached to the LAHC head groups. During the condensation process, the ACA molecules are detached from the modified TIPT molecules due to the presence of a transparent yellow liquid layer above the gel. Then, the gel is dried at 80°C to obtain dried gel (zero gel) of TiO_2. Finally, the dried gel is calcined at equal to or above 500°C, which is a sufficiently high temperature for both the LAHC surfactant removal from the gel network and the photocatalyst crystallization process, to yield the mesoporous-

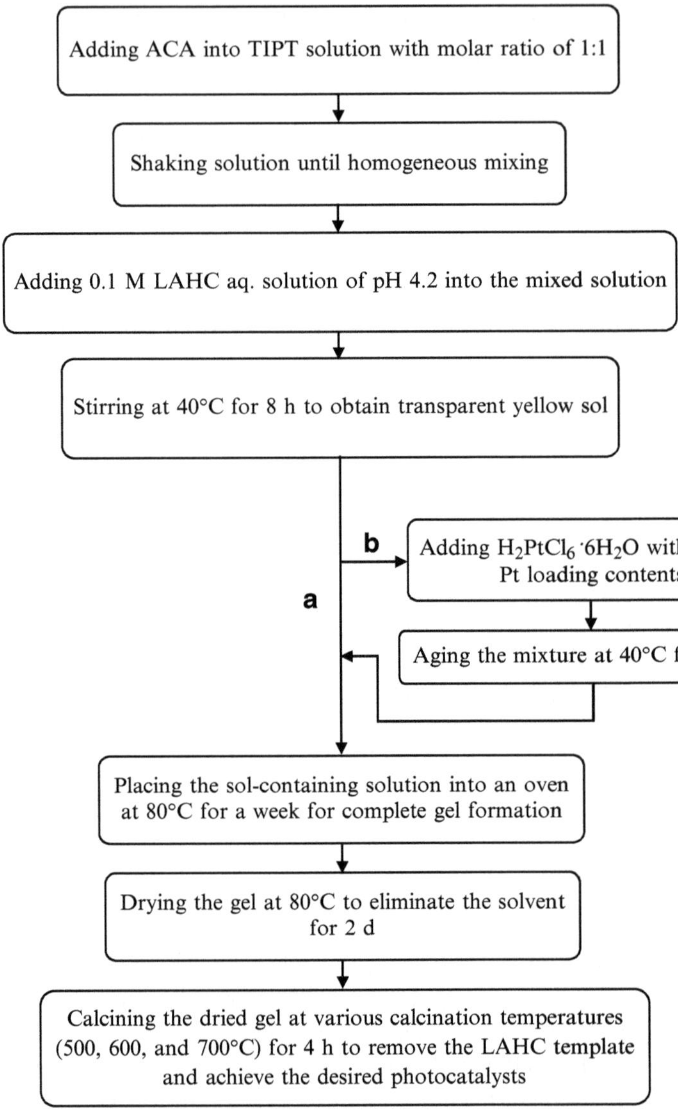

Fig. 6.5 Synthesis procedure for mesoporous-assembled TiO_2 nanocrystal photocatalysts (**a**) without and (**b**) with Pt loading by single-step sol-gel method

assembled TiO_2 nanocrystal photocatalysts. In case of platinum (Pt) loading on the surface of mesoporous-assembled TiO_2 nanocrystal photocatalyst by a single-step sol-gel method, a Pt precursor (e.g., $H_2PtCl_6 \cdot 6H_2O$) is added to the obtained transparent yellow sol prior to the gelation step. The mixture is then homogenized at 40°C, followed by the gelation, gel drying, and calcination steps in the same manner as explained above to yield the Pt-loaded mesoporous-assembled TiO_2 nanocrystal photocatalyst. The synthesis procedure for the mesoporous-assembled TiO_2 photocatalysts without and with Pt loading by the single-step sol-gel method is shown in Fig. 6.5.

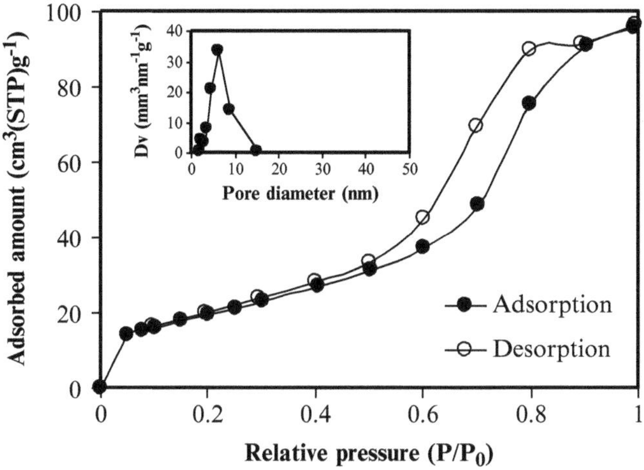

Fig. 6.6 N_2 adsorption-desorption isotherms and pore size distribution of the mesoporous-assembled TiO_2 nanocrystal photocatalyst calcined at 500°C (*Inset*: Pore size distribution) [87]

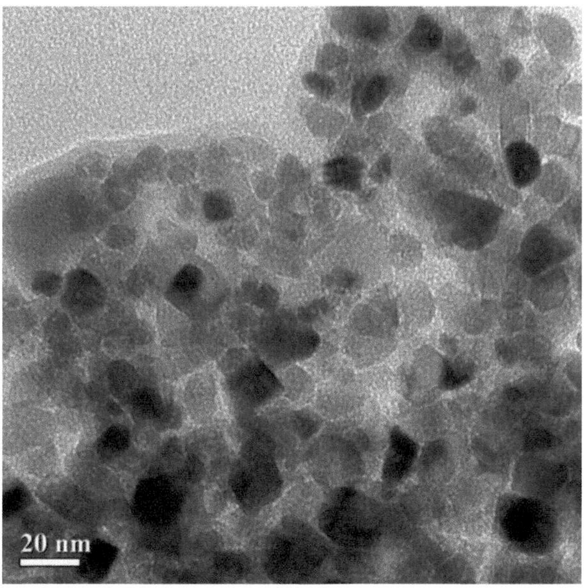

Fig. 6.7 TEM micrograph of the mesoporous-assembled TiO_2 nanocrystal photocatalyst calcined at 500°C [87]

The synthesized TiO_2 nanocrystal photocatalysts possess very narrow and monomodal pore size distribution in the mesoporous range (i.e., pore diameter between 2 and 50 nm), as verified by the N_2 adsorption-desorption isotherm and pore size distribution analyses in Fig. 6.6. The particle size of the synthesized mesoporous TiO_2 is also very uniform in a nanosized range, as observed by the TEM analysis in Fig. 6.7.

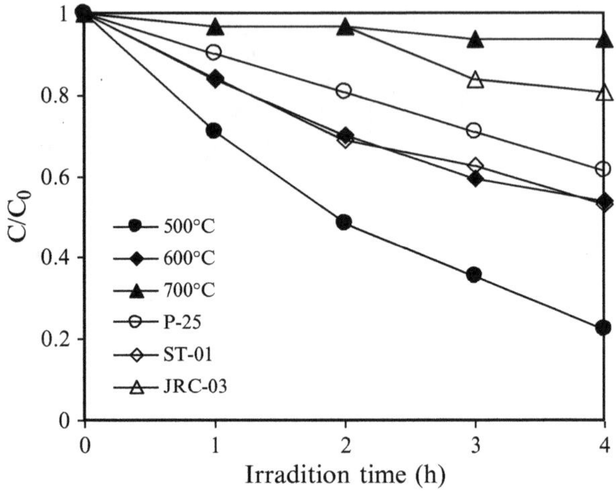

Fig. 6.8 Photocatalytic degradation of MO dye in terms of relative concentration (C/C_0) over the synthesized mesoporous-assembled TiO_2 nanocrystal photocatalysts calcined at various temperatures (500°C, 600°C, and 700°C) and commercial TiO_2 photocatalysts as a function of irradiation time (photocatalyst dosage = 2 g/l; reaction volume = 80 ml; initial MO concentration = 5 mg/l; and UV lamp power = 44 W) [87]

The synthesized mesoporous-assembled TiO_2 nanocrystal can be employed as an efficient photocatalyst for degrading various kinds of azo dyes, e.g., methyl orange (MO) monoazo dye, acid yellow (AY) monoazo dye, and acid black (AB) diazo dye. As the first instance, the photocatalytic degradation of MO dye over the synthesized mesoporous-assembled TiO_2 nanocrystal photocatalysts calcined at various temperatures is shown in Fig. 6.8, as compared to that over various commercially available non-mesoporous TiO_2 powders (i.e., Degussa P-25 TiO_2, Ishihara ST-01 TiO_2, and Ishihara JRC-03 TiO_2) [87]. It is clearly revealed that the mesoporous-assembled TiO_2 nanocrystal photocatalyst calcined at a suitable temperature of 500°C provides the highest photocatalytic MO dye degradation efficiency, which is considerably higher than that of the commercial TiO_2 powders. It can therefore be inferred that the mesoporosity with uniform pore size distribution of the synthesized TiO_2 nanocrystal photocatalyst is a prime and very important factor in achieving its acceptably high photocatalytic dye degradation performance (this will also be confirmed for the $SrTiO_3$ and $SrTi_xZr_{1-x}O_3$ nanocrystal photocatalysts, as shown in the next section). Moreover, a suitable Pt loading content of 0.6 wt.% can improve the photocatalytic MO dye degradation performance of the synthesized mesoporous-assembled TiO_2 nanocrystal photocatalyst [88], as shown in Fig. 6.9. This is because of the electron transfer-mediating capability of the loaded Pt particles that prevents the electron-hole recombination at both surface and bulk traps.

As the synthesized mesoporous-assembled TiO_2 nanocrystal photocatalyst calcined at 500°C shows a superior photocatalytic activity, its photocatalytic performance for simultaneously degrading a mixture of AY and AB dyes with more complex molecular structures is also evaluated [89], as shown in Fig. 6.10.

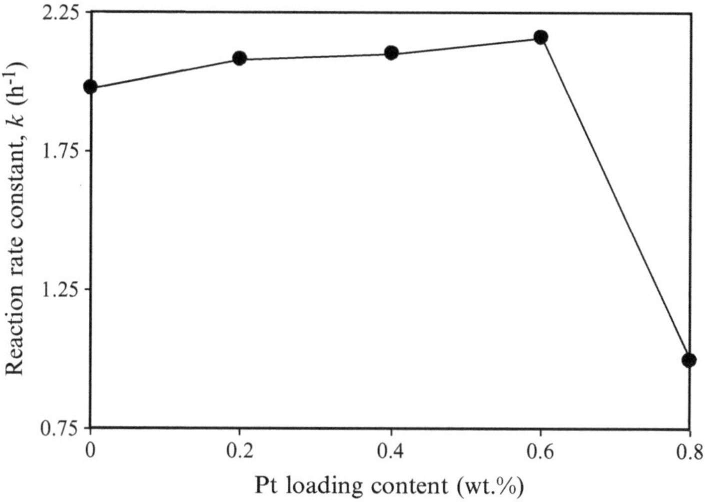

Fig. 6.9 Photocatalytic degradation of MO dye in terms of pseudo-first-order rate constant over the synthesized Pt-loaded mesoporous-assembled TiO_2 nanocrystal photocatalysts prepared by single-step sol-gel method and calcined at 500°C as a function of Pt loading content (photocatalyst dosage = 2 g/l; reaction volume = 80 ml; initial MO concentration = 5 mg/l; and UV lamp power = 44 W) [88]

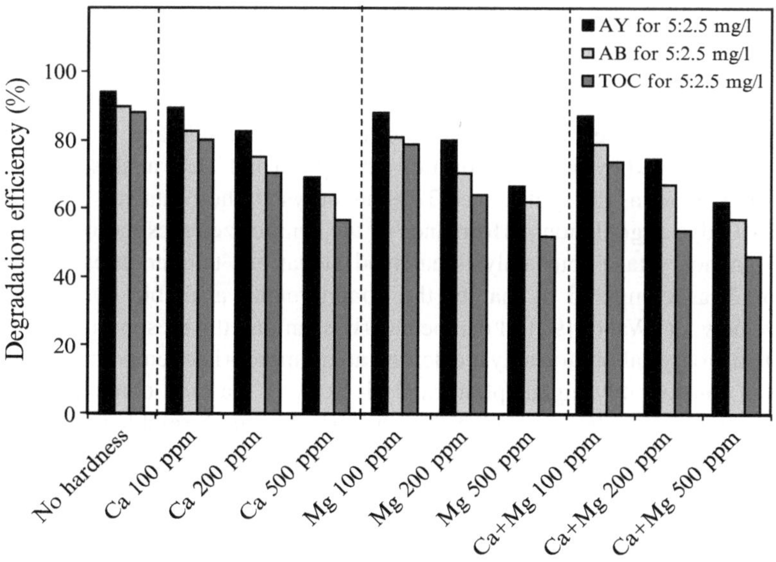

Fig. 6.10 Photocatalytic degradation of a mixture of AY and AB dyes in the presence of hardness in terms of degradation efficiency over the synthesized mesoporous-assembled nanocrystal TiO_2 photocatalysts calcined at 500°C as functions of type and concentration of hardness (photocatalyst dosage = 10 g/l; reaction volume = 80 ml; initial AY concentration = 5 mg/l; initial AB concentration = 2.5 mg/l; irradiation time = 60 min; and UV lamp power = 44 W) [89]

It is apparent that even in the presence of hardness found in tap water (normally found at a concentration of approximately 100–200 ppm in the form of either Ca, Mg, or Ca-Mg mixture up to 500 ppm, which represents an extremely high hardness concentration), the synthesized mesoporous-assembled TiO_2 nanocrystal photocatalyst still exhibits a considerably high photocatalytic degradation activity for both dyes present in the aqueous mixture. This verifies that the mesoporous-assembled TiO_2 nanocrystal photocatalyst possesses an exceptional photocatalytic property and can be efficiently employed for the azo dye removal from wastewaters.

6.6.2 Mesoporous-Assembled $SrTiO_3$ and $SrTi_xZr_{1-x}O_3$ Nanocrystal Photocatalysts

The aforementioned surfactant-assisted templating sol-gel process, with some modifications, can also be adapted to synthesize the mesoporous-assembled $SrTiO_3$ and $SrTi_xZr_{1-x}O_3$ nanocrystal photocatalysts by using strontium titanate ($Sr(NO_3)_2$) as the Sr precursor, tetraisopropyl orthotitanate (TIPT) as the Ti precursor, zirconium butoxide (ZRB, 80 wt.% in 1-butanol) as the Zr precursor, acetylacetone (ACA) as the modifying agent, laurylamine hydrochloride (LAHC) as the structure-templating surfactant, and ethanol (EtOH) as the co-solvent [90, 91]. The synthesis procedure for the mesoporous-assembled $SrTiO_3$ and $SrTi_xZr_{1-x}O_3$ nanocrystal photocatalysts without and with Pt loading by the single-step sol-gel method is illustrated in Fig. 6.11.

The synthesized $SrTiO_3$ and $SrTi_xZr_{1-x}O_3$ nanocrystal photocatalysts also possess narrow pore size distribution in the mesoporous range and uniform particle sizes, as confirmed by the N_2 adsorption-desorption and TEM analyses [90, 91], in the same manner as the synthesized TiO_2 nanocrystal photocatalyst. The photocatalytic MO dye degradation performance of the synthesized mesoporous-assembled $SrTiO_3$ nanocrystal photocatalysts calcined at various temperatures is shown in Fig. 6.12, as compared to that of the commercially available non-mesoporous $SrTiO_3$ powder (Wako) [90]. It can be clearly seen that the mesoporous-assembled $SrTiO_3$ nanocrystal photocatalyst calcined at an appropriate temperature of 700°C shows a considerably higher photocatalytic activity than the commercial $SrTiO_3$ powder. It can then be concluded that the mesoporous-assembled structure with a uniform pore size distribution of $SrTiO_3$ plays the most important role that governs the photocatalytic activity of the $SrTiO_3$ photocatalyst. The photocatalytic activity of the mesoporous-assembled $SrTiO_3$ nanocrystal photocatalyst can be further greatly enhanced by partly substituting Ti atoms in the $SrTiO_3$ photocatalyst by a suitable quantity of Zr atoms to obtain the $SrTi_{0.9}Zr_{0.1}O_3$ photocatalyst [91], as shown in Fig. 6.13 for the photocatalytic degradation of AB

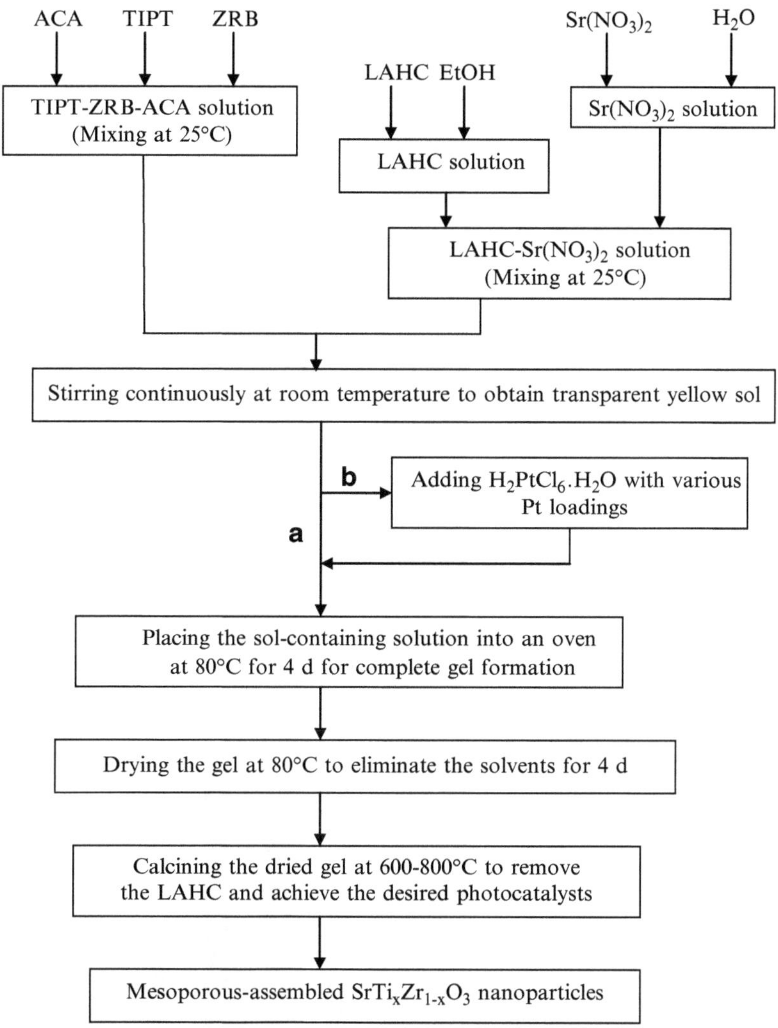

Fig. 6.11 Synthesis procedure for mesoporous-assembled $SrTiO_3$ and $SrTi_xZr_{1-x}O_3$ nanocrystal photocatalysts (**a**) without and (**b**) with Pt loading by single-step sol-gel method

dye. Moreover, the Pt loading with an appropriate content of 1 wt.% can additionally improve the photocatalytic AB dye degradation performance of the mesoporous-assembled $SrTi_{0.9}Zr_{0.1}O_3$ nanocrystal photocatalyst, as shown in Fig. 6.14. Hence, the surfactant-assisted templating sol-gel process can be a promising and effective means for synthesizing very highly photoactive mesoporous-assembled photocatalysts for azo dye degradation application.

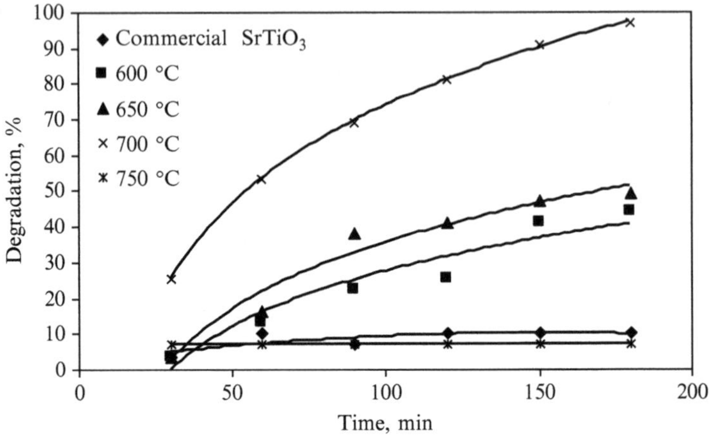

Fig. 6.12 Photocatalytic degradation of MO dye in terms of degradation efficiency over the synthesized mesoporous-assembled $SrTiO_3$ nanocrystal photocatalysts calcined at various temperatures (600°C, 650°C, 700°C, and 750°C) and commercial $SrTiO_3$ photocatalyst as a function of irradiation time (photocatalyst dosage = 5 g/l; reaction volume = 15 ml; initial MO concentration = 10 mg/l; and UV lamp power = 15 W) [90]

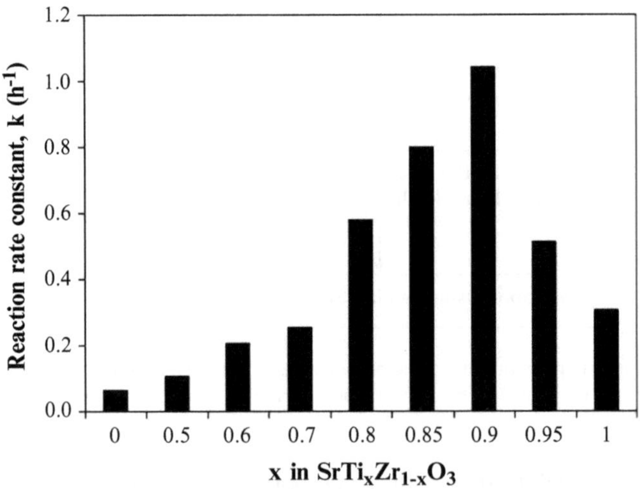

Fig. 6.13 Photocatalytic degradation of AB dye in terms of pseudo-first-order rate constant over the synthesized mesoporous-assembled $SrTi_xZr_{1-x}O_3$ nanocrystal photocatalysts calcined at 700°C as a function of Ti content (x) in $SrTi_xZr_{1-x}O_3$ (photocatalyst dosage = 5 g/l; reaction volume = 100 ml; initial AB concentration = 15 mg/l; irradiation time = 4 h; and UV lamp power = 44 W) [91]

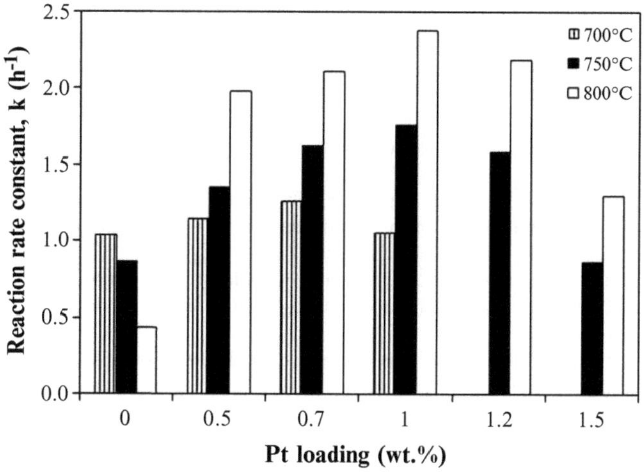

Fig. 6.14 Photocatalytic degradation of AB dye in terms of pseudo-first-order rate constant over the synthesized Pt-loaded mesoporous-assembled $SrTi_{0.9}Zr_{0.1}O_3$ nanocrystal photocatalysts prepared by single-step sol-gel method and calcined at various temperatures as a function of Pt loading content (photocatalyst dosage = 5 g/l; reaction volume = 100 ml; initial AB concentration = 15 mg/l; irradiation time = 4 h; and UV lamp power = 44 W) [91]

Acknowledgment The Asahi Glass Foundation, Japan; the Research Grant for Mid-Career University Faculty (RMU) co-funded by the Thailand Research Fund (TRF), the Commission on Higher Education, Thailand, and Chulalongkorn University, Thailand; the Ratchadapisek Somphot Endowment Fund, Chulalongkorn University, Thailand; the Research Unit of Applied Surfactants for Separation and Pollution Control under the Ratchadapisek Somphot Endowment Fund, Chulalongkorn University, Thailand; the Research Unit of Petrochemical and Environmental Catalysis under the Ratchadapisek Somphot Endowment Fund, Chulalongkorn University, Thailand; the Sustainable Petroleum and Petrochemicals Research Unit, Center for Petroleum, Petrochemicals, and Advanced Materials, Chulalongkorn University, Thailand; and the Petroleum and Petrochemical College, Chulalongkorn University, Thailand are acknowledged.

References

1. Sokolowsak-Gajda J, Freeman HS, Reife A (1996) Synthetic dyes based on environmental considerations. Part 2: iron complexed formazan dyes. Dye Pigment 30:1–20
2. Hsu TC, Chiang CS (1997) Activated sludge treatment of dispersed dye factory wastewater. J Environ Sci Health A Environ Sci Eng Tox Hazard Subst Control A 32:1921–1932
3. Kabdach I, Tunay O, Orhon D (1999) Wastewater control and management in a leather tanning district. Water Sci Technol 40:261–267
4. Slampova A, Smela D, Vondrackova A, Jancarova I, Kuban V (2001) Determination of synthetic colorants in foodstuffs. Chem Listy 95:163–168
5. Wrobel D, Boguta A, Ion RM (2001) Mixtures of synthetic organic dyes in a photoelectronic cell. J Photochem Photobiol A Chem 138:7–22
6. Wang Y (2000) Solar photocatalytic degradation of eight commercial dyes in TiO_2 suspension. Water Res 34:990–994

7. Guillard C, Lachheb H, Houas A, Ksibi M, Elaloui E, Herrmann JM (2003) Influence of chemical structure of dyes, of pH and of inorganic salts on their photocatalytic degradation by TiO_2 comparison of the efficiency of powder and supported TiO_2. J Photochem Photobiol A Chem 158:27–36
8. Netpradit S, Thiravetyan P, Towprayoon S (2004) Adsorption of three azo reactive dyes by metal hydroxide sludge: effect of temperature, pH, and electrolytes. J Colloid Interface Sci 270:255–261
9. Yang Y, Wu Q, Guo Y, Hu C, Wang E (2005) Efficient degradation of dye pollutants on nanoporous polyoxotungstate–anatase composite under visible-light irradiation. J Mol Catal A Chem 225:203–212
10. Marugan J, Lopez-Munoz MJ, van Grieken R, Aguado J (2007) Photocatalytic decolorization and mineralization of dyes with nanocrystalline TiO_2/SiO_2 materials. Ind Eng Chem Res 46:7605–7610
11. Khataee AR, Kasiri MB (2010) Photocatalytic degradation of organic dyes in the presence of nanostructured titanium dioxide: influence of the chemical structure of dyes. J Mol Catal A Chem 328:8–26
12. Robinson T, McMullan G, Marchant R, Nigam P (2001) Remediation of dyes in textile effluent: a critical review on current treatment technologies with a proposed alternative. Bioresour Technol 77:247–255
13. Fu Y, Viraraghavan T (2001) Fungal decolorization of dye wastewaters: a review. Bioresour Technol 79:251–262
14. Forgacsa E, Cserhatia T, Oros G (2004) Removal of synthetic dyes from wastewaters: a review. Environ Int 30:953–971
15. Konstantinou IK, Albanis TA (2004) TiO_2-assisted photocatalytic degradation of azo dyes in aqueous solution: kinetic and mechanistic investigations: a review. Appl Catal B Environ 49:1–14
16. van der Zee FP, Villaverde S (2005) Combined anaerobic–aerobic treatment of azo dyes – a short review of bioreactor studies. Water Res 39:1425–1440
17. http://www.colour-index.org/. (Accessed on 30 April 2012)
18. http://www.sdc.org.uk/. (Accessed on 30 April 2012)
19. http://www.aatcc.org/. (Accessed on 30 April 2012)
20. Mary HG (1991) Encyclopedia of chemical technology. Wiley-Interscience, New York
21. Banat IM, Nigam P, Singh D, Marchant R (1996) Microbial decolorization of textile dye-containing effluents: a review. Bioresour Technol 58:217–227
22. Ravi Kumar MNV, Sridhari TR, Bhavani KD, Dutta PK (1998) Trends in color removal from textile mill effluents. Colorage 40:25–34
23. O'Neill C, Hawkes FR, Hawkes DL, Lourenco ND, Pinheiro HM, Delee W (1999) Colour in textile effluents–sources, measurement, discharge consents and simulation: a review. J Chem Technol Biotechnol 74:1009–1018
24. Sun Q, Yang L (2003) The adsorption of basic dyes from aqueous solution on modified peat-resin particle. Water Res 37:1535–1544
25. Dabrowski A (2001) Adsorption, from theory to practice. Adv Colloid Interface Sci 93: 135–224
26. Gogate PR, Pandit AB (2004) A review of imperative technologies for wastewater treatment I: oxidation technologies at ambient conditions. Adv Environ Res 8:501–551
27. McMullan G, Meehan C, Conneely A, Kirby N, Robinson T, Nigam P, Banat IM, Marchant R, Smyth WF (2001) Microbial decolourisation and degradation of textile dyes. Appl Microbiol Biotechnol 56:81–87
28. Marco A, Esplugas S, Saum G (1997) How and why to combine chemical and biological processes for wastewater treatment. Water Sci Technol 35:231–327
29. Oller I, Malato S, Sánchez-Pérez JA (2011) Combination of advanced oxidation processes and biological treatments for wastewater decontamination – a review. Sci Total Environ 409: 4141–4166

30. Lin Y, Yuan G, Liu R, Zhou S, Sheehan SW, Wang D (2011) Semiconductor nanostructure-based photoelectrochemical water splitting: a brief review. Chem Phys Lett 507:209–215
31. Robertson PKJ (1996) Semiconductor photocatalysis: an environmentally acceptable alternative production technique and effluent treatment process. J Cleaner Prod 4:203–212
32. Moreira ML, Andrs J, Longo VM, Li MS, Varela JA, Longo F (2009) Photoluminescent behavior of $SrZrO_3/SrTiO_3$ multilayer thin films. Chem Phys 473:293–298
33. Hoffmann MR, Martin ST, Choi W, Bahnemann DW (1995) Environmental applications of semiconductor photocatalysis. Chem Rev 95:69–96
34. Linsebigler AL, Lu G, Yater JT Jr (1995) Photocatalyst on TiO_2 surfaces: principles, mechanisms, and selected results. Chem Rev 95:735–758
35. Mills A, Le Hunte S (1997) An overview of semiconductor photocatalysis. J Photochem Photobiol A Chem 108:1–35
36. Carp O, Huisman CL, Reller A (2004) Photoinduced reactivity of titanium dioxide. Prog Solid State Chem 32:33–177
37. Deng X, Yue Y, Gao Z (2002) Gas phase photo-oxidation of organic compounds over nanosized TiO_2 photocatalysts by various preparations. Appl Catal B Environ 39:135–147
38. Mugglie DS, Ding L (2001) Photocatalytic performance of sulfated TiO_2 and Degussa P-25 TiO_2 during oxidation of organics. Appl Catal B Environ 32:181–188
39. Mills A, Lee SK, Lepre A (2003) Photodecomposition of ozone sensitised by a film of titanium dioxide on glass. J Photochem Photobiol A Chem 155:199–205
40. Tanaka H, Misono M (2001) Advances in designing perovskite catalysts. Curr Opin Solid State Mater Sci 5:381–387
41. Fang CM, Ahuja R (2006) Structures and stability of ABO_3 orthorhombic perovskites at the Earth's mantle conditions from first-principles theory. Phys Earth Planet Inter 157:1–7
42. Ashokkumar M (1998) An overview on semiconductor particulate systems for photoproduction of hydrogen. Int J Hydrog Energy 23:427–438
43. Chen L, Zhang S, Wang L, Xue D, Yin S (2008) Photocatalytic activity of Zr:$SrTiO_3$ under UV illumination. J Cryst Growth 311:735–737
44. Kamat PV (1995) Tailoring nanostructured thin films. Chemtech
45. Chen X, Mao SS (2007) Titanium dioxide nanomaterials: synthesis, properties, modifications, and applications. Chem Rev 107:2891–2959
46. Tanaka K, Padermpole K, Hisanaga T (2000) Photocatalytic degradation of commercial azo dyes. Water Res 34:327–333
47. Galindo C, Jacques P, Kalt A (2000) Photodegradation of the aminoazobenzene Acid Orange 52 by three advanced oxidation processes: UV/H_2O_2, UV/TiO_2 and VIS/TiO_2: comparative mechanistic and kinetic investigations. J Photochem Photobiol A Chem 130:35–47
48. Bianco-Prevot A, Baiocchi C, Brussino MC, Pramauro E, Savarino P, Augugliaro V, Marci G, Palmisano L (2001) Photocatalytic degradation of Acid Blue 80 in aqueous solutions containing TiO_2 suspensions. Environ Sci Technol 35:971–976
49. Houas A, Lachheb H, Ksibi M, Elaloui E, Guillard C, Hermann JM (2001) Photocatalytic degradation pathway of methylene blue in water. Appl Catal B Environ 31:145–157
50. Daneshvar N, Salari D, Khataee AR (2003) Photocatalytic degradation of azo dye Acid Red 14 in water: investigation of the effect of operational parameters. J Photochem Photobiol A Chem 157:111–116
51. Vinodgopal K, Bedja I, Hotchandani S, Kamat PV (1994) A photocatalytic approach for the reductive decolorization of textile azo dyes in colloidal semiconductor suspensions. Langmuir 10:1767–1771
52. Vinodgopal K, Wynkoop D, Kamat P (1996) Environmental photochemistry on semiconductor surfaces: photosensitized degradation of a textile azo dye, Acid Orange 7, on TiO_2 particles using visible light. Environ Sci Technol 30:1660–1666
53. Zhang F, Zhao J, Shen T, Hidaka H, Pelizzetti E, Serpone N (1998) TiO_2-assisted photodegradation of dye pollutants II. Adsorption and degradation kinetics of eosin in TiO_2 dispersions under visible light irradiation. Appl Catal B Environ 15:147–156

54. Chen F, Xie Y, Zhao J, Lu G (2001) Photocatalytic degradation of dyes on a magnetically separated photocatalyst under visible and UV irradiation. Chemosphere 44:1159–1168
55. Epling GA, Lin C (2002) Photoassisted bleaching of dyes utilizing TiO_2 and visible light. Chemosphere 46:561–570
56. Ishizaki K, Komarneni S, Nanko M (1988) Porous materials: process technology and applications. Kluwer Academic, London
57. Rouquerol F, Rouquerol J, Sing K (1999) Adsorption by powders and porous solid: principle, methodology, and applications. Academic, San Diego
58. Stein A (2003) Advanced materials progress report on advances in microporous and mesoporous solids–highlights of recent progress. Adv Mater 15:763–775
59. Kresge CT, Leonowicz ME, Roth WJ, Vartuli JC, Beck JS (1992) Ordered mesoporous molecular sieves synthesized by a liquid-crystal template mechanism. Nature 359:710–712
60. Beck JS, Vartuli JC, Roth WJ, Leonowicz ME, Kresge CT, Schmitt KD, Chu TTW, Olson DH, Sheppard EW, McCullen SB, Higgins JB, Schlenker JL (1992) A new family of mesoporous molecular sieves prepared with liquid crystal templates. J Am Chem Soc 114:10834–10843
61. Wang CC, Ying J (1999) Sol-gel synthesis and hydrothermal processing of anatase and rutile titania nanocrystals. Chem Mater 11:3113–3120
62. Wu JC, Chen CH (2004) A visible-light response vanadium-doped titania nanocatalyst by sol-gel method. J Photochem Photobiol A Chem 163:509–515
63. Sreethawong T, Suzuki Y, Yoshikawa S (2005) Synthesis, characterization, and photocatalytic activity for hydrogen evolution of nanocrystalline mesoporous titania prepared by surfactant-assisted templating sol-gel process. J Solid State Chem 178:329–338
64. Sreethawong T, Susuki Y, Yoshikawa S (2005) Photocatalytic evolution of hydrogen over mesoporous TiO_2 supported NiO photocatalyst prepared by single-step sol-gel process with surfactant template. Int J Hydrog Energy 30:1053–1062
65. Sreethawong T, Yoshikawa S (2005) Comparative investigation on photocatalytic hydrogen evolution over Cu-, Pd-, and Au-loaded mesoporous TiO_2 photocatalysts. Catal Commun 6:661–668
66. Sreethawong T, Ngamsinlapasathian S, Suzuki Y, Yoshikawa S (2005) Nanocrystalline mesoporous Ta_2O_5-based photocatalysts prepared by surfactant-assisted templating sol-gel process for photocatalytic H_2 evolution. J Mol Catal A Chem 235:1–11
67. Sreethawong T, Yoshikawa S (2006) Enhanced photocatalytic hydrogen evolution over Pt supported on mesoporous TiO_2 prepared by single-step sol-gel process with surfactant template. Int J Hydrog Energy 31:786–796
68. Sreethawong T, Laehsalee S, Chavadej S (2008) Comparative investigation of mesoporous- and non-mesoporous-assembled TiO_2 nanocrystals for photocatalytic H_2 production over N-doped TiO_2 under visible light irradiation. Int J Hydrog Energy 33:5947–5957
69. Sreethawong T, Laehsalee S, Chavadej S (2009) Use of Pt/N-doped mesoporous-assembled nanocrystalline TiO_2 for photocatalytic H_2 production under visible light irradiation. Catal Commun 10:538–543
70. Puangpetch T, Sreethawong T, Yoshikawa S, Chavadej S (2009) Hydrogen production from photocatalytic water splitting over mesoporous-assembled $SrTiO_3$ nanocrystal-based photocatalysts. J Mol Catal A Chem 312:97–106
71. Sreethawong T, Junbua C, Chavadej S (2009) Photocatalytic H_2 production from water splitting under visible light irradiation using Eosin Y-sensitized mesoporous-assembled Pt/TiO_2 nanocrystal photocatalyst. J Power Sources 190:513–524
72. Puangpetch T, Sreethawong T, Chavadej S (2010) Hydrogen production over metal-loaded mesoporous-assembled $SrTiO_3$ nanocrystal photocatalysts: effects of metal type and loading. Int J Hydrog Energy 35:6531–6540
73. Puangpetch T, Sommakettarin P, Chavadej S, Sreethawong T (2010) Hydrogen production from water splitting over Eosin Y-sensitized mesoporous-assembled perovskite titanate nanocrystal photocatalysts under visible light irradiation. Int J Hydrog Energy 35: 12428–12442

74. Puangpetch T, Chavadej S, Sreethawong T (2011) Mesoporous-assembled V_2O_5 nanosheet synthesized via a surfactant-modified sol-gel technique and its photocatalytic H_2 production activity under visible light irradiation. Powder Technol 208:37–41
75. Onsuratoom S, Chavadej S, Sreethawong T (2011) Hydrogen production from water splitting under UV light irradiation over Ag-loaded mesoporous-assembled TiO_2-ZrO_2 mixed oxide nanocrystal photocatalysts. Int J Hydrog Energy 36:5246–5261
76. Sakulkhaemaruethai S, Sreethawong T (2011) Synthesis of mesoporous-assembled TiO_2 nanocrystals by a modified urea-aided sol-gel process and their outstanding photocatalytic H_2 production activity. Int J Hydrog Energy 36:6553–6559
77. Wongwisate P, Chavadej S, Gulari E, Sreethawong T, Rangsunvigit P (2011) Effects of monometallic and bimetallic Au-Ag supported on sol-gel TiO_2 on photocatalytic degradation of 4-chlorophenol and its intermediates. Desalination 272:154–163
78. Puangpetch T, Chavadej S, Sreethawong T (2011) Hydrogen production over Au-loaded mesoporous-assembled $SrTiO_3$ nanocrystal photocatalyst: effects of molecular structure and chemical properties of hole scavengers. Energy Convers Manag 52:2256–2261
79. Brinker CJ, Scherer GW (1989) Sol-gel science. Academic, New York
80. Murakami Y, Matsumoto T, Takasu Y (1999) Salt catalysts containing basic anions and acidic cations for the sol–gel process of titanium alkoxide: controlling the kinetics and dimensionality of the resultant titanium oxide. J Phys Chem B 103:1836–1840
81. Israelachvili JN, Mitchell DJ, Ninham BW (1976) Theory of self-assembly of hydrocarbon amphiphiles into micelles and bilayers. J Chem Soc Faraday Trans II 72:1525–1568
82. Nagarajan R (2002) Molecular packing parameter and surfactant self-assembly: the neglected role of the surfactant tail. Langmuir 18:31–38
83. Zheng JY, Pang JB, Qiu KY, Wei Y (2001) Synthesis and characterization of mesoporous titania and silica-titania materials by urea templated sol-gel reaction. Microporous Mesoporous Mater 49:189–195
84. Oliver S, Kuperman A, Coombs N, Lough A, Ozin GA (1995) Lamellar aluminophosphates with surface patterns that mimic diatom and radiolarian microskeletons. Nature 378:47–50
85. Tanev PT, Pinnavaia TJ (1996) Biomimetic templating of porous lamellar silicas by vesicular surfactant assemblies. Science 271:1267–1269
86. Yuan ZY, Liu JQ, Peng LM, Su BL (2000) Morphosynthesis of vesicular mesostructured calcium phosphate under electron irradiation. Langmuir 18:2450–2452
87. Jantawasu P, Sreethawong T, Chavadej S (2009) Photocatalytic activity of nanocrystalline mesoporous-assembled TiO_2 photocatalyst for degradation of methyl orange monoazo dye in aqueous wastewater. Chem Eng J 155:223–233
88. Jantawasu P (2008) Photocatalytic decomposition of methyl orange azo dye using nanostructured TiO_2 photocatalyst. M.S. thesis, The Petroleum and Petrochemical College, Chulalongkorn University, Thailand
89. Wongkalasin P, Chavadej S, Sreethawong T (2011) Photocatalytic degradation of mixed azo dyes in aqueous wastewater using mesoporous-assembled TiO_2 nanocrystal synthesized by a modified sol-gel process. Colloid Surf A Physocochem Eng Asp 384:519–528
90. Puangpetch T, Sreethawong T, Yoshikawa S, Chavadej S (2008) Synthesis and photocatalytic activity in methyl orange degradation of mesoporous-assembled $SrTiO_3$ nanocrystals prepared by sol-gel method with the aid of structure-directing surfactant. J Mol Catal A Chem 287:70–79
91. Khunrattanaphon P, Chavadej S, Sreethawong T (2011) Synthesis and application of novel mesoporous-assembled $SrTi_xZr_{1-x}O_3$-based nanocrystal photocatalysts for azo dye degradation. Chem Eng J 170:292–307

Chapter 7
Microwave-Assisted Organic Pollutants Degradation

Ackmez Mudhoo

7.1 Water Pollutants and Environmental Concerns

The global environment is under much stress due to fast urbanization, industrialization, and rapid population growth which in turn set a limit in the usage of the limited natural resources. The problems are compounded by drastic changes that have been taking place in the lifestyle and habits of people. The nature and the magnitude of the environmental problems are ever changing, bringing new challenges and creating a constant need for developing newer and more appropriate technologies helpful in the remediation of polluted strata.

The problems of the environment can be classified as follows and may be traced to be one or more of the following either directly or indirectly: Waste generation (sewage, wastewaters, kitchen wastes, industrial solid wastes and effluents, agricultural wastes, and food wastes) and the use of chemicals for various purposes in the form of insecticides, pesticides, herbicides, chemical fertilizers, toxic products, and by-products from chemical industries. Waste generation is a side effect of consumption and anthropogenic production activities and tends to increase with economic advance. What is of concern is the increased presence of toxic organic chemicals such as halogen aliphatics, aromatics, polychlorinated biphenyls (PCBs) and other organic and inorganic pollutants which may reach air, water, or soil and affect the environment in several ways, ultimately threatening the self-regulating capacity of the biosphere [1, 2]. These pollutants may be present in high levels at their respective points of discharge or may remain low or fixed in some chemical form, but be still highly toxic for the receiving bodies as a result of bioaccumulation up the food chain, wherein their concentrations increase as they pass through the food chain [3, 4].

A. Mudhoo (✉)
Department of Chemical and Environmental Engineering, Faculty of Engineering, University of Mauritius
e-mail: ackmezchem@yahoo.co.uk; a.mudhoo@uom.ac.mu

S.K. Sharma and R. Sanghi (eds.), *Advances in Water Treatment and Pollution Prevention*, DOI 10.1007/978-94-007-4204-8_7,

Organic pollutant inputs have increased in recent decades, and the result has been degradation of water quality in many rivers, lakes, and coastal oceans. This degradation shows up in the disruption of natural aquatic ecosystems and the consequent loss of their component species as well as the amenities that these ecosystems once provided to society [5]. The most common impairments of surface waters is eutrophication caused by excessive inputs of phosphorus (P) and nitrogen (N), and severe cellular damage and mutation caused to the existing fauna in these waters due to assimilation of toxic contaminants. Impaired waters are defined as those that have been contaminated with organic and/or inorganic pollutants and have thus become unsuitable for designated uses such as drinking, irrigation, industry, recreation, or fishing [5].

7.2 Remediation Approaches

7.2.1 Steps in Pollutant Remediation

There are three main approaches in dealing with contaminated sites. The first step consists in the identification of the problem, followed by the assessment of the nature and degree of the hazard, and finally selecting and implementing the best choice of remedial action. The need to remediate sites which have been contaminated with organic (and inorganic pollutants) has led to the development of a number of (new) technologies that emphasize the degradation, detoxification, and/or destruction of the organic pollutants [6, 7] rather than the conventional approach of end of pipe disposal. All the more, rapid developments in understanding activated sludge processes and wastewater remediation have paved the way in research and applied research for the exploitation and design of different strategies for studying the degradation of such pollutants.

7.2.2 Conventional Technologies Versus Bioremediation

The remediation technologies for organic pollutants may be broadly lumped in two major categories. These are conventional remediation and bioremediation options. Some of these conventional techniques for organic pollutants remediation are advanced oxidation processes [8], electrokinetic remediation [9], UV-assisted photocatalytic degradation, adsorption on activated carbon, coagulation/flocculation processes; nanofiltration membrane technology [10], high-performance catalytic air oxidation, ozonation, and photoelectrocatalysis [11]. On the other hand, bioremediation techniques which are more cost effective as compared to the conventional treatment options embrace the concept of Green Technology and Environmental Sustainability. Bioremediation is in principle the engineered use of microorganisms

Table 7.1 Selected studies on the bioremediation of organic pollutants

Pollutant(s)	Bioremediation technique	Reference
Phenanthrene (PHEN), and 1,2,4,5-tetrachlorobenzene (TeCB)	Biosorption by *Escherichia coli*	[17]
Pentachlorophenol (PCP)	Biosorption by *Aspergillus niger* biomass	[18]
Anionic reactive dyes were C.I. Reactive Black 8, C.I. Reactive Brown 9, C.I. Reactive Green 19, C.I. Reactive Blue 38, and C.I. Reactive Blue 3	Biosorption by *Aspergillus niger*, *Aspergillus japonica*, *Rhizopus nigricans*, *Rhizopus arrhizus*, and *Saccharomyces cerevisiae* biomass	[19]
Direct brown MR (DBMR) azo dye in the textile industry	Biodegradation of dye by *Acinetobacter calcoaceticus* NCIM 2890	[20]
Methyl red, orange II, G-Red (FN-3G), basic cationic, and basic fuchsin	Biodegradation by *Chlorella vulgaris*, *Lyngbya lagerlerimi*, *Nostoc lincki*, *Oscillatoria rubescens*, *Elkatothrix viridis*, and *Volvox aureus*	[21]
Phenanthrene (Phen), anthracene (Anth), and benzo(a)pyrene (BaP)	Vermicomposting using *Eisenia fetida*	[22]
4-Nonylphenol (known endocrine disruptor)	Degradation during biosolids composting	[23]
Fungicides and triazoles	Dissipation of pesticides during composting	[24]
Pirimiphos-methyl	Degradation of pollutant during composting of greenhouse tomato plant residues	[25]
Simazine, terbuthylazine, cyanazine and prometryn	Degradation of pollutant in soil amended with olive cake, compost and vermicompost of olive cake	[26]
Agricultural pesticides 2,4-dichlorophenoxyacetic acid (2,4-D) and isoproturon	Biodegradation potential during anaerobic digestion process	[27]
Isoproturon and 2,4-D	Biodegradability of herbicides in sequencing batch reactors (anaerobic systems)	[28]
Pentachlorophenol and methanol	Anaerobic biological treatment of pentachlorophenol and methanol in a horizontal-flow anaerobic immobilized biomass (HAIB) reactor	[29]

and microbial processes to degrade and detoxify environmental contaminants [12, 13]. Vidali [14] has proposed the following classification of microorganisms involved in bioremediation processes: Aerobic microbes, anaerobic bacteria, ligninolytic fungi, and methylotrophs. Some of the biological remediation technologies/techniques such as biosorption, bioaugmentation, biostimulation, biopulping, biodeterioration, biobleaching, bioaccumulation, biotransformation, bioattenuation, vermicomposting and composting [15] anaerobic digestion biotechnology and phytoremediation are being actively studied [16] for the degradation of organic contaminants for pollutant media remediation. Table 7.1 lists some recent studies which have employed bioremediation options for the degradation/detoxification of organic pollutants.

7.3 Novel Green Remediation Options

Green Technology, emanating directly from Green Chemistry may be described as the utilization of a set of principles that reduces or eliminates the use or generation of hazardous substances in the design, manufacture, and application of chemical products [30]. In practice, Green Chemistry is taken to cover a broader range of issues than the definition itself may imply.

7.3.1 Green Technology Principles

Anastas and Warner [31] have developed "The Twelve Principles of Green Chemistry" that serve as valuable benchmark guidelines for practicing chemists, researchers, and engineers in developing and assessing how green a synthesis, compound, process, or technology is. These principles are related to the concepts of prevention, atom economy, less hazardous chemical syntheses, designing safer chemicals, safer solvents and auxiliaries, design for energy efficiency, use of renewable feedstocks, reduce derivatives, catalysis, design for degradation, real-time analysis for pollution prevention, and inherently safer chemistry for accident prevention. Green chemistry is an essential part of green engineering. The definitions of green chemistry and green engineering share many commonalities, and the application of both chemistry and engineering principles is needed to advance the goals of environmental sustainability [32]. A practical definition of green engineering proposed by Kirchhoff [32] is the design, commercialization, and use of processes and products that are feasible and economical while minimizing pollution at the source and risk to human health and the environment. While green chemistry focuses on the design of chemical products and processes that reduce or eliminate the use and generation of hazardous substances, it also lays down the ground plan for the design of the green engineering technologies needed to implement sustainable products, processes, and systems [32]. Sonochemistry and microwave irradiation are two crucial branches of green chemistry, and their application in organic pollutant remediation is increasingly being studied. While the focus of this chapter shall be on microwave irradiation and pollutants degradation, a succinct outline of the nature and application of sonochemistry is herein made for the nonexpert. Sonochemistry is chemistry that deals with sonic waves on chemical reactions.

7.3.2 A Window on Sonochemistry

Decomposition and removal of anthropogenic organic contaminants from surface waters, groundwaters, sediments, wastewaters, and soils are also very important in environmental remediation. The unique mechanism of ultrasound combines simultaneous oxidation, thermolysis, shear degradation of shock waves, microjets pitting, and enhanced mass transfer and mixing together. As a result, sonication is a very attractive and interesting technique in environmental remediation, especially for decontamination of recalcitrant and hazardous compounds. Cavitation is the formation, growth, and the implosion of bubbles in a liquid by ultrasound. Ultrasound refers to inaudible sound waves with frequencies in the range of 16 kHz to 500 MHz, greater than the upper limit of human hearing. Like any sound wave, ultrasound is propagated via a series of compression and rarefaction waves induced in the molecules of the medium through which it passes. Compression cycles push molecules together, while expansion cycles pull them apart. At sufficiently high power, the rarefaction cycle may exceed the attractive forces of the molecules of the liquid and cavitation bubbles will form.

According to Feng and Aldrich [33], the likely mechanism of ultrasonic desorption can be explained by considering the different effects of ultrasound in heterogenous media. First, the high temperatures in localized hot spots enhance the breaking of physical bonds between the adsorbate (contaminants) and the adsorbent surface. Second, acoustic cavitation produces high-speed microjets and high-pressure shock waves that impinge on the surface and erode the adsorbate [34, 35]. Eventually, ultrasound produces acoustic vortex microstreaming within the pores of the solid particles, as well as the solid–liquid interface and this phenomenon arises by the increase in momentum brought about as the liquid absorbs energy from the propagating sound waves, even in the absence of cavitation [36]. Ultrasonication not only assists the desorption of the contaminants from the soil but also promotes the formation of the strong oxidant, •OH radical [37]. Ultrasonication can hence destroy the contaminants through oxidation by free radicals and pyrolysis processes. There are over 100 anthropogenic contaminants that have been studied by sonolytic degradation [38–40]. Due to limited spaces, only selected studies on the sonolytic degradation of organic contaminants are summarized in Table 7.2.

The next sections substantiate the essential characteristics of microwave power and discuss the research and application of microwave irradiation as a green remediation technique employed for the degradation of a number of organic pollutants including textile dyes, pesticides, herbicides, xenobiotics, PAHs, PCBs, and other persistent organic pollutants for the remediation of contaminated waters and wastewaters.

7.4 Microwave Power: Essentials and Applications

7.4.1 Characteristics of Microwave Power

Microwaves (frequencies of 0.3–300 GHz and wavelengths of 1 m to 1 mm) lie between radio wave frequencies and infrared frequencies in the electromagnetic spectrum. Domestic and industrial microwave ovens generally operate at a

Table 7.2 Ultrasound-assisted degradation of contaminants in real and synthetic effluents

Contaminant(s)	Degradation conditions	Degradation performance	Reference
Polycyclic aromatic hydrocarbons (PAHs) in a petrochemical industry wastewater	Ambient conditions (25°C), increasing sonication time (0–150 min), temperature (30–60°C), dissolved oxygen (DO, 2–10 mg L^{-1}) and hydrogen peroxide (H_2O_2, 100–2,000 mg L^{-1}) concentrations tested on the removal of PAHs	Maximum PAH removals were 80.2%, 91%, 98.5%, and 98% at 25°C, 60°C, DO concentration of 6 mg L^{-1} and H_2O_2 concentration of 2,000 mg L^{-1}, respectively, after 150 min sonication Sonication alone provides PAH removals varying between 88% and 92% without DO PAH sonodegradation appeared to be pseudo-first order in PAHs naphthalene (NAP), acenaphthylene (ACL), phenanthrene (PHE), pyrene (PY), and benz[*b*] fluoranthene	[41]
Ethylenediamine-tetraacetic acid (EDTA)–copper wastewater	Electrochemical performance of ultrasound integrated with an electrodeposition technique for the reclamation of wastewater	The ultrasound + electrodeposition technique could effectively remove copper (95.6%, w/w) and decompose the EDTA (84% COD removal) from the wastewater containing EDTA–copper complex compounds under certain operational control	[42]
Ammonia-contaminated simulated wastewater from strippers at petroleum refinery	High ultrasound frequency sonication at 2.4 and 1.7 MHz conducted to study the effect on removal of ammonia	Sonication could remove ammonia with 5% concentration to meet the local standard of treated wastewater within less than 2 h for 0.080 L solution	[43]
Nitrobenzene (NB) in wastewater	Removal of NB by slag containing titania under ultrasound irradiation	Slag titania removed NB catalytically under sonication with 18 g L^{-1} high titania slag, and removal efficiency of NB could reach 100% alter 140 min under 45 kHz ultrasonic irradiation	[44]

(continued)

Table 7.2 (continued)

Contaminant(s)	Degradation conditions	Degradation performance	Reference
Phenol in wastewater	Ultrasound alone and with ultrasound associated with electrolysis	With a 20 kHz sonication, the electrochemical oxidation of phenol in NaCl media allows the conversion of 75% of initial phenol within 10 min of treatment	[45]
		At 500 kHz a conversion of 95% of the initial phenol was obtained within the same treatment time, and final products of degradation were acetic and chloroacrylic acids	
4-Chlorophenol, 2,4-dichlorophenol, [aryl-$^2H_3$]2,4-dichlorophenol, 4-chloro-3,5-dimethylphenol, 4-fluorophenol, 2,4,6-trinitrotoluene, 2-amino-4,6-dinitrotoluene and 4-amino-2,6-dinitrotoluene in dilute aqueous solution	Ultrasonic mineralization studied as functions of substrate structure and concentration, bulk phase temperature, pH, and the presence of co-solutes such as detergents and humic acids	All substrates were found to degrade sonochemically, as evidenced by the release of Cl^- and NO_3^-, respectively	[46]
		Chloride release from chlorophenols was approximately proportional to substrate total chlorine content, irrespective of structural differences, and reached 80% of the theoretical limit. Fluoride release from 4-fluorophenol was ca. tenfold lower than that of chloride from 4-chlorophenol	
2,4-Dibromophenol in aqueous solutions	Combined effect of sonication and MW in a new flow reactor	Complete degradation within 6 h, at which time organic compounds were no longer detectable	[47]
	Degradation of 2,4-dibromophenol (0.1 g L^{-1} in water) by Fenton's reagent	Even if no Fenton's reagent was added, about 50% of 2,4-dibromophenol was degraded after 3 h irradiation	

frequency of 2.45 GHz corresponding to a wavelength of 12.2 cm and energy of 1.02×10^{-5} eV [48]. Microwaves can be reflected, transmitted, and/or absorbed. The absorbed MW energy is converted into heat within the material, resulting in an increase in temperature. Gases, liquids, and solids can interact with microwaves and be heated. Under certain conditions, gases can be excited by microwaves to form plasmas that also can be useful for processing.

An important characteristic of microwave heating is the phenomenon of "hotspot" formation, whereby regions of very high temperature form due to nonuniform heating [49]. This thermal instability arises because of the nonlinear dependence of the electromagnetic and thermal properties of the material on temperature [50]. The formation of standing waves within the microwave cavity results in some regions being exposed to higher energy than others. This results in an increased rate of heating in these higher energy areas due to the nonlinear dependence. Cavity design is an important factor in the control, or the utilization of this hotspot phenomenon.

Their absorption may be increased by adding absorbing constituents (e.g., silicon carbide, carbon, and organic binders), altering their microstructures and defect structures, changing their form (e.g., bulk vs. powder), or changing the frequency of the incident radiation. The latter often is not feasible due to the relative unavailability of equipment. Fortunately, absorption of poorly absorbing ceramics also can be improved by increasing their temperatures. This phenomenon has led to the development of hybrid heating, where the ceramics are heated initially using conventional methods. Hybrid heating can be achieved either by using an independent heat source, such as a gas or electric furnace in combination with microwaves, or through the use of an external susceptor that couples with the microwaves. In the latter, the material is exposed simultaneously to microwave energy and radiant conventional heat produced by the susceptor. Microwaves can be absorbed by ceramics either through polarization or conduction processes. Polarization involves short-range displacement of charge through formation and rotation of electric dipoles (or magnetic dipoles, if present). Conduction requires long-range (compared to rotation) transport of charge. Both processes give rise to absorption losses in certain frequency ranges.

7.4.2 Applications and Merits of Microwave Irradiation

Microwave energy has been widely used in several domestic, industrial, and medical applications such as food sterilization, organic/inorganic syntheses, polymerization, dehydration, analyses and extraction, and biological destruction [51]. The potential applications of microwave energy as remedial alternatives for various types of wastes and diverse contamination of soils, sludge, or wastewaters have also been growing field of research during the last decade [52]. The application of microwave energy has proven to be superior to the use of conventional heating on accelerating rates of biochemical/chemical reactions, improving yields, selectively activating or suppressing reaction pathways [51], allowing for greater control of the heating or drying process and reducing equipment size and wastes, and environmental remediation.

7.5 Organic Contaminants in Water and Wastewaters

Wastewaters are waterborne solids and liquids discharged into sewers that represent the wastes of community life [53] and a variety of industrial and manufacturing processes. Wastewater includes dissolved and suspended organic solids, which are normally biologically decomposable to some extent.

7.5.1 Wastewater Categories

Two general categories of wastewaters, not entirely separable, are recognized: domestic and industrial. Wastewater treatment is a process in which the solids and contaminants in wastewater are partially removed and partially changed by decomposition from highly complex, putrescible, organic solids to mineral or relatively stable organic solids [53]. Wastewater generally contains a high load of oxygen-demanding wastes, pathogenic or disease-causing agents, organic materials, nutrients that stimulate plant growth, inorganic chemicals, and minerals and sediments. It may also contain toxic compounds. The most commonly detected organic pollutants in wastewater are dyes, pesticides, herbicides, xenobiotics, pharmaceuticals and personal care products, PAHs, PCBs, and other persistent organic pollutants. These are discussed in more detail in the following sections.

7.5.2 Organic Pollutants Discharges

Chemical inputs of organic pollutants to rivers, lakes, and oceans originate either from point or nonpoint sources. Point sources include effluent pipes from municipal sewage treatment plants and factories. Pollutant discharges from such sources tend to be continuous, with little variability over time, and often they can be monitored by measuring discharge [5] and chemical concentrations periodically at a single place. Consequently, point sources are relatively simple to monitor and regulate, and can often be controlled by treatment at the source. Nonpoint inputs can also be continuous, but are more often intermittent and linked to seasonal agricultural activity such as planting and plowing or irregular events such as heavy rains or major construction. Nonpoint inputs often arise from various activities across extensive stretches of the landscape, and materials enter receiving waters as overland flow, underground seepage, or through the atmosphere. Consequently, nonpoint sources are difficult to measure and control [5]. Data in Table 7.3 and the discussions below trace down the main categories of organic pollutants detected in water and wastewaters.

Table 7.3 Organic pollutants detected in contaminated waters and wastewaters

Pollutant(s)	Contaminated media	Reference
Textile dye Acid Blue 25	Textile wastewater	[54]
Methylene blue, red basic dye, blue basic dye, nonylphenol, and octylphenol	Synthetic contaminated aqueous solutions	[55]
Vat Green 01	Textile wastewater	[56]
C.I. Acid Red 14 (AR14)	Synthetic solutions in batch mode testing	[57]
Malachite green	Synthetic aqueous solutions	[58]
PAHs	Urban lakes, water, sediment and soil in drinking water resource, sediments in marshes, river water, subsurface water samples	[59–61]
PCBs, polybrominated diphenyl ethers (PBDEs)	Water and sediments from estuaries, municipal sewage treatment plant effluents, river water, subsurface water samples, air and seawater, soils	[62–65]
Pharmaceuticals and personal care products (PPCPs)	River waters, sewage treatment plant effluents	[66]
Dioxins and furans (PCDD/Fs)	Soils/water in natural park in urban sites	
Organochlorine and organophosphate pesticides	River waters, sediments in rivers, soils, ground water	[67–69]
Dichlorodiphenyltrichloroethane (DDTs)	Wastewater	[64]
Estrogens, progestogens, and polar pesticides	Sewage treatment plant effluents	[66]
Agricultural pesticides and herbicides (atrazine, alachlor, trifluralin, metazachlor, metamitron and sulcotrione)	Surface water and groundwater, rainwater	[70–76]
Pharmaceuticals, personal care products, antibiotics	Wastewater streams, surface waters	[77–82]
Hormones	Wastewater streams, surface waters	[83–87]
Petroleums	Wastewater streams, surface waters	[88–91]

7.5.2.1 Dyes

Wastewaters from the textile industry is a complex mixture of many polluting substances ranging from residual dyestuffs to heavy metals associated with the dyeing and printing process [92, 93]. More than 110,000 commercially available dyes are known [94]. Furthermore, the color and chemical composition of textile effluents are usually subject to daily and seasonal variations dictated by the production routine. The presence of even trace concentrations of dyes in effluent is highly visible and undesirable [95]. The release of colored wastewater in the ecosystem is a remarkable source of esthetic pollution, eutrophication, and disruptions in aquatic life. Dye effluent usually contains chemicals, including dye itself, that are toxic, carcinogenic, mutagenic, or teratogenic to various microbiological and fish species [96].

It has been estimated that more than 10% of the total dyestuff used in dyeing processes is released into the environment [97]. Azo dyes are the largest group of dyes used in textile industry [98] constituting 60–70% of all dyestuff produced [99]. They have one or more azo groups (R_1–N=N–R_2) having aromatic rings mostly substituted by sulfonate groups. These complex aromatic substituted structures make conjugated system and are responsible for intense color, high water solubility, and resistance to degradation of azo dyes under natural conditions [100]. Various chemical and physical processes such as precipitation, adsorption, air stripping, flocculation, reverse osmosis, and ultrafiltration can be used for color removal from textile effluents [101]. However, these techniques are nondestructive, since they only transfer the nonbiodegradable matter into sludge, giving rise to new type of pollution, which needs further treatment [102, 103].

7.5.2.2 Xenobiotics

The term "xenobiotic" has a range of meanings and implications revolving around the chemical industry, its products and by-products [104]. Xenobiotics in the strict sense of the word are defined as man-made molecules, foreign to life, and should have never been encountered by bacterial populations before their introduction by man. Examples are polychlorinated biphenyls (PCBs), PAHs, and various pesticides. In this respect, a more wide-ranging definition of xenobiotics includes "all compounds that are released in any compartment of the environment by the action of man and thereby occur in a concentration that is higher than natural" [105].

Many synthetic organic chemicals (e.g., organochlorines, organophosphates, PAHs, and organometals) are of growing environmental concern, because of their high toxicity and high persistence in the environment and in biological systems [106]. Furthermore, the high lipophilicity of many of these xenobiotics greatly enhances their bioconcentration/biomagnification, thereby posing potential health hazards on predators at higher trophic levels (including human beings) [106]. Persistent xenobiotic compounds have been found in every part of the ocean: from arctic to antarctic, and from intertidal to abyssal [106]. Most xenobiotic compounds occur only at very low concentrations in the environment, and their threats to marine life and public health may be detected. However, sublethal effects of these compounds over long-term exposure may cause significant damage to marine populations through the impairment in reproduction functions of animals while others may be carcinogenic, mutagenic, or teratogenic [106].

PAHs are a class of organic compounds that have accumulated in the natural environment mainly as a result of anthropogenic activities such as the combustion of fossil fuels [107, 108]. The increasing use of fossil fuels and their combustion products by human beings during the past two centuries raises several questions about PAHs hazards for living organisms. Since most PAHs are highly hydrophobic

[109], their pathways of transfer through geological and biological media are not well understood. Also, explicit correlations between PAH sources and carcinogenic effects have been reported only for intense exposure to PAHs. Polychlorinated dibenzodioxins and polychlorinated dibenzofurans (dioxins) consist of 210 different compounds that have similar chemical properties [110]. This class of compounds is persistent, toxic, and bioaccumulative. They are generated as by-products during incomplete combustion of chlorine containing wastes such as municipal solid waste, sewage sludge, and hospital and hazardous wastes [110]. PCBs were widely used in the past and now contaminate many industrial and natural areas.

According to the United States Environmental Protection Agency, the term "pesticide" is a broad nonspecific term covering a large number of substances including, insecticides, herbicides, and fungicides, "though often misunderstood to refer only to insecticides." Chemical pesticides have consistently demonstrated their merit by increasing the global agricultural productivity [111], reducing insect-borne, endemic diseases and protecting plantations, forests, and harvested wood [112]. As of date, pesticides are more valued in developing countries, particularly those in tropical regions seeking to enter the global economy by providing off-season fresh fruits and vegetables to countries in more temperate climates [111]. However, the continuous use of pesticides has caused severe irreversible damage to the environment, caused human ill-health, negatively impacted on agricultural production, and reduced agricultural sustainability [113].

7.5.2.3 Pharmaceuticals

The term "pharmaceutical" covers a wide-ranging class of compounds with substantial variability in structures, function, behavior, and activity [114]. Pharmaceutical compounds and their metabolites are emerging contaminants that have been frequently detected at low levels in environmental samples [110, 115], biota, and human tissues. The persistence of pharmaceutical contaminants in the environment has been attributed to human consumption of drugs (not including natural and synthetic hormones) and subsequent discharges from sewage treatment plants, as well as veterinary use of drugs, and nonpoint discharges from agricultural runoff [15]. The use of some pesticides may have experienced a fall in recent years as new laws have been introduced to minimize their use [116], but even if they should prove problematic, pharmaceuticals are unlikely to be restricted in this way, due to their beneficial human health effects and economic importance [117]. Indeed, their use is expected to grow with the increasing average age of the population [118], and consequently, they and their metabolites are much likely to be detected in the environment [119]. Drugs may be degraded during sewage treatment processes [117] but many pharmaceuticals are not thermally stable [120] and so may be expected to break down during processes such as composting due to microbial heat generated during the process.

Medicinal compounds are generally excreted after being partially or completely converted to water-soluble metabolites [121, 122] but a significant amount of the original substance may also be excreted unchanged [123]. This has previously been regarded as inconsequential because of the dilution received in the sewerage system. However, recent studies on pharmaceutical residues (primarily in Germany) have demonstrated that elimination of high to medium polar pharmaceuticals in municipal STPs is often incomplete, ranging between 60% and 90% [124, 125]. One of the most comprehensive studies of this type was performed by Kolpin et al. [115] who chronicled the detection of over 95 organic chemicals in US streams and rivers.

7.5.2.4 Hormones

During recent years, estrogenic compounds such as nonylphenol, octylphenol, nonylphenol polyethoxylates, dihydrofolliculin (β-estradiol), estrone, estriol, and ethynylestradiol (α-estradiol) have increasingly been found in treated domestic wastewater effluents [126]. These estrogenic compounds are of great concern because of their potential in altering the normal endocrine function and physiological status of animals and humans [127]. Natural estrogens line β-estradiol and α-estradiol are excreted by both humans and animals [128]. Many of these known estrogenic compounds end up in the aquatic environment via sewage, the discharge of municipal and/or industrial effluents, and agricultural runoff. Chang et al. [129] monitored five estrogens, nine androgens, nine progestogens, six glucocorticoids, and one mineralocorticoid in samples from urban rivers. Androgens were the dominant steroids detected (total concentrations up to 480 ng/L), followed by glucocorticoids (up to 52 ng/L), progestogens (up to 50 ng/L), and estrogens (up to 9.8 ng/L). The summed concentration for each class of detected hormones in the 13 discharging site samples was higher than that in river samples, up to 1,887 ng/L for androgens, 390 ng/L for glucocorticoids, 75 ng/L for progestogens, and 25 ng/L for estrogens. Chang et al. [129] found that 62.7% of the mean summed hormones were contributed by freshly discharged untreated sewage, 29.4% by treated sewage and/or naturally attenuated untreated sewage, and 7.9% by an unknown source, possibly pharmaceutical manufacturing plants.

7.5.2.5 Petroleums

Total petroleum hydrocarbons (TPH) describe a broad family of several hundred chemical compounds that originally come from crude oil. Crude oils can vary depending on their chemical constituents, and so can the petroleum products that are made from crude oils. Because modern society uses so many petroleum-based products (gasoline, kerosene, fuel oil, mineral oil, and asphalt), contamination of the environment by them is potentially widespread. Contamination caused by petroleum products contains a variety of these hydrocarbons. Because they are found in a complex mixture, it is not usually practical to measure each one individually and treat them separately with complete remediation.

7.6 Microwave-Assisted Degradation Processes for Organic Pollutants

Microwave power has been applied widely in the remediation of contaminated soils, treatment of wastes, oil and water separation, reduction of nitrogen oxides and sulfur dioxide, polymer synthesis, organic reactions, and regeneration of activated carbon. Up to now, two main pathways of using microwave energy as an emerging technology for the treatment of organic pollutants' contaminated waters and wastewaters have been adopted. The first one is that the contaminants are adsorbed on the surfaces of carbonaceous material or soil from aqueous media, which then is irradiated by microwave and the adsorbed pollutants are decomposed simultaneously. The second approach is that the effluent is irradiated directly by microwave and contaminants are degraded in the aqueous phase. The treatment of industrial and municipal wastewaters is always among the most severe obstacles faced by all countries.

Research carried out so far in this respect and with regard to the applicability of microwave irradiation has shown that microwave technology may be effective in treating wastewaters at the various stages of the treatment process, and hence improving the degradation of a variety of organic pollutants. A survey of the literature shows that there are three major types of research areas in the use of microwave irradiation for the degradation of organic pollutants. These are standalone microwave-assisted degradation processes, combined microwave-assisted degradation processes, and microwave-assisted catalytic degradation processes. In stand-alone processes, the microwave irradiation is the only kind of treatment conditions imposed on the contaminants. In the combined process (hybrid process), microwave irradiation is coupled or performed in the presence of other degradation processes like ozonation, electrokinetic remediation, advanced oxidation, ultraviolet illumination, or sonolysis. In the microwave-assisted catalytic degradation processes, the contaminants are degraded in the presence of a catalyst, either hetereogeneous or homogeneous. Most of the studies in this category have reported the use of hetereogeneous catalysts coupled with photocatalysis. Table 7.4 presents the main results of a representative set of studies on the microwave-assisted degradation of organic contaminants in wastewaters, synthetic effluents, and other surface waters.

Other researchers have also reported some salient findings and inferences on the advantages and possible mechanisms involved in the use of microwave irradiation for organic pollutant degradation. Liu et al. [135, 136] concluded that adopting microwave irradiation resulted in rapid degradation rates, low cost, no residual intermediates, and no secondary pollution in actual application. In their work, Zhihui et al. [137] analyzed the synergistic effects of several microwave-assisted advanced oxidation processes (MW/AOPs) for the degradation of 4-chlorophenol (4-CP). Their results showed that the synergistic effects between MW and H_2O_2, UV/H_2O_2, TiO_2 photocatalytic oxidation (PCO) resulted in a higher degradation efficiency for 4-CP. All the more, Park et al. [138] found that the photocatalytic

Table 7.4 Comparison of microwave-assisted degradation processes for organic contaminants

Contaminant(s)	Test media	Degradation conditions	Degradation performance	Reference
2,4-Dichlorophenoxy-acetic acid	Agricultural land	Coupled photocatalytic/microwave method in an aqueous TiO_2 dispersion Aqueous dispersion was contained in a high-pressure Teflon batch (TB) reactor that also integrated a double glass cylindrical plasma lamp (DGCPL) as the source of the UV-Vis radiation	Rates of degradation were 2×10^{-3} mM min^{-1} (photocatalytic/microwave method (PD/MW)) and 1.1×10^{-3} mM min^{-1} (photocatalytic method (PD)). The coupled PD/MW method was about ten times more efficient than the PD method alone	[130]
Methylene blue	Aqueous suspensions	Photocatalytic degradation with microwave power from electrodeless discharge lamps (EDLs) in TiO_2 suspensions	45% mineralization of dye achieved after 15 min of irradiation	[131]
Phenol	Aqueous solutions and batch mode analyses	Photocatalytic degradation in the presence of microwave-assisted prepared zinc oxide	15 min microwave-irradiated sample shows 88% phenol (0.6 g L^{-1} dose) degradation at pH 5.0 and 4 h reaction time under sunlight. Zinc oxide samples prepared by microwave irradiation and calcined at 300°C exhibited highest surface area, acid sites and lowest crystallite sizes and showed highest activity toward photocatalytic oxidation of phenol	[132]

(continued)

Table 7.4 (continued)

Contaminant(s)	Test media	Degradation conditions	Degradation performance	Reference
Phenol	Aqueous solutions with continuous bubbling of air through the liquid phase	Microwave-enhanced catalytic degradation (MECD) of phenol were investigated with mix-valenced nickel oxide catalyst, NiO_x, prepared from nickel nitrate aqueous solution through a precipitation with sodium hydroxide and an oxidation by sodium hypochlorite	The introduction of microwave irradiation could greatly shorten the time of phenol degradation	[133]
4-Chlorophenol (4-CP)	Aqueous solutions with continuous bubbling of air through the liquid phase	Microwave-enhance catalytic degradation (MECD) with mix-valenced nickel oxide catalyst, NiO_x, prepared from nickel nitrate aqueous solution through a precipitation with sodium hydroxide and an oxidation by sodium hypochlorite	4-CP was degraded completely by MECD method within 20 min under pH 7, $T = 40°C$ and $C = 200$ g dm^{-3} over catalyst	[134]

degradation of bromothymol blue increased with the TiO_2 particle dosages and microwave intensity in a microwave/UV/TiO_2/ozone/H_2O_2 hybrid process system. When an auxiliary oxidant such as ozone or hydrogen peroxide was added to the microwave-assisted photocatalysis, however, a synergetic effect that enhanced the reaction rate considerably was equally observed.

With respect to the proposed mechanisms of microwave-mediated contaminant degradation, Horikoshi et al. [139] have indicated that a greater efficacy of aqueous TiO_2 dispersions using an integrated microwave/UV-illumination technique appeared to be the result of enhanced formation of reactive oxygen species (•OH radicals), and due to the activity of bulk water or the TiO_2 particle surface. In their work on the microwave-assisted photocatalysis of atrazine on TiO_2 nanotubes, Gao et al. [140] indicated that atrazine had been completely degraded in 5 min and the mineralization efficiency was up to 98.5% in 20 min. Gao et al. [140] equally attributed this enhanced degradation performance to the intense UV radiation generated by electrodeless discharge lamps under microwave irradiation, the increased number of •OH, additional defect sites on TiO_2 under the irradiation of microwave, and larger specific surface area of TiO_2 nanotubes which could adsorb more organic substances to degrade than TiO_2 nanoparticles.

7.7 Conclusions

This chapter has been synthesized after over 400 research papers were read through and scrutinized. The first sections of this chapter have discussed the environmental issues and concerns that are occasioned in a cascade of impacts through the contamination of water bodies and wastewaters by a wide variety of organic pollutants (xenobiotics, hormones, and petroleums). The subsequent sections have presented the findings of several research studies on the application of microwave irradiation in assisting the degradation and detoxification of a number of these organic pollutants in contaminated waters. The major inferences come out to be that microwave-assisted degradation of organic pollutants seems to achieve a relatively cleaner, faster, and almost complete degradation of many organic pollutants. In this regard, microwave power demonstrates an initial promising momentum for further applied research in upscaling the application of microwave power for organic pollutant remediation.

Acknowledgements The author expresses his gratitude to all the researchers whose valuable data have been of significance to this chapter, and is also grateful to other colleagues for their constructive criticisms and suggestions. Kind regards are also due to the editors, Prof. S.K. Sharma and Dr. R. Sanghi, for granting the opportunity to contribute a chapter in this comprehensive book on green technology and wastewater remediation.

References

1. Prasad MNV, Freitas H, Fraenzle S, Wuenschmann S, Markert B (2010) Knowledge explosion in phytotechnologies for environmental solutions. Environ Pollut 158:18–23
2. Beltrame MO, De Marco SG, Marcovecchio JE (2010) Effects of zinc on molting and body weight of the estuarine crab *Neohelice granulata* (Brachyura: Varunidae). Sci Total Environ 408:531–536
3. Kelly BC, Ikonomou MG, Blair JD, Morin AE, Gobas FAPC (2007) Food web–specific biomagnification of persistent organic pollutants. Science 317:236–239
4. Takeuchi I, Miyoshi N, Mizukawa K, Takada H, Ikemoto T, Omori K, Tsuchiya K (2009) Biomagnification profiles of polycyclic aromatic hydrocarbons, alkylphenols and polychlorinated biphenyls in Tokyo Bay elucidated by $\delta^{13}C$ and $\delta^{15}N$ isotope ratios as guides to trophic web structure. Mar Pollut Bull 58:663–671
5. Carpenter S, Caraco NF, Correll DL, Howarth RW, Sharpley AN, Smith VH (1998) Nonpoint pollution of surface waters with phosphorus and nitrogen. Issue Ecol 3:1–12
6. Kulkarni PS, Crespo JG, Afonso CAM (2008) Dioxins sources and current remediation technologies — a review. Environ Int 34:139–153
7. Busca G, Berardinelli S, Resini C, Arrighi L (2008) Technologies for the removal of phenol from fluid streams: a short review of recent developments. J Hazard Mater 160:265–288
8. Mascolo G, Ciannarella R, Balest L, Lopez A (2008) Effectiveness of UV–based advanced oxidation processes for the remediation of hydrocarbon pollution in the groundwater: a laboratory investigation. J Hazard Mater 152:1138–1145
9. Wang JY, Huang XJ, Kao JCM, Stabnikova O (2007) Simultaneous removal of organic contaminants and heavy metals from kaolin using an upward electrokinetic soil remediation process. J Hazard Mater 144:292–299
10. Kim JH, Park PK, Lee CH, Kwon HH (2008) Surface modification of nanofiltration membranes to improve the removal of organic micro–pollutants (EDCs and PhACs) in drinking water treatment: graft polymerization and cross–linking followed by functional group substitution. J Membr Sci 321:190–198
11. Nissen S, Alexander BD, Dawood I et al (2009) Remediation of a chlorinated aromatic hydrocarbon in water by photoelectrocatalysis. Environ Pollut 157:72–76
12. Margesin R, Hämmerle M, Tscherko D (2007) Microbial activity and community composition during bioremediation of diesel–Oil–contaminated soil: effects of hydrocarbon concentration, fertilizers, and incubation. Microb Ecol 53:259–269
13. Zhao B, Poh CL (2008) Insights into environmental bioremediation by microorganisms through functional genomics and proteomics. Proteomics 8:874–881
14. Vidali M (2001) Bioremediation. An overview. Pure Appl Chem 73:1163–1172
15. Mudhoo A, Mohee R (2010) Composting as a bioremediation technique for hazardous organic contaminants. In: Sharma SK, Mudhoo A (eds) Green chemistry for environmental sustainability. Taylor & Francis, Boca Raton, pp 215–247
16. Whiteley CG, Lee D-J (2006) Enzyme technology and biological remediation. Enzyme Microb Technol 38:291–316
17. Xiao L, Qu X, Zhu D (2007) Biosorption of nonpolar hydrophobic organic compounds to *Escherichia coli* facilitated by metal and proton surface binding. Environ Sci Technol 41:2750–2755
18. Mathialagan T, Viraraghavan T (2009) Biosorption of pentachlorophenol from aqueous solutions by a fungal biomass. Bioresour Technol 100:549–558
19. Kumari K, Abraham TE (2007) Biosorption of anionic textile dyes by nonviable biomass of fungi and yeast. Bioresour Technol 98:1704–1710
20. Ghodake G, Jadhav S, Dawkar V, Govindwar S (2009) Biodegradation of diazo dye Direct brown MR by Acinetobacter calcoaceticus NCIM 2890. Int Biodeterior Biodegrad 63: 433–439

21. El-Sheekh MM, Gharieb MM, Abou-El-Souod GW (2009) Biodegradation of dyes by some green algae and cyanobacteria. Int Biodeterior Biodegrad 63:699–704
22. Contreras-Ramos SM, Álvarez-Bernal D, Dendooven L (2009) Characteristics of earthworms (*Eisenia fetida*) in PAHs contaminated soil amended with sewage sludge or vermicompost. Appl Soil Ecol 41:269–276
23. Das KC, Xia K (2008) Transformation of 4–nonylphenol isomers during biosolids composting. Chemosphere 70:761–768
24. Kupper T, Bucheli TD, Brändli RC, Ortelli D, Edder P (2008) Dissipation of pesticides during composting and anaerobic digestion of source–separated organic waste at full–scale plants. Bioresour Technol 99:7988–7994
25. Ghaly AE, Alkoaik F, Snow A (2007) Degradation of pirimiphos–methyl during thermophilic composting of greenhouse tomato plant residues. Can Biosyst Eng 49:6.1–6.11
26. Delgado-Moreno L, Peña A (2009) Compost and vermicompost of olive cake to bioremediate triazines–contaminated soil. Sci Total Environ 407:1489–1495
27. Elefsiniotis P, Li W (2008) Biodegradation behavior of agricultural pesticides in anaerobic batch reactors. J Environ Sci Health B 43:172–178
28. Celis E, Elefsiniotis P, Singhal N (2008) Biodegradation of agricultural herbicides in sequencing batch reactors under aerobic or anaerobic conditions. Water Res 42:3218–3224
29. Baraldi EA, Damianovic MHRZ, Manfio GP, Foresti E, Vazoller RF (2008) Performance of a horizontal–flow anaerobic immobilized biomass (HAIB) reactor and dynamics of the microbial community during degradation of pentachlorophenol (PCP). Anaerobe 14:268–274
30. Kidwai M, Mohan R (2005) Green chemistry: an innovative technology. Found Chem 7:269–287
31. Anastas PT, Warner JC (1998) Green chemistry, theory and practice. Oxford University Press, Oxford
32. Kirchhoff MM (2003) Promoting green engineering through green chemistry. Environ Sci Technol 37:5349–5353
33. Feng D, Aldrich C (2000) Sonochemical treatment of simulated soil contaminated with diesel. Adv Environ Res 4:103–112
34. Suslick KS, Casadonte DJ, Green MLH, Thompson ME (1987) Effects of high intensity ultrasound on inorganic solids. Ultrasonics 25:56–61
35. Stephanis CG, Hariris JG, Mourmouras DE (1997) Process (mechanism) of erosion of soluble brittle materials caused by cavitation. Ultrason Sonochem 4:269–271
36. Ley SV, Low CMR (1989) Ultrasound in synthesis. Springer, Berlin, Ch. 2
37. Flores R, Blass G, Dominguez V (2007) Soil remediation by an advanced oxidative method assisted with ultrasonic energy. J Hazard Mater 140:399–402
38. Adewuyi YG (2005) Sonochemistry in environmental remediation. Combinative and hybrid sonophotochemical oxidation processes for the treatment of pollutants in water. Environ Sci Technol 39:3409–3420
39. Chowdhury P, Viraraghavan T (2009) Sonochemical degradation of chlorinated organic compounds, phenolic compounds and organic dyes – a review. Sci Total Environ 407: 2474–2492
40. Pham TD, Shrestha RA, Virkutyte J, Sillanpaa M (2009) Recent studies in environmental applications of ultrasound. Can J Civil Eng 36:1849–1858
41. Sponza DT, Oztekin R (2010) Removals of PAHs and acute toxicity via sonication in a petrochemical industry wastewater. Chem Eng J 162:142–150
42. Chang JH, Ellis AV, Yan CT, Tung CH (2009) The electrochemical phenomena and kinetics of EDTA–copper wastewater reclamation by electrodeposition and ultrasound. Sep Purif Technol 68:216–221
43. Matouq MA-D, Al-Anber ZA (2007) The application of high frequency ultrasound waves to remove ammonia from simulated industrial wastewater. Ultrason Sonochem 14:393–397
44. Kang Y, Xue X, Yang H, Liu J (2009) Degradation of nitrobenzene in wastewater by slag titania with ultrasound. J Chin Ceram Soc 37:536–542

45. Trabelsi F, Aït-Lyazidi H, Ratsimba B et al (1996) Oxidation of phenol in wastewater by sonoelectrochemistry. Chem Eng Sci 51:1857–1865
46. Goskonda S, Catallo WJ, Junk T (2004) Sonochemical degradation of aromatic organic pollutants. Waste Manag 22:351–356
47. Cravotto G, Di Carlo S, Curini M, Tumiatti V, Roggero C (2007) A new flow reactor for the treatment of polluted water with microwave and ultrasound. J Chem Technol Biotechnol 82:205–208
48. Jacob J, Chia LHL, Boey FYC (1995) Review—thermal and non–thermal interaction of microwave radiation with materials. J Mater Sci 30:5321–5327
49. Hill JM, Marchant TR (1996) Modelling microwave heating. Appl Math Model 20:3–15
50. Reimbert CG, Minzoni AA, Smyth NF (1996) Effect of radiation losses on hotspot formation and propagation in microwave heating. IMA J Appl Math 57:165–179
51. Wu T-N (2008) Environmental perspectives of microwave applications as remedial alternatives: review. Pract Period Hazard Tox Radioact Waste Manag 12:102–115
52. Nüchter M, Ondruschka B, Bonrath W, Gum A (2004) Microwave assisted synthesis – a critical technology overview. Green Chem 6:128–141
53. Sonune A, Ghate R (2004) Developments in wastewater treatment methods. Desalination 167:55–63
54. Mahmoodi NM, Arami M (2009) Degradation and toxicity reduction of textile wastewater using immobilized titania nanophotocatalysis. J Photochem Photobiol B Biol 94:20–24
55. Kurinobu S, Tsurusaki K, Natui Y, Kimata M, Hasegawa M (2007) Decomposition of pollutants in wastewater using magnetic photocatalyst particles. J Magn Magn Mater 310:1025–1027
56. Schrank SG, Ribeiro N, dos Santos J, Santos Souza D, Santos Souza EE (2007) Decolourisation effects of Vat Green 01 textile dye and textile wastewater using H_2O_2/UV process. J Photochem Photobiol A Chem 186:125–129
57. Aleboyeh A, Daneshvar N, Kasiri MB (2008) Optimization of C.I. Acid Red 14 azo dye removal by electrocoagulation batch process with response surface methodology. Chem Eng Process Process Intensif 47:827–832
58. Khattri SD, Singh MK (2009) Removal of malachite green from dye wastewater using neem sawdust by adsorption. J Hazard Mater 167:1089–1094
59. Li J, Cheng H, Zhang G, Qi S, Li X (2009) Polycyclic aromatic hydrocarbon (PAH) deposition to and exchange at the air–water interface of Luhu, an urban lake in Guangzhou, China. Environ Pollut 157:273–279
60. Zhu L, Chen Y, Zhou R (2008) Distribution of polycyclic aromatic hydrocarbons in water, sediment and soil in drinking water resource of Zhejiang Province, China. J Hazard Mater 150:308–316
61. Liang Y, Tse MF, Young L, Wong MH (2007) Distribution patterns of polycyclic aromatic hydrocarbons (PAHs) in the sediments and fish at Mai Po Marshes Nature Reserve, Hong Kong. Water Res 41:1303–1311
62. Cailleaud K, Forget-Leray J, Souissi S, Hilde D, LeMenach K, Budzinski H (2007) Seasonal variations of hydrophobic organic contaminant concentrations in the water–column of the Seine Estuary and their transfer to a planktonic species *Eurytemora affinis* (Calanoïda, copepoda). Part 1: PCBs and PAHs. Chemosphere 70:270–280
63. Magnusson K, Magnusson M, Östberg P, Granberg M, Tiselius P (2007) Bioaccumulation of ^{14}C–PCB 101 and ^{14}C–PBDE 99 in the marine planktonic copepod *Calanus finmarchicus* under different food regimes. Mar Environ Res 63:67–81
64. Chak S, Kluskens B, Ford D (2010) Determination of polychlorinated biphenyl (PCBs) and dichlorodiphenyltrichloroethane (DDTs) in sediments in Boeng Cheung Ek, Phnom Penh, Cambodia. Asia J Water Environ Pollut 7:3–11
65. Vane CH, Harrison I, Kim AW (2007) Polycyclic aromatic hydrocarbons (PAHs) and polychlorinated biphenyls (PCBs) in sediments from the Mersey Estuary, U.K. Sci Total Environ 374:112–126

66. Kuster M, López J, de Alda M, Hernando MD, Petrovic M, Martín-Alonso J, Barceló D (2008) Analysis and occurrence of pharmaceuticals, estrogens, progestogens and polar pesticides in sewage treatment plant effluents, river water and drinking water in the Llobregat river basin (Barcelona, Spain). J Hydrol 358:112–123
67. Zhou R, Zhu L, Yang K, Chen Y (2006) Distribution of organochlorine pesticides in surface water and sediments from Qiantang River, East China. J Hazard Mater 137:68–75
68. Leong KH, Tan LLB, Mustafa AM (2002) Contamination levels of selected organochlorine and organophosphate pesticides in the Selangor River, Malaysia between 2002 and 2003. Chemosphere 66:1153–1159
69. Qiu S, Zhu T, Wang F, Hu J (2008) Air–water Gas exchange of organochlorine pesticides in Taihu Lake, China. Environ Sci Technol 42:1928–1932
70. Iwakuma T, Shiraishi H, Nohara S, Takamura N (1993) Runoff properties and change in concentrations of agricultural pesticides in a river system during a rice cultivation period. Chemosphere 27:677–691
71. Knee KL, Gossett R, Boehm AB, Paytan A (2010) Caffeine and agricultural pesticide concentrations in surface water and groundwater on the north shore of Kauai (Hawaii, USA). Mar Pollut Bull 60:1376–1382
72. Schwab AP, Splichal PA, Banks MK (2006) Persistence of atrazine and alachlor in ground water aquifers and soil. Water Air Soil Pollut 171:203–235
73. Chen J-Z, Meng S-L, Hu G-D, Qu J-H (2009) Bioaccumulation of herbicide atrazine in *Carassius auratus*. J Agro-Environ Sci 28:1313–1318
74. Navarro S, Vela N, José Giménez M, Navarro G (2004) Persistence of four s–triazine herbicides in river, sea and groundwater samples exposed to sunlight and darkness under laboratory conditions. Sci Total Environ 329:87–97
75. Scheyer A, Morville S, Mirabel P, Millet M (2007) Pesticides analysed in rainwater in Alsace region (eastern France): comparison between urban and rural sites. Atmos Environ 41:7241–7252
76. Mamy L, Barriuso E, Gabrielle B (2005) Environmental fate of herbicides trifluralin, metazachlor, metamitron and sulcotrione compared with that of glyphosate, a substitute broad spectrum herbicide for different glyphosate–resistant crops. Pest Manag Sci 61:905–916
77. Bartelt-Hunt SL, Snow DD, Damon T, Shockley J, Hoagland K (2009) The occurrence of illicit and therapeutic pharmaceuticals in wastewater effluent and surface waters in Nebraska. Environ Pollut 157:786–791
78. Fatta D, Achilleos A, Nikolaou A, Meriç S (2007) Analytical methods for tracing pharmaceutical residues in water and wastewater. TrAC Trend Anal Chem 26:515–533
79. Shariati FP, Mehrnia MR, Salmasi BM et al (2010) Membrane bioreactor for treatment of pharmaceutical wastewater containing acetaminophen. Desalination 250:798–800
80. Badawy MI, Wahaab RA, El-Kalliny AS (2009) Fenton–biological treatment processes for the removal of some pharmaceuticals from industrial wastewater. J Hazard Mater 167: 567–574
81. O'Connell DW, Birkinshaw C, O'Dwyer TF (2008) Persistence and fate of highly soluble pharmaceutical products in various types of municipal wastewater treatment plants. WIT Trans Ecol Environ I:799–807
82. Sim WJ, Lee JW, Oh JE (2010) Occurrence and fate of pharmaceuticals in wastewater treatment plants and rivers in Korea. Environ Pollut 158:1938–1947
83. Plósz BG, Leknes H, Liltved H, Thomas KV (2010) Diurnal variations in the occurrence and the fate of hormones and antibiotics in activated sludge wastewater treatment in Oslo, Norway. Sci Total Environ 408:1915–1924
84. Ikehata K, Gamal El-Din M, Snyder SA (2008) Ozonation and advanced oxidation treatment of emerging organic pollutants in water and wastewater. Ozone: Sci Eng 30:21–26
85. Schlüsener MP, Bester K (2008) Behavior of steroid hormones and conjugates during wastewater treatment – a comparison of three sewage treatment plants. CLEAN – soil, Air. Water 36:25–33

86. Pedersen JA, Soliman M, Suffet IL (2005) Human pharmaceuticals, hormones, and personal care product ingredients in runoff from agricultural fields irrigated with treated wastewater. J Agric Food Chem 53:1625–1632
87. Nghiem LD, Manis A, Soldenhoff K, Schäfer AI (2004) Estrogenic hormone removal from wastewater using NF/RO membranes. J Membr Sci 242:37–45
88. Saien J, Nejati H (2007) Enhanced photocatalytic degradation of pollutants in petroleum refinery wastewater under mild conditions. J Hazard Mater 148:491–495
89. Antić MP, Jovancicevic B, Vrvić MM, Schwarzbauer J (2006) Petroleum pollutant degradation by surface water microorganisms. Environ Sci Pollut Res 13:320–327
90. El-Naas MH, Al-Zuhair S, Abu Alhaija M (2010) Removal of phenol from petroleum refinery wastewater through adsorption on date–pit activated carbon. Chem Eng J 162:997–1005
91. Meng J-B, Lu S-M, Yang L (2008) A test of emergency treatment of petroleum pollution in source water by powder activated carbon. Ind Water Wastewater 39:22–25
92. Correia VM, Stephenson T, Judd SD (1994) Characterisation of textile wastewaters – a review. Environ Technol 15:917–929
93. Hood EE (2002) From green plants to industrial enzymes. Enzyme Microb Technol 30: 279–283
94. Selvam K, Swaminathan K, Keo-Sang C (2003) Microbial decolorization of azo dyes and dye industry effluent by Fomes lividus. World J Microbiol Biotechnol 19:591–593
95. Hao OJ, Kim H, Chiang P-C (2000) Decolorization of wastewater. Crit Rev Environ Sci Technol 30:449
96. Daneshvar N, Ashassi-Sorkhabi H, Tizpar A (2003) Decolorization of orange II by electrocoagulation method. Sep Purif Technol 31:153
97. Maguire RJ (1992) Occurrence and persistence of dyes in a Canadian river. Water Sci Technol 25:265–270
98. Maximo C, Amorim MTP, Costa-Ferreira M (2003) Biotransformation of industrial reactive azo dyes by *Geotricum* sp. CCMI 1019. Enzyme Microb Technol 32:145–151
99. Carliell CM, Barclay SJ, Buckley CA (1995) Microbial decolourizationof a reactive azo dye under anaerobic conditions. Water SA 21:61–69
100. Rajaguru P, Kalaiselvi K, Palanivel M, Subburam V (2000) Biodegradation of azo dyes in a sequential anaerobiceaerobic system. Appl Microbiol Biotechnol 54:268–273
101. Robinson TF, McMullan G, Marchant R, Nigam P (2001) Remediation of dyes in textile effluent: a critical review on current treatment technologies with a proposed alternative. Bioresour Technol 77:247
102. Arslan I, Balcioglu IA, Tuhkanen T, Bahnemann D (2000) H_2O_2/UV–C and Fe^{2+}/H_2O_2/UV–CversusTiO_2/UV–A treatment for reactive dye wastewater. J Environ Eng 126:903
103. Stock N, Peller J, Vinodgopal K, Kamat PV (2000) Combinative sonolysis and photocatalysis for textile dye degradation. Environ Sci Technol 34:1747
104. Hutzinger O, Veerkamp W (1981) Xenobiotic chemicals with pollution potential. In: Leisinger T, Cook AM, Hiitter R, Naesch J (eds) Microbial degradation of xenobiotics and recalcitrant compounds. Academic, London, pp 3–45
105. Leisinger T (1983) Microorganisms and xenobiotic compounds. Experientia 39:1183–1220
106. Wu RSS (1999) Eutrophication, water borne pathogens and xenobiotic compounds: environmental risks and challenges. Mar Pollut Bull 39:11–22
107. Bamforth SM, Singleton I (2005) Bioremediation of polycyclic aromatic hydrocarbons: current knowledge and future directions. J Chem Technol Biotechnol 80:723–736
108. Johnsen AR, Wick LY, Harms H (2005) Principles of microbial PAH–degradation in soil. Environ Pollut 133:71–84
109. Wild SR, Jones KC (1992) Organic chemicals entering agricultural soils in sewage sludges: screening for their potential to transfer to crop plants and livestock. Sci Total Environ 119: 85–119
110. Bhandari A, Xia K (2005) Hazardous organic chemicals in biosolids recycled as soil amendments. In: Handbook of environmental chemistry, vol 5/Part F. Springer, Berlin/Hiedelberg, pp 217–239

111. Ecobichon DJ (2001) Pesticide use in developing countries. Toxicology 160:27–33
112. Ecobichon DJ (2000) Our changing perspectives on beneft and risks of pesticides: a historical overview. Neurotoxicology 21:211–218
113. Wilson C, Tisdell C (2001) Why farmers continue to use pesticides despite environmental, health and sustainability costs. Ecol Econ 39:449–462
114. Derksen JGM, Rijs GBJ, Jongbloed RH (2004) Diffuse pollution of surface water by pharmaceutical products. Water Sci Technol 49:213–221
115. Kolpin DW, Furlong ET, Meyer MT, Thurman EM, Zaugg SD, Barber LB, Buxton HT (2002) Pharmaceuticals, hormones, and other organic wastewater contaminants in U.S. streams, 1999–2000: a national reconnaissance. Environ Sci Technol 36:1202–1211
116. Daughton CG (2002) Environmental stewardship and drugs as pollutants. Lancet 360: 1035–1036
117. Jones OAH, Voulvoulis N, Lester JN (2005) Human pharmaceuticals in wastewater treatment processes. Crit Rev Environ Sci Technol 35:401–427
118. Daughton CG, Ternes TA (1999) Pharmaceuticals and personal care products in the environment: agents of subtle change? Environ Health Perspect 107:907–942
119. Sedlak DL, Pinkston KE, Gray JL, Kolodziej EP (2003) Approaches for quantifying the attenuation of wastewater–derived contaminants in the aquatic environment. Chimia 57: 567–569
120. Ternes TA (2001) Analytical methods for the determination of pharmaceuticals in aqueous environmental samples. Trends Anal Chem 20:419–434
121. Grahame-Smith DG, Aronson JK (2002) Oxford textbook of clinical pharmacology and drug therapy. Oxford University Press, Oxford
122. Rang HP, Dale M, Ritter JM (1999) Pharmacology. Churchill Livingstone, St. Louis
123. Hirsch R, Ternes T, Haberer K, Kratz KL (1999) Occurrence of antibiotics in the aquatic environment. Sci Total Environ 225:109–118
124. Ternes TA (1998) Occurrence of drugs in German sewage treatment plants and rivers. Water Res 32:3245–3257
125. Ternes T (2000) Pharmaceuticals and metabolites as contaminants of the aquatic environment: an overview. Abstr Pap Am Chem Soc 219:30-ENVR
126. Johnson A, Jurgens M (2003) Endocrine active industrial chemicals: release and occurrence in the environment. Pure Appl Chem 75:1895–1904
127. Hoffmann B, Landeck A (1999) Testicular endocrine function, seasonality and semen quality of the stallion. Anim Reprod Sci 57:89–98
128. Meng Z, Chen W, Mulchandani A (2005) Removal of estrogenic pollutants from contaminated water using molecularly imprinted polymers. Environ Sci Technol 39:8958–8962
129. Chang H, Wan Y, Ghu J (2009) Determination and source apportionment of five classes of steroid hormones in urban rivers. Environ Sci Technol 43:7691–7698
130. Horikoshi S, Hidaka H, Serpone N (2004) Environmental remediation by an integrated microwave/UV illumination technique: VI. A simple modified domestic microwave oven integrating an electrodeless UV–Vis lamp to photodegrade environmental pollutants in aqueous media. J Photochem Photobiol A Chem 161:221–225
131. Hong J, Sun C, Yang SG, Liu YZ (2006) Photocatalytic degradation of methylene blue in TiO_2 aqueous suspensions using microwave powered electrodeless discharge lamps. J Hazard Mater 133:162–166
132. Parida KM, Parija S (2006) Photocatalytic degradation of phenol under solar radiation using microwave irradiated zinc oxide. Solar Energy 80:1048–1054
133. Lai TL, Lee CC, Wu KS, Shu YY, Wang CB (2006) Microwave–enhanced catalytic degradation of phenol over nickel oxide. Appl Catal B Environ 68:147–153
134. Lai TL, Liu JY, Yong KF, Shu YY, Wang CB (2008) Microwave–enhanced catalytic degradation of 4–chlorophenol over nickel oxides under low temperature. J Hazard Mater 157:496–502

135. Liu Y, Yang S, Hong J, Sun C (2007) Low–temperature preparation and microwave photocatalytic activity study of TiO_2–mounted activated carbon. J Hazard Mater 142: 208–215
136. Liu X, Zhang Q, Zhang G, Wa R (2008) Application of microwave irradiation in the removal of polychlorinated biphenyls from soil contaminated by capacitor oil. Chemosphere 72: 1655–1658
137. Zhihui A, Peng Y, Xiaohua L (2005) Degradation of 4–Chlorophenol by microwave irradiation enhanced advanced oxidation processes. Chemosphere 60:824–827
138. Park Sh, Kim SJ, Seo SG, Jung SC (2010) Assessment of microwave/UV/O_3 in the photo–catalytic degradation of bromothymol blue in aqueous nano TiO_2 particles dispersions. Nanoscale Res Lett 5:1627–1632
139. Horikoshi S, Hidaka S, Serpone N (2002) Environmental remediation by an integrated microwave/UV–illumination method. 1. Microwave–assisted degradation of Rhodamine–B dye in aqueous TiO_2 dispersions. Environ Sci Technol 36:1357–1366
140. Gao Z, Yang S, Ta N, Sun C (2007) Microwave assisted rapid and complete degradation of atrazine using TiO_2 nanotube photocatalyst suspensions. J Hazard Mater 145:424–430

Chapter 8
Direct Flocculation Process for Wastewater Treatment

Mei Fong Chong

Notations

A_p, A_m, A_s	Hamaker constant of the solids, solvent and polymer respectively, J
a_m	Effective monomer size, nm
b_R	Fitting parameter in Eq. 8.10, dimensionless
C_L	Aggregate structure prefactor, dimensionless
d_{am}	Arithmetic mean floc diameter, μm
d_F	Fractal dimension, dimensionless
$\overline{d_i}$	Arithmetic mean diameter of flocs in size class i, μm
D	Impeller diameter, m
D_{sc}	Scaling length, nm
e	Elementary charge, C
E_f	Fluid collection efficiency of an aggregate, dimensionless
G	Global average fluid velocity gradient or shear rate, s^{-1}
h_o	Minimum separation distance between particle surfaces, nm
$H_{(x,y)}$	Unretarded geometric functions, dimensionless
K	Debye-Hückel parameter, J/m^3C
K_B	Boltzmann constant, J/K
m_i	Salt concentration, mol/m^3
n	Number concentration of particles or aggregates, m^{-3}
N	Rotational speed, rpm
N_{AV}	Avogadro's number, mol^{-1}

M.F. Chong (✉)
Department of Chemical and Environmental Engineering, Faculty of Engineering, The University of Nottingham, Malaysia Campus, Jalan Broga, 43500 Semenyih, Selangor, Malaysia
e-mail: MeiFong.Chong@nottingham.edu.my

S.K. Sharma and R. Sanghi (eds.), *Advances in Water Treatment and Pollution Prevention*, DOI 10.1007/978-94-007-4204-8_8,

N_i	Number concentration of particles or aggregates in section i, m^{-3}
N_P	Power number of impeller, W
r_{ci}, r_{cj}	Floc collision radius, μm
r_i	Particle radius at size class i, μm
r_o	Composite radius of a particle with adsorbed polymer layers, μm
r_{oi}, r_{oj}	Primary particles radii, μm
$\overline{r}$	Mass mean aggregate radius, μm
s	Distance between particles centers, nm
S	Specific rate constant of fragmentation, s^{-1}
t	Flocculation time, s
T	Suspension temperature, K
v, u	Particle or aggregate volumes, m^3
V	Volume of the suspension, J
V_T	Net interaction energy between two primary particles, J
V_{edl}	Electrical double layer repulsion, J
V_s	Energy of steric repulsion or bridging attraction, J
V_{vdw}	Van der Waals energy, J
z_c	Valence of counterion, dimensionless
z_i	Valence of electrolyte ions, dimensionless

Greek Letters

α	Collision efficiency factor, dimensionless
α_{Sc}	Numerical constant, dimensionless
β	Collision frequency factor, m^3/s
γ	Breakage distribution function, dimensionless
ε_o, ε_r	Dielectric constant of a vacuum and the solvent, C/mV
$\overline{\varepsilon}$	Average turbulent energy dissipation rate, m^2/s^3
ρ	Density of the suspension, kg/m^3
ρ_p	Particle density, kg/m^3
μ	Fluid dynamic viscosity, kg/ms
v	Kinematic viscosity, m^2/s
υ_i	Floc volume fraction in size class i, dimensionless
δ	Adsorbed polymer layer thickness, nm
λ_R	Characteristic wavelength of interaction, nm
ψ_{oi}, ψ_{oj}	Surface potential, mV
Γ	Total amount of polymer adsorbed on a single surface
Γ_o	Adsorbed amount at saturation
Φ_{so}	Polymer volume fraction at a single saturated surface, dimensionless

8.1 Introduction

Chemical treatment is one of the most utilized treatment methods in any water and wastewater treatment process. Chemical treatment, often being termed as primary treatment, involves a process of coagulation and flocculation. In this process, the colloidal particles brought about are destabilized while the soluble constituents are precipitated, both into microflocs by the addition of a chemical reagent called as coagulant. This is then followed by flocculation where the destabilized particles agglomerate and form bulky floccules, which can be settled, called flocs. The addition of another reagent called flocculant or a coagulant aid may promote the formation of the flocs [1]. The aim of applying coagulation and flocculation treatment is generally to remove the colloidal matters such as suspended solids present in the wastewater.

The most common coagulants used are hydrolyzable metal cations such as lime, aluminum sulfate (alum), ferric chloride, and ferrous sulfate whereas polymers are employed as flocculants. These coagulants and flocculants are employed extensively in water and wastewater treatment [2–4]. Although inorganic coagulants are inexpensive and readily available, their usage requires high chemical cost due to high dosage. It also generates excessive volumes of phytotoxic sludge and cannot be readily disposed. A large amount of caustic soda is needed to alter the solution pH to achieve its isoelectric point and coupling with flocculation is needed to improve the efficiency [5].

The flocculants of organic macromolecules polymers offer significant advantages in coagulation-flocculation process. The concentrations needed are only a few milligrams per liter and they generate small quantity of nonhazardous sludge for easy disposal. On a price-per-weight basis, they are much more expensive than inorganic coagulants, but overall operating cost is lower because of a reduced dosage, elimination of pH-adjusting chemicals, and reduced sludge disposal costs due to lower sludge volumes [6].

With the wide availability of flocculants or more precisely polyelectrolytes at variety of type, charge density, and molecular size, complete elimination of coagulation by using direct flocculation is gaining its popularity and importance. The major reasons to its gaining application are its biodegradability, simplicity, and it is also inexpensive. This chapter provides some useful information on direct flocculation which consists of the detailed description on the basic principle of flocculation, the difference between conventional coagulation-flocculation and direct flocculation, process modeling and simulation for flocculation process based on Population Balance Model (PBM), type of flocculants and their applications, industrial applications of direct flocculation, and finally, a special case study on the adsorption-flocculation for boron removal from wastewater.

8.2 Basic Principle of Flocculation

Coagulation and flocculation are often used interchangeably and ambiguously, as in reality most coagulants often enable the formation of agglomerates by bridging and thus helping to flocculate. At the same time, the cationic or anionic charge carried by the flocculants will simultaneously destabilize the colloidal particles and bridge them together. However, in this chapter, the terms coagulation and flocculation carry distinguish meanings as:

1. Coagulation is the process of destabilization of colloidal particles brought about or precipitation of soluble constituents as complex metal hydroxide by the addition of a chemical reagent called as coagulant which is mostly metal salts, e.g., alum, ferric chloride, and polyaluminum chloride (PAC). The destabilization of colloidal particles takes place when the DLVO (Derjaguin, Landau, Verwey, and Overbeek) energy barrier is effectively eliminated thus lowering the energy barrier.
2. Flocculation is defined as the agglomeration of destabilized particles due to bridging, by the addition of long chains polymers called flocculants, when they are driven toward each other by hydraulic shear forces in the rapid mix, into bulky, visible floccules which can be settled as flocs.

8.2.1 Stabilized Colloidal Suspensions

The agglomeration of destabilized colloidal suspensions by polymer flocculants, which undergo an irreversible process controlled by hydrodynamic and physicochemical conditions, is gaining its popularity in terms of scientific interest and industrial importance [7]. The understanding of the mechanisms of flocculation is the key to successful treatment performance while successful flocculation should start with the understanding of the type of contaminants/constituents present in the wastewater. Industrial wastewater treatment experts always relate flocculation process to water clarification as a means of suspended solids removal or turbidity reduction. In other words, the contaminants/constituents present in the wastewater must be in the form of stabilized insoluble colloidal suspension so that effective flocculation can be initialized. Contaminants in soluble form should be preconditioned into precipitated microflocs by means of coagulation and pH adjustment prior to flocculation.

The insoluble stabilized colloidal suspensions especially range from 0.01 to 5 μm in size, which contributing to water turbidity, pose great challenge in wastewater treatment due to their unsettle-ability. The stability of the colloidal suspensions is strongly influenced by their electrokinetic or simply surface charge, which in nature is usually negative. The surface charge, as shown in Fig. 8.1, causes the adjacent particles to repel each other and as a result, they tend to remain discrete, dispersed, and in suspension [6].

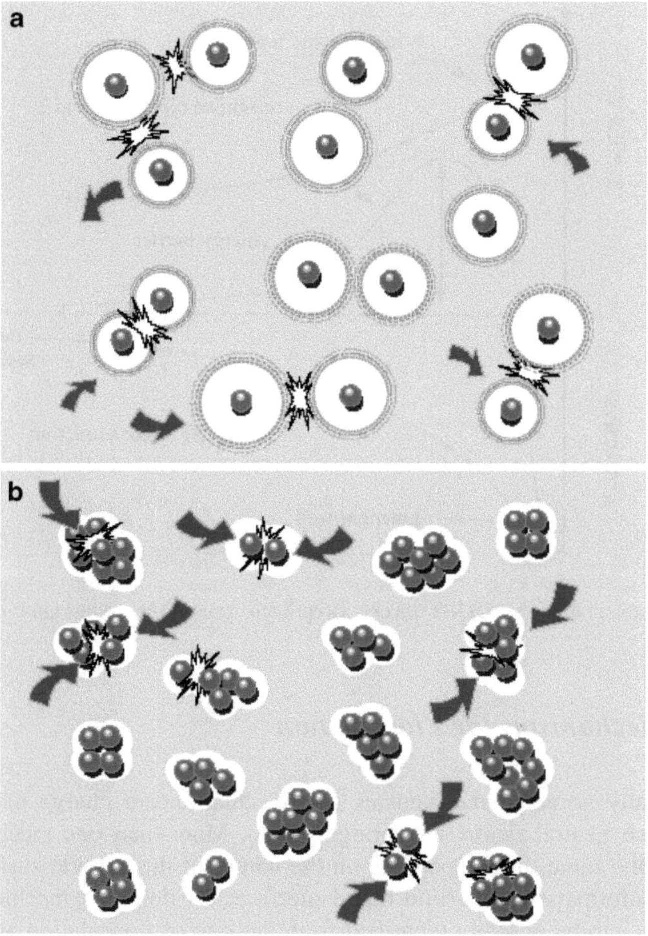

Fig. 8.1 (**a**) Stabilized particles which repel each other due to the surface charge. (**b**) Uncharged particles which tend to agglomerate during collisions (Reprinted from [6]. With kind permission of © Zeta-Meter Inc)

The tendency of the colloids to destabilize from their stabilized form is dependent on the balance between two opposing forces of electrostatic repulsion and van der Waals attraction. The classical DLVO theory [8, 9] stated that the net interaction energy of the colloids is equal to the sum of van der Waals attraction and electrical double layer repulsion. The net interaction can be attractive or repulsive depending on the distance between the colloidal particles and if it falls at the repulsive section, then the region is called the energy barrier as shown in Fig. 8.2. In order to agglomerate, the energy barrier has to be lowered or completely removed so that the net interaction energy is always attractive.

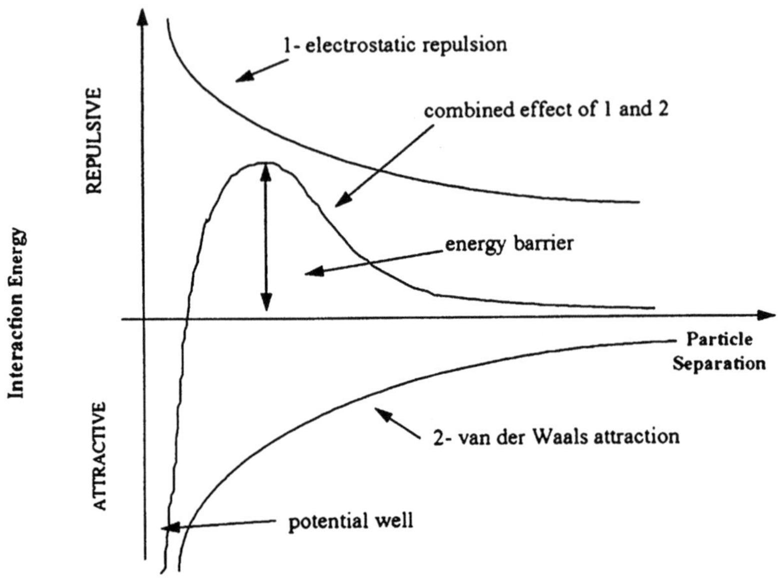

Fig. 8.2 Representation of DLVO theory (Reprinted from [10]. With kind permission of © Elsevier)

8.2.2 *Mechanisms for Flocculation*

The generally accepted mechanisms for flocculation are charge neutralization, charge patch neutralization, and bridging [11]. More than one mechanism may operate at the same time depending on the nature of the particle surface and the polymer conformation at the solid-liquid interface. The dominant mechanism during flocculation can be possibly identified from the rate of flocculation as the rate of flocculation by bridging is several orders of magnitude higher than that by either charge-patch neutralization or simple charge neutralization. The rate obtained from charge patch neutralization is about two to three times greater than that by simple charge neutralization [12].

Bridging flocculation takes place when the polymer with long chains adsorbs onto the surface of the colloidal particles following several elementary processes which occur simultaneously under turbulent flow as schematically illustrated in Fig. 8.3, which are [7]:

1. Dilution of the flocculation into homogeneous solution.
2. Collision between colloidal particles.
3. Transportation of polymer flocculant toward the surface of colloidal particles.
4. Reconformation of adsorbed polymer on the surface of colloidal particles.
5. Formation of a bond (or bridge) between colloidal particles.
6. Rearrangement and breakup of the structure of a floc.

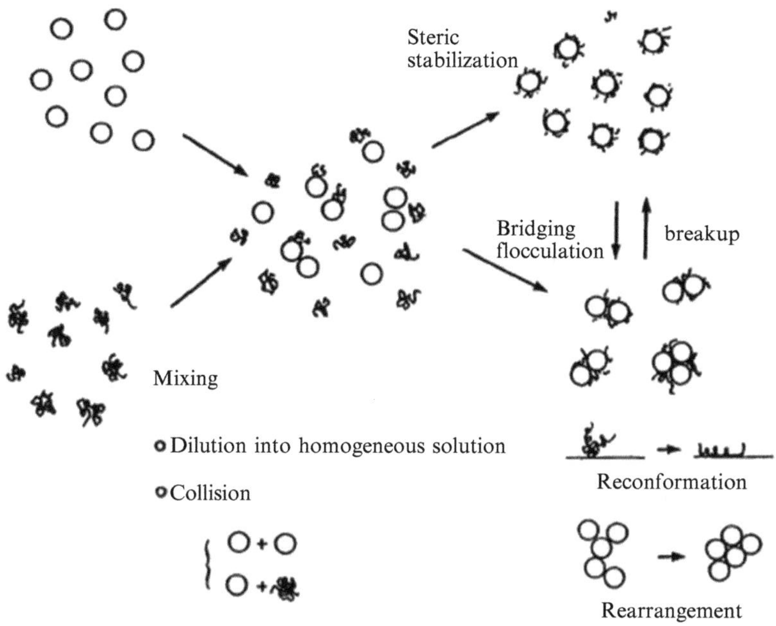

Fig. 8.3 Schematic diagram of adsorbing polymer on colloidal particles (Reprinted from [7]. With kind permission of © Elsevier)

Depending on the affinity of the polymer to the surface, polymer chains can have any one or a combination of the three conformations: trains or thin layers (starched flat on the surface), loops or coils, and dangling tails that are starched into the solution, at some angle to the surface [13, 14]. Figure 8.4 shows a schematic representation of the three different polymer conformations at the solid–liquid interface.

When the long chains polymer adsorbing on the colloidal particles conveys a charge, especially in most of the occasions where polyelectrolyte is used, the factor of charge neutralization must be considered. Only simple charge neutralization will be considered in this chapter and it often applies to cationic polymer besides the typical inorganic coagulants due to the opposite surface charge of the colloidal particles. Charge neutralization is simply lowering the DLVO energy barrier when a positively charged polymer adsorbs onto the surface of the colloidal particles. The positively charged coating neutralizes the negative charge of the colloidal particles, resulting in a near isoelectric point or zero net interaction energy.

8.3 Coagulation-Flocculation Versus Direct Flocculation

The conventional chemical pretreatment for wastewater often involves coagulation-flocculation process where inorganic coagulants (metal salts) are first added into the system to alter the physical state of dissolved and suspended solids to obtain

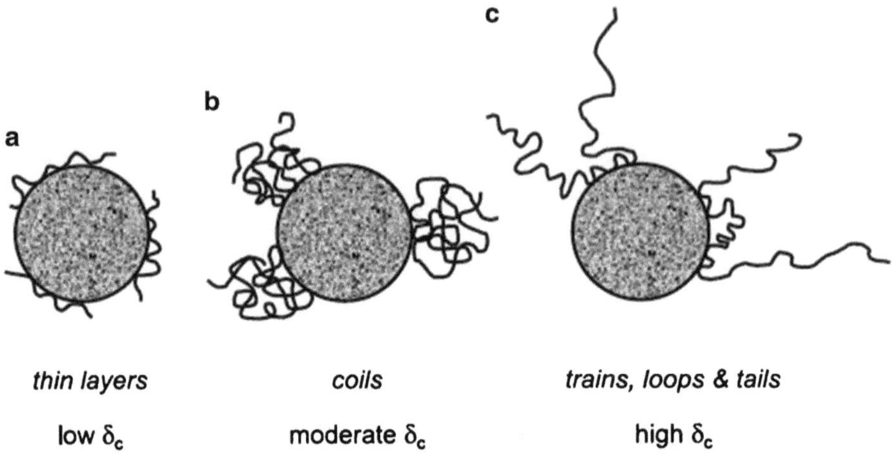

Fig. 8.4 Schematic representation of the three different polymer conformations at the solid–liquid interface. (**a**) thin layers, (**b**) coils, and (**c**) trains, loops, and tails. δ_C is adsorbed layer thickness (Reprinted from [15]. With kind permission of © Elsevier)

complex precipitates of metal hydroxides at the desired pH to facilitate sedimentation. This is followed by the addition of flocculants or often being termed as coagulant aids to enhance the treatment efficiency and sedimentation rate by the mechanical bringing of the microflocs into visible, dense, and rapid settling flocs. In this case, long chains nonionic or anionic polymers are usually used and the major flocculation mechanism involved is bridging.

The direct flocculation (i.e., without addition of coagulants) using cationic and/or anionic polymers offers the possibility to completely replace the inorganic coagulants with water-soluble organic polymers in chemical pretreatment under a constant applied shear. The direct flocculation process can be classified into the single and dual polymer systems. The single polymer system utilizes the medium charge density with high molecular weight cationic polymer. The cationic polymer serves as double acting polymer by first neutralizing the negative charge of the particles (charge neutralization) and then visible flocs formation by bridging. The dual polymer system is employed when the single polymer system failed to achieve the desired flocculation. In this system, the cationic polymer is first added for charge neutralization and bridging. The bridging-type long chain anionic polymer is then added to further enhance the bridging effects by mechanically bridging the flocs into larger and rapid settling flocs.

Although direct flocculation can be used to completely replace the inorganic coagulants without the need of pH alteration and with less sludge generation, which eventually lead to lower operational cost; the conventional coagulation-flocculation process still remains for its attractiveness especially for the wastewater with dissolved inorganic constituents which can only be precipitated as metal hydroxide at the present of suitable coagulants. Thus, the selection between conventional coagulation-flocculation and direct flocculation is highly depending on the type of wastewater and the understanding of the mechanism of flocculation is also important to ensure successful treatment. Table 8.1 provides a brief overview on the differences between coagulation-flocculation and direct flocculation which is useful as a preliminary selection guideline between the two processes.

In this following section, the pretreatment of palm oil mill effluent (POME) by using conventional coagulation-flocculation and direct flocculation is analyzed as a case study to evaluate their differences. The direct flocculation of single and dual polymer systems are also discussed to provide further insight on how dual polymer system can improve the treatment efficiency of the single polymer system. A preliminary cost analysis was also conducted to give a direct cost comparison between the two systems.

8.3.1 Coagulation-Flocculation for POME Pretreatment

POME is a colloidal suspension of 95–96% water, 0.6–0.7% oil, and 4–5% total solids including 2–4% suspended solids originating from the mixture of a sterilizer condensate, separator sludge, and hydrocyclone wastewater [36]. In the coagulation-flocculation for POME pretreatment, alum (Envifloc 40L) and flocculant (Envifloc 20S), obtained from Envilab Sdn. Bhd., Malaysia, were used throughout the study conducted by Ahmad et al. [16].

The coagulation results obtained from Ahmad et al. [16], which were plotted in Fig. 8.5, shows that water recovery and the supernatant turbidity decreased at the increasing pH from 4.5 to 9 when a constant coagulant dosage of 15,000 mg/L was used. It is interesting to note that even though minimum turbidity value of the supernatant at pH 9 was found, higher volume of sludge was generated due to the weak flocs formation which led to the poor sedimentation and water recovery. This indicates that further flocculation was required to enhance dense flocs formation for improved sedimentation and water recovery.

Figure 8.6 depicts the effects of flocculant dosage on the coagulated POME with the alum dosage of 15,000 mg/L at the pH of 6.5. The results show that the flocculation enhanced the treatment efficiency by increasing the water recovery with at least 10%. However, the water recovery and turbidity had a slight decrease with an increasing flocculant dosage.

8.3.2 Direct Flocculation for POME Pretreatment

In the study of the direct flocculation process for POME pretreatment, the performance of different type of cationic and anionic polymers at different dosage was evaluated based on the selected operating parameters of temperature, stirring speed, and stirring time. The efficiency of the direct flocculation process was evaluated

Table 8.1 Differences between coagulation-flocculation and direct flocculation

Process	Coagulation-flocculation	Direct flocculation
Ability	Able to remove highly soluble solids and metals from the wastewater besides the insoluble portion	Able to remove the suspended and stabilized colloidal particles. It is only suitable when the removal of the soluble portion is not the major concern
Type of wastewater	More suitable for inorganic wastewater, e.g., wastewater from semiconductor, metal plating, mining industries, though it is also applicable to organic based wastewater. Also suitable for portable water treatment where the water source can be from river or underground basins	Only suitable for organic-based wastewater with considerate suspended solids concentration, e.g., food, paper, and pulp, dyestuff, slaughtering house
Type of chemicals for the initial stage	Addition of coagulants (metal salts) to precipitate the dissolved constituents into complex metal hydroxide and to neutralize the surface charge of the suspended particles. Relatively weak flocs will form	Addition of flocculants (usually cationic polymers depending on the surface charge of the particles) to neutralize the surface charge and mechanical bridging of the suspended particles. Strong and dense flocs can be readily formed at this stage
Type of chemical for the subsequent stage	Addition of flocculants (usually anionic polymers) for bringing of the microflocs into visible, dense, and rapid settling flocs	Sometimes (quite rare) it requires the addition of opposite charge polymers (usually anionic polymers) to further enhance the dense flocs formation by bridging
Other requirements	Requires additional chemicals of caustic and acid for pH adjustment due to the acidic behavior of the coagulants added and the desired isoelectric point for precipitation	No requirement for pH adjustment due to the near neutral behavior of the flocculants and its dependant on charge neutralization without undergoing precipitation
Chemicals	Inorganic, inexpensive (per weight basis), and readily available	Organic, more expensive (per weight basis) and readily available
Sludge generated	Generates excessive volumes of phytotoxic sludge and cannot be readily disposed	Less sludge generation and readily for disposal as the polymers are organic in nature. However, it still depends on the type of wastewater treated
Overall treatment cost	More expensive due to the high dosage of coagulants and phytotoxic sludge disposal at excessive volume	Less expensive due to the low dosage of flocculants and nontoxic sludge disposal at small volume
Operation	More complicated as it involves more chemicals (coagulants and flocculants), handling of corrosive chemicals (acids and caustics), and the need for an exact pH adjustment	Easy as it usually involves only dosing of one chemical (cationic polymer)

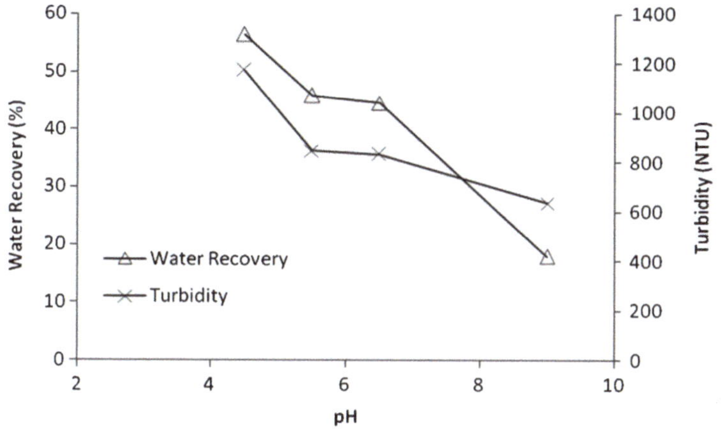

Fig. 8.5 Effects of pH on the water recovery and supernatant turbidity for the coagulation of POME (Reprinted from [16]. With kind permission of © Elsevier)

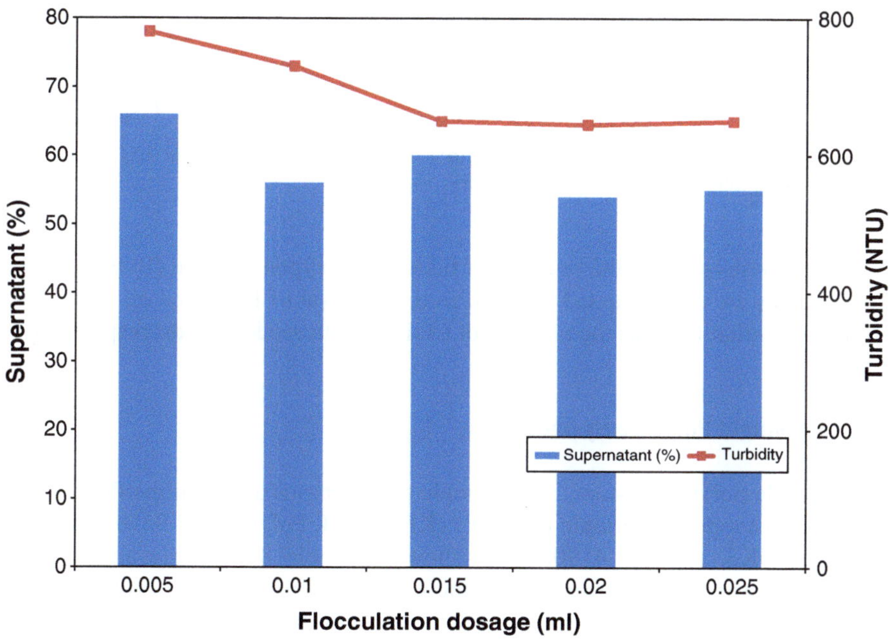

Fig. 8.6 Effect of flocculant dosage to supernatant turbidity and water recovery (Reprinted from [16]. With kind permission of © Elsevier)

based on the important responses in terms of suspended solids removal, chemical oxygen demand (COD) removal, ratio of suspended solids concentration in the filtrate to the supernatant, and water recovery. The study of these responses is adequate to evaluate the overall performance of the direct flocculation process.

Table 8.2 The supplier and price of cationic polymer

Type of cationic polymer	Supplier	Price* (RM/kg)
KP1200H	Euro Chemo Pharma Sdn. Bhd.	19.80
Polyfloc KP9650	Dia-Chemical Sdn. Bhd.	13.50
KP7000	Euro Chemo Pharma Sdn. Bhd.	33.00
Envifloc 70KS	Envilab Sdn. Bhd.	16.50
FO 4190SH	Exotic Chemical Sdn. Bhd.	16.00

*The prices quoted as per private communications in 2004 are for reference only

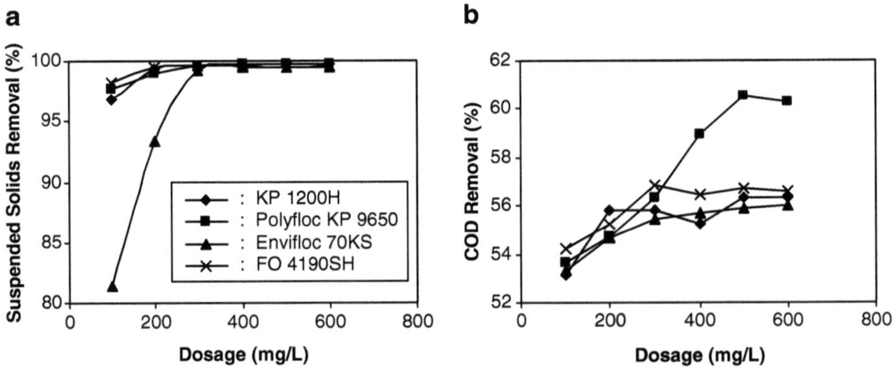

Fig. 8.7 Effect of cationic polymer dosage on (**a**) suspended solids removal and (**b**) COD removal of POME for different type of cationic polymers

Other responses such as the removal of biological oxygen demand (BOD), and oil and grease were not discussed in detail as the removal of BOD and oil and grease showed a similar trend as the removal of COD and suspended solids, respectively.

8.3.2.1 Single Polymer System

In the single polymer system, only cationic polymer was added in the flocculation process. Five types of cationic polymers (KP 1200H, Polyfloc KP 9650, KP 7000, Envifloc 70KS, and FO 4190SH) as shown in Table 8.2 were evaluated. The experiment was done on-site so that the experimental data were representative. The performance of all the cationic polymers was evaluated at the dosage range of 100–600 mg/L while the other parameters of temperature (71°C, the on-site temperature) and pH (4.1, the on-site pH) remained constant. The POME was stirred at 150 rev/min for 1 min after the addition of cationic polymer.

Figure 8.7a shows the effect of dosage for the cationic polymers on the suspended solids removal. For all the cationic polymers, the highest suspended solids removal (>99.4%) with the concentration less than 100 mg/L was achieved at the dosage of 300–600 mg/L. However, Envifloc 70KS gave poorer performance in terms of suspended solids removal compared to other polymers. The COD removal as shown

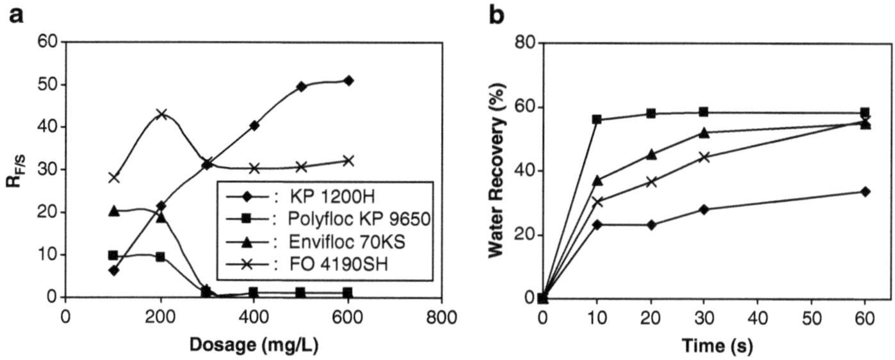

Fig. 8.8 Effect of cationic polymer dosage on (**a**) $R_{F/S}$ and (**b**) water recovery of POME for different type of cationic polymers

in Fig. 8.7b was highest for all cationic polymers at the dosage of 300–600 mg/L except for Polyfloc KP 9650. For Polyfloc KP 9650, the COD removal increased as the dosage increased to 500 mg/L and slightly decreased at 600 mg/L. However, the range of COD removal was small as it varied only from 53% to 61% for all the cationic polymers.

The flocculation efficiency based on filtration study is measured in terms of the water recovery and the ratio of suspended solids concentration in the filtrate to the supernatant, $R_{F/S}$. An efficient flocculation should have $R_{F/S} \approx 1$. If $R_{F/S} > 1$, this indicates that some fine flocs have been carried over to the filtrate, the flocs obtained are easy to break, and thus it is not suitable to be dewatered.

Figure 8.8a shows the effect of cationic polymers dosage on the $R_{F/S}$ value. For all the polymers, the filtrate-suspended solids concentration at the dosage of 200 mg/L and below increased tremendously with $R_{F/S}$ from 6 to 61. The filtrate-suspended solids concentration increased tremendously at all dosages for KP 1200H and FO 4190SH. This indicates that flocs generated were soft, weak, easy to break, and were not suitable for dewatering process. Thus, KP 1200H and FO 4190SH were not suitable in this system. For Polyfloc KP 9650 and Envifloc 70KS, the $R_{F/S}$ value at the dosage of 300 mg/L and above was close to unity (filtrate suspended solids remained below 100 mg/L) and this indicates that flocs breakage did not occur during filtration. This shows that the flocs were dense and therefore, suitable for dewatering. Based on the observed results, the dosage of 300 mg/L was the optimum dosage for the cationic polymer flocculation.

Figure 8.8b shows the water recovery for the cationic polymers at the dosage of 300 mg/L. Polyfloc KP 9650 gave the highest percentage of water recovery (56%) at 10 s. The Envifloc 70KS and FO 4190SH achieved the maximum water recovery of 56% after 60 s. This shows that Polyfloc KP 9650 had the best dewatering capability followed by Envifloc 70KS. The FO 4190SH should not be considered as it generated weak flocs as discussed in Fig. 8.8a. Between Polyfloc KP 9650 and Envifloc 70KS, Polyfloc KP 9650 was chosen for the cationic polymer flocculation

Table 8.3 The supplier and price of anionic polymer

Type of anionic polymer	Supplier	Price* (RM/kg)
AN 350M	Euro Chemo Pharma Sdn. Bhd.	16.00
Polyfloc AP 8350	Dia-Chemical Sdn. Bhd.	12.70
Polyfloc AP 8300	Dia-Chemical Sdn. Bhd.	12.70

*The prices quoted as per private communications in 2004 are for reference only

system as it offered lower price. Though Polyfloc KP 9650 demonstrated the best dewatering capability, the water recovery of 56% was still very low compared to the desired water recovery of 75%. In addition, though the maximum suspended solids removal of >99.4% with the concentration of <100 mg/L was achieved, the performance was still poor compared to the desired suspended solids concentration of <50 mg/L. Therefore, addition of anionic polymer flocculation was needed to enhance the suspended solids removal and dewatering capability of the flocculation system.

8.3.2.2 Dual Polymer System

In the dual polymer system, the cationic polymer was first added and followed by anionic polymer to enhance the performance of flocculation process. The flocculation of POME by using anionic polymer was aimed to achieve the suspended solids concentration of less than 50 mg/L and water recovery of more that 75% to meet the physical constraint of the subsequent treatments. The anionic polymers of AN 350M, Polyfloc AP 8350, and Polyfloc AP 8300 as shown in Table 8.3 were evaluated. This experiment was also done on-site so that the experimental data were representative. The performance of the anionic polymers was evaluated at the dosage range of 10–60 mg/L at the fixed parameters of cationic polymer dosage (300 mg/L), cationic polymer type (Polyfloc KP9650), temperature (71°C, on-site temperature), and pH (4.1, on-site pH). The POME was stirred at 150 rev/min for 1 min after the addition of cationic polymer and 50 rev/min for 1 min after the addition of anionic polymer.

Figure 8.9a shows that the suspended solids removal increased when the anionic polymer dosage increased. The suspended solids concentration of less than 50 mg/L with the removal of more than 99.7% was achieved at the dosage of 50–60 mg/L for AN 350M and Polyfloc AP 8350. Polyfloc AP 8300 showed the lowest performance in terms of suspended solids removal and failed to achieve the desired concentration of 50 mg/L for all the dosage. Figure 8.9b shows that the COD removal increased when the anionic polymer dosage increased with the Polyfloc AP 8350; it also gave the highest COD removal. However, the range of COD removal was small as it was varied only from 53% to 61% for all the anionic polymers.

Figure 8.10a shows that the $R_{F/S}$ remained at the value close to unity for the dosage of 50–60 mg/L. This indicates that the floc breakage during filtration at the dosage of 50 mg/L and above was very minimal. It shows that, the floc formed

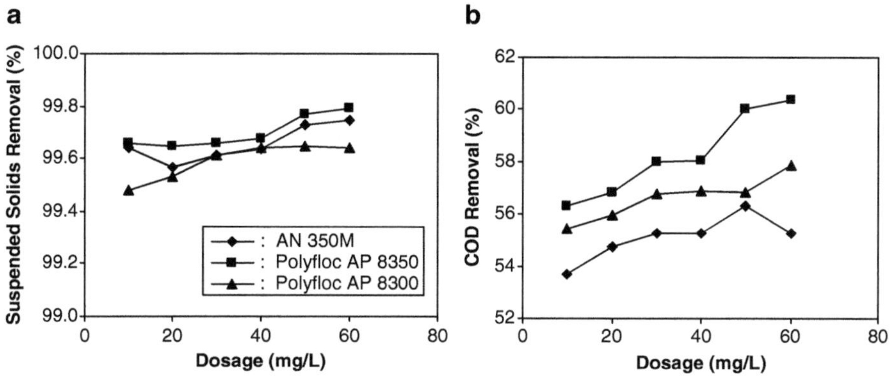

Fig. 8.9 Effect of anionic polymer dosage on (**a**) suspended solids removal and (**b**) COD removal of POME for different type of anionic polymers

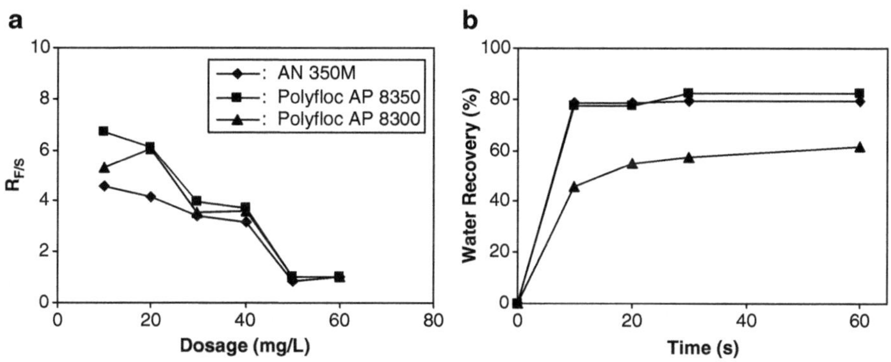

Fig. 8.10 Effect of anionic polymer dosage on (**a**) $R_{F/S}$ and (**b**) water recovery of POME for different type of anionic polymers

was dense and suitable for the dewatering process. Based on these findings, the anionic polymer dosage of 50 mg/L was recommended. Figure 8.10b shows the water recovery of the anionic polymers at the dosage of 50 mg/L. Both AN 350M and Polyfloc AP 8350 gave the highest water recovery with 78% achieved in just 10 s while Ployfloc AP 8300 gave poor water recovery. Based on the price of anionic polymers shown in Table 8.3, the Polyfloc AP 8350 was cheaper than the AN 350M. Therefore, the anionic polymer chosen was Polyfloc AP 8350.

8.3.3 Preliminary Cost Analysis

A preliminary cost analysis presented in Table 8.4 was carried out to evaluate and compare the treatment costs between the direct flocculation of POME and

Table 8.4 Comparison of the treatment costs between the direct flocculation of POME and the conventional coagulation-flocculation process

	Parameters	Direct flocculation	Conventional pretreatment[a]
1st stage	Type of coagulant/flocculant	Cationic polymer	Alum
	Dosage of coagulant/flocculant	300 mg/L POME treated	15,000 mg/L POME treated
	Unit cost of coagulant/flocculant	RM 15.50/kg	RM 1.00/kg
	Total cost of coagulant/flocculant	RM 4.65/m^3 POME treated	RM 15.00/m^3 POME treated
2nd stage	pH adjustment	Not needed	Needed
3rd stage	Type of flocculant	Anionic polymer	Cationic polymer
	Dosage of flocculant	50 mg/L POME treated	300 mg/L POME treated
	Unit cost of flocculant	RM 9.00/kg	RM 11.00/kg
	Total cost of flocculant	RM 0.45/m^3 POME treated	RM 3.30/m^3 POME treated
	Total treatment cost	RM 5.10/m^3 POME treated	RM 18.30/m^3 POME treated
	Suspended solids removal	99.66%	>99%
	COD removal	55.79%	>50%
	Oil and grease removal	99.66%	>99%
	Water recovery	80.78%	78%

[a]The literature data obtained from Ahmad et al. [16, 17]

the conventional coagulation-flocculation process [16, 17]. The cost estimates were based on the current market price in Malaysia for all the materials as quoted by suppliers. Based on the comparison of treatment efficiency between direct flocculation and conventional pretreatment of POME in terms of water recovery, suspended solids, COD, oil and grease removal, the direct flocculation showed comparable treatment efficiency if it was not better. However, without even considering the cost of chemical used in pH adjustment for the conventional pretreatment, the total treatment cost of conventional pretreatment was 3.6 times higher than the total treatment cost of direct flocculation. Therefore, direct flocculation was more cost effective than the conventional pretreatment of POME.

8.4 Process Modeling and Simulation for Flocculation Process

PBM of Smoluchowski [18] is commonly used for modeling aggregation phenomena in colloidal suspensions. In the aggregation-fragmentation processes, fragmentation is generally assumed to take place only due to fluid stress and not due to collisions between different aggregates, though it is also possible. The incorporation of both aggregation and fragmentation kinetics in the PBM is given by the following partial integral-differential equation [18]:

$$\frac{\partial n\,(\upsilon,t)}{\partial t} = -\int_0^{\infty} \alpha\,(\upsilon,u)\;\beta\,(\upsilon,u)\;\;n\,(\upsilon,t)\;\;n\,(u,t)\;\;du$$
$$+\frac{1}{2}\int_0^{\nu} \alpha\,(\nu-u,u)\;\beta\,(\nu-u,u)\;\;n\,(\nu-u,t)\;\;n\,(u,t)\;\;du$$
$$-\,S(v)\,n\,(v,t) + \int_{\nu}^{\infty} S(u)\;\gamma\,(v,u)\;\;n\,(u,t)\;\;du \tag{8.1}$$

where n is the number concentration of particles or aggregates, v and u are particle or aggregate volumes, t is flocculation time, α is collision efficiency factor, β is collision frequency factor, S is the specific rate constant of fragmentation, and γ is breakage distribution function. The first term in the Eq. 8.1 accounts for loss or disappearance of particles or aggregates of size v due to their interaction with primary particles or aggregates belonging to all sizes. The second term represents the growth of aggregates due to the interaction between primary particles and aggregates belonging to smaller size classes. The third term accounts for the loss of aggregation due to fragmentation while the last term represents generation of primary particles or smaller aggregates due to breakage or erosion of larger aggregates.

Equation 8.1 is a stochastic model. It is necessary to employ numerical solution after discretizing the equation with respect to size into a set of nonlinear ordinary differential equation (ODE). Based on the geometric discretization techniques [19], the rate of change of particle or aggregate number concentration during the simultaneous aggregation and fragmentation is given by the following discretized and lumped PBM:

$$\frac{\mathrm{d}N_i}{\mathrm{d}t} = \frac{1}{2}\,\alpha_{i-1,i-1}\beta_{i-1,i-1}N_{i-1}^2 + N_{i-1}\sum_{j=1}^{i-2} 2^{j-i+1}\alpha_{i-1,j}\beta_{i-1,j}N_j$$
$$-\,N_i\sum_{j=1}^{i-1} 2^{j-i}\alpha_{i,j}\beta_{i,j}N_j - N_i\sum_{j=i}^{\max_1}\alpha_{i,j}\beta_{i,j}N_j - S_iN_i + \sum_{j=i}^{\max_2}\gamma_{i,j}\;S_j\;N_j \tag{8.2}$$

where N_i is number concentration of particles or aggregates in section i, $\max_1$ is maximum number of sections used to represent the complete aggregate size spectrum, and $\max_2$ corresponds to the largest section from which flocs in the current section are produced by fragmentation. The first and second terms on the right of Eq. 8.2 account for growth, while the third and fourth terms account for loss of aggregates by aggregation, fifth term accounts for the loss of aggregates due to fragmentation, and the last term account for generation of smaller aggregates due to breakage or erosion. The flocculation can be visualized as a three-step process: aggregate transport represented by the collision frequency factor, $\beta_{i,j}$, attachment given by the collision efficiency, $\alpha_{i,j}$ and aggregates breakage represented by the specific rate constant of fragmentation, S_i, and breakage distribution function $\gamma_{i,j}$.

8.4.1 Collision Frequency

In the wastewater treatment system, orthokinetic aggregation (aggregation due to applied shear) is often preferred. Shear is applied by the stirring motion of impeller to accelerate aggregation process. The collision frequency factor of orthokinetic [15] is given by:

$$\beta_{i,j}^{\mathrm{sh}} = \frac{4}{3} G\left(\sqrt{E_{\mathrm{f}i}}\, r_{\mathrm{c}i} + \sqrt{E_{\mathrm{f}j}}\, r_{\mathrm{c}j}\right)^3 \tag{8.3}$$

where G is the global average fluid velocity gradient or shear rate, $r_{\mathrm{c}i}$ or $r_{\mathrm{c}j}$ is the floc collision radius, and E_{f} is the fluid collection efficiency of an aggregate. The E_{f} is in the range of $0 \leq E_{\mathrm{f}} \leq 1$. The collision frequency factor for permeable aggregates can be reverted to those for rigid spheres (rectilinear model) by setting E_{f} equal to 1. If the E_{f} is less than 1, it is curvilinear model [20, 21]. The shear rate, G can be obtained as [22]:

$$G = \sqrt{\frac{N_{\mathrm{P}} \rho N^3 D^5}{V \mu}} = \sqrt{\frac{\overline{\varepsilon}}{\nu}} \tag{8.4}$$

where N_{P} is the power number of impeller, ρ is density of the suspension, N is rotational speed, D is the impeller diameter, V is the volume of the suspension, μ is the fluid dynamic viscosity, $\overline{\varepsilon}$ is the average turbulent energy dissipation rate, and ν is the kinematic viscosity. The flocs collision radius of an aggregate, $r_{\mathrm{c}i}$ containing n_{o} primary particles is given by [23]:

$$r_{\mathrm{c}i} = r_{\mathrm{o}} \left(\frac{n_{\mathrm{o}i}}{C_L}\right)^{1/d_{\mathrm{F}}} \tag{8.5}$$

where C_L is the aggregate structure prefactor, r_{o} is composite radius of a particle with adsorbed polymer layers, and d_{F} is the fractal dimension. The collision frequency of Eq. 8.3 can be computed by assigning appropriate values of d_{F}. The fractal dimension is an indirect indicator of the flocs structure and its openness. The fractal dimension is used to incorporate the qualitatively analysis of flocs structure into the PBM which is quantitative (number and radii of flocs).

8.4.2 *Collision Efficiency*

The collision efficiency factor is computed as the reciprocal of the Fuchs' stability ratio, W between the primary particles [15]:

$$W_{i,j} = \left(r_{oi} + r_{oj}\right) \int\limits_{r_{oi}+r_{oj}}^{\infty} \frac{\exp\left(V_T/K_B T\right)}{s^2}\, ds \tag{8.6}$$

where V_T is net interaction energy between two primary particles of radii r_{oi} and r_{oj}, s is the distance between particles centers ($s = r_{oi} + r_{oj} + h_o$), h_o is the minimum separation distance between particle surfaces, K_B is the Boltzmann constant, and T is the suspension temperature.

In the DLVO (Derjaguin Landau Verwey Overbeek) theory [15, 24, 25], the net interaction energy between two primary particles, V_T is equal to the sum of Van der Waals energy V_{vdw}, electrical double layer repulsion V_{edl}, and energy of steric repulsion or bridging attraction V_s.

$$V_T = V_{vdw} + V_{edl} + V_s \tag{8.7}$$

8.4.2.1 Van der Waals Energy

In the case where inorganic coagulant is used as the coagulant, the van der Waals energy of attraction between bare particles, V_{vdw}^{H} should be considered [15].

$$V_{vdw}^{H} = -\frac{A}{6}\left\{\frac{2r_{oi}r_{oj}}{s^2 - \left(r_{oi} + r_{oj}\right)^2} + \frac{2r_{oi}r_{oj}}{s^2 - \left(r_{oi} - r_{oj}\right)^2} + \ln\left[\frac{s^2 - \left(r_{oi} + r_{oj}\right)^2}{s^2 - \left(r_{oi} - r_{oj}\right)^2}\right]\right\} \tag{8.8}$$

where A is the Hamaker constant of solids across the solvent medium. However, in the case where polymer is used, the adsorbed polymer layers on the particles should be considered. The expression for the Van der Waals energy for the case of two solids of the same kind with equal adsorbed polymer layer thickness is [15, 26]:

$$\begin{aligned} -12V_{vdw}^{V} = {} & H_{si\,sj}\left(A_{si}^{1/2} - A_{m}^{1/2}\right)\left(A_{sj}^{1/2} - A_{m}^{1/2}\right) + H_{pi\,pj}\left(A_{pi}^{1/2} - A_{si}^{1/2}\right) \\ & \times \left(A_{pj}^{1/2} - A_{sj}^{1/2}\right) + H_{pi\,sj}\left(A_{pi}^{1/2} - A_{si}^{1/2}\right)\left(A_{sj}^{1/2} - A_{m}^{1/2}\right) \\ & + H_{pj\,si}\left(A_{pj}^{1/2} - A_{sj}^{1/2}\right)\left(A_{si}^{1/2} - A_{m}^{1/2}\right) \end{aligned} \tag{8.9}$$

where A_p, A_m, A_s is the Hamaker constant of the solids, solvent, and polymer respectively across vacuum, $H_{(x,y)}$ is the unretarded geometric functions. The retardation effect is incorporated by multiplying the unretarded van der Waals energy between the polymer coated particles with a correction function, $f_{R(h_o)}$ [27].

$$f_{R(h_o)} = 1 - \frac{b_R h_o}{\lambda_R} \ln\left(1 + \frac{\lambda_R}{b_R h_o}\right) \tag{8.10}$$

λ_R is the characteristic wavelength of interaction, 100 nm and b_R is a fitting parameter of 5.32.

8.4.2.2 Electrical Double Layer Repulsion

The interaction energy due to the electrical double layers between two spheres of radii, (r_{oi} and r_{oj}) and surface potentials, ψ_{oi} and ψ_{oj} is given by [28]:

$$V_{edl} = 64\pi\varepsilon_o\varepsilon_r\left(\frac{K_B T}{z_c e}\right)^2 \left(\frac{r_{oi} r_{oj}}{r_{oi} + r_{oj}}\right) \tanh\left(\frac{z_c e \psi_{oi}}{4K_B T}\right) \tanh\left(\frac{z_c e \psi_{oj}}{4K_B T}\right) \exp(-\mathrm{K} h_o) \tag{8.11}$$

where e is the elementary charge, z_c is the valence of counter ion, ε_o, ε_r are the dielectric constant of vacuum and the solvent, and K is the Debye-Hückel parameter which is defined as [29]:

$$\mathrm{K} = \frac{N_{AV}\, e^2 \sum_i m_i z_i^2}{\varepsilon_o \varepsilon_r K_B T} \tag{8.12}$$

where N_{AV} is the Avogadro's number, m_i is salt concentration, and z_i is valence of electrolyte ions. The i in Eq. 8.12 refers to an electrolyte species in solution.

8.4.2.3 Bridging Attraction

The interaction energy due to bridging attraction (V_s) is dependent on the adsorbed polymer layers. It is important to understand the electrochemical changes brought by the adsorbed polymer on particle surfaces. The scaling theory [30, 31] is used to compute forces due to adsorbed polymer layers. It was chosen because it permits derivation of analytical formulas for interaction between spherical particles. The scaling theory is based on minimization of a surface free energy functional subject to the constraint that total amount of polymer adsorbed is fixed in the region between two surfaces having adsorbed layers. The interaction energy due to bridging attraction between two unequal polymer-coated spheres can be computed as [25]:

$$V_s = \left(\frac{2\pi r_{oi} r_{oj}}{r_{oi} + r_{oj}}\right)\left(\frac{\alpha_{Sc} K_B T}{a_m^3}\right)\Phi_{so}^{9/4} D_{Sc} \times \left\{-\frac{16\Gamma D_{Sc}}{\Gamma_o}\ln\left(\frac{2\delta}{h_o}\right) + \frac{4D_{Sc}^{5/4}}{2^{5/4}}\left(\frac{8\Gamma}{\Gamma_o}\right)^{9/4}\left[\frac{1}{h_o^{1/4}} - \frac{1}{(2\delta)^{1/4}}\right]\right\} \tag{8.13}$$

where δ is adsorbed polymer layer thickness, α_{Sc} is numerical constant which can be obtained from osmotic pressure and light scattering experiments on polymer solution, a_m is effective monomer size, Γ is total amount of polymer adsorbed on a single surface, D_{sc} is the scaling length, Φ_{so} is polymer concentration at a single saturated surface, and Γ_o is adsorbed amount at saturation.

The proposed scaling theory of Eq. 8.13 is based on flocculation in the absence of shear. When shear is applied, there will be lateral sliding and friction forces besides the normal forces. Due to the lateral shear stress, the adsorbed polymer chains may swell or stretch because of the fluid velocity gradients and osmotic pressure can increase as the fluid tries to squeeze out of the gap between particle surfaces. The chain may get desorbed if the shear rate is too high. However, shear forces between surfaces covered with adsorbed polymers have not been studied, either theoretically or experimentally [25]. Due to current limitations, the scaling theory is being applied in the modeling of flocculation under applied shear by ignoring the sliding and friction forces on the adsorbed layer.

8.4.3 Aggregates Breakage

The increase of collision frequency by applying shear does not necessary result in high rates of flocculation. Aggregates breakage or fragmentation often occurs due to shear stress especially at high stirring speed. The parameters used to compute the aggregates breakage is the specific rate constant of fragmentation, S_i and the breakage distribution function, $\gamma_{i,j}$. The specific rate constant of fragmentation, S_i is defined as [15, 23]:

$$S_i = \left(\frac{4}{15\pi}\right)^{\frac{1}{2}}\left(\frac{\bar{\varepsilon}}{\nu}\right)^{\frac{1}{2}}\exp\left(\frac{-\varepsilon_{b,i}}{\bar{\varepsilon}}\right) \tag{8.14}$$

where $\varepsilon_{b,i}$ is the critical turbulent energy dissipation rates at which floc breakage takes place. The breakage distribution function, $\gamma_{i,j}$ is a fitting parameter.

8.4.4 Dynamic Scaling

As aggregation proceeds, flocs develop as porous object with highly irregular and open structures. PBM represent aggregation process quantitatively (number and radii of flocs); however, it does not compute the aggregates quality in terms of flocs compactness. Flocs quality, besides having large aggregate radii, and compactness of the flocs is also crucial to assist sedimentation and liquid-solid separation. The fractal dimension, d_F is a simple parameter to represent the complex structure of aggregates. The fractal dimension falls in the range of $1 \leq d_F \leq 3$ [32]. The formed aggregates are more compact when the system has high fractal dimension. The fractal dimension of polymer flocculated flocs ranges from 1.7 to 2.5 [21]. The fractal dimension can be computed by using the scaling law for mean aggregate size as a function of time [33]:

$$\bar{r} \propto t^{1/d_F} \tag{8.15}$$

where $\bar{r}$ is the mass mean aggregate radius. A plot of log $(\bar{r})$ against log (t) will be a straight line and d_F can be obtained as the reciprocal of the slope.

8.4.5 *Solution of the Model*

Equation 8.2 of discretized PBM forms the governing equation for the flocculation process. The time evolution floc size distribution data can be predicted from Eq. 8.2 once the parameters of collision frequency factor, $\beta_{i,j}$, collision efficiency, $\alpha_{i,j}$, and specific rate constant of fragmentation, S_i are specified. The collision frequency factor, $\beta_{i,j}$ can be calculated from the Eqs. 8.3, 8.4, and 8.5 once the value of fractal dimension, d_F is obtained from the dynamic scaling based on Eq. 8.15. The collision efficiency, $\alpha_{i,j}$ can be obtained from the Eqs. 8.6, 8.7, 8.8, 8.9, 8.10, 8.11, 8.12, and 8.13 which accounts the effects of van der Waals energy, V_{vdw}^{V}, electrical double layer repulsion, V_{edl} and bridging attraction, V_{s}. The specific rate constant of fragmentation, S_i can be obtained from Eq. 8.14. The discretized PBM of Eq. 8.2 forms a set of nonlinear ODEs and can be solved numerically by the orthogonal collocation technique [34]. The Eq. 8.2 coupled with Eqs. 8.3, 8.4, 8.5, 8.6, 8.7, 8.8, 8.9, 8.10, 8.11, 8.12, 8.13, and 8.14 can be solved using any computer software package, e.g., Matlab 7.0 following the algorithm as shown in Fig. 8.11.

8.5 Type of Flocculants and Their Applications

It is always a challenging task to select an appropriate polymer for a specific wastewater treatment due to its wide availability from the manufacturers. The characteristics in terms of type of charge, molecular weight, molecular structure, and charge density are always used for the polymer's classification. The knowledge on the polymer characteristics as shown in Table 8.5 will aid in polymer selection; however, preliminary bench testing of polymers by using standard jar tests is the most important part of the selection process to identify the specific polymer, its dosage, and mixing requirements.

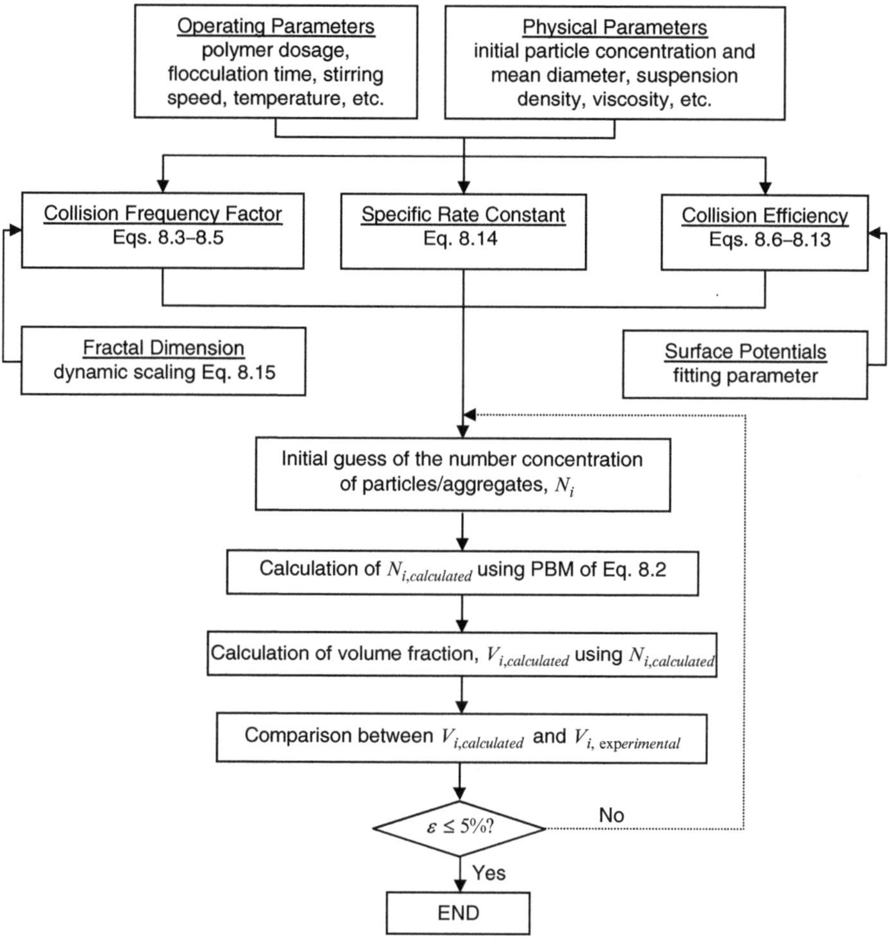

Fig. 8.11 Flow diagram of the algorithm for solution of the governing equations

To date, there are some world's leading manufacturers of flocculants for water and wastewater treatment, which include SNF, Inc., a subsidiary of SNF Floeger, France, Ciba Specialty Chemicals Corporation, Germany, BASF-The Chemical Company, Germany, Dia-Nitrix Co., Ltd., Japan, Stockhausen, Inc., Germany, Sanyo Chemical Industries, Ltd., Japan, Mitsui Chemical Aqua Polymer Inc., Japan, CYTEC Industries, USA, Kolon Industries, Inc., and Korea. There are also many more other manufacturers especially from China and Korea. Table 8.6 provides some general rules on polymer selection based on the experience obtained from SNF Floerger. Looking at the great diversity of the flocculants, this could offer some limited helps to narrow the flocculants selection prior to the jar test. Even so, there are no specific rules of thumb to give systematic guidelines on flocculants selection but cumulated experience in hands-on flocculation experiments is the most important key to instant selection of the right flocculants.

Table 8.5 Characterizing polymers based on their classification

Type	
Nonionic	
Very low charge density, typically known as polyacrylamide, flocculates through bridging	
Anionic	
Negatively charged, and normally used for bridging	
Cationic	
Positively charged, *double acting* ability: charge neutralization and bridging	
Amphoteric	
Exhibit both cationic and anionic behavior	
Physical form[a]	
Powder	
Advantage: Contains 100% active matter	
Liquid	
Advantage: Simplicity of use	
Emulsion	
Advantage: Easy to use and an increased efficiency on certain applications due to the specific molecular structures	
Beads	
Advantage: Absence of dust and rapid dissolution	
Dispersion	
Advantage: Ability to be directly fed inline without any expensive makeup equipment or aging time. They have also been found efficient in phase separation of flotation processes	
Molecular weight	
Very high	>10,000,000 g/mol
High	1,000,000–10,000,000 g/mol
Medium	200,000–1,000,000 g/mol
Low	100,000–200,000 g/mol
Very low	50,000–100,000 g/mol
Very, very low	<50,000 g/mol
Charge density	
Very high	
High	
Medium	
Low	
Very low	

[a]Data obtained from SNF FLOERGER

8.6 Industrial Applications of Direct Flocculation

Ever since the introduction of flocculants, industrial applications of direct flocculation in water and wastewater treatment is very minimal compare to the conventional coagulation-flocculation though it is widely applied in sludge conditioning and dewatering. One of the major reasons is the inadequate knowledge and understanding of the water chemistry, colloidal particles surface behaviors,

Table 8.6 General rules for polymer selection

Industries	Nonionic flocculants	Anionic flocculants	Cationic flocculants
Agri foodstuff			
Sludge treatment			√
Water treatment		√	√
Dyeing			
Sludge treatment			√
Water treatment		√	
Paper making			
Sludge treatment		√	√
Water treatment		√	√
Chemical industry			
Sludge treatment		√	√
Water treatment	√	√	√
Effluent with oil			√
Mechanical industry			
Sludge treatment	√	√	√
Water treatment	√	√	√
Effluent with oil			√
Municipal effluent			
Sludge treatment		√	√
Water treatment	√	√	√

Note: data obtained from SNF Floerger

and flocculation mechanisms. In some cases, the wastewater treatment plants are designed, constructed, and commissioned together with the processing plants by a team of engineers who are not environmental engineers.

Tables 8.7 and 8.8 provide some examples of successful cases where the conventional coagulation-flocculation processes were completely replaced by direct flocculation processes in Malaysia. The treatment efficiencies were evaluated by focusing only on suspended solids reductions as the major role of direct flocculation was to remove the stabilized colloidal particles. Table 8.7, which summarizes the industrial applications of direct flocculation for single polymer system, shows that cationic polymers with very high molecular weights indicating long chains type were used in majority to flocculating the suspended solids regardless on their influent concentrations. As a general rule, the higher influent suspended solids concentration, the higher cationic polymer dosage required. However, it is not always applicable as the polymer dosage required still depends highly on the wastewater characteristic. In certain occasions, dual polymer system as in Table 8.8 was used to enhance the treatment efficiency, where cationic polymers were always added for charge neutralization and bridging, followed by anionic polymers for only mechanical bridging. It is important to note that the direct flocculation was able to achieve high suspended solids removal with more than 90% at reasonable cost reductions as compared to the conventional coagulation-flocculation process.

Table 8.7 Industrial applications of direct flocculation-single polymer system

Industry	Core products	Flowrate (m^3/h)	Polymer type	Molecular weight (10^6 g/mol)	Charge density (%)	Dosage (mg/L)	Suspended solids (mg/L)		Cost reduction (%)
							Influent	Effluent	
Ceramic tiles	Sanitary ware	5	Cationic	12	80	60	4,500	1	30
	Homogenous porcelain tiles	6	Cationic	12	80	1,500	11,400	2	20
	Fine porcelain tableware	0.04	Cationic	12	80	1,000	36,700	1	20
	Modern ceramic sanitary ware	5	Cationic	12	80	45	36,700	7	20
	Sanitary ware	21	Cationic	12	80	30	3,670	24	20
	Porcelain tiles	100	Anionic	16	50	50	122	1	20
Food	Dairy products	20	Cationic	0.5	40	15	77	1	–
	Dairy products	35	Cationic	12	80	10	25	1	–
	Processed chicken products	3	Cationic	12	20	1	600	1	–
	Rice vermicelli	2	Cationic	12	40	20	400	10	–
	Pasteurized liquid egg	6	Cationic	12	20	0.5	633	1	–
Oleo-chemical	Fatty acids, glycerin, triacetin	42	Cationic	12	80	15	108	3	–
	Fatty alcohol	70	Cationic	3	100	15	278	1	–
Paper mill and packaging	Boxes for packaging	1	Cationic	12	20	200	15,800	8	50
	Papers for household usage	100	Cationic	12	40	4	985	22	5
Petrochemical	Various chemical products	25	Cationic	14	80	15	330	1	–
	Various chemical products	25	Cationic	12	40	15	775	1	85
	Purified telepthalic acids	250	Cationic	14	40	5	385	1	–
	Olefins and polyolefins	18	Cationic	12	40	15	25	1	–
Textile	Automotive textile	0.2	Cationic	12	20	1,000	2,850	5	–
	Polyester staple fiber	20	Cationic	12	20	10	77	1	–
Slaughtering house	Fresh dressed chicken	8	Cationic	12	40	12	640	9	90
	Fresh dressed chicken	4.5	Cationic	12	40	15	1,885	7	75
	Fresh dressed chicken	300	Cationic	12	20	3	640	39	–
	Fresh dressed chicken	40	Cationic	12	20	10	1,122	20	–

Table 8.8 Industrial applications of direct flocculation-dual polymer system

			Single polymer				Dual polymer				Suspended solids (mg/L)		
Industry	Core products	Flowrate (m^3/h)	Polymer type	Molecular weight (10^6 g/mol)	Charge density (%)	Dosage (mg/L)	Polymer type	Molecular weight (10^6 g/mol)	Charge density (%)	Dosage (mg/L)	Influent	Effluent	Cost reduction (%)
Oleochemical	Soap, glycerin	2	Cationic	12	40	2	Anionic	16	30	2	1,210	2	–
	Biodiesel	0.5	Cationic	12	40	5	Anionic	16	30	20	302	1	60
Paper mill	Brown grade paper	5	Cationic	12	40	500	Anionic	16	30	50	55,600	61	–
Rubber and latex	Natural rubber concen-trates	0.6	Cationic	12	40	20	Anionic	16	30	25	12,500	2	–

8.7 Adsorption-Flocculation for Boron Removal from Wastewater

A novel adsorption-flocculation method proposed for the removal of boron and clarification of ceramic wastewater is an innovative approach for direct flocculation application. In ceramic industry, the wastewater is highly turbid due to the existence of fine solid particles (clay) besides having high boron concentration. The palm oil mill boiler (POMB) bottom ash was used as an alternative adsorbent and after boron adsorption on POMB bottom ash, the suspended particles as well as the bottom ash were flocculated by using the long chain polymer flocculant [35].

The optimum operating conditions for boron removal by using adsorption-flocculation process was obtained following a standard jar test. The optimum operating conditions and the quality of the treated wastewater are shown in Tables 8.9 and 8.10 respectively. At the proposed optimum operating condition as shown in Table 8.9, the boron concentration was reduced from 15 to 3 mg/L which was lower than the legislation requirement by Malaysia Department of Environment (DOE) in Standard B, 4 mg/L (Environmental Quality Act 1974). Standard B is classified corresponding to the location of the industrial area, which is the downstream region of the water reservoir. The suspended solids concentration of the wastewater was also greatly reduced. The suspended solids concentration of the wastewater was reduced to less than 5 mg/L. This was well below the DOE requirement in Standard B of 100 mg/L.

Table 8.9 Optimum operating conditions for adsorption-flocculation

Process	Parameter	Optimum conditions	
Adsorption	pH	8	
	Dosage (g bottom ash/ 300 mL wastewater)	40	
	Residence time (h)	1	
	Mixing speed (rpm)	100	
Flocculation	Flocculant type[a]	KP 1200 B	AP 120 C
	Dosage (mg flocculant/L wastewater)	100	50
	Mixing speed (rpm)	200	
	Mixing time (min)	1	

Reprinted from [35]. With kind permission of © Elsevier
[a]Supplied from Dia-Chemical Sdn. Bhd.

Table 8.10 Boron, suspended solids, and COD values before and after treatment compared to DOE standards

Parameter	Raw wastewater	Treated wastewater	DOE Standard B
Boron (mg/L)	15	3	4
Suspended solids (mg/L)	2,000	5	100
COD (mg/L)	46	22	100
pH	6	9	5–9

Source: Chong et al. [35]

The COD concentration and pH level of the treated wastewater was tested in order to ensure that the treated wastewater was in the allowable range for discharge. The COD concentration was 22 mg/L with the pH of 9.0. Thus, no further treatment for COD was required but a final pH adjustment was needed.

Nonetheless, looking at the optimum dosage of 40 g of bottom ash/300 ml of wastewater, this will result in applying 133 kg of bottom ash to purify every cubic meter of wastewater. This implies that the proposed treatment scheme is only suitable for the industries having small volume or low concentration of boron contaminated wastewater. The bottom ash is dosed directly into the adsorption tank or readily available tank without major (or any) modification in their existing treatment plant and this is the major benefit of the proposed adsorption-flocculation mechanism. In the case of high volume or concentration of boron contaminated wastewater, column operations should be considered.

References

1. Degrémont (1979) Water treatment handbook, 5th edn. Halstead Press, New York
2. Vilgé-Ritter A, Masion A, Boulangé T, Rybacki D, Bottero JY (1999) Removal of natural organic matter by coagulation-flocculation: a pyrolysis-GC-MS study. Environ Sci Technol 33:3027–3032
3. Lee JD, Lee SH, Jo MH, Park PK, Lee CH, Kwak JW (2000) Effect of coagulation conditions on membrane filtration characteristics in coagulation-microfiltration process for water treatment. Environ Sci Technol 34:3780–3788
4. Wang D, Tang H, Gregory J (2006) Coagulation behavior of aluminum salts in eutrophic water: significance of Al_{13} species and pH control. Environ Sci Technol 40:325–331
5. Sarika R, Kalogerakis N, Mantzavinos D (2005) Treatment of olive mill effluents Part II. Complete removal of solids by direct flocculation with poly-electrolytes. Environ Int 31:297–304
6. Ravina L (1993) Everything you want to know about coagulation and flocculation, 4th edn. Zeta-Meter Inc, Staunton
7. Adachi Y (1995) Dynamic aspects of coagulation and flocculation. Adv Colloid Interface Sci 56:1–31
8. Derjaguin BV, Landau LD (1941) A theory of the stability of strongly charged lyophobic sols and of the adhesion of strongly charged particles in solutions of electrolytes. Acta Physiochim USSR 14:633
9. Verwey EJW, Overbeek JThG (1948) Theory of the stability of lyophilic colloids. Elsevier, Amsterdam
10. Thomas DN, Judd SJ, Fawcett N (1999) Flocculation modeling: a review. Water Res 33:1579–1592
11. Levine S, Friesen WI (1987) Flocculation of colloids particles by water-soluble polymers. In: Attia YA (ed) Flocculation in biotechnology and separation systems. Elsevier, Amsterdam
12. Gregory J (1973) Rates of flocculation of latex particles by cationic polymers. J Colloid Interface Sci 42:448–456
13. Tjipangandjara K, Huang YB, Somasundaran P, Turro NJ (1990) Correlation of alumina flocculation with adsorbed polyacrylic acid conformation. Colloid Surf A 44:229–236
14. Fleer GJ, Cohen Stuart MA, Scheutjens JMHM, Cosgrove T, Vincent B (1993) Polymers at interfaces. Chapman and Hall, New York
15. Somasundaran P, Runkana V (2003) Modeling flocculation of colloidal mineral suspensions using population balances. Int J Miner Process 72:33–55

16. Ahmad AL, Ismail S, Bhatia S (2003) Water recycling from palm oil mill effluent (POME) using membrane technology. Desalination 157:87–95
17. Ahmad AL, Ismail S, Bhatia S (2005) Optimization of coagulation-flocculation process for palm oil mill effluent using response surface methodology. Environ Sci Technol 39:2828–2834
18. Kumar S, Ramkrishna D (1996) On the solution of population balance equations by discretization: I. A fixed pivot technique. Chem Eng Sci 51:1311–1332
19. Spicer PT, Pratsinis SE (1996) Coagulation and fragmentation: universal steady-state particle size distribution. AIChE J 42:1612–1620
20. Thill A, Moustier S, Aziz J, Wiesner MR, Bottero JY (2001) Flocs restructuring during aggregation: experimental evidence and numerical simulation. J Colloid Interface Sci 243:171–182
21. Bushell GC, Yan YD, Woodfield D, Raper J, Amal R (2002) On techniques for the measurement of the mass fractal dimension of aggregates. Adv Colloid Interface Sci 95:1–50
22. Hopkins DC, Ducoste JJ (2003) Characterizing flocculation under heterogeneous turbulence. J Colloid Interface Sci 264:184–194
23. Flesch JC, Spicer PT, Pratsinis SE (1999) Laminar and turbulence shear-induced flocculation of fractal aggregates. AIChE J 45:1114–1124
24. Runkana V, Somasundaran P, Kapur PC (2004) Mathematical modeling of polymer-induced flocculation by charge neutralization. J Colloid Interface Sci 270:347–358
25. Runkana V, Somasundaran P, Kapur PC (2006) A population balance model for flocculation of colloidal suspensions by polymer bridging. Chem Eng Sci 61:182–191
26. Vincent B (1973) The van der Waals attraction between colloid particles having adsorbed layers II. Calculation of interaction curves. J Colloid Interface Sci 42:270–285
27. Gregory J (1981) Approximate expressions for retarded van der Waals interaction. J Colloid Interface Sci 83:138–145
28. Bell GM, Levine S, McCartney LN (1970) Approximate methods of determining the double-layer free energy of interaction between two charged colloidal spheres. J Colloid Interface Sci 33:335–359
29. Israelachvili JN (1991) Intermolecular and surface forces, 2nd edn. Academic, New York
30. Genes PG (1981) Polymer solutions near an interface. 1. Adsorption and depletion layers. Macromolecules 14:1637–1644
31. Genes PG (1982) Polymer solutions near an interface. 2. Interaction between two plates carrying adsorbed polymer layers. Macromolecules 15:492–500
32. Sterling MC, Bonner JS, Ernest ANS, Page CA, Autenrieth RL (2005) Application of fractal flocculation and vertical transport model to aquatic sol-sediment systems. Water Res 39:1818–1830
33. Amal R, Coury JR, Raper JA, Walsh WP, Waite TD (1990) Structure and kinetics of aggregating colloidal hematite. Colloid Surf A 46:1–19
34. Constantinides A, Mostoufi N (2000) Numerical methods for chemical engineers with MATLAB applications. Prentice Hall PTR, Upper Saddle River, pp 502–522
35. Chong MF, Lee KP, Chieng HJ, Ramli IIS (2009) Removal of boron from ceramic industry wastewater by adsorption–flocculation mechanism using palm oil mill boiler (POMB) bottom ash and polymer. Water Res 43:3326–3334
36. Ma AN (2000) Environment management for the palm oil industry. Palm Oil Developments 30:1–10

Chapter 9
Combined Macromolecular Adsorption and Coagulation for Improvement of Membrane Separation in Water Treatment

Mohammed Al-Abri, Chedly Tizaoui, and Nidal Hilal

Nomenclature

C_p	Permeate concentration (mg/l)
C_b	Bulk concentration (mg/l)
C_w	Wall concentration (mg/l)
R_m	Hydraulic membrane resistance (m^{-1})
ΔP	Trans-membrane pressure (bar)
J_0	Pure water flux ($L/m^2.s^1$)
J_v	Permeate flux ($L/m^2.s$)
J_i	Pure water flux after 30 min backwash ($L/m^2.s$)
k	Mass transfer coefficient ($L/m^2.s$)
d_h	Hydraulic diameter of the filtration channel (m)
D	Bulk diffusivity of solute (m^2/s)
R_g	Gel layer resistance (m^{-1})
C_g	Gel concentration (mg/l)
R_c	Concentration polarization resistance (m^{-1})
R_a	Adsorption resistance (m^{-1})
R_{a1}	Weak adsorption resistance (m^{-1})
R_{a2}	Strong adsorption resistance (m^{-1})
η	Dynamic viscosity (kg/m/s) or (Pa.s)

M. Al-Abri
Petroleum and Chemical Engineering Department, College of Engineering,
Sultan Qaboos University, Muscat, Oman
e-mail: alabri@squ.edu.om

C. Tizaoui • N. Hilal (✉)
Centre for Water Advanced Technologies and Environmental Research (CWATER),
College of Engineering, Swansea University, Swansea, UK
e-mail: c.tizaoui@swansea.ac.uk; N.Hilal@Swansea.ac.uk

S.K. Sharma and R. Sanghi (eds.), *Advances in Water Treatment and Pollution Prevention*, DOI 10.1007/978-94-007-4204-8_9,

9.1 Introduction

Desalination is the process of removing contaminants and reducing the dissolved salts content in a water source to produce water suitable for a particular application. Several technologies were developed to achieve this and these can be categorized as thermal or membrane processes. Reverse osmosis (RO) is the most commonly used membrane technology for water desalination. In the marketplace, when judged by installed capacity, the RO process leads with 44% of total capacity, closely followed by the multistage flash with 40% of total capacity. The remaining 16% are divided between other processes [1]. The estimated water cost for the multistage flash process is $0.52/m^3, whereas the RO cost is $0.45/m^3 of desalinated seawater [2]. The expected costs for the RO desalination of brackish water are between $0.20 and $0.35 per m^3 of produced water [3]. The cost figures indicate that RO will be the dominating desalination technology in the near future.

Fouling is the biggest obstacle facing the operation of RO desalination plants. Ideal RO membranes should be resistant to chemical and microbial attack, where their separation and mechanical characteristics should not change after a long-term operation. Unfortunately, seawater contains many foulants that foul RO membranes such as suspended particles, natural organic matter (NOM), micro-organisms, and heavy metals. RO process is restricted to certain operating conditions. The typical RO elements have limitations with respect to temperature (45°C), pH value (2–10), silt density index (<3 SDI), chlorine (dechlorination mandatory), and several other parameters [4]. Seawater and brackish water contain different composition of pollutants, depending on the source of collection. These pollutants cause different deteriorating effects on the RO desalination process; thus, many pretreatment methods have been proposed to obtain the desired results. Extensive pretreatment is required to increase the water recovery ratio of 30:50% for RO seawater desalination [5].

Different processes such as coagulation, flocculation, acid treatment, pH adjustment, addition of anti-scalant, and media filtration have been used as conventional pretreatment for years [6]. Nowadays, the trend is moving in the direction of integrated membrane systems pretreatment. Reasons are mainly feasibility, process reliability, plant availability, modularity, relative insensitivity in the case of raw water, and lower operating costs. Integrated membrane systems are expected to offer 14% reduction in RO treatment plant compared to conventional pretreatment [7]. This is due to the substantial drop in membrane prices in the last few years, and further reduction is expected in the coming years [3].

9.2 Seawater Composition

9.2.1 *Humic Substances*

Typical natural water contains natural organic matter (NOM), mono- and multivalent ions, micro-organisms, and organic and inorganic colloids [8]. These contaminants are divided into soluble (<1 nm), colloidal (1 nm to 1 μm), and particulate fractions (>1 μm) [9]. Natural organic matter is present in all water sources and is a complex mixture of compounds formed from the breakdown of plant and animal material in the environment. It has no readily identifiable structure. It consists of small, low molecular weight species such as carboxylic, amino acids, and proteins, through to larger and higher molecular weight humic substances (HS). Natural organic matter is one of the major fouling agents during membrane filtration [10]. They bind particles to each other and to the membrane surface [11].

Humic substances constitute a major part of the NOM present in ground water [12]. They occur as long linear chains at high pH and low ionic strength due to charge repulsion of functional groups, and as coiled, spherical molecules at low pH and high ionic strength; hence, they are more soluble at high pH [8]. It is well established that HS are amphiphilic molecules, containing at the same time hydrophilic and hydrophobic groups. Representing the hydrophilic groups in HS are those containing oxygen, nitrogen, phosphorous, and sulfur, while hydrophobic groups are aliphatic, aromatic, and cyclic hydrocarbons [13]. HS are refractory anionic macromolecules of low to moderate molecular weight. They contain both aromatic and aliphatic components with primarily carboxylic and phenolic functional groups. Carboxylic functional groups account for 60–90% of all functional groups. As a result, humic substances are negatively charged at the pH range of natural waters [14].

Humic substances are usually defined as substances containing humic acids (HA) and fulvic acids (FA) plus other components. Humic acids are acidic components that are soluble in bases but precipitate in acids. Fulvic acids are the acidic components that are soluble in both bases and acids [15]. To better understand the different behavior of HA and FA, it is necessary to consider their different molecular weight as well. Fulvic acids have a lower molecular weight and a higher percentage of carboxylic groups than HA, and that increases their hydrophilicity. Therefore, FA are soluble also at low pH values (pH ~2), where humic acids precipitate. Thus, FA have less affinity for binding hydrophobic compounds, while they often show a high metal complexion capacity; otherwise, the high molecular weight and basic hydrophobicity of the HA favor the formation of micelle-like structures [16].

The molecular weight and molecular weight distribution of humic substances differ from site to site. The average molecular weight varies from 0.5 to 10 kDa for aquatic FA [17] to a value as high as 40–300 kDa for soil-extracted HA [18]. These values are far greater than the values reported earlier, typically, 0.5–2 kDa for aquatic FA, 1.5–5 kDa for aquatic HA, and 4–6 kDa for soil HA and FA [19–21].

9.2.2 Heavy Metals

Many metals exist in natural water. Their concentration varies significantly with water source [22]. River and anthropogenic inputs, and biological and geochemical cycling may influence the concentration of metals in estuarine and coastal seawater to a much greater extent than that occurring in open ocean water. Transport of metals from rivers and estuaries is dependent on the partitioning of metals between dissolved and particulate phases. The partitioning of trace metals between dissolved and particulate phases is a function of several factors including specific metal ion, metal concentration, nature of particles, particle concentration, pH, salinity, and dissolved oxygen [23, 24]. In a study of trace elements, three groups had been distinguished according to their behavior during filtration and association with colloids. Heavy metals like Co, Ni, Zn, Cu, and Cd were grouped as elements of 1–10 kDa fraction that have a tendency to form inorganic or organic complexes [22].

Although the toxicity of trace metals to humans is fairly low, bioaccumulations of even small amounts in the body may be lethal [25]. In addition to trace metal toxicity, they are of particular concern in desalination due to their effects on RO membrane fouling [26]. The conventional and most commonly used method for the treatment of industrial wastewater containing heavy metals is chemical precipitation method. However, this process requires a large amount of treatment chemicals. However, nanofiltration (NF) and RO membranes can achieve >97% and >99% removal, respectively [27].

9.2.3 Colloids

The term "colloids" is still subject to an operational definition, although there is a general consensus that colloids fall largely in the nanometer to sub-micrometer size range and occur in relatively large concentrations (more than 10^9 colloids per liter). Colloids are ubiquitous; they occur in natural waters, seawater, groundwater, and interstitial soil water. Colloids may include clay minerals, oxides, or hydroxides of iron and aluminum, colloidal silica, organic matter, and biocolloids, including viruses and bacteria [28]. In suspensions, colloids interact with each other via electrostatic double-layer repulsion and/or van der Waals attraction and/or short-range repulsion (born or hydration forces) [29].

Colloids have a large specific surface area and a large number of reactive surface functional groups per unit mass because of their small size. Therefore, they are efficient sorbents for contaminants such as heavy metals, nonpolar organic compounds, and radionuclides and can potentially enhance contaminant mobility [30]. Eyrolle and Benaim [31] studied Amazonian surface water using sequential ultrafiltration (UF). Several fractions defined by the nominal molecular weight cut-off (MWCO) of the UF membranes were obtained: <5 kDa, 5–20 kDa, 20–100 kDa, 100 kDa to 0.2 μm, and 0.2–0.45 μm. The metal binding capacity of these fractionated samples revealed that the 20–100 kDa fraction was the most efficient fraction for copper binding. In addition, the UF experiments conducted by Pokrovsky and Schott [22] showed that several transition metals including Cu, Zn, Ni, Co, and Cd can be partially transported by low molecular weight (<1–10 kDa) organic acids. A full understanding of metal chemistry in natural waters and their membrane filtration needs to take into account the competitive reactions of metals with humic matter. The binding ability of HS toward heavy metals is known to depend on ion size [32] as well as HS–metal ratio [28]. Therefore, although trace metals have molecular sizes less than the pore sizes of ultrafiltration membranes, they can still be rejected in the presence of HS using UF membranes due to HS–metal complex formation [28, 33].

9.3 Ultrafiltration Membranes

Ultrafiltration membranes have pore sizes in the range of 10–500 nm and are capable of retaining species of molecular weight higher than 1 kDa. Typical rejected species include sugars, biomolecules, polymers, and colloidal particles. Most UF membranes are described by their nominal molecular weight cut-off (MWCO), which is usually defined as the smallest molecular weight species for which the membrane has more than 90% retention. Ultrafiltration can provide high-quality feedwater with low SDI and free from all suspended solids and microorganisms [34]. UF process is strongly dependent on membrane material, operating conditions, and raw water quality, but it is insensitive to high salinity [35].

9.3.1 Retention Mechanisms

Membranes and membrane suitability for a particular separation is measured by their retention among other factors. Single solute retention is defined as [36]:

$$R_{obs} = 1 - \frac{C_p}{C_b} \tag{9.1}$$

where R_{obs} is solute retention, C_p is permeate concentration, and C_b is bulk (feed) concentration.

However, during actual separation, membrane performance changes with time, depending on many factors including membrane fouling. Real retention of solutes differs with different membrane-fouling phenomena. There are different retention models that have been developed to approximate solutes' real retention instead of observed retention such as concentration-polarization model and gel-layer model. These fouling phenomena are discussed in Sect. 9.3.2.

During any membrane separation process, the concentration of solute at the membrane surface is higher than the bulk concentration due to the gradual increase of the rejected solute near the surface. This well-known phenomenon is concentration polarization. Concentration-polarization retention model is used when higher solute concentration at the membrane surface forms a thin boundary layer above the membrane of thickness (δ). Real retention is calculated using [36]:

$$\frac{C_w}{C_b} = R_{real}.e^{\left(\frac{J_v}{k}\right)} + 1 - R_{real} \tag{9.2}$$

where C_w is membrane surface concentration, R_m is hydraulic membrane resistance, J_v is permeate flux, k is mass transfer coefficient, and R_{real} is solute real retention. The mass transfer coefficient (k) is related to Sherwood number:

$$Sh = \frac{kd_h}{D} = a\text{Re}^b Sc^c \left(\frac{d_h}{L}\right)^d \tag{9.3}$$

where Sh is the Sherwood number (kd_h/D), d_h is the hydraulic diameter of the filtration channel, D is the bulk diffusivity of the solute, Re is the Reynolds number (ud_h/v), Sc is the Schmidt number (v/D), L is the length of the membrane channel, and a, b, c, and d are constants. The value of k depends strongly on the hydrodynamics of the system and, therefore, can be optimized.

Concentration polarization can be very severe in membrane filtration, especially if membrane flux is high and the diffusivity of the macromolecules is low, leading to high membrane retention. This may result in the formation of a high concentration gel layer at the surface of the membrane. The difference between concentration-polarization and gel-layer resistances is that the former is a thermodynamic modification of the pressure driving force, and the latter is the viscous resistance for flow through highly concentrated (precipitated or gelled) solutes [37]. Gel-layer concentration depends on the size, shape, chemical structure, and degree of salvation [38]. In the gel-layer region, k is expressed using the equation:

$$J_v = k \ln \frac{(C_g)}{(C_b)} \tag{9.4}$$

where C_g is gel concentration, assuming that the gel concentration remains constant across the layer.

9.3.2 Membrane Fouling

Fouling often associated with accumulation of substances on the membrane surface or within the membrane pore structure, worsens membrane performance and, ultimately, reduces membrane efficiency and shortens membrane life. These include dissolved and macromolecular organic substances, sparingly soluble inorganic compounds, colloidal and suspended particles, and microorganisms [39–41]. The following fouling mechanisms may play a role: colloidal or particulate fouling, organic fouling, biological fouling (biofouling), and scaling. To prevent RO membrane fouling, organic, colloidal, and biological matter need to be removed from the feedwater to the RO system [42]. Different fouling mechanisms require different pretreatment strategies. Biological fouling, defined as accumulation of microorganisms at the surface of the membrane, is reduced by a biological process in the pretreatment stage. It can be controlled by reducing the degradable components in the feed using slow sand filtration or by soil passage or using membrane filtration and effective cleaning procedures [39]. Scaling can be dealt with by reducing membrane recovery and by adding chemicals, e.g., acid and anti-scalants.

Colloidal matter in natural water includes sulfur, silica, and ferric and aluminum hydroxides. Their residual colloids form membrane deposits with a high hydraulic resistance contributing to the permeate flux decline due to their small size. They can also precipitate at the membrane surface forming gel layer and enhancing the osmotic pressure [43, 44]. Natural organic matter is considered a major cause of fouling in membrane filtration of natural waters. HS is the most detrimental NOM foulant, causing fouling by membrane adsorption and pore plugging [4, 45–47]. Organic fouling of membranes is dependent on the chemical properties of the NOM molecules and/or chemical transformations that they undergo at the membrane surface [48, 49]. Colloidal and organic fouling pretreatment includes conventional pretreatment such as coagulation and sedimentation and membrane pretreatment.

Two kinds of fouling can be distinguished: reversible and irreversible fouling. Reversible fouling is caused by the physical separation mechanisms that induce concentration-polarization phenomena, deposit formation, and plugging of the pores. The phenomena induce important variations of the system permeability. These variations can be regulated by enhancing the hydrodynamic shear stresses close to the membrane such as cross-flow circulation, back flushing, and momentary filtration stopping. Irreversible fouling is linked to interactions between compounds in solution and membrane material. Chemical cleaning is necessary to detach these compounds from the membrane surface. However, in water treatment, this kind of fouling does not cause very important variation of membrane permeability [50]. Polysaccharides usually foul the membranes irreversibly [51].

There are different causes of flux decline. In general, the flux decline is caused by a decreased driving force and/or an increased resistance [52]. The fouling process may be attributed to a number of mechanisms including pore blocking by solutes that are of similar diameter to the pores, formation of a cake from rejected solutes, precipitation or gelation of inorganic and organic particulates at the membrane surface as a result of the localized high concentrations that occur at the membrane/solution interface and reversible and irreversible physical changes to the membrane (e.g., compression). A schematic diagram of membrane fouling mechanisms is found in Fig. 9.1 [52]. These mechanisms can be defined using a resistance in series model [37]:

$$J_{\mathrm{v}} = \frac{\Delta P}{\eta\left(R_{\mathrm{m}} + R_{\mathrm{c}} + R_{\mathrm{g}} + R_{\mathrm{a}}\right)} \tag{9.5}$$

where J_{v} is flux through the membrane, ΔP is trans-membrane pressure, η is dynamic viscosity, R_{m} is membrane hydraulic resistance, R_{c} is concentration polarization resistance, R_{g} is gel layer resistance, R_{a} is adsorption resistance.

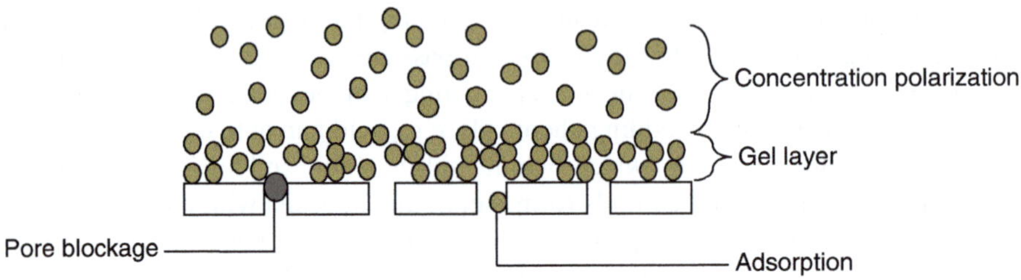

Fig. 9.1 Schematic diagram of membrane fouling mechanisms

9.4 Pretreatment Prior to Desalination

9.4.1 Conventional Pretreatment

A conventional water treatment plant consists of a multistep process applying screen filtration, ozonation, coagulation and flocculation, sedimentation, sand filtration, and usually disinfection as a last step. Each step of this process has to be controlled to get an optimal performance of the overall process, which results in a complex control system [53, 54]. Pretreatment chemicals may include ozone, flocculants, coagulants, hydrogen peroxide, lime, chlorine, sulfuric acid, sodium metabisulfiate, and anti-scalants. The use of such chemicals requires special precautions for safety purposes [55]. A conventional pretreatment is an extremely complicated chemical process. Due to its complexity and inefficiency compared to membrane pretreatment, it frequently fails in providing the required seawater quality [56]. Conventional filtration is also susceptible to inconsistent filtrate water quality when the feedwater quality varies. In addition, chemical addition and changing water conditions can lead to an operator intensive plant [57].

9.4.2 Membrane Pretreatment

The failure of conventional methods to produce treated seawater capable of reducing the fouling tendencies of RO plants and the high cost of specific operations of such conventional methods are the main factors that lead to the use of membrane filtration as a pretreatment to RO [4]. Membrane technology is widely accepted as a means of producing various qualities of water from surface water, brackish water, and seawater. Over the past years, such processes have been widely adopted by different industries. The advantages of membrane processes as compared to conventional process include [53, 58]:

- Little conditioning or ripening time after backwashing.
- Limited or no chemical requirement. No chemicals are used apart from those for cleaning the membranes, i.e., substantially reduced residues and by-products.

- Ease of operation and minimal operator involvement. The membrane plant is easier to automate owing to its simple design and greater flexibility owing to its modular construction (e.g., later extension of capacity).
- Reduction of sludge and water volume.

Desirable membranes must have suitable porosity to ensure high volume flux, low pore sizes, and a narrow pore-size distribution to meet the separation requirements, essential mechanical strength to withstand the pressure employed in the operation, and also have high chemical and biological resistance [59]. Application of UF pretreatment has the benefits of a better/safer pretreatment compared to conventional pretreatment. It has the potential to reduce the cost of seawater desalination [60]. UF systems require significantly less space and often have less power and chemical consumption than conventional pretreatment systems. As a result, UF membrane systems are a practical solution to seawater reverse osmosis (SWRO) pretreatment needs, especially for locations with limited space or variable seawater quality [57]. UF pretreatment also allows the economical utilization of RO membranes in areas where membrane desalination has not been considered as the appropriate technology due to difficult raw water conditions. UF, unlike conventional pretreatment technologies, provides a physical barrier to microorganisms, particulates, and colloids, which ensures that the RO plants can operate on a continuous basis, at high and stable fluxes, at higher recovery rates, and also allow a better control of contaminants limit values [61].

Water Supply Company of North Holland and Kiwa combined UF/ultralow pressure RO for the treatment of surface water. Measurements showed >95% removal of haloacetic acids [62]. Removal of colloidal silica and bacteria can be significant using UF membrane treatment [63]. In addition, removal of fouling constituents of seawater is more efficient using UF pretreatment compared to conventional pretreatment [51]. A study by Van Hoof et al. [64] showed a reduction of turbidity from 3 to 4 NTU to around 0.3 NTU using UF. UF has the potential of complete turbidity removal [65]. UF membrane pretreatment is a reliable technology capable of providing consistently excellent quality water independently of the fluctuation in raw water quality [66, 67]. Installation of UF prior to RO may increase the output of RO by >25% [68].

9.4.3 Fouling During Membrane Filtration

The loss of membrane flux due to fouling is a major impediment to the development of membrane processes for use in drinking water treatment [69]. The general mechanistic view of NOM interactions with membranes which plays a major role in the filtration process are [63, 70]:

- The NOM mixture has an intrinsic chemical nature (aromaticity, polarity, ionizable groups, etc.) and molecular size. The actual charge, configuration, and chemical potential of the NOM in solution depend on the current solution environment (pH, ionic strength, ion compositions, temperature, pressure, etc.), which varies throughout the filtration process.

- The combination of (a) the operating conditions of the filtration process (transmembrane pressure and hydrodynamic mass transfer at the membrane/feed interface); (b) the membrane geometry (porosity and pore size distribution), and (c) the membrane's retention characteristics toward the NOM controls the NOM's concentration at the membrane surface and in the pores.
- The chemical nature of the NOM; its concentration at the membrane fluid–solid interfaces and the chemical and geometrical nature of the membrane (under the given solution conditions) control the amount (and degree) of gel or precipitate formation and reversible and irreversible adsorption that occurs.
- The NOM's interfacial concentration, the interfacial solution's viscosity, and the mass and porosity of the adsorbed layer influence the hydrodynamic aspects of flux decline and the change in the filtration process's apparent retention of the NOM through both porous media and physical property aspects.

Increasing the feed flow increases both permeate flow and recovery up until an optimum feed flow is reached, then recovery begins to decline [71]. According to Teng et al. [63] low flux–high recovery operation is more suitable than high flux–low recovery approach for direct application on seawater UF. Intending to increase the production capacity with higher flux has to be weighed carefully against the increase in chemical cost of backwashing. Brehant et al. [51] showed that combined MF/UF cannot operate at high-flux rates when treating highly fouling surface water because of severe membrane fouling and plugging of fibers.

Membranes may be fouled by relatively hydrophilic and/or hydrophobic NOM components, depending on NOM characteristics, membrane properties and operating conditions [72]. The major contribution to fouling was attributed by the NOM fraction comprising small, neutral, hydrophilic compounds. The NOM fractions comprising HA and FA made only a minor contribution to fouling [73]. Yiantsios and Karabelas [74] studied the effect of colloid stability on membrane fouling. The rate of fouling was relatively low for hydrophobic fractions of NOM and for the charged hydrophilic fraction. However, in the case of the neutral hydrophilic fraction, the rate of fouling was considerably faster. Colloidal material may cause fouling by forming a cake on the membrane surface, while dissolved material may cause fouling by precipitating at the membrane surface or adsorption within the membrane pore space [73]. In the study by Jarusutthirak et al. [75], colloidal fractions showed high flux decline on UF membranes primarily due to the effects of pore blockage. The hydrophobic and transphilic fractions exhibited less fouling than the colloids due to their molecular size as well as electrostatic repulsion between organic acids and the membrane surface.

9.5 Coagulation

In general, very small particles are difficult to remove from water using conventional separation methods such as sedimentation and filtration. Those separation techniques work very effectively for larger particles. The goal of coagulation and flocculation is to coalesce the smaller particles into larger and more easily removed particles (flocs). Removal of NOM by chemical coagulation requires physical/chemical incorporation into flocs, which are subsequently removed by a solid–liquid separation process [76]. Coagulation/flocculation processes are mainly used for colloidal material removal, which cause color and turbidity [77]. HS interact strongly with cationic additives, especially hydrolyzing metal coagulants and cationic polyelectrolytes [78]. Kinetics of coagulant reactions is very fast and is normally complete within the first few seconds after coagulant addition [79].

Due to the sensitivity of coagulation processes, different optimal doses were obtained for different water and NOM qualities. Variations in the daily and seasonal raw water quality and chemistry issues such as pH, alkalinity, NOM, and temperature determine the optimum coagulant dose [79]. Coagulation has an optimum dosage since a higher dose may result in an increase of the residual turbidity of the settled water; and, at low dosages, a substantial reduction of residual turbidity may be observed [80]. There exists a threshold coagulant concentration value below which there is deleterious effect on plant operation. This implies that floc growth needs to proceed to a certain critical floc size prior to challenging membrane filtration, which otherwise is, apparently, partially irreversibly clogged by the flocculants solids [81].

Humic substances with higher apparent molecular weight and lower carboxylic acidities are preferentially removed by alum coagulation [76]. Enhancement of coagulation is more obvious when the organic content in water is relatively high [80] and the multivalency of the coagulant could conceivably lead to large aggregates [82]. Experimental results indicate that the removal of organics by coagulation is directly related to molecular weight. The interaction of alum with HS involves complexation, charge neutralization, precipitation, and adsorption. As the dosage is increased, the major mechanism of HS removal can be expected to shift from complexation–charge neutralization–precipitation to adsorption [83].

Jar tests done by Bolto et al. [84] on reconstituted water with alum and/or cationic polyelectrolyte showed synergistic benefits from combinations of the two. The more hydrophobic NOM fractions were the most easily removed by the polymer. The performance of cationic polymers improved significantly with increasing charge density and molecular weight. In a study by Kam and Gregory [78], they disagreed in that there was no systematic effect of molecular weight in removing dissolved organic matter, but agreed in that the most highly charged polyelectrolyte showed good removal of HA (up to around 90%) over a fairly narrow range of dosages. As the charge density of the polyelectrolyte decreases, the zone of good removal becomes broader and the removal becomes progressively worse (to about 75%).For cationic polymers with a charge density of around 3 meq/g or greater, there seems to be a simple one-to-one charge interaction, so that charge neutralization occurs with a certain amount of added polymer charge, irrespective of charge density or molecular weight.

9.6 Integration of Coagulation with Membrane Filtration

Seawater pretreatment may constitute up to one fourth of the total costs of a membrane desalting facility [85]. Pretreatment may consist of chlorination, coagulation, sand filtration, acidification, anti-scalant, micro-filtration, and dechlorination [86]. Combination of different water treatment processes depend on the feedwater quality, feedwater composition, and product requirements [7, 34, 51, 71]. The use of membrane separation as opposed to conventional clarification techniques permits a much reduced flocculation time and, thus, a more compact plant [81]. According to Kampa et al. [87], coagulation, sedimentation, and filtration (CSF) pretreatment led to unacceptable colloidal fouling, which could not be improved by an additional (in line) filtration step. Additional pretreatment by UF showed a superior particle removal. In addition, combined use of CSF/UF reduced biological parameters significantly; thereby, reducing the risks of a severe biofouling. Therefore, the treatment scheme CSF-UF-RO was pursued for a practical application.

The literature contains many contradictory results on the effects of coagulation on membrane fouling. This contradiction may be attributed to the type of water treated, type of coagulants and membranes used, and operational conditions. According to Abdessemed and Nezzal [88], coagulation improves the permeate flux in a very significant way. The coupling makes it possible to modify the structure of the deposit on the surface of the membrane by the formation of larger particles of flocs. Results obtained by Maartens et al. [82] indicate that a reduction in NOM concentration by coagulation before filtration cannot reduce or prevent membrane fouling. In fact, irreversible membrane fouling was much worse, indicating that more substances were adsorbed onto the membranes from the pretreated water, probably because metal-ions that remained in solution formed complexes with the NOM in the feed solution. On the contrary, Low et al. [89] stated that it is possible to protect polypropylene membranes from fouling occurring inside the membrane pores as well as the formation of a cake layer using a two-step coagulation using ferric chloride with anionic polyelectrolyte in the first step and aluminum coagulant in the second one.

Most studies available in the literature used inorganic coagulants combined with membrane filtration as pretreatment to seawater desalination. Recently, the industry introduced polyelectrolyte coagulants as an aid to membrane pretreatment. Unfortunately, all available research studied the use of polyelectrolytes under low salinity feedwater conditions. The aim of this study is to investigate the benefits of combined polyelectrolyte coagulation and membrane filtration as pretreatment to brackish water and seawater using synthetic water resembling their salinity.

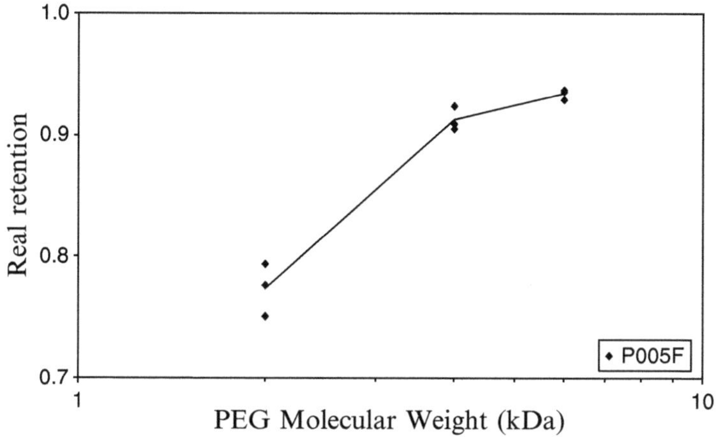

Fig. 9.2 P005F membrane real retention using polyethylene glycols (PEG)

9.7 P005F Membrane Properties

P005F polyethersulfone membrane used during the study was provided by Nadir filtration GmbH (Germany). It has an MWCO of 5 kDa as specified by the manufacturer although a real retention study illustrated in Fig. 9.2 showed that the membrane has an MWCO of 3.6–3.8 kDa. In addition, atomic force microscopy study showed that the membrane has a mean pore size of 10.0 nm and porosity of 17.3%, as illustrated in Fig. 9.3. More details are provided in Hilal et al. [90].

9.8 Membrane Retention

9.8.1 Agglomerates of Humic Substance and Heavy Metals

9.8.1.1 Humic Substance Retention

The influence of salinity, HS concentration, and heavy metals concentration on HS retention is studied using P005F-UF membrane. HS retention decreased when salinity level increased from 10,000 to 25,000 ppm NaCl, but no further increase in retention was experienced when salinity level was further increased to 35,000 ppm NaCl, as shown in Fig. 9.2. HS retention using P005F operating at 3 bar and initial feed concentration of 10 mg/l HS and 5 mg/l heavy metals was 0.77 and 0.73 at salinity of 10,000 and 25,000 ppm NaCl, respectively, as shown in Fig. 9.4.

Cation sodium ions (Na^+) are attracted to the area surrounding the anionic HS particles, forming a layer called the diffuse layer. As the salinity increases,

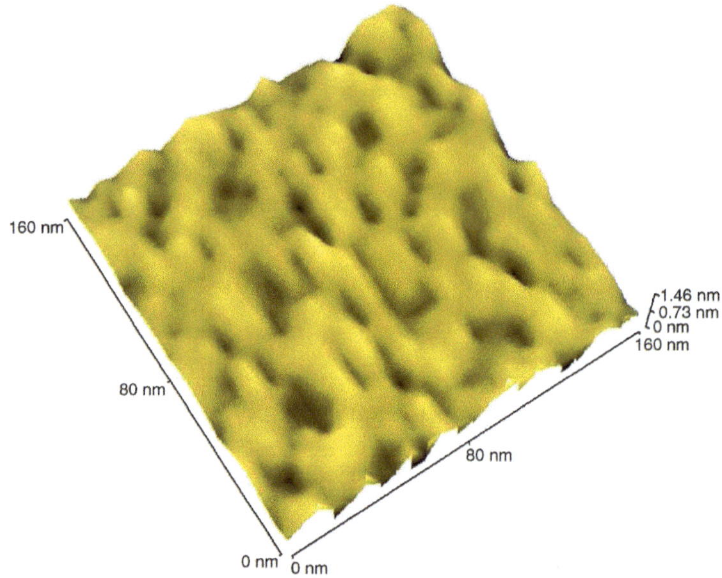

Fig. 9.3 3D topographic AFM image of P005F membrane under air environments

more counterions are added to the suspension and being attracted toward the HS particles causes the diffuse layer to compress [91]. Ionic strength has an effect on HS retention due to the increased coiling of HS molecules with increasing ionic strength. At low salinity level, HS functional groups are stretched. When salinity level is increased, these groups curl up and aggregate causing reduction in size of colloids, thus increasing HS permeability, as shown in Fig. 9.5 [8]. This phenomenon is called double-layer compression [91]. Discussion of the effects of high-level salinity (35,000 ppm NaCl) is provided in Sect. 9.8.2.

HS retention slightly decreased with increasing HS concentration, as shown in Fig. 9.4. HS retention using P005F operating at 3 bar and initial feed concentration of 25,000 ppm NaCl and 10 mg/l heavy metals was 0.73, 0.71, and 0.68 at HS concentration of 10, 20, and 30 mg/l, respectively. As HS concentration increases, the number of HS species settling at the surface of membrane will increase. This settling will enhance concentration polarization at the surface of the membrane and reduce HS retention.

Figure 9.4 shows the effect of heavy metals concentration on HS retention. Heavy metals concentration did not affect HS retention. Retention of HS using P005F operating at 3 bar and initial feed concentration of 25,000 ppm NaCl and 10 mg/l HS was 0.72, 0.73, and 0.73 at heavy metals concentration of 0, 5, and 10 mg/l, respectively. The same trend was experienced by Alpatova et al. [28] and Verbych et al. [92] using UF membranes.

Divalent ions affect HS removal through double-layer compression similar to monovalent ion effect. Divalent ions also affect HS removal through chemical

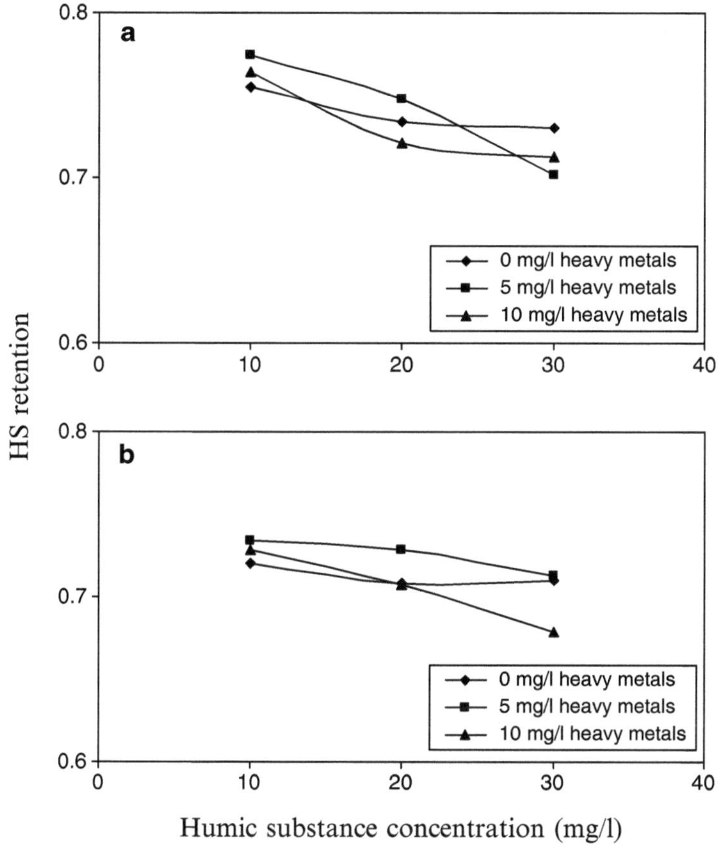

Fig. 9.4 Effect of salinity, HS concentration and heavy metals concentration on HS retention; P005F membrane operating at 3 bar. (**a**) 10,000 ppm NaCl and (**b**) 25,000 ppm NaCl

association. It is well known that HS offers binding and mobility sites for heavy metals in aquatic water and soil [93]. It may be hypothesized that the effect of double-layer compression (reducing HS colloids' size) is opposed by the chemical aggregation of heavy metals with HS (increasing the size of resultant colloids). Thus, the aggregation effects of heavy metals and HS did not affect HS retention using the tested P005F membrane. More details are presented elsewhere [94, 95].

9.8.1.2 Cu and Zn Retention

The influence of salinity, HS concentration, and heavy metals concentration on heavy metals retention was studied using P005F membrane. Heavy metal solutions consists of equal concentration of Cu, Co, Ni, Zn, and Cd, Co, Ni, and Cd retention will not be discussed in this study.

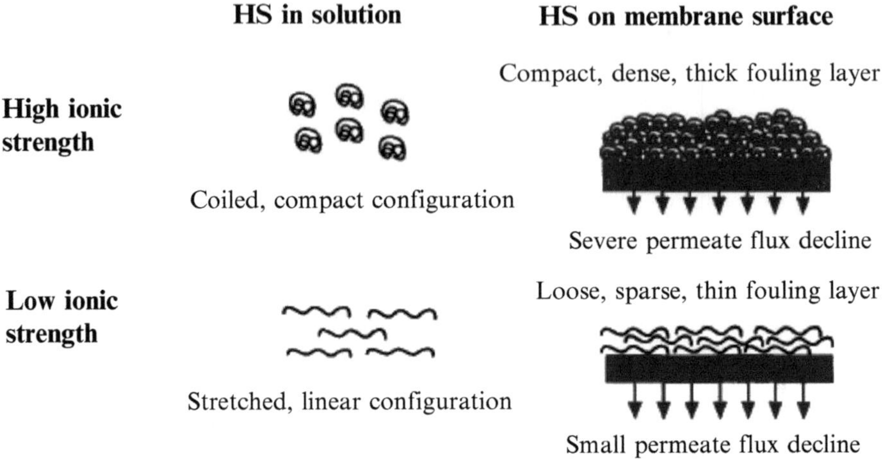

Fig. 9.5 Schematic description of the effect of ionic strength on the conformation of HS in the solution and on the membrane surface and the resulting effect on membrane permeate flux [8]

Cu and Zn retention decreased with increasing salinity level as shown in Fig. 9.6. Cu retention using P005F operating at 3 bar with initial feed concentration of 10 mg/l HS and 5 mg/l heavy metals was 0.92 and 0.85 at salinity levels of 10,000 and 25,000 ppm NaCl, respectively. The reduction in heavy metal retention with increase in ionic strength is due to the effect of salinity on HS structure, as discussed previously. As the ionic strength increases, HS molecules coil up and their surface area is reduced. Heavy metals interact with the surface of the colloidal HS molecules and the reduction in the surface area of the HS molecules will decrease the removal efficiency of the heavy metals. This results in an increase in the quantity of nonassociated metal ions that pass through the membrane pores decreasing their retention [96, 97].

Figure 9.6 shows the effect of HS concentration on Cu and Zn retention. Heavy metals retention increased with HS concentration. Cu retention using P005F operating at 3 bar with initial feed concentration of 25,000 ppm NaCl and 5 mg/l heavy metals was 0.92, 0.94, and 0.98 at HS concentration of 10, 20, and 30 mg/l, respectively. HS have a great capacity for interaction with heavy metal ions forming soluble complexes, colloidal, and/or insoluble substances because of the HS colloidal character and their high number of surface functional groups. This mechanism involves the complexation of metallic ions with soluble HS [93, 98]. Increase in HS concentration increases HS complexation because more HS molecules are available as binding sites for the heavy metals [99]. In previous experiments on fresh water undertaken by Alpatova et al. [28] and Verbych et al. [92], a limit was reached where an increase in HS concentration does not improve the removal of the heavy metals; HS–metal ratio of 2:1 was optimalfor maximum metal removal. Present

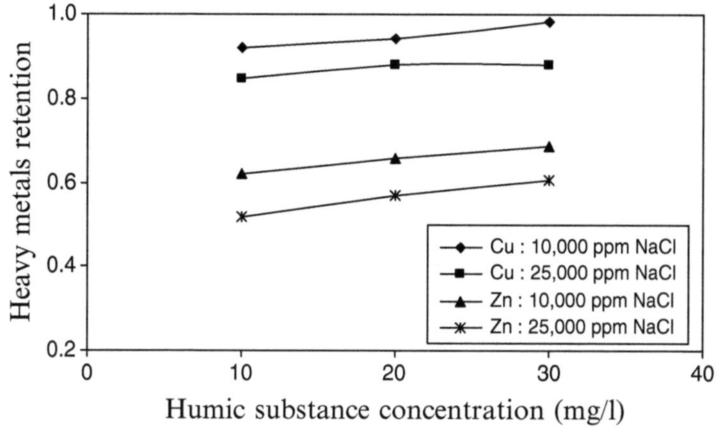

Fig. 9.6 Effect of salinity and HS concentration on heavy metals retention using P005F membrane operating at 3 bar; Initial feed concentration of 5 mg/l heavy metals

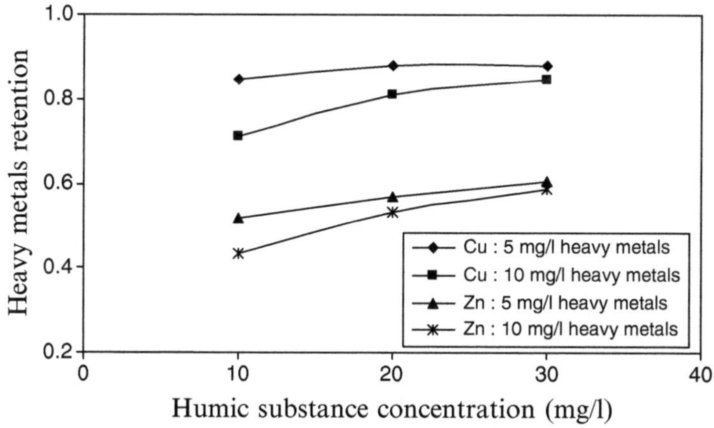

Fig. 9.7 Effect of heavy metals concentration on heavy metals retention using P005F membrane operating at 3 bar; Initial feed concentration of 25,000 ppm NaCl

results show a higher ratio; between 2:1 and 4:1. This limit was not experienced in this study due to the effect of salinity on the aggregation of heavy metals with HS molecules. Salinity compresses HS colloidal structure and reduces their active sites, as discussed previously. This will increase the amount of HS molecules required to obtain the same heavy metal removal.

Figure 9.7 shows the effect of heavy metal concentration on heavy metal retention. Heavy metal retention decreased with increasing heavy metal concentration. Cu retention using P005F operating at 3 bar with initial feed concentration of 25,000 ppm NaCl and 10 mg/l heavy metals was 0.84 and 0.71 at heavy metals concentration of 5 and 10 mg/l, respectively. Retention of heavy metals decreases when heavy metal concentration is increased due to the increase in concentration polarization at the surface of the membrane. When heavy metals concentration is increased, more metal ions are available in the bulk solution. These ions are transported to the surface of the membrane, where they accumulate and increase concentration polarization, thus reducing retention. In addition, increasing heavy metals concentration will increase the amount of dissociated heavy metals in solution due to limitation of heavy metals binding with HS.

Heavy metals retention is in the range of Cu (II) > Zn (II) using P005F, as shown in Fig. 9.7. Heavy metals retention is dependent on metal ions size, their stability constants, and the reactivity and selectivity of HS [28, 92]. The molecular weights of the studied heavy metals are 63.4 and 65.4 Da for Cu (II) and Zn (II), respectively. As the metals' molecular weight increases, their binding with HS colloids decreases. In addition, metals with higher stability constants form more stable complexes with HS. Stability constants (Log K values) of the studied heavy metals are 4.6 and 3.1 for Cu (II) and Zn (II), respectively [100].

9.8.2 *Humic Substance Coagulation*

The influence of salinity, HS concentration, and polyelectrolyte type and concentration on HS retention are investigated using P005F membrane. HS retention decreased when salinity level increased from 10,000 to 25,000 ppm NaCl, but no further increase in retention was experienced when salinity level was further increased to 35,000 ppm NaCl, as shown in Fig. 9.7. HS retention using P005F operating at 3 bar with initial feed concentration of 10 mg/l HS and 1 mg/l PDADMAC was 0.88, 0.82, and 0.83 at salinity levels of 10,000, 25,000, and 35,000 ppm NaCl, respectively.

Salinity effects on HS retention with polyelectrolyte coagulation follow the same behavior as the effect of salinity on HS retention with heavy metals agglomeration, as discussed previously. At low salinity level, HS functional groups are stretched. As salinity increases, these groups curl up and aggregate causing reduction in size of colloids due to double-layer compression [91]. The increase in salinity from 25,000 to 35,000 ppm did not exert any change in HS retention due to the very high salinity of the feed; salinity is no longer an important factor. The stability of HS is a function of the attraction and the repulsive forces. The attraction force is independent of the ionic strength; whereas, the repulsive force is affected by ionic strength. Repulsive force affects the stability of HS through the factors $\sum \left(n_i z_i^2\right)^{-1}$ and $e^{-\sum \left(n_i z_i^2\right)}$, where z_i is the valency and n_i is the number of ions of the *i*th species per unit volume. Due to the negative exponential factor, the increase in double-layer compression will rapidly become smaller as the ionic strength grows. Therefore, at a certain point HS colloids can be considered fully destabilized and any additional increase in ionic strength will not affect the HS colloidal structure [101].

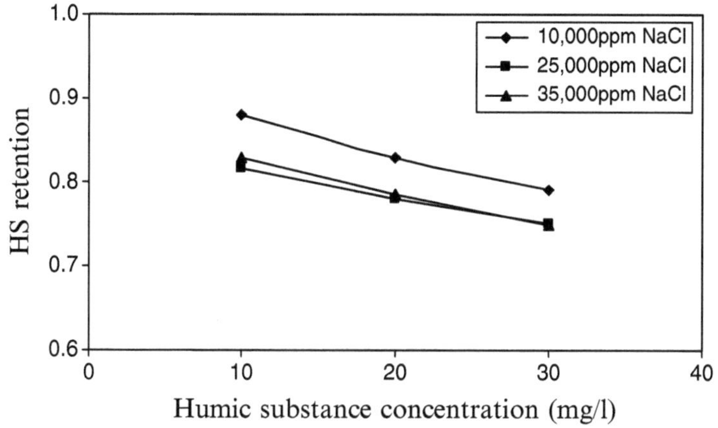

Fig. 9.8 Effects of salinity and HS on HS retention. P005F operating at 3 bar with initial feed concentration of 1 mg/l PDADMAC

Humic substance retention decreased with HS concentration, as shown in Fig. 9.7. HS retention using P005F operating at 3 bar with initial feed concentration of 25,000 ppm NaCl and 1 mg/l PDADMAC was 0.88, 0.83, and 0.97 at HS concentration of 10, 20, and 30 mg/l, respectively. These results follow the same behaviour as described in Sect. 9.8.1.1. As the HS concentration increases, the number of HS species settling at the surface of membrane will increase. This settling will enhance concentration polarization at the surface of the membrane and reduce HS retention.

Figure 9.8 shows the effect of polyelectrolyte type and concentration on HS retention. HS retention using PDADMAC was higher than CoAA. HS retention using P005F operating at 3 bar with initial feed concentration of 25,000 ppm NaCl and 10 mg/l HS was 0.92 and 0.88 using 2 mg/l PDADMAC and CoAA, respectively. HS retention also increased with polyelectrolyte concentration. HS retention using P005F operating at 3 bar with initial feed concentration of 25,000 ppm NaCl and 10 mg/l HS was 0.72, 0.82, and 0.92 at PDADMAC concentrations of 0, 1, and 2 mg/l, respectively.

Polyelectrolytes strengthen the linkage between colloidal HS, which enlarge the size of the colloids, thus increasing their retention using the tested membranes. Furthermore, the polyelectrolytes' molecular weights are very high (>250 kDa) and HS particles might be adsorbed onto the polyelectrolyte flocs forming polymer–floc complexes, leading to higher HS retention [78, 102]. Coagulation performance exhibited by the various types of polyelectrolytes is influenced by the charge density of the polyelectrolyte rather than the molecular weight. PDADMAC produces higher retention of HS because PDADMAC has charge density of 100% compared to 85% charge density of CoAA. Increasing polyelectrolyte concentration will provide more cationic species to interact with the HS, thus increasing HS neutralization and HS retention.

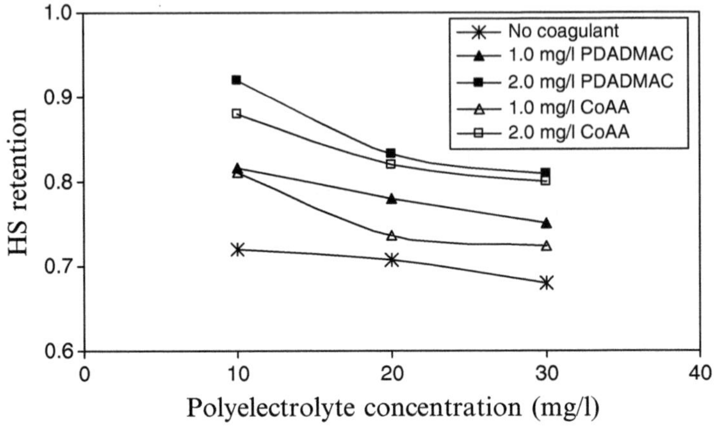

Fig. 9.9 Effects of polyelectrolyte type and concentration on HS retention. P005F operating at 3 bar with initial feed concentration of 25,000 ppm NaCl

9.8.3 Humic Substance and Heavy Metals Coagulation

9.8.3.1 Humic Substance Retention

The influence of salinity, HS concentration, heavy metals concentration and polyelectrolyte type and concentration on HS retention was studied using P005F membrane. Humic substance retention decreased with increasing salinity level using P005F membrane at low polyelectrolyte concentration. No effect of salinity on HS retention was experienced at high polyelectrolyte concentration, although highest HS retention was achieved at lower polyelectrolyte concentration using 10,000 ppm NaCl, as shown in Fig. 9.9. HS retention was 0.86 and 0.81 at 10,000 and 25,000 ppm NaCl, respectively, at initial feed concentration of 2 mg/l PDADMAC, while HS retention was 0.98 and 0.99 at 10,000 and 25,000 ppm NaCl, respectively, at initial feed concentration of 2 mg/l PDADMAC. Maximum HS retention was reached at 5 and 7 mg/l PDADMAC at 10,000 and 25,000 ppm NaCl, respectively. These results show the same behavior as previous results discussed in Sect. 9.8.1. Due to the coiling of HS structure and compression of the double layer with increased ionic strength, HS colloidal size decreases; thus, a reduction in HS retention is experienced. At high polyelectrolytes concentration, salinity does not play a major role in HS retention since maximum HS retention is achieved due to the addition of high polyelectrolyte dosage.

HS retention decreased with increasing HS concentration using P005F membranes, as shown in Fig. 9.10. HS retention using P005F operating at 3 bar at was 0.91, 0.85, and 0.81 at HS concentration of 10, 20, and 30 mg/l, respectively. These results follow a similar behavior as previous results discussed in Sect. 9.8.1. As HS concentration increases, the number of HS species settling at the surface of

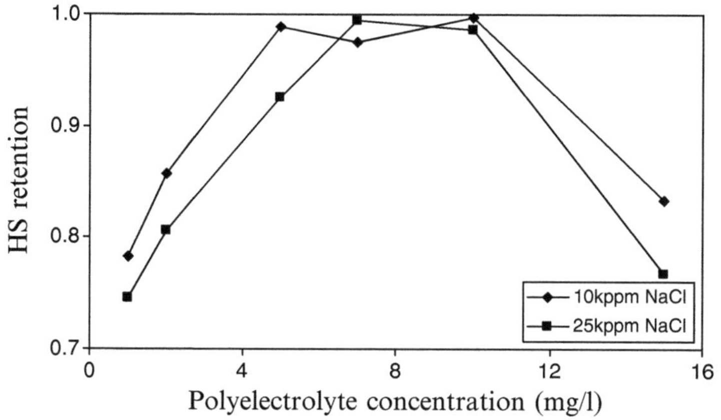

Fig. 9.10 Effects of membrane type and salinity on HS retention using P005F membrane operating at 3 bar. Initial feed concentration of 30 mg/l HS, 5 mg/l heavy metals and using PDADMAC

membrane will increase. This settling will enhance concentration polarization at the surface of the membrane and reduces HS retention. At high polyelectrolyte concentration, similar maximum HS retention was achieved at all studied HS concentrations, although a higher polyelectrolyte concentration was required to reach the maximum achievable HS retention at higher HS concentrations, as shown in Fig. 9.10. Maximum HS retention reached using P005F was 0.99 and was achieved at PDADMAC dose of 3, 5, and 7 mg/l at HS concentration of 10, 20, and 30 mg/l, respectively.

Comparison of HS coagulation without heavy metals results, Sect. 9.8.1, with HS and heavy metals coagulation shows that the addition of heavy metals does not increase HS retention at the same conditions, as illustrated in Fig. 9.11. HS retention using P005F was 0.81 and 0.82 at 5 and 10 mg/l heavy metals, respectively. As stated previously, the effect of double-layer compression (reducing HS colloids' size) is opposed by the chemical aggregation of heavy metals with HS (increasing the size of the resultant colloids).

Figure 9.11 shows the effect of polyelectrolyte type and concentration on HS retention using P005F membrane. Typical to the results shown in Sect. 9.8.2, PDADMAC produced higher HS retention compared to CoAA at low polyelectrolyte concentration. HS retention was 0.81 and 0.77 for PDADMAC and CoAA, respectively, at 2 mg/l polyelectrolyte. Coagulation performance exhibited by the various types of polyelectrolytes is influenced by the charge density of the polyelectrolyte rather than the molecular weight. PDADMAC produces higher retention of HS because PDADMAC has a higher charge density of 100% compared to 85% charge density of CoAA. Furthermore, no difference in HS retention using PDADMAC and CoAA was observed at high polyelectrolyte concentration. HS retention was 0.99 for both polyelectrolytes at 7 mg/l. Increase in polyelectrolyte concentration increased HS retention until the maximum retention was achieved.

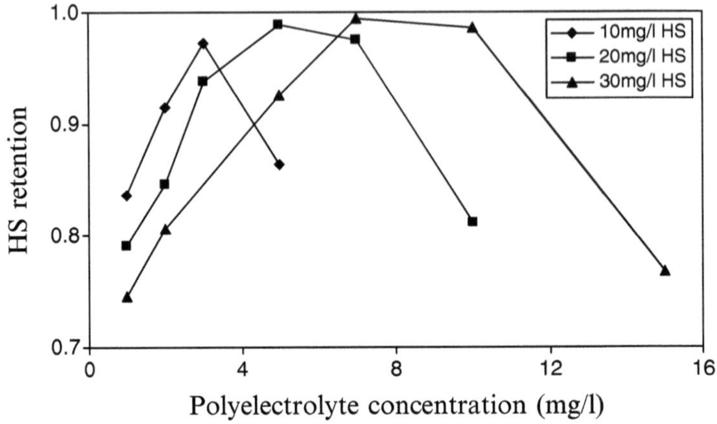

Fig. 9.11 Effects of HS concentration on HS retention using P005F membrane operating at 3 bar. Initial feed concentration of 25,000 ppm NaCl, 5 mg/l heavy metals, and using PDADMAC

Further increase in polyelectrolyte concentration deteriorated HS retention, as shown in Fig. 9.11. HS retention was 0.75, 0.81, 0.93, 0.99, 0.99, and 0.77 at PDADMAC concentration of 1, 2, 5, 7, 10, and 15 mg/l, respectively. Increasing polyelectrolyte concentration will provide more cationic species to interact with the HS, which increases HS neutralization, thus increasing HS retention. At polyelectrolyte concentration higher than the optimum dosage, HS colloids will be surrounded by excess polyelectrolyte molecules leading to high electrostatic and steric repulsion thus reducing HS retention.

9.8.3.2 Cu and Zn Retention

The influence of salinity, HS concentration, heavy metals concentration, and polyelectrolyte type and concentration on Cu and Zn retention was studied. Cu and Zn retention decreased slightly with increasing salinity at low polyelectrolyte concentration as shown in Fig. 9.12. Cu retention was 0.99 and 0.96 at salinity levels of 10,000 and 25,000 ppm NaCl, respectively. Reduction in heavy metal retention with increase in ionic strength is due to the effect of salinity on HS structure, as explained earlier in Sect. 9.8.1.2. As the ionic strength increase, HS molecules coil up reducing their colloidal surface area. Contraction of HS double layer will, thereby, reduce the probability of HS–metal interaction. This results in an increase in the quantity of nonassociated metal ions that pass through the membrane pores and decreases retention.

Figure 9.13 shows the effect of HS concentration on Cu and Zn retention. Cu and Zn retention increased with HS concentration. Cu retention was 0.86, 0.92, and 0.96 at HS concentration of 10, 20, and 30 mg/l, respectively. These results follow

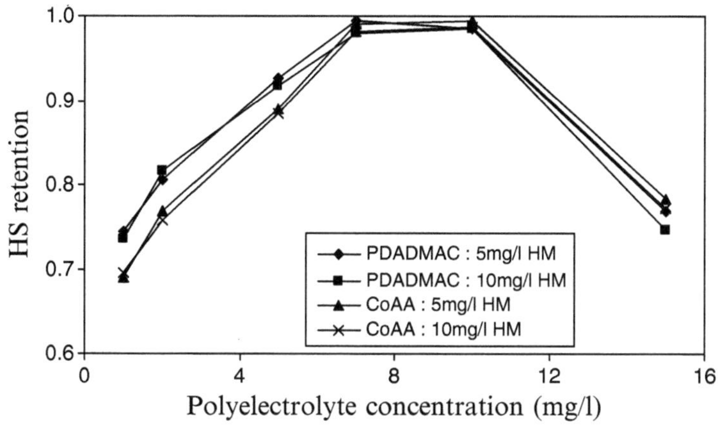

Fig. 9.12 Effects of heavy metals (HM) concentration on HS retention using P005F membrane operating at 3 bar. Initial feed concentration of 25,000 ppm NaCl and 30 mg/l HS

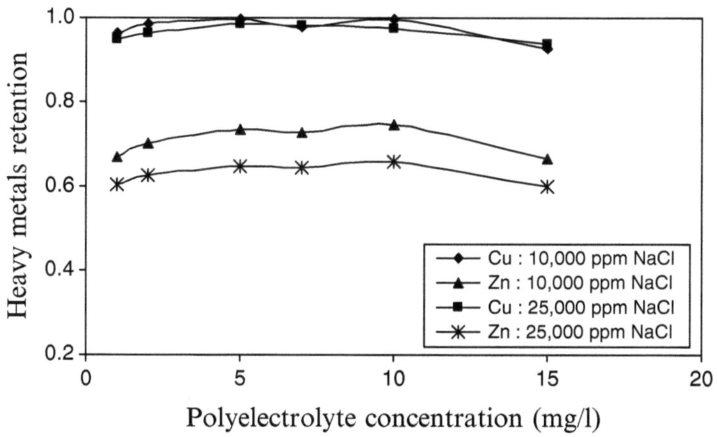

Fig. 9.13 Effect of salinity on Cu and Zn retention using P005F membrane operating at 3 bar. Initial feed concentration of 30 mg/l HS, 5 mg/l heavy metals and using PDADMAC

a similar trend to previous experiments discussed in Sect. 9.8.1.2. As explained previously, due to properties of HS colloidal surface, HS have a great capacity for interaction with heavy metal ions. This mechanism involves the complexation of metallic ions with soluble HS. Increase in HS concentration increases humic substance complexation because more HS molecules are available as binding sites for the heavy metals.

Figure 9.14 shows the effect of heavy metal concentration on Cu and Zn retention. Heavy metal retention decreased with increasing heavy metal concentration.

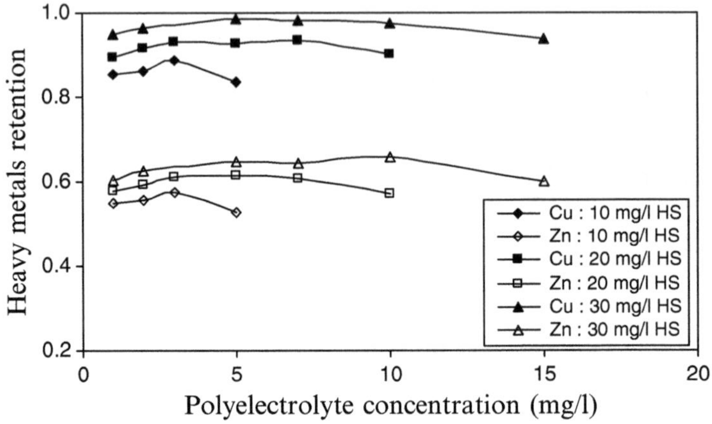

Fig. 9.14 Effect of HS concentration on Cu and Zn retention using P005F membrane operating at 3 bar. Initial feed concentration of 25,000 ppm NaCl, 5 mg/l heavy metals, and using PDADMAC

Cu retention was 0.95 and 0.89 at heavy metals concentration of 5 and 10 mg/l, respectively. These results follow a similar trend to previous results discussed in 9.8.1.2. Retention of heavy metals decreases when heavy metal concentration is increased due to the increase in concentration polarization at the surface of the membrane. When heavy metals concentration is increased, more metal ions are available in the bulk solution. These ions are transported to the surface of the membrane, where they accumulate and increase concentration polarization, thus reducing retention. In addition, increasing heavy metals concentration will increase the amount of dissociated heavy metals in solution due to limitation of heavy metals binding with HS.

Figure 9.15 shows the effect of polyelectrolyte type and concentration on Cu and Zn. Cu and Zn retention using PDADMAC was higher than CoAA. Cu retention was 0.96 and 0.93 for PDADMAC and CoAA, respectively. PDADMAC higher retention of heavy metals is due to its higher retention of HS. Metal retention is dependent on HS retention using P005F membrane as discussed previously. As HS retention increase, metals retention increase.

Increasing polyelectrolyte concentration increased Cu and Zn retention until maximum retention is achieved. Further increase in polyelectrolyte concentration reduced metal retention. Cu retention was 0.95, 0.96, 0.98, 0.98, 0.98, and 0.94 at 1, 2, 5, 7, 10, and 15 mg/l PDADMAC, respectively. As explained previously, metals retention using UF membranes is dependent on HS retention. An increase or decrease in HS retention due to the effect of polyelectrolyte will increase metals retention.

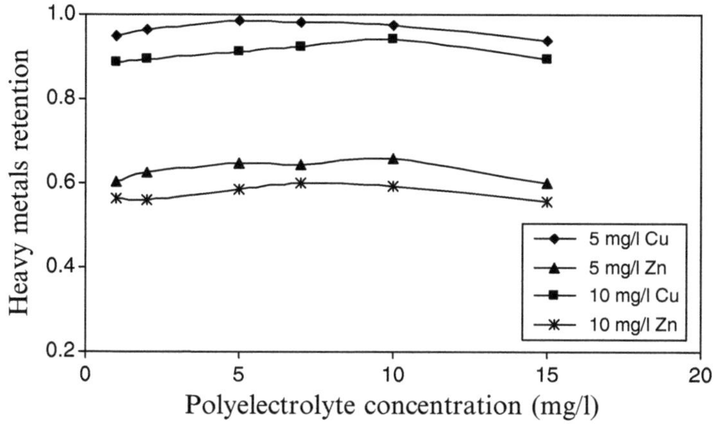

Fig. 9.15 Effect of heavy metals concentration on Cu and Zn retention using P005F membrane operating at 3 bar. Initial feed concentration of 25,000 ppm NaCl, 30 mg/l HS, and using PDADMAC

9.9 Membrane Fouling

9.9.1 Agglomerates of Humic Substance and Heavy Metals

Many factors affect membrane fouling such as membrane's surface charge, properties, and composition of the bulk solution and hydrodynamic conditions. Different membranes respond differently to any changes in these factors. The influence of salinity, HS concentration, and heavy metals concentration on P005F UF membrane fouling were studied. Figure 9.16 shows the effect of salinity on membrane fouling. Increasing salinity level increased membrane fouling. P005F fouling at 3 bar and initial feed concentration of 10 mg/l HS and 5 mg/l heavy metals were 27.1% and 32.3% at salinity levels of 10,000 and 25,000 ppm NaCl, respectively. As explained in Sect. 9.8.1, increasing the ionic strength compresses the diffuse layer of HS colloids. Double-layer compression will reduce HS colloidal charge and the electrostatic repulsion between HS colloidal macromolecules. The reduction in HS interchain repulsion will coil up HS colloids forming as a result a more densely packed deposition of HS at the surface of the membrane and formation of a compact fouling layer, as shown in Fig. 9.5.

Figure 9.16 shows the effect of HS concentration on membrane fouling. Increasing HS concentration increased membrane fouling. P005F fouling at 3 bar and initial feed concentration of 25,000 ppm NaCl and 5 mg/l heavy metals were 32.3%, 35.6%, and 37.9% at HS concentration of 10, 20, and 30 mg/l, respectively. As HS concentration increases, number of HS molecules settling at the surface of membrane increases. This settling enhances concentration polarization at the surface of the membrane and increases membrane fouling.

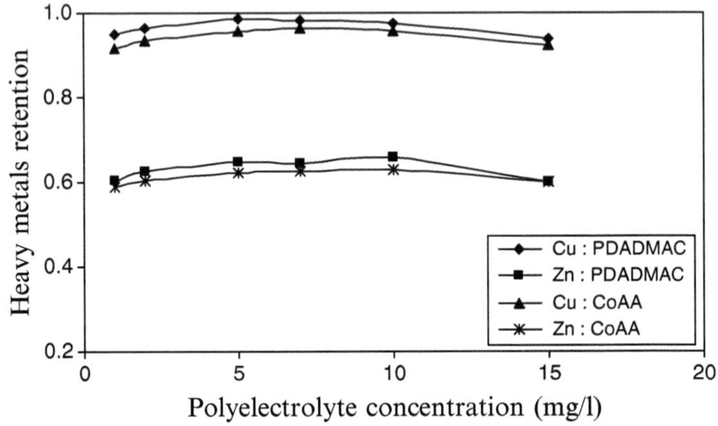

Fig. 9.16 Effect of polyelectrolyte type and concentration on Cu and Zn retention using P005F membrane operating at 3 bar. Initial feed concentration of 25,000 ppm NaCl, 30 mg/l HS, and 5 mg/l heavy metals

Figure 9.16 shows the effect of heavy metals concentration on membrane fouling. Membrane fouling increased with heavy metals concentration. P005F fouling at 3 bar and initial feed concentration of 25,000 ppm NaCl and 10 mg/l HS were 31.2%, 32.3%, and 35.9% at heavy metals concentration of 0, 5, and 10 mg/l, respectively. Unlike Na^+ cations, divalent cations interact specifically with humic carboxyl functional groups; thus, substantially reducing HS charge and the electrostatic repulsion between the HS macro-molecules. The reduction in HS interchain repulsion results in increased HS deposition at the membrane surface and the formation of a densely packed fouling layer.

9.9.2 Humic Substance Coagulation

The influence of salinity, HS concentration, and polyelectrolyte type and concentration on membrane fouling was studied using P005F membrane. Figure 9.17 shows the effect of salinity level on membrane fouling. Increasing salinity level from 10,000 to 25,000 ppm NaCl increased fouling, but no further increase in fouling was experienced when salinity was increased to 35,000 ppm NaCl. P005F fouling at 3 bar and initial feed concentration of 10 mg/l HS and 2 mg/l PDADMAC were 30.1%, 36.0%, and 35.2% at salinity levels of 10,000, 25,000, and 35,000 ppm NaCl, respectively. Increasing the ionic strength compresses the diffuse layer of HS colloids. Double-layer compression reduces the HS colloidal charge and the electrostatic repulsion between HS colloidal macro-molecules. The reduction in HS interchain repulsion will coil up HS colloids forming as a result a more densely packed deposition of HS at the surface of the membrane and formation of a compact

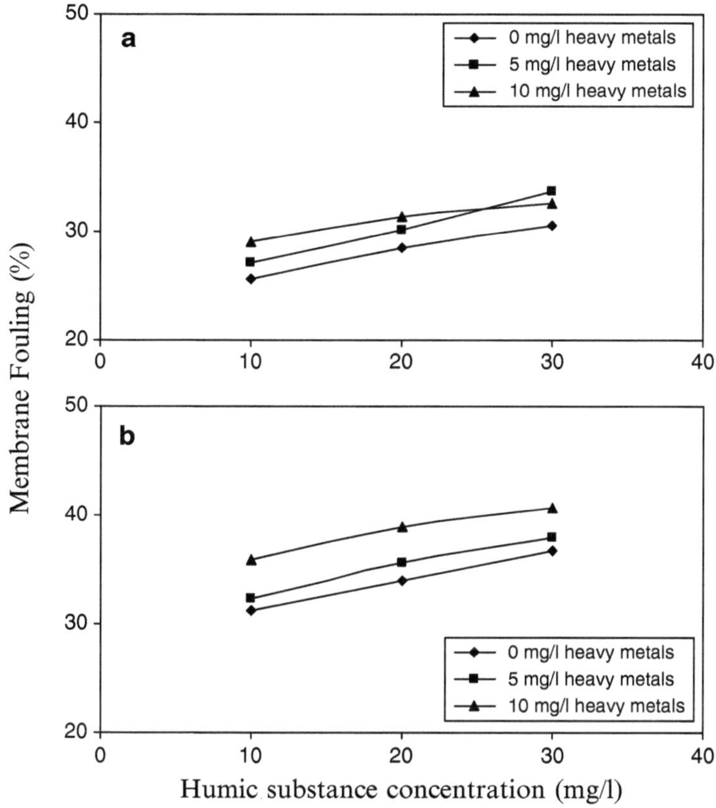

Fig. 9.17 Effect of membrane type, salinity, and HS concentration on membrane fouling using P005F operating at 3 bar. (**a**) 10,000 ppm NaCl and (**b**) 25,000 ppm NaCl

fouling layer. The increase in salinity from 25,000 to 35,000 ppm did not exert any change in HS colloids; due to the very high salinity of feed, salinity is no longer an important factor, as explained in Sect. 9.8.2.

Figure 9.17 shows the effect of HS concentration on membrane fouling. Membrane fouling increased with HS concentration. P005F fouling at 3 bar and initial feed concentration of 25,000 ppm NaCl and 2 mg/l PDADMAC were 36.0%, 39.2%, and 40.6% at HS concentration of 10, 20, and 30 mg/l, respectively. These results follow a similar trend to previous results discussed in Sect. 9.9.1. As HS concentration increases, the number of HS molecules settling at the surface of membrane will increase. This settling will enhance concentration polarization at the surface of the membrane and increase membrane fouling.

Figure 9.18 shows the effect of polyelectrolyte type and concentration on membrane fouling. No difference in fouling was experienced using PDADMAC and CoAA coagulants. P005F fouling at 3 bar and initial feed concentration of 25,000 ppm NaCl and 10 mg/l HS were 33.5% and 34.0% using 1 mg/l PDADMAC and CoAA, respectively. In addition, membrane fouling increased with increasing

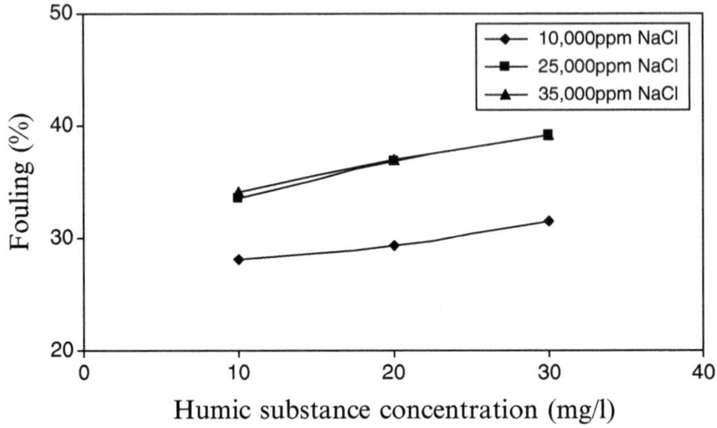

Fig. 9.18 Effects of salinity and HS on membrane fouling using P005F operating at 3 bar. Initial feed concentration of 2 mg/l PDADMAC

polyelectrolyte concentration. P005F fouling at 3 bar and initial feed concentration of 25,000 ppm NaCl and 10 mg/l HS were 31.2%, 33.5%, and 36.0% at PDADMAC concentration of 0, 1, and 2 mg/l, respectively. The increase in fouling with increase in polyelectrolyte concentration is due to the increase in polyelectrolyte and colloids settling at the surface of membrane. An increase in polyelectrolyte concentration will increase the amount of polyelectrolyte molecules settling at the surface of the membrane inducing concentration. The addition of polyelectrolyte also changes the structure of the HS molecules enabling them to foul the membrane more extensively.

9.9.3 Humic Substance and Heavy Metals Coagulation

The influence of salinity, HS concentration, heavy metals concentration, and polyelectrolyte type and concentration on membrane fouling were studied using P005F membrane. Figure 9.19 shows the effect of salinity on membrane fouling. Increasing salinity level increased membrane fouling. P005F fouling was 36.0 and 41.7% at salinity levels of 10,000 and 25,000 ppm NaCl, respectively. As explained previously, increasing the ionic strength compresses the diffuse layer of HS colloids. Double-layer compression will reduce HS colloidal charge and the electrostatic repulsion between HS colloidal macromolecules. The reduction in HS interchain repulsion will coil up HS colloids forming as a result a more densely packed deposition of HS at the surface of the membrane and formation of a compact fouling layer as shown in Fig. 9.5.

Figure 9.20 shows the effect of HS concentration on membrane fouling. Increasing HS concentration increased fouling. P005F fouling was 37.0%, 39.0%, and 41.7% at HS concentration of 10, 20, and 30 mg/l, respectively. These results show similar trend to previous results discussed in Sect. 9.9.1. As HS concentration in-

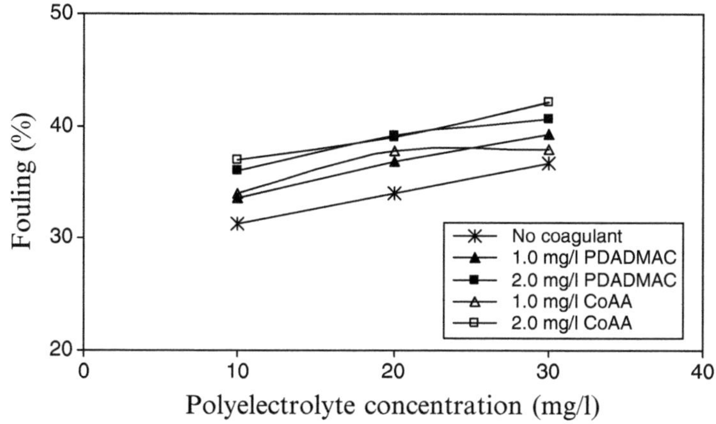

Fig. 9.19 Effects of polyelectrolyte type and concentration on membrane fouling using P005F operating at 3 bar. Initial feed concentration of 25,000 ppm NaCl and 10 mg/l HS

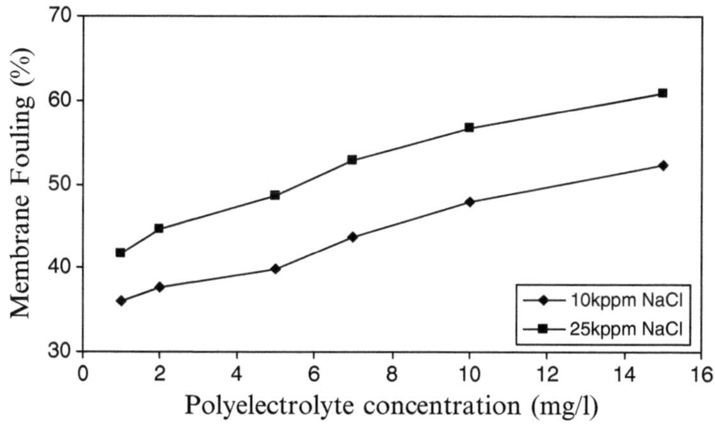

Fig. 9.20 Effect of salinity on membrane fouling using P005F membrane operating at 3 bar. Initial feed concentration of 30 mg/l HS, 5 mg/l heavy metals, and using PDADMAC

creases, the number of HS molecules settling at the surface of membrane increases. This settling enhances concentration polarization and increases membrane fouling.

Figure 9.21 shows the effect of heavy metals concentration on membrane fouling. Membrane fouling increased with heavy metals concentration following similar trend as previous results discussed in Sect. 9.9.1. P005F fouling was 44.7% and 45.9% at heavy metals concentration of 5 and 10 mg/l, respectively. As explained previously, divalent cations interact specifically with humic carboxyl functional groups thus, substantially reducing HS charge and the electrostatic repulsion between the HS macromolecules. The reduction in HS interchain repulsion results in increased HS deposition at the membrane surface and the formation of a densely packed fouling layer.

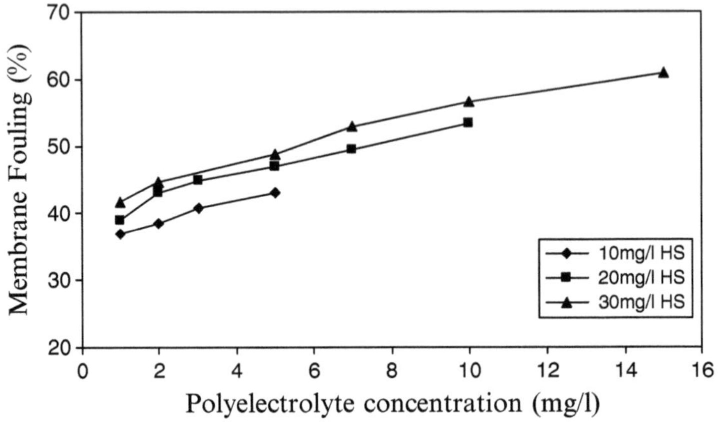

Fig. 9.21 Effect of HS concentration on membrane fouling using P005F membrane operating at 3 bar. Initial feed concentration of 25,000 ppm NaCl, 5 mg/l heavy metals, and using PDADMAC

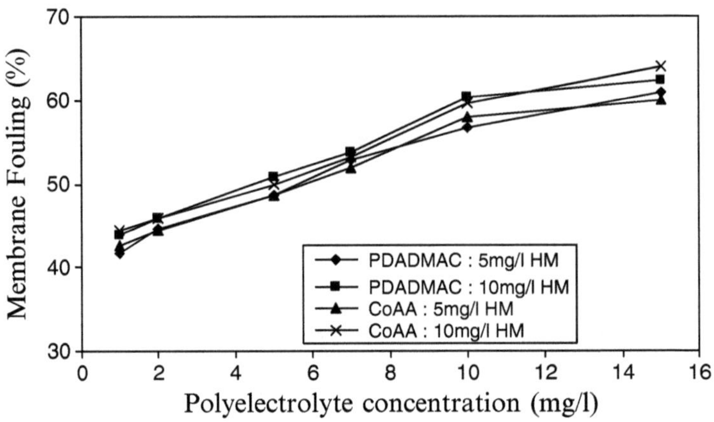

Fig. 9.22 Effect of Heavy metals concentration and polyelectrolyte type and concentration on membrane fouling using P005F membrane operating at 3 bar. Initial feed concentration of 25,000 ppm NaCl and 30 mg/l HS

Figure 9.21 also shows the effect of polyelectrolyte type and concentration on membrane fouling. No difference in fouling was experienced using PDADMAC and CoAA coagulants. P005F fouling was 44.7% and 45.9% using 2 mg/l PDADMAC and CoAA, respectively. In addition, fouling increased with increasing polyelectrolyte concentration. P005F fouling was 41.7%, 44.7%, 48.7%, 52.9%, 56.6%, and 60.9% at PDADMAC concentration of 1, 2, 5, 7, 10, and 15 mg/l, respectively. These results show similar behaviour to previous results discussed previously in Sect. 9.9.2. Increase in fouling with increasing polyelectrolyte concentration is due to the increase in polyelectrolyte and colloids settling at the surface of membrane. An increase in polyelectrolyte concentration will increase the amount of polyelectrolyte molecules settling at the surface of the membrane inducing concentration. The addition of polyelectrolyte also changes the structure of the HS molecules enabling them to foul the membrane more extensively (Fig. 9.22).

Acknowledgment The authors would like to thank the Middle East Desalination Research Center (MEDRC) for funding this work (project number 03-AS-02).

References

1. Ebensperger U, Isley P (2005) Review of the current state of desalination. Water Policy Working Paper, 2005–008
2. Borsani R, Rebagliati S (2005) Fundamentals and costing of MSF desalination plants and comparison with other technologies. Des J 182:29–37. doi:10.1016/j.desal.2005.03.007
3. Semiat R (2000) Desalination: present and future. IWRA 25:54–65
4. Nicolaisen B (2003) Developments in membrane technology for water treatment. Des J 153:355–360. doi:10.1016/S0011-9164(02)01127-X
5. Ghabayen S, McKee M, Kemblowski M (2004) Characterization of uncertainties in the operation and economics of the proposed seawater desalination plant in the Gaza Strip. Des J 161:191–201. doi:10.1016/S0011-9164(04), 90054-9
6. Sikora J, Hansson C, Ericsson B (1989) Pretreatment and desalination of mine drainage water in a pilot plant. Des J 75:363–373. doi:10.1016/0011-9164(89), 85022-2
7. Redondo J (2001) Brackish-, sea- and wastewater desalination. Des J 138:29–40. doi:10.1016/S0011-9164(01), 00241-7
8. Schaefer A, Schwicker U, Fischer M, Fane A, Waite T (2000) Microfiltration of colloids and natural organic matter. J Membr Sci 171:151–172. doi:10.1016/S0376-7388(99), 00286-0
9. Wilkinson K, Negre J, Buffle J (1997) Coagulation of colloidal material in surface water: the role of natural organic matter. J Contaminant Hydrol 26:229–243. doi:10.1016/S0169-7722(96)00071-X
10. Costa A, De Pinho M (2002) The role of membrane morphology on ultrafiltration for natural organic matter removal. Des J 145:299–304. doi:10.1016/S0011-9164(02), 00426-5
11. Xu W, Chellam S, Clifford D (2004) Indirect evidence for deposit rearrangement during dead-end microfiltration of iron coagulated suspensions. J Membr Sci 239:243–254. doi:10.1016/j.memsci.2004.03.039
12. Hong S, Elimelech M (1997) Chemical and physical aspects of natural organic matter (NOM) fouling of nanofiltration membranes. J Membr Sci 132:159–181. doi:10.1016/S0376-7388(97), 00060-4
13. Raspor B (1989) Adsorption of humic substances from seawater at differently charged surfaces. Sci Total Environ 81–82:319–328. doi:10.1016/0048-9697(89), 90139-3
14. Aiken G, McKnight D, Wershaw R, MacCarthy E (1985) Humic substances in soil, sediment, and water. Wiley, New York
15. Ruohomaki K, Vaisanen P, Metsamuuronen S, Kulovaara M, Nystrom M (1998) Characterization and removal of humic substances in ultra- and nanofiltration. Des J 118:273–283. doi:10.1016/S0011-9164(98), 00147-7
16. De Paolis F, Kukkonen J (1997) Binding of organic pollutants to humic and fulvic acids: influence of pH and the structure of humic material. Chemosphere J 34:1693–1704. doi:10.1016/S0045-6535(97)00026-X
17. Amy G, Collins M, Kuo C, King P (1987) Comparing gel permeation chromatography and ultrafiltration for the molecular weight characterization of aquatic organic matter. AWWA 79:43–49

18. Rao L, Choppin G (1995) Thermodynamic study of the complexation of neptunium(V) with humic acids. Radiochim Acta 69:87–95
19. Aiken G, Malcolm R (1987) Molecular weights of aquatic fulvic acids by vapor pressure osmometry. Geochim Cosmochim Acta 51:2177–2184
20. Beckett R, Zhang J, Giddings J (1987) Determination of molecular weight distributions of fulvic and humic acids, using flow field-flow fractionation. Environ Sci Technol 21:289–295
21. Reid P, Wilkinson A, Tipping E, Jones M (1990) Determination of molecular weights of humic substances by analytical (UV scanning) ultracentrifugation. Geochim Cosmochim Acta 54:131–138
22. Pokrovsky O, Schott J (2002) Iron colloids/organic matter associated transport of major and trace elements in small boreal rivers and their estuaries (NW Russia). Chem Geol 190:141–179
23. Munksgaard N, Parry D (2001) Trace metals, arsenic and lead isotopes in dissolved and particulate phases of North Australian coastal and estuarine seawater. Mar Chem 75:165–184
24. Gavriil A, Angelidis M (2005) Metal and organic carbon distribution in water column of a shallow enclosed Bay at the Aegean Sea Archipelago: Kalloni Bay, Island of Lesvos, Greece. Estuar Coast Shelf Sci 64:200–210
25. Yates D, Joyce K, Heaney P (1998) Complexation of copper with polymeric silica in aqueous solution. Appl Geochem 13:235–241
26. Goosen M, Sablani S, Al-Hinai H, Al-Obeidani S, Al-Belushi R, Jackson D (2005) Fouling of reverse osmosis and ultrafiltration membranes: a critical review. Sep Sci Tech 39:2261–2297
27. Abu Qdais H, Moussa H (2004) Removal of heavy metals from wastewater by membrane processes: a comparative study. Des J 164:105–110
28. Alpatova A, Verbych S, Bryk M, Nigmatullin R, Hilal N (2004) Ultrafiltration of water containing natural organic matter: heavy metal removing in the hybrid complexation–ultrafiltration process. Sep Pur Tech 40:155–162
29. Harmant P, Aimar P (1998) Coagulation of colloids in a boundary layer during cross-flow filtration. Colloids Surf A Physicochem Eng Asp 138:217–230
30. Kretzschmar R, Sticher H (1998) Colloid transport in natural porous media: Influence of surface chemistry and flow velocity. Phys Chem Earth 23:133–139
31. Eyrolle F, Benaim J (1999) Metal available sites on colloidal organic compounds in surface waters (Brazil). Water Res 33:995–1004
32. Tipping E, Lofts S, Lawlor A (1998) Modelling the chemical speciation of trace metals in the surface waters of the Humber system. Sci Total Environ 210–211:63–77
33. Martell A (1957) The chemistry of metal chelates in plant nutrition. Soil Sci 84:13–26
34. Pervov A, Andrianov A, Efremov R, Desyatov A, Baranov A (2003) A new solution for the Caspian Sea desalination: low-pressure membranes. Des J 157:377–384
35. Teuler A, Glucina K, Laine J (1999) Assessment of UF pretreatment prior RO membranes for seawater desalination. Des J 125:89–96
36. Schafer A, Fane A, Waite T (1998) Nanofiltration of natural organic matter: removal, fouling and the influence of multivalent ions. Des J 118:109–122
37. Schaefer A, Fane A, Waite T (2000) Fouling effects on rejection in the membrane filtration of natural waters. Des J 131:215–224
38. Agashichev S (2006) Enhancement of concentration polarization due to gel accumulated at membrane surface. J Membr Sci 285:96–101
39. Vrouwenvelder H, Van Paassen J, Folmer H, Hofman A, Nederlof M, Kooij D (1998) Biofouling of membranes for drinking water production. Des J 118:157–166
40. Vrouwenvelder J, Kappelhof J, Heijman S, Schippers J, Kooij D (2003) Tools for fouling diagnosis of NF and RO membranes and assessment of the fouling potential of feed water. Des J 157:361–365
41. Van Der Bruggen B, Braeken L, Vandecasteele C (2002) Evaluation of parameters describing flux decline in nanofiltration of aqueous solutions containing organic compounds. Des J 147:281–288
42. Isaias N (2001) Experience in reverse osmosis pretreatment. Des J 139:57–64

43. Luo M, Wang Z (2001) Complex fouling and cleaning in-place of a reverse osmosis desalination system. Des J 141:15–22
44. Chua K, Hawlader M, Malek A (2003) Pretreatment of seawater: results of pilot trials in Singapore. Des J 159:225–243
45. Kaiya Y, Itoh Y, Fujita K, Takizawa S (1996) Study on fouling materials in the membrane treatment process for potable water. Des J 106:71–77
46. Clark M, Lucas P (1998) Diffusion and partitioning of humic acid in a porous ultrafiltration membrane. J Membr Sci 143:13–25
47. Jones K, O'Melia C (2000) Protein and humic acid adsorption onto hydrophilic membrane surfaces: effects of pH and ionic strength. J Membr Sci 165:31–46
48. Chang Y, Benjamin M (2003) Formation of natural organic matter fouling layer on ultrafiltration membranes. Envir Eng J 129:25–32
49. Lee S, Lee C (2006) Microfiltration and ultrafiltration as a pretreatment for nanofiltration of surface water. J Sep Sci Tech 41:1–23
50. Choksuchart P, Heran M, Grasmick A (2002) Ultrafiltration enhanced by coagulation in an immersed membrane system. Des J 145:265–272
51. Brehant A, Bonnelye V, Perez M (2002) Comparison of MF/UF pretreatment with conventional filtration prior to RO membranes for surface seawater desalination. Des J 144:353–360
52. Derradji A, Taha S, Dorange G (2005) Application of the resistances in series model in ultrafiltration. Des J 184:1357–1364
53. Clever M, Jordt F, Knauf R, Rabiger N, Rudebusch M, Hilker-Scheibel R (2000) Process water production from river water by ultrafiltration and reverse osmosis. Des J 131:325–336
54. Bottino A, Capannelli C, Del Borghi A, Colombino M, Conio O (2001) Water treatment for drinking purpose: ceramic microfiltration application. Des J 141:75–79
55. Stover R, Ameglio A, Khan P (2005) The ghalilah SWRO plant an overview of the solutions adopted to minimize energy consumption. Des J 184:1197–1201
56. Gille D, Czolkoss W (2005) Ultrafiltration with multi bore membranes as seawater pretreatment. Des J 182:295–301
57. Halpern D, McArdle J, Antrim B (2005) UF pretreatment for SWRO: pilot studies. Des J 182:317–326
58. Kothari N, Taylor J (1998) Pilot scale microfiltration at Manitowoc. Des J 119:93–102
59. Wang K, Matsuura T, Chung T, Guo W (2004) The effects of flow angle and shear rate within the spinneret on the separation performance of poly(ethersulfone) (PES) ultrafiltration hollow fiber membranes. J Membr Sci 240:67–79
60. Cote P, Siverns S, Monti S (2005) Comparison of membrane-based solutions for water reclamation and desalination. Des J 182:251–257
61. Wolf P, Siverns S, Monti S (2005) UF membranes for RO desalination pretreatment. Des J 182:289–296
62. Kruithof J, Schippers J, Kamp P, Folmer H, Hofman J (1998) Integrated multi-objective membrane systems for surface water treatment: pretreatment of reverse osmosis by conventional treatment and ultrafiltration. Des J 117:37–48
63. Teng C, Hawlader M, Malek A (2003) An experiment with different pretreatment methods. Des J 156:51–58
64. Van Hoof S, Hashim A, Kordes A (1999) The effect of ultrafiltration as pretreatment to reverse osmosis in wastewater reuse and seawater desalination applications. Des J 124:231–242
65. Schafer A, Richards B (2005) Field testing of a hybrid membrane system for groundwater desalination. Des J 183:55–62
66. Glueckstern P, Priel M, Wilf M (2002) Field evaluation of capillary UF technology as a pretreatment for large seawater RO systems. Des J 147:55–62
67. Hofman J, Beumer M, Baars E, Van Der Hoek J, Koppers H (1998) Enhanced surface water treatment by ultrafiltration. Des J 119:113–125
68. Murrer J, Rosberg R (1998) Desalting of seawater using UF and RO – results of a pilot study. Des J 118:1–4

69. Speth T, Gusses A, Summers R (2000) Evaluation of nanofiltration pretreatments for flux loss control. Des J 130:31–44
70. Assemi S, Newcombe G, Hepplewhite C, Beckett R (2004) Characterization of natural organic matter fractions separated by ultrafiltration using flow field-flow fractionation. Water Res 38:1467–1476
71. Hassan A, Farooque A, Jamaluddin A, Al-Amoudi A, Al-Sofi M, Al-Rubaian A, Kither N, Al-Tisan I, Rowaili A (2000) A demonstration plant based on the new NF-SWRO process. Des J 131:157–171
72. Lee H, Amy G, Cho J, Yoon Y, Moon S, Kim I (2001) Cleaning strategies for flux recovery of an ultrafiltration membrane fouled by natural organic matter. Water Res 35:3301–3308
73. Carroll T, King S, Gary S, Bolto B, Booker N (2000) The fouling of microfiltration membranes by NOM after coagulation treatment. Water Res 34:2861–2868
74. Yiantsios S, Karabelas A (1998) The effect of colloid stability on membrane fouling. Des J 118:143–152
75. Jarusutthirak C, Amy G, Croue P (2002) Fouling characteristics of wastewater effluent organic matter (EfOM) isolates on NF and UF membranes. Des J 145:247–255
76. Kuo C, Amy G (1988) Factors affecting coagulation with aluminium sulphate-II: dissolved organic matter removal. Water Res 22:863–872
77. Al-Mutairi N, Hamoda M, Al-Ghusain I (2004) Coagulant selection and sludge conditioning in a slaughterhouse wastewater treatment plant. Bioresource Tech 95:115–119
78. Kam S, Gregory J (2001) The interaction of humic substances with cationic polyelectrolytes. Water Res 35:3557–3566
79. Vickers J, Thompson M, Kelkar U (1995) The use of membrane filtration in conjunction with coagulation processes for improved NOM removal. Des J 102:57–61
80. Ma J, Liu W (2002) Effectiveness of ferrate (VI) preoxidation in enhancing the coagulation of surface waters. Water Res 36:4959–4962
81. Judd S, Hillis P (2001) Optimisation of combined coagulation and microfiltration for water treatment. Water Res 35:2895–2904
82. Maartens A, Swart A, Jacobs E (1999) Feed-water pretreatment: methods to reduce membrane fouling by natural organic matter. J Membr Sc 163:51–62
83. Huang C, Shiu H (1996) Interactions between alum and organics in coagulation. Colloids Surf A: Physiochem Eng Aspects 113:155–163
84. Bolto B, Abbt-Braun G, Dixon D, Eldridge R, Frimmel F, Hesse S, King S, Toifl M (1999) Experimental evaluation of cationic polyelectrolytes for removing natural organic matter from water. Water Sci Technol 40:71–79
85. Gabelich C, Yun T, Coffey B, Mel Suffet I (2002) Effects of aluminum sulfate and ferric chloride coagulant residuals on polyamide membrane performance. Des J 150:15–30
86. Zidouri H (2000) Desalination in Morocco and presentation of design and operation of the Laayoune seawater reverse osmosis plant. Des J 131:137–145
87. Kampa P, Kruithof J, Folmer H (2000) UF/RO treatment plant Heemskerk: from challenge to full scale application. Des J 131:27–35
88. Abdessemed D, Nezzal G (2002) Treatment of primary effluent by coagulation-adsorption-ultrafiltration for reuse. Des J 152:367–373
89. Low S, Han H, Jin W (2004) Characteristics of a vibration membrane in water recovery from fine carbon-loaded wastewater. Des J 160:83–90
90. Hilal N, Al-Abri M, Al-Hinai H (2007) Characterization and retention of membranes using PEG, HS and polyelectrolytes. Des J 206:568–578. doi:10.1016/j.desal.2006.02.077
91. O'Melia C, Becker W, Au K (1999) Removal of humic substances by coagulation. Water Sci Technol 40:47–54
92. Verbych S, Bryk M, Alpatova A, Chornokur G (2005) Ground water treatment by enhanced ultrafiltration. Des J 179:237–244
93. Alvarez-Puebla R, Valenzuela-Calahorro C, Garrido J (2004) Retention of Co(II), Ni(II), and Cu(II) on a purified brown humic acid. Modeling and characterization of the sorption process. Langmuir J 20:3657–3664

94. Hilal N, Al-Abri M, Moran A, Al-Hinai H (2008) Effects of heavy metals and polyelectrolytes in humic substance coagulation under saline conditions. Des J 220:85–95. doi:10.1016/j.desal.2007.01.024
95. Hilal N, Al-Abri M, Al-Hinai H, Somerfield C (2008) Combined humic substance and heavy metals agglomeration, and membrane filtration under saline conditions. Sep Sci Tech J 43:1488–1506. doi:10.1080/01496390801941091
96. Spark K, Wells J, Johnson B (1997) Sorption of heavy metals by mineral-humic acid substrates. Aust J Soil Res 35:113–122
97. Spark K, Wells J, Johnson B (1997) The interaction of a humic acid with heavy metals. Aust J Soil Res 35:89–101
98. Alvarez-Puebla R, Valenzuela-Calahorro C, Garrido J (2004) Cu(II) retention on a humic substance. J Colloid Interface Sci 270:47–55
99. Zhou P, Yan H, Gu B (2005) Competitive complexation of metal ions with humic substances. Chemosphere J 58:1327–1337
100. Fukushima M, Nakayasu K, Tanaka S, Nakamura H (1995) Chromium(III) binding abilities of humic acids. Anal Chim Acta 317:195–206
101. Alaerts G, Van Haute A (1981) Flocculation of brackish water from a tidal river. Water Res 15:517–523
102. Chang E, Chiang P, Tang W, Chao S, Hsing H (2005) Effects of polyelectrolytes on reduction of model compounds via coagulation. Chemosphere J 58:1141–1150

Chapter 10
Hybrid Sonochemical Treatment of Contaminated Wastewater: Sonophotochemical and Sonoelectrochemical Approaches. Part I: Description of the Techniques

B. Neppolian, M. Ashokkumar, I. Tudela, and J. González-García

10.1 Introduction

The chemical and biological effects of ultrasound were reported by Loomis et al. about 80 years ago [1, 2]. Ultrasound as an initiator of chemical reactions and physical transformations of material has emerged as a potential technology within last few decades (from 1970). There are widespread applications of ultrasound in many industrial fields. Ultrasound frequencies in the range 20 kHz–1 MHz have been used in sonochemistry [3]. The physical and chemical effects of ultrasound have been widely used in numerous applications such as cleaning, degassing, polymerization, emulsification, drilling, cutting, flow measurements, imaging, biomedical, automotive, food preservation, drug delivery, and dairy industry [4, 5]. Ultrasound irradiation has found use in plastic welding, drying, air scrubbing, and nebulization. It is used in manufacturing industries such as synthetic textiles and reinforced composite materials [5]. Over the last two decades, new opportunities have emerged for utilizing ultrasound for chemical reactions such as controlled synthesis of nanomaterials and oxidation of organic pollutants (wastewater treatment) [6–22].

B. Neppolian
SRM Research Institute, SRM University, Kattankulathur, Chennai 603203, India
e-mail: b_neppolian@yahoo.com

M. Ashokkumar
School of Chemistry, University of Melbourne, Melbourne, Australia
e-mail: masho@unimelb.edu.au

I. Tudela • J. González-García (✉)
Grupo de Nuevos Desarrollos Tecnológicos en Electroquímica: Sonoelectroquímica y Bioelectroquímica, Departamento de Química Física e Instituto de Electroquímica, Universidad de Alicante, Alicante, Spain
e-mail: i.tudela@ua.es; jose.gonzalez@ua.es

S.K. Sharma and R. Sanghi (eds.), *Advances in Water Treatment and Pollution Prevention*, DOI 10.1007/978-94-007-4204-8_10,

This chapter is concerned with the applications of ultrasound for the treatment of water and wastewater in environmental remediation. The focus of this chapter is to introduce different experimental techniques that are commonly used for the degradation of organic pollutants present in an aqueous environment.

10.2 Equipment

Sonochemical degradation of organic pollutants is a very simple process that can be easily adapted to any working environment. Ultrasound waves are generated by transducers that convert electrical energy into mechanical energy. The transducers may be magnetostrictive- or piezoelectric-type materials [23]. Among the two types, piezoelectric transducers are commonly used in ultrasonic instruments. In lab-scale studies, bath- and horn-type ultrasonicators are commonly used. Most experimental studies to date have used low-frequency (~20 kHz) horn type or high-frequency plate type (100–600 kHz) sonicators (200–500 mL reactor capacity). In lab-scale experiments, reactors are fabricated mostly in Pyrex glass of about 1 L capacity. For industrial or pilot-scale studies, stainless steel reactors that can hold many high-frequency piezoelectric transducers of different capacity are used. Calorimetery is the most common method used to measure the acoustic power delivered into the liquid.

10.3 Operational Variables

Sonochemistry is strongly influenced by many operational variables of ultrasound such as frequency, power delivered to the liquid, different types of transducers, namely, horn type (low frequency, ~20 kHz) and plate type (high frequency, ~100 kHz–1 MHz), different modes of operation (pulse and normal mode), solution temperature, dual frequency configuration, geometry, and size of the reactor [24–29]. These are important considerations in the effective application of ultrasound for any particular type of chemical reaction. All operational variables play a significant role in the direction and efficiency of the sonochemical oxidation reactions that take place in the reactor [24, 25, 27]. For example, the choice of frequency has a major impact on the sonochemical reactivity. Low-frequency ultrasound (20–100 kHz) generates a large bubble and as a result violent cavitational collapse occurs, producing higher localized temperatures and pressures at cavitation sites as well as strong shockwaves [23]. This cavitation effect is commonly used in industry for cleaning particles bound to a surface and for wastewater treatment. Whereas, at higher frequencies (100 kHz–1 MHz), more cavitation events occur per unit time but the smaller bubbles tend to collapse less violently, producing lower temperatures and pressures [29]. Hence, high-frequency ultrasound is used in the electronics industry to clean components such as silicon wafers [3]. The production of free radicals is much higher at higher frequencies and, hence, the rate of chemical reactions is relatively faster.

Another important variable of ultrasound is acoustic power. Most of the commercially available ultrasonic equipments usually operate at a fixed frequency with variable power levels [8, 9, 25, 30]. Depending on the nature of reactions, one needs to optimize the acoustic power to be delivered to the medium. Many reports have demonstrated the importance of using optimal acoustic power in sonochemical reactions. The use of higher acoustic power results in stronger cavitation events. The rate of oxidation of organic compounds, e.g., is greatly enhanced with an increase in acoustic power [8].

10.4 Lab-Scale Reactors Used in Environmental Remediation

10.4.1 Low-Frequency Horn-Type Ultrasonic Reactors

10.4.1.1 Sonochemical-Oxidative Degradation of Organic Compounds

There have been many reports on utilizing low-frequency horn-type ultrasonicators for the degradation of either volatile or nonvolatile organic pollutants. Simple glass reactor vessels with a double wall for cooling water circulation have been used in lab-scale studies. As an example, a low-frequency horn-type ultrasonicator (20 kHz) is used for the degradation of highly volatile organic compounds such as methyl tert-butyl ether (MTBE) that is widely used in petroleum industry. MTBE is considered to be a potential carcinogen. The total degradation of MTBE was achieved by ultrasound irradiation within a short period of time in the presence of other oxidizing agents [8, 31]. MTBE is a highly volatile organic compound and, hence, the reactor was fabricated in such a way that it remains air tight and could hold the volatile compound without releasing it into the atmosphere. Figure 10.1 shows the reactor setup used for the effective degradation of volatile organic compounds by ultrasound, using a low-frequency horn-type ultrasonicator. The reactor assembly is made up of borosilicate glass of 150 mL equipped with an ultrasonicator (VCX-400 vibracell). The O-ring with metal collar connects the glass reactor to the stainless steel probe. Sampling and gas purging ports are sealed with Teflon valves and covered with rubber septa. The temperature of the reactor is maintained by circulating cooling water between the double walls of the reactor, as shown in Fig. 10.1. The rate of degradation of MTBE steadily increased with an increase in power density of the ultrasonicator from 22 to 76 W and also with a rise in reactor solution temperature from 10°C to 30°C. In the presence of an oxidizing agent, potassium persulfate, the sonolytic rate of degradation of MTBE was enhanced substantially. The intermediate products formed such as *tert*-butyl formate and acetone were also completely degraded by this process [8].

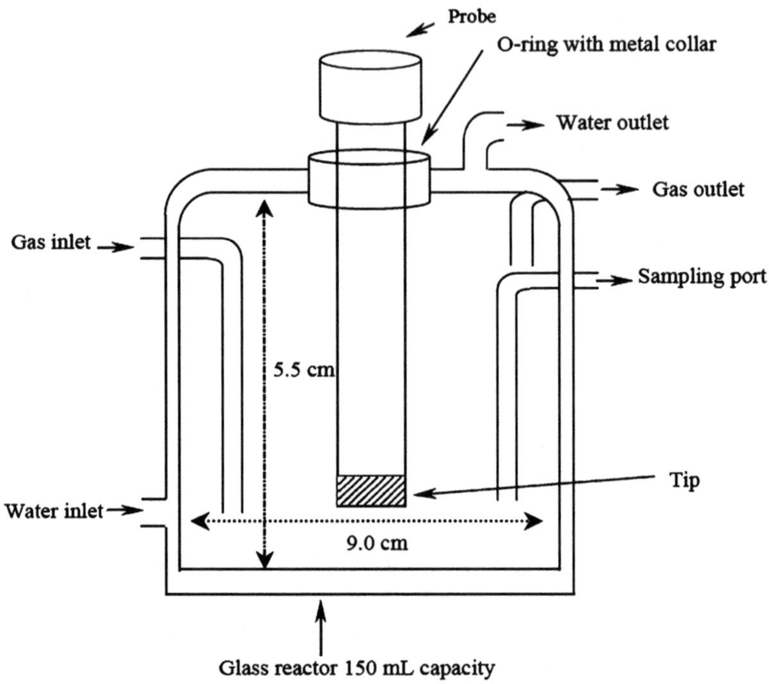

Fig. 10.1 Schematic diagram of a gas-tight glass reactor with a metal collar for attachment to stainless-steel sonication probe. This type of reactor design is recommended for the degradation of volatile compounds (Reproduced from Ref. [8]. With kind permission of © Elsevier (2002))

10.4.1.2 Sonochemical Oxidation of Highly Toxic As(III) Metalloid to Less Toxic As(V)

Ultrasound irradiation can also be used for the oxidation of metalloids, such as As(III) to As(V). As(III) is highly toxic and more mobile than As(V). Once As(III) is oxidized to As(V), the latter can be easily removed from the water and wastewater systems by using conventional adsorption methods. Many researchers have used different techniques for the oxidation of As(III) to As(V) [32–37]. Recently, Neppolian et al. reported on the effect of the sonochemical oxidation of As(III) to As(V) using a low-frequency horn-type ultrasonicator [24, 25]. OH radicals generated during acoustic cavitation oxidize As(III) to As(V). The physical forces generated during acoustic cavitation at low ultrasound frequency (20 kHz) substantially enhanced the As(III) oxidation rate. Sonication in a pulse mode has been found to show a profound effect on the rate of the oxidation of As(III) to As(V) in comparison with the continuous mode of operation. An almost two times higher rate of As(III) oxidation was observed in the pulse mode of ultrasonication. Further, the reaction rate was remarkably faster for the oxidation of As(III) to As(V) by ultrasound in combination with other oxidizing agents such as peroxydisulfate [24]. Figure 10.2 shows the percentage oxidation of As(III) to As(V) using low-frequency

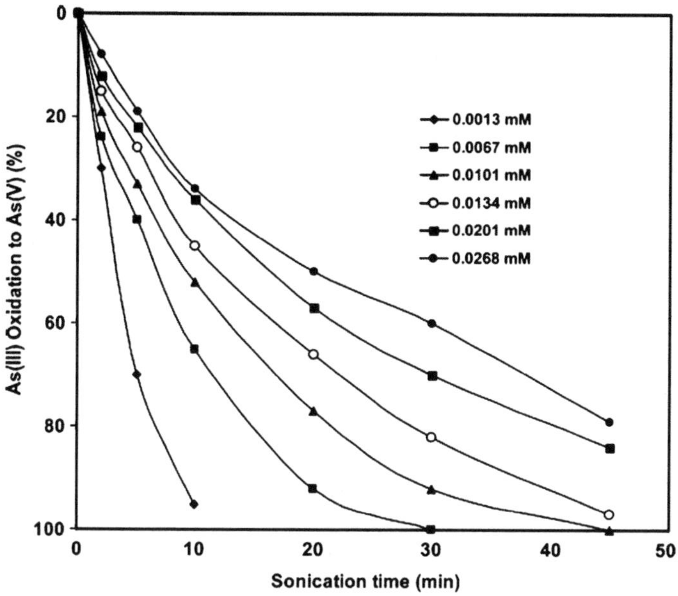

Fig. 10.2 Effect of the different initial concentrations of As(III) on the rate of oxidation of As(III) to As(V). Experimental conditions: pH = 7, tip diameter = 19 mm and power = 36 W (Reproduced from Ref. [25]. With kind permission of © The American Chemical Society (2009))

Table 10.1 Amount of As(III) oxidized into As(V) after Sonication

Initial [As] (ppb)	Amount oxidized in 10 min (ppb)	Amount oxidized in 30 min (ppb)
500	325	500
750	390	690
1000	430	820
1500	680	1050
2000	690	1200

Reproduced from Ref. [25]. With kind permission of © The American Chemical Society (2009)

(20 kHz) sonication. It can be seen in this figure that the percentage of sonochemical oxidation of As(III) increases with an increase in As(III) concentration. The data shown in Fig. 10.2 are processed for a better understanding of the effect of [As(III)] and presented in Table 10.1. It is clearly seen that the amount of oxidation of As(III) increases with an increasing concentration of As(III) [25]. For example, within 10 min of sonication time, 325 ppb of As(III) was oxidized into As(V) when the initial concentration of As(III) was 500 ppb, at the same time, 680 ppb of As(III) could be oxidized for the initial concentration of 1,500 or 2,000 ppb As(III). The same trend was observed for 30 min of reaction time (Table 10.1). Thus, the overall oxidation was higher for the higher concentration of As(III). This is a great advantage of using the sonochemical method for oxidizing high concentration of As(III) [25].

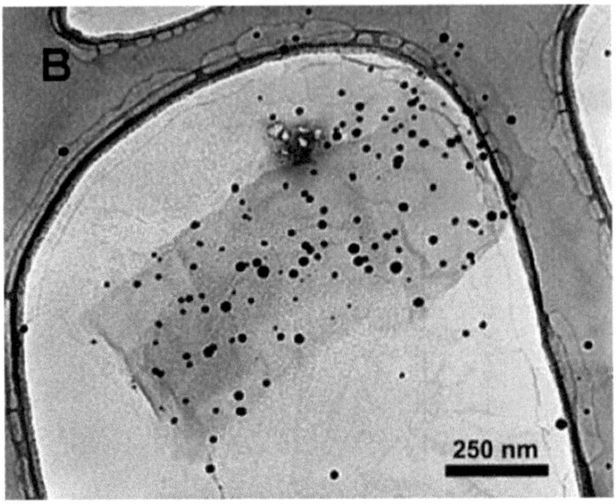

Fig. 10.3 TEM image of Au loaded-graphene oxide prepared by ultrasonic irradiation (Reproduced from Ref. [50]. With kind permission of © The American Chemical Society (2010))

10.4.2 Ultrasonic Reactors with Plate-Type Transducers

High-frequency ultrasound in the range 100–700 kHz is suitable for degradation of organic pollutants. The production of OH radicals is generally high with high frequency of the ultrasonicator [4]. Ashokkumar et al. reported that the amount of OH radicals obtained from a 200 kHz frequency ultrasonicator is about ten times higher than that generated by using a 20 kHz horn sonicator [4]. Many reports demonstrated the application of high-frequency ultrasound for the oxidation of organic compounds, such as different surfactants that are commonly used in domestic as well as industrial applications. The use of high-frequency ultrasound efficiently degrades surfactants without the addition of any external chemicals or catalysts [11, 13, 14]. Various other pollutants such as phenols and textile dyes were also successfully degraded by high frequency of ultrasound [38–44].

Recently, catalyst assemblies, based on metal nanoparticles deposited on single or few layers of graphene sheets, have gained considerable attention in a variety of applications ranging from photocatalysts to fuel cells and sensors to storage batteries [45–53]. Recently, Vinodgopal et al. [50] reported that high-frequency ultrasound at 211 kHz was very effective for the simultaneous reduction of metallic precursor to pure metal as well as converting graphene into reduced graphene oxide. The sonochemically prepared metal particles along with graphene oxide sheets were found to be very stable over a long period of time. With high-frequency ultrasound, graphene oxide sheets consisting of few layers of (~2–3) could be obtained by this method. Further, due to the physical forces generated from the collapse of acoustic cavitation during ultrasound irradiation, high levels of metal nanoparticles were deposited on the sheets of graphene oxide, as shown in the following TEM image (Fig. 10.3). Thus, ultrasound is useful for reducing graphene very effectively as well as for the effective loading of metal catalysts onto the sheets.

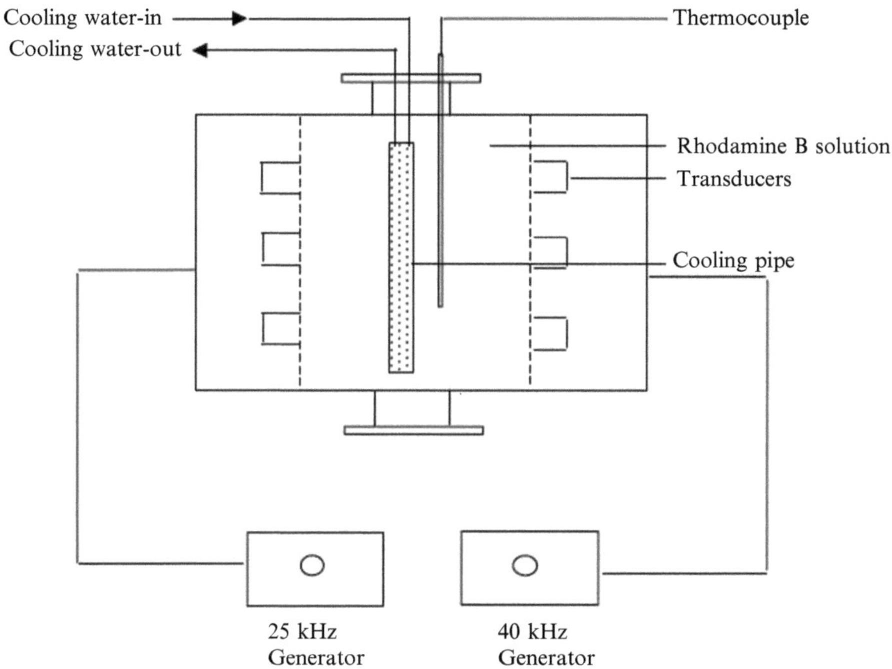

Fig. 10.4 Schematic diagram of dual frequency bath reactor (Reproduced from Ref. [58]. With kind permission of © Elsevier (2001))

10.4.3 Dual-Frequency Ultrasonic Reactors

It is generally reported that when using a single frequency transducer, the chemical reaction zone is limited to the volume of propagation or transmission of ultrasonic waves in a liquid medium. For large-scale industrial wastewater or water treatment, flow-through reactors attached with many transducers are required in order to achieve maximum volume for reaction. Few reports described the application of dual frequencies for different sonochemical processes [54–57]. Sivakumar and Pandit [58] fabricated a reactor, which could be operated for both continuous as well as batch type reactions. Their study involved the degradation of rhodamine blue dye as a model pollutant to check the feasibility of the reactor for environmental remediation. The reactor was made up of a stainless steel vessel and fitted with 20 and 40 kHz frequency piezoelectric transducers, as shown in Fig. 10.4. This reactor can work either with single frequency or dual frequency. A cooling pipe located inside the reactor (Fig. 10.4) was used for the circulation of water from a thermostated bath in order to maintain the temperature at around 25°C. The reactor was operated at three modes: 20, 40, and 20 + 40 kHz. The results showed that there was no synergetic or detrimental effect on the degradation of the dye with dual frequency. Only an additive effect of the separate frequencies was observed in the dual frequency mode. Despite the absence of a synergy effect, this work showed that flow-through reactors are suitable for the effective treatment of large-scale volumes of wastewater containing organic pollutants.

10.5 Large Size Reactors

10.5.1 Multifrequency Reactors

Utilizing ultrasound for industrial-scale studies is a challenging issue with limited information available in the literature. One has to study the effect of various important parameters of ultrasound for a large-scale reactor. Gogate et al. [59] reported the effect of various operational variables of ultrasound such as frequency of irradiation, use of multiple frequencies and power dissipation into the system on the extent of degradation of Rhodamine B as a model pollutant with a 7.5 L capacity reactor. The configuration of hexagonal reactor setup is shown in Fig. 10.5, which

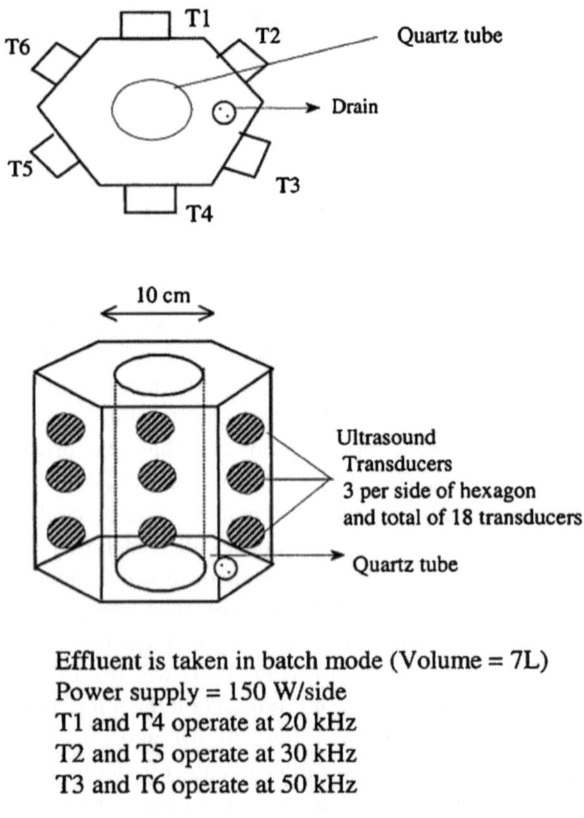

Fig. 10.5 Schematic representation of triple frequency flow cell used in the experimental work (Reproduced from Ref. [59]. With kind permission of © Elsevier (2004))

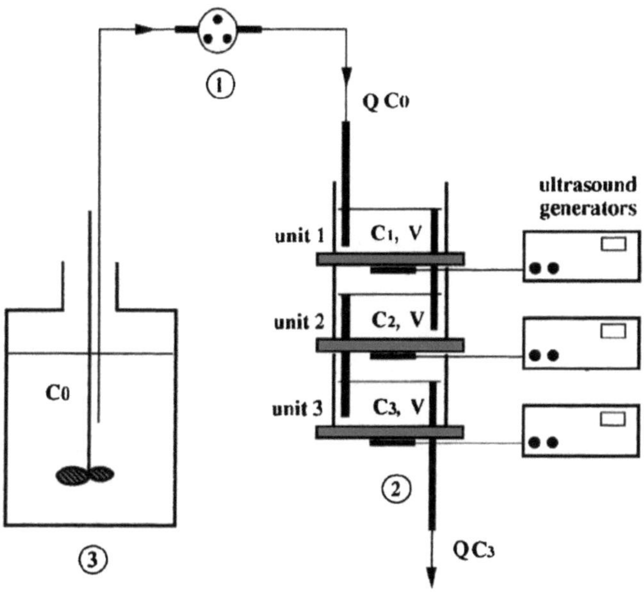

Fig. 10.6 Scheme of the experimental setup: *1*. Peristaltic pump; *2*. three-stage sonochemical reactor; *3*. continuously stirred vessel (Reproduced from Ref. [60]. With kind permission of © Elsevier (1999))

consists of three different frequency transducers mounted on each side. A total of 18 transducers with equal power rating of 50 W are present in the reactor. This reactor was able to operate both batch as well as continuous mode of operation. There were seven different configurations of frequencies (20, 30, 50, 20 + 30, 20 + 50, 30 + 50, and 20 + 30 + 50 kHz) that could be operated to control the effect of power and frequency for a particular reaction. The maximum power dissipation was 900 W when all transducers were in use with a combination of 20 + 30 + 50 kHz. This study revealed that the rate of degradation of Rhodamine B strongly depended on the cavitation intensity of a particular application. Aeration and the presence of solid particles could effectively enhance the rate of degradation of the dye. The above information is useful with respect to designing large- and appropriate-scale reactors for sonochemically assisted wastewater treatment.

10.5.2 Flow-Through Reactors

As a part of scale-up of the ultrasonic process for a large-scale wastewater treatment, Petrier and coworkers [60] first reported a three stage sonochemical reactor operating in continuous flow mode for the degradation of pentachlorophenol (PCP) as a model pollutant. Figure 10.6 shows the three stage sonochemical reactor

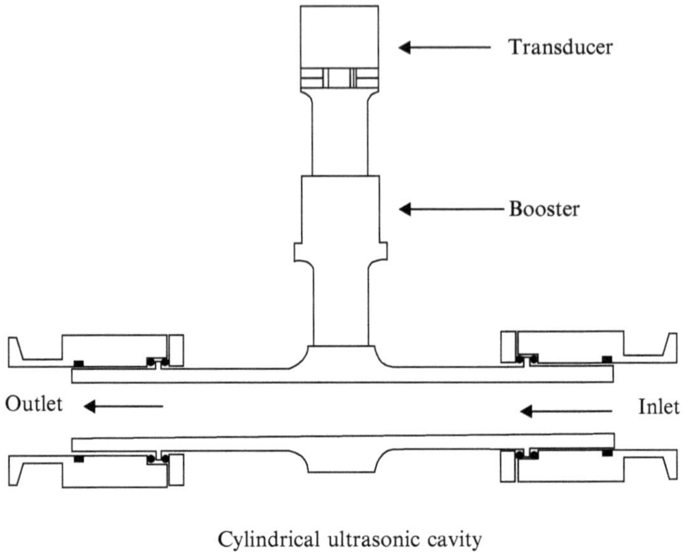

Fig. 10.7 Cylindrical reactor of 35 kHz (Reproduced from Ref. [61]. With kind permission of © Elsevier (2003))

setup. Each unit consists of a 500 kHz piezoelectric disk with a variable power supply in the range of 0–100 W. The temperature of the reaction solution was maintained at 20°C using a cooling unit. The study involved the influence of different operating variables such as ultrasonic power, volume of the reactor, and volumetric feed flow rate on the reactor performance. The rate of PCP degradation increased with an increase in the acoustic power of the ultrasonicator. At the same time, the reaction decreased with an increase in the flow rate of the solution. The rate of oxidation of phenol with the flow-type reactor mainly depended on acoustic power as well as flow rate of the solution, similar to other batch type reactors. According to theoretical as well as experimental results and interpretations, many challenging factors need to be taken into consideration when scaling up this method for industrial-scale application is desired.

10.5.3 Cylindrical Reactors

Generally, geometry and size of reactors play a significant role in determining the rate of a particular reaction. Many reports have described batch or semi-batch type reactors with square or rectangular shape for ultrasonic reactions. Petrier [61] was able to fabricate a cylindrical flow-through reactor with three different frequencies (20, 35, and 500 kHz), operated in a continuous flow mode for the degradation of phenol in aqueous solutions. Figure 10.7 shows the reactor assembly in which the reactor can be fitted with one of the transducers at a particular time. A titanium tube

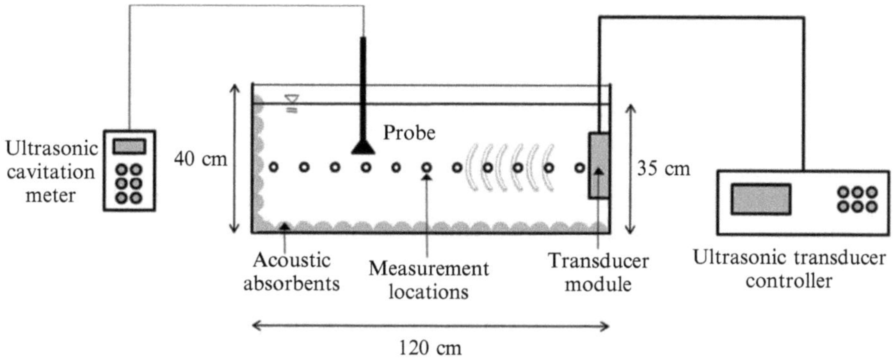

Fig. 10.8 Schematic of the large-scale sonoreactor and the location of measurement of cavitation energy (*side view*) (Reproduced from Ref. [62]. With kind permission of © Elsevier (2009))

was used for transmitting ultrasound waves at 20 kHz, whereas a piezoelectric disk fixed on a Pyrex plate emits ultrasound at 500 kHz. The degradation of phenol was carried out using a continuous flow mode of operation within a reaction volume of 350 mL. The rate of degradation of phenol was higher at 500 kHz than that at 20 or 35 kHz. However, the cylindrical type flow reactor enhanced the rate of phenol oxidation at lower frequency (35 kHz) than that of higher frequency (500 kHz) in the presence of oxidizing agent H_2O_2 and copper sulfate. Further, the intermediate products formed during the degradation of phenol were easily degraded by 35 kHz frequency. This is a simple and highly efficient way of degradation organic pollutants by using cylindrical flow type reactor.

10.5.4 Large-Scale Reactors

It is necessary to study the effect of operational variables of ultrasound for large-scale reactors (>100 L capacity) during organic compound degradation. This type of study might give sufficient information to build a pilot-plant reactor for industrial applications using ultrasound. Son et al. [62] described the performance and the cavitation energy distribution in a large-scale reactor (250 L capacity). They were able to design a large-scale reactor with variable frequencies from 35 to 170 kHz. Figure 10.8 shows a schematic representation of the reactor. The large-scale reactor consists of an acrylic bath (L 1.20 m × W 0.60 m) and ultrasonic transducer module placed at the face center of the reactor as shown in Fig. 10.8. The module has nine PZT transducers and can produce ultrasound at 35, 72, 110, and 170 kHz frequencies. The temperature of the reactor was maintained around 18°C. Among the various frequencies of the ultrasonicator, 72 kHz provided both the maximum and uniform energy distribution throughout the large-scale reactor (Table 10.2). Many research reports suggested that high frequencies in the range of

Table 10.2 Average cavitation energy in the whole sonoreactor

Frequency (kHz)	35	72	110	170
Average cavitation energy (W)	53.50	83.87	6.41	0.10

200–1,700 kHz were more efficient for complete and relatively faster degradation of pollutants in small-scale reactors. This is due the formation of large number of bubbles at high frequencies that lead to more radical production. However, in large-scale reactors, the energy distribution was found to be very high and uniform only for low-frequency sonication (Table 10.2) because sound waves over 100 kHz do not propagate as effectively for long distances in a reactor as those of lower frequency. Such information is highly useful in designing a large-scale reactor for industrial wastewater treatment.

10.6 Ultrasound and Other AOTs as Examples of Hybrid Sonochemical Treatments

10.6.1 Sonolysis in Combination with Photocatalysis

Advanced oxidation technologies (AOTs) such as photocatalysis, sonolysis, ozonolysis, the Fenton's process, and photochemical oxidation have been extensively used for the treatment of contaminated water and wastewater. These AOT processes can produce OH radicals as a primary nonselective oxidant, which assists the oxidation process. Many researchers reported the effect of combining two or more AOTs to increase the rate of degradation of organics by combining the advantages of individual process [4, 8, 9, 30, 31, 38–45, 63–70]. For example, during TiO_2 photocatalysis, adsorbed reaction intermediates on the TiO_2 surface slowly decrease the activity of TiO_2. The fluid flow generated by ultrasound acoustic bubbles enhances mass transport of pollutants as well as prevent particles settling at the bottom of the reactor. In combining both these AOTs, adsorption of reaction products on the surface of catalysts can be prevented as well as make the catalyst's surface readily available for reaction. Using this approach, Stock et al. [70] reported the effect of combining photocatalysis and sonolysis for the degradation of the textile azo dye naphthol blue black (NBB). They used 600 kHz high-frequency ultrasonicator for this study. The reaction cell had a volume capacity of 600 mL and was placed 4 cm above the transducer (Fig. 10.9). The temperature of the reactor during sonolysis was maintained around 30°C. The reaction solution was sparged with oxygen throughout the experiment. The results revealed that sonolysis was an effective method for the degradation of the parent compound (NBB); whereas, photocatalysis was highly effective for complete mineralization of NBB and the reaction intermediates. Thus, this combined method result in complete mineralization and an additive effect on the degradation of NBB.

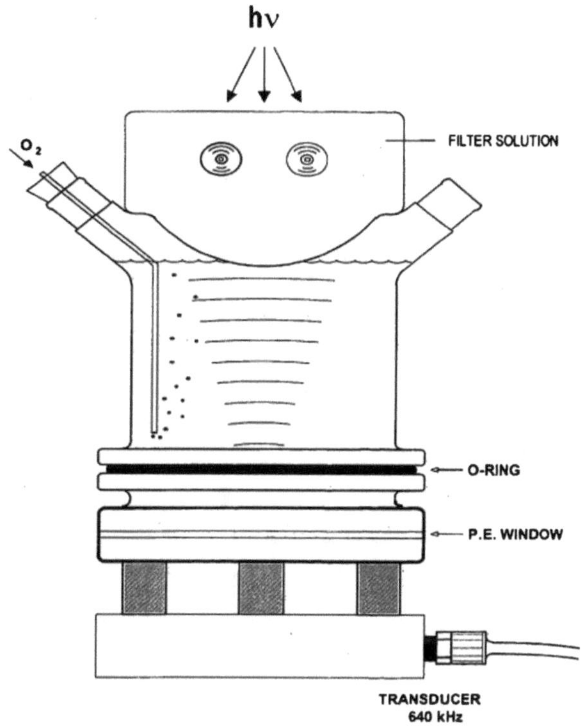

Fig. 10.9 High-frequency ultrasound sample cell. The transducer is separated from the sample solution by a polyethylene film (Reproduced from Ref. [70]. With kind permission of © The American Chemical Society (2000))

10.6.2 Degradation of Organic Pollutant by Ultrasonic Irradiation Along with Ozone

The generation of high temperatures during the acoustic cavitation bubble collapse provides the conditions with which to decompose not only water molecules but also oxidants like ozone very efficiently [71–73]. Ozone readily decomposes in the gas phase of the cavitation bubbles in a process that generates molecular oxygen and oxygen atoms (Reactions 10.1 and 10.2). Oxygen atoms react with water molecules and produce OH radicals. In addition to this, many other reactions occur between ozone and other free radicals in this coupled method, which facilitates the fast and rapid mineralization of organic pollutants. Kang and Hoffmann [71] applied ultrasound irradiation coupled with ozonation for the oxidation of methyl-*tert*-butyl ether (MTBE). The rate of oxidation of MTBE was remarkably high using the coupled method of oxidation of MTBE than the individual methods of oxidation, as shown in Table 10.3. The ozone-US coupled method enhanced the rate of MTBE oxidation by a factor of 1.5–4, depending on the initial concentration of MTBE. Similarly, the coupled method showed a very rapid mineralization of phenol into CO_2. On the other hand, ozone itself could not degrade phenol completely and led to intermediate products such as formate and oxalate, which accounted for the residual

Table 10.3 Comparison of various pseudo-first order rate constants for the sonochemical degradation of MTBE for various initial reaction conditions

Conditions[a]	$[O_3]$ or $[O_2]$ (mM)	$[MTBE]_o$ (mM)	Power (WL^{-1})	k_o (s^{-1})	$t_{1/2}$(min)
O_2 + US	0.25	0.05	200	8.7×10^{-4}	13.3
O_3 + US	0.31	0.05	200	8.7×10^{-4}	3.7
O_2 + no US	0.25	0.70	NA[b]	0	NA
O_3 + no US	0.25	0.31	NA	8.7×10^{-4}	192

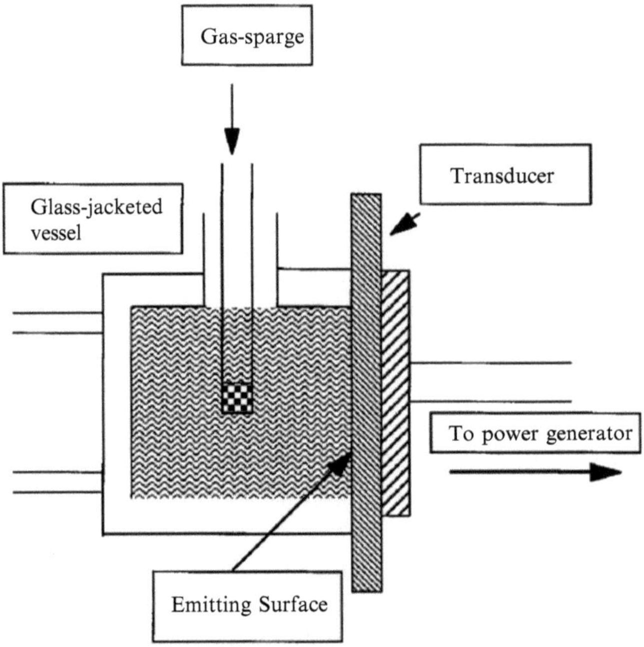

Fig. 10.10 Ultrasonic reactor configuration for sonication at 513 kHz (Figure reproduced from Ref. [72]. With kind permission of © The American Chemical Society (1997))

total organic carbon (TOC). A 205 kHz ultrasonic transducer was used in this study and mounted with a glass reactor, as shown in Fig. 10.10. The reactor capacity was 500 mL with a double walled construction for water cooling during ultrasonic irradiation. The reactor had four ports, which were used for gas inlet, gas outlet, to withdraw samples, and for adding other chemicals.

$$O_3 \xrightarrow{)))\Delta} O_2 + O \quad (10.1)$$

$$O + H_2O \rightarrow 2OH^{\bullet} \quad (10.2)$$

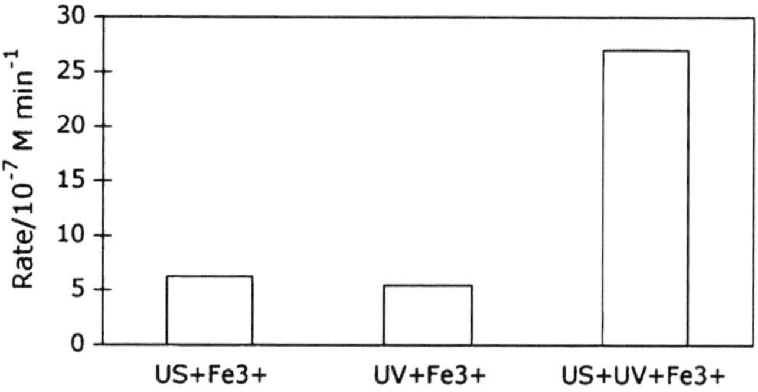

Fig. 10.11 A comparison of degradation rates of AR88 dye in the presence of Fe^{3+} using different processes (Reproduced from Ref. [41]. With kind permission of © Elsevier (2010))

10.6.3 Ultrasound-Assisted Homogeneous Photocatalysis

Ultrasound irradiation is also used as a source for the production of H_2O_2 as well as to accelerate homogenous photocatalytic degradation of organic pollutants. Fenton's reaction (Fe^{2+}/H_2O_2) and modified Fenton's like oxidation (Fe^{3+}/H_2O_2, Fe^{3+}/oxone and etc.) in the presence of UV light irradiation is known as homogeneous photocatalysis (reaction 10.3). This method has also been reported as a method for the effective degradation aqueous organic pollutants. Fenton's reaction produces OH radicals through splitting of H_2O_2 by iron salts (reaction 10.4). Ultrasound assists water splitting reaction to generate hydrogen peroxide by pyrolysis. H_2O_2 produced from ultrasound is a source for OH radicals that react with Fe^{2+} ions; and considered to be one of the main advantages of combining this Fenton's process along with ultrasound, since the external addition of H_2O_2 is not required.

$$Fe^{3+} + H_2O + h\upsilon \rightarrow Fe^{2+} + HO^{\bullet} + H^{+} \tag{10.3}$$

$$Fe^{2+} + H_2O_2 \rightarrow Fe^{3+} + HO^{\bullet} + H_2O \tag{10.4}$$

Madhavan et al. [43] have developed a method for the degradation of a monoazo textile dye acid-red 88 (AR88) using this combined process. Hydrated iron(III) nitrate was used as a source for Fe^{3+} ion. A 213 kHz frequency ultrasonicator was used to produce cavitation. The reactor volume had a capacity of 200 mL. Water was circulated around the reactor to maintain the constant temperature at around 25°C. The study demonstrated that there was a synergistic effect on the degradation of textile dye with a synergistic index of 2.3 (Fig. 10.11) compared with individual processes. UV light irradiation played a major role in reducing Fe^{3+} to Fe^{2+}, which instigated the Fenton's reaction for the dye degradation. The regeneration of catalyst could be achieved with OH radicals by continuous UV light irradiation of Fe^{3+}. This is another simple and highly efficient method for the degradation of textile dyes.

10.6.4 Ultrasound-Assisted Fenton-Like Oxidation

Goethite (FeOOH) is solid iron oxide, which reacts with H_2O_2 at low pH values (~3) and produce OH radicals. It is, generally, known as Fenton-like oxidation. This process has been used in the past few years to decompose organic contaminants present in wastewater as well as in soil. Neppolian et al. [9] studied the effect of ultrasound coupled with Fenton's like oxidation using FeOOH particles for the oxidative degradation of p-chlorobenzoic acid (p-CBA) with a 20 kHz horn type ultrasonicator. The reactor setup is described in Sect. 10.4.1.1. During sonochemical reactions, the physical forces of ultrasound can enhance mass transport of FeOOH particles and H_2O_2. As a result, there was enhanced decomposition of H_2O_2 observed in the presence of FeOOH, which led to a higher oxidation rate of organics present in this system. Figure 10.12 shows the influence of ultrasound on the decomposition efficiency of H_2O_2 under this coupled method of oxidation. The high rate of decomposition of H_2O_2 and organic compounds could be attributed to the indirect chemical effects associated with continuous cleaning and activation of FeOOH surfaces and the enhanced mass transport resulting from the turbulent effects of cavitation in the system.

10.6.5 Ultrasound-Assisted Preparation of Highly Active TiO_2 Photocatalysts for Water Treatment

Many researchers have reported the application of ultrasound during the synthesis of semiconductor nanoparticles [7, 74–80]. Neppolian et al. [7] applied ultrasound during the pH swing method of preparation of TiO_2 photocatalysts and observed a reduction in the particle sizes of the TiO_2 particles than in the absence of ultrasound. The authors also reported that during ultrasonic irradiation, the fragmentation of nanoparticles and increased high velocity interparticle collisions prevented the formation of larger particles. The catalytic activity of prepared TiO_2 showed a relatively better performance than the P-25 commercially available TiO_2. Figure 10.13 shows the application of ultrasound and its effect on the particle size during the pH swing method of synthesizing TiO_2 nanoparticles. It was also found that not only particle size of TiO_2 could be controlled by ultrasound but also other important parameters of TiO_2 such as surface area, ratio of anatase, rutile, and brookite, aggregated pore volume as well as the pore size of TiO_2 particles very effectively just by altering the ultrasound irradiation time [80]. Both horn- (20 kHz) as well as bath-type ultrasonicators (a cleaning bath, 40 kHz) showed almost the same trend on the physicochemical characteristics of TiO_2 particles produced.

Figure 10.14 shows the surface area of TiO_2 particles, steadily increasing with increasing ultrasound time and then decreasing after 100 min. The decreasing size

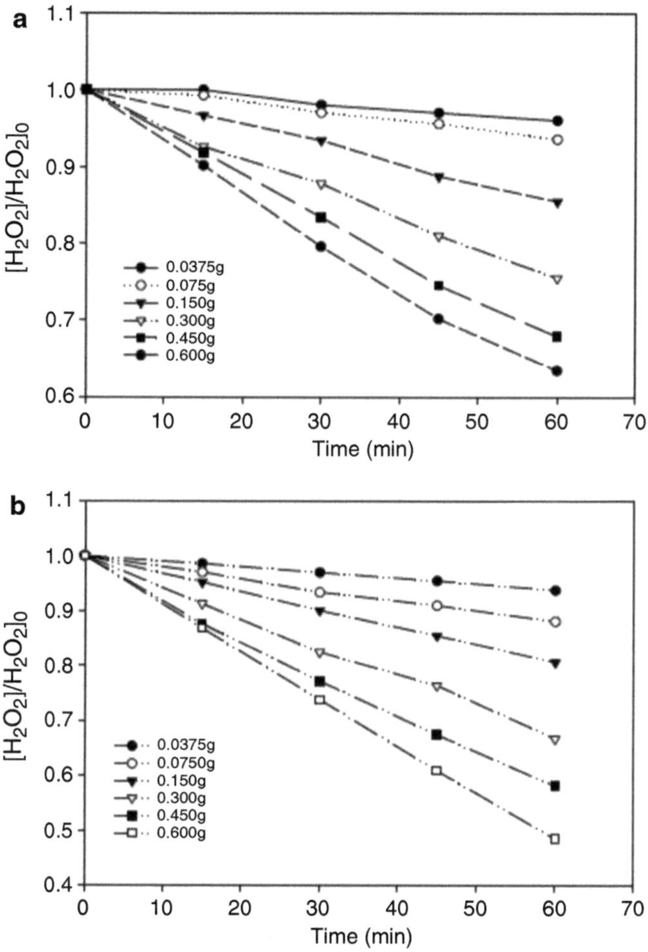

Fig. 10.12 Effect of FeOOH concentration in Fenton-like method on H_2O_2 decomposition: without using ultrasound (**a**); with ultrasound (**b**) (Reproduced from Ref. [9]. With kind permission of © Elsevier (2004))

of TiO_2 after some time was due to the formation of large size rutile particles. During ultrasound assisted preparation, the important operating variables of an ultrasonicator such as power density, ultrasound irradiation time, initial solution temperature, and different types of ultrasonic sources (horn and bath types) play a significant role in controlling the properties of TiO_2 nanoparticles. Thus, ultrasound is a useful technique to control many properties of synthesized semiconductor nanoparticles.

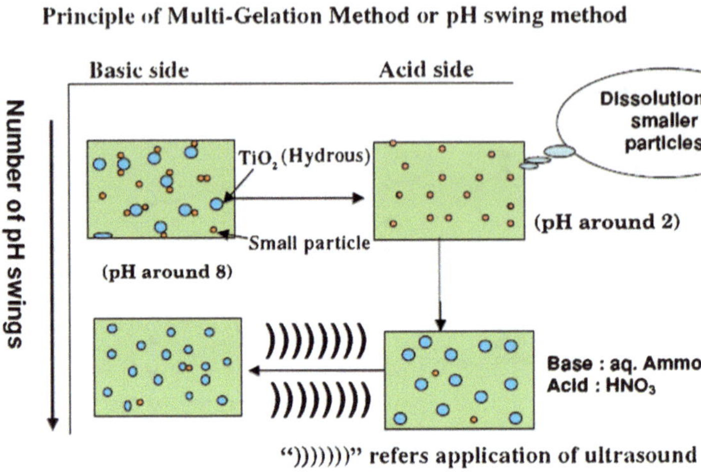

Fig. 10.13 Schematic representation of pH swing method with ultrasound (Reproduced from Ref. [7]. With kind permission of © Springer (2008))

Fig. 10.14 Surface area of TiO_2 prepared by (**a**) bath-type ultrasonicator; (**b**) horn-type ultrasonicator (Reproduced from Ref. [80]. With kind permission of © Elsevier (2008))

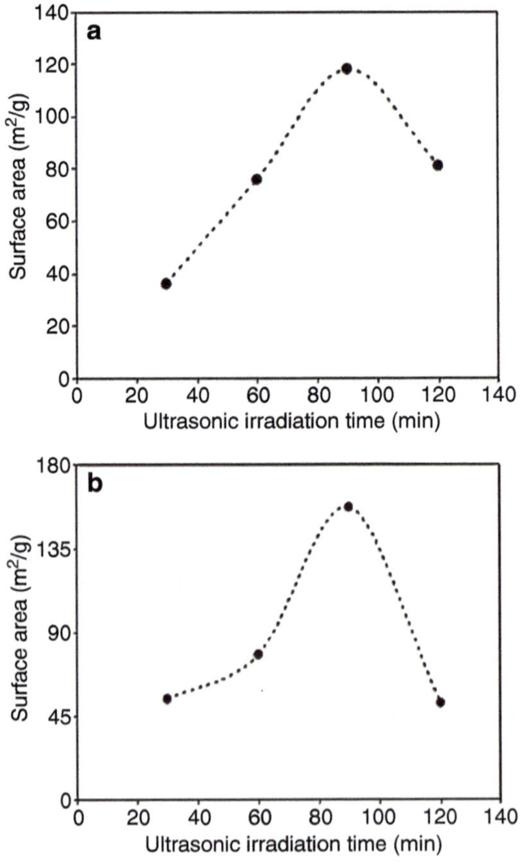

10.7 Ultrasound and Electrochemistry as Hybrid Sonochemical Treatment

The combination of an electrical field with an ultrasound field, namely, sonoelectrochemistry has been studied since 1930s [81] as an interesting way to generate chemical reactions. This is due to the high-energy chemical microenvironment provided by the combination of both fields. These benefits have been routinely highlighted in the literature [82–84], where several reviews [85–88] summarized different applications of this discipline. Among these applications, the approach to the environmental remediation can be considered as a hybrid methodology for the treatment of contaminated wastewater [89]. There are several special features of the sonoelectrochemical technique, coming from the electrochemical contribution, which justify a specific section in this chapter. One of them is the possibility to carry out reduction and oxidation processes at the same time in different regions of the reactor, and with the same operating cost. Another aspect is the enhancement of the kinetics of reactions, as will be discussed later. Finally, the avoidance of dangerous and harmful reagents and solvents remains as a remarkable feature of the sonoelectrochemical treatment of wastewater from an environmental point of view.

10.7.1 Different Adaptations of the Sonoreactors in Sonoelectrochemical Devices

The mechanical nature of the ultrasound field opens different possibilities in the application of electrochemistry. In this way, while the electrical field is established directly in the working solution with the introduction of the electrodes, the ultrasound field can be (or not) applied to an external medium that can be transmitted through the wall of the electrochemical cell. Therefore, different experimental setups can be found for the development of sonoelectrochemical reactions, which are discussed below.

10.7.1.1 Physical Separation Configuration

The most easy and convenient experimental setup is the immersion of the electrochemical cell in an ultrasonic bath [90], as shown in Fig. 10.15. This configuration is still in use, despite the commercial availability of ultrasonic horns [91, 92], due to its low cost and an easier control of the electrochemical system from an electric point of view. As we can find low- and high-frequency devices in an ultrasound bath configuration, it is possible to develop sonoelectrochemical experiments in a wide range of frequencies. However, drawbacks such as its low reproducibility and the fact that the power transmitted inside the cell is low and the dependence of the efficiency on the position of the cell inside the ultrasonic field could prevent

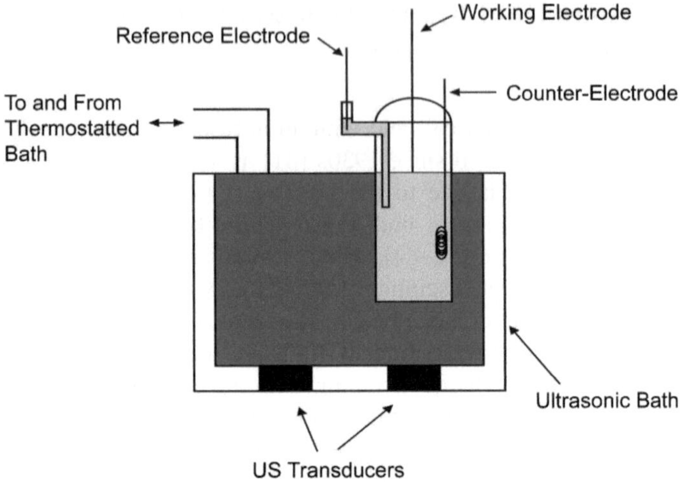

Fig. 10.15 Electrochemical cell dipped into an ultrasonic bath (Reproduced from Ref. [90]. With kind permission of © The Royal Society of Chemistry (RSC) for the Centre National de la Recherche Scientifique (CNRS) and the RSC (1998))

its use for systematic analysis. Due to these practical issues, sonoelectrochemists have developed more sophisticated designs with high-pressure conducting liquid [93], trying to overcome its drawbacks. Better reproducibility has been achieved, although technical issues with the overpressure remain to be improved further [94]. In spite of the fact that there is no reason for not working in flow-through mode, this physical separation configuration has been routinely used in lab-scale, batch mode with low volume of working solution [95].

10.7.1.2 Electrode-Apart-Transducer Configurations

In this configuration, the electrodes and the ultrasonic horn are directly dipped into the working solution of the sonochemical cell. The configuration design depends on the transducer shape and geometry and, therefore, different experimental arrangements can be found. The simplest configuration is the introduction of the electrodes and ultrasonic horn from the top of the cell, as shown in Fig. 10.16 [96].

The other one is the introduction of the electrodes and horn from opposite sides: (1) the electrodes from the bottom and the ultrasound tip from the top (Fig. 10.17 [96]) and (2) the ultrasound horn from the bottom and the electrodes from the top (Fig. 10.18 [97]).

All previous configurations are normally used for high-power low-frequency ultrasound field, due to the fact that there are many different commercially available equipments in the market. Besides, its technology is further developed with low

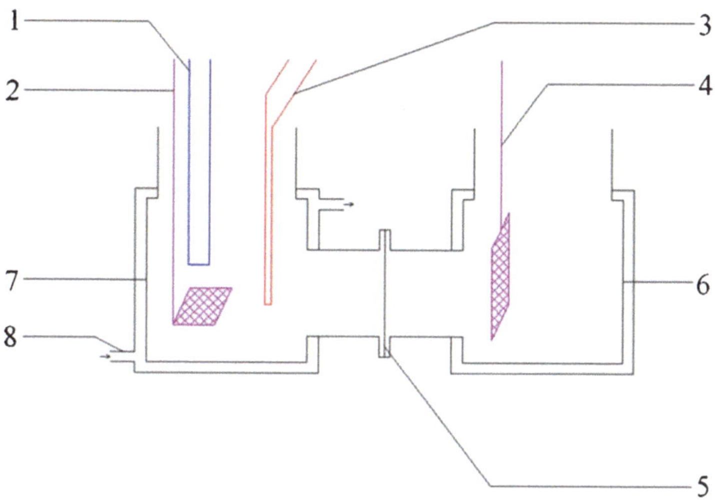

Fig. 10.16 Experimental setup for sonoelectrolysis in a batch scale in divided configuration: (*1*) ultrasound horn, (*2*) working electrode (or cathode), (*3*) luggin capillary, (*4*) counter-electrode, (*5*) separator (membrane), (*6*) glass cell, (*7*) cooling jacket, and (*8*) port for the cooling fluid (Reproduced from Ref. [96]. With kind permission of © Elsevier (2010))

Fig. 10.17 Sonovoltammeric cell: (*1*) luggin capillary, (*2*) glass cell, (*3*) gas port, (*4*) counter-electrode, (*5*) working electrode, and (*6*) ultrasound horn (Reproduced from Ref. [96]. With kind permission of © Elsevier (2010))

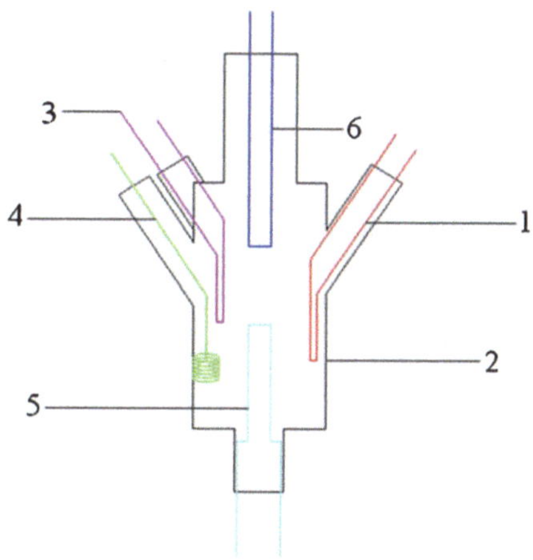

power, high frequencies. At low-power, high-frequency ultrasound fields, the configuration is basically limited to the flat plate configuration (see Fig. 10.19).

With all these different kinds of configurations, we can find in the literature that studies have been carried out in batch and flow-through operation modes [87], which have provided a wide range of database for the performance of the sonoelectrochemical technology from the technical, economical, and environmental point of views.

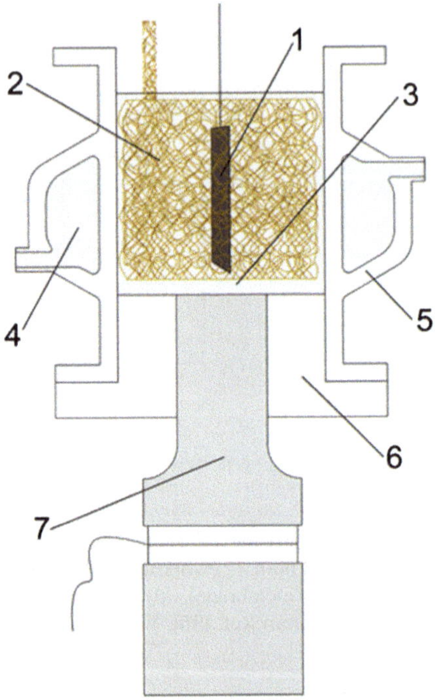

Fig. 10.18 20 kHz sonoreactor adapted as a sonoelectrochemical device: (*1*) lead dioxide electrode, (*2*) copper counter-electrode (foam geometry), (*3*) working solution, (*4*) cooling jacket, (*5*) glass cell, (*6*) Teflon holder, (*7*) ultrasonic transducer (Reproduced from Ref. [97]. With kind permission of © Elsevier (2011))

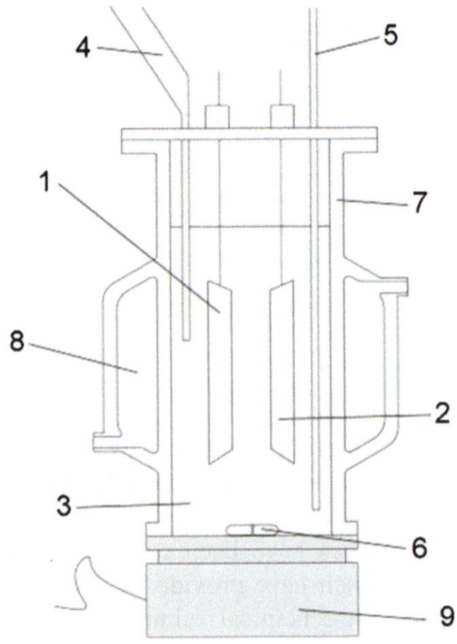

Fig. 10.19 Scheme showing a high-frequency sonoelectrochemical reactor: (*1*) working electrode, (*2*) counterelectrode, (*3*) working solution, (*4*) luggin capillary, (*5*) temperature probe, (*6*) stirring bar, (*7*) glass reactor, (*8*) cooling fluid, (*9*) ultrasonic transducer

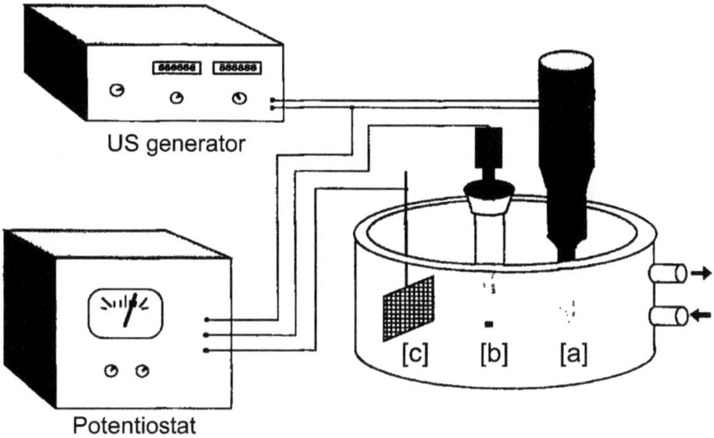

Fig. 10.20 Schematic diagram of a sonoelectrochemical reactor in the sonoelectrode configuration: [*a*] Ti sonotrode (working electrode), [*b*] saturated sulfate electrode in a Tacussel bridge (reference electrode), [*c*] Pt grid (counter-electrode) (Reproduced from Ref. [98]. With kind permission of © Elsevier (1996))

10.7.1.3 Sonotrode or Sonoelectrode Configuration

Finally, in this case, the working electrode acts as an ultrasound tip, see Fig. 10.20 [98]. This configuration provides a much localized microenvironment for the reaction at the surface of the sonoelectrode, which normally works at low frequency and high-power ultrasound.

In these conditions, not only acoustic streaming by cavitation [86, 99] occurs but also other effects of cavitation such as microjetting [94] and shock waves [100] occur at the electrode surface. Therefore, the presence of ultrasound in an electrochemical process enhances various processes such as depassivation [101], electrochemical reaction mechanism modification [102], surface activation [103], decrease of adsorption phenomena [104], and the mass transport [105]. This idea has also been used in scale-up processes at flow-through operation mode. The equipment is commercially provided by Advanced Sonic Processing Systems (Fig. 10.21 [106]), where the flat electrodes are also the ultrasound emitters. Each electrode oscillates at different frequencies and this provides a well-defined cavitation pattern in the electrode chambers.

10.7.2 Sonoelectrochemical Treatment of Contaminated Wastewater

As we have discussed above, a sonoelectrochemical process presents a wide variety of strategies. It can be carried out with or without separation between the

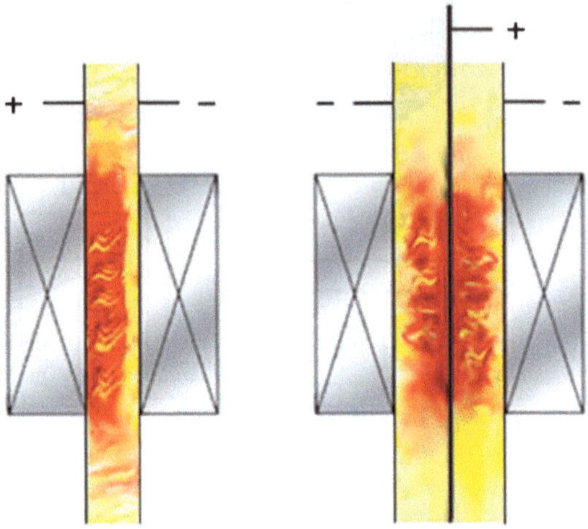

Fig. 10.21 Schematic diagram of the electro-acoustical reaction cell from advanced sonics (Reproduced from Ref. [106]. With kind permission of © Advanced Sonic Processing System (2004))

anodic (oxidation) and the cathodic (reduction) reactions. This possibility allows environmental researchers not only to study the degradation of the pollutant by oxidation, reduction or by forcing the pollutants (and their by-products) to undergo both electrons transfer possibilities, but also to investigate the relation among them by modifying the reaction mechanisms with the formation of new intermediates. In addition, we can apply both electrical and ultrasonic energy fields simultaneously in the same reactor or in different chambers. Therefore, extensive study of the sonoelectrochemical treatment of wastewater could be developed. However, the chemical nature of the pollutant can normally allow the choice among the anodic, cathodic, or dual treatment. This suggestion can be supported on the preliminary analysis carried out using sonovoltametric techniques [90] that may also provide information about the electrocatalysis of the different electrodic materials to be used or the need for the separation between cathodic and anodic compartment in a sonoelectrochemical reactor. After this study, the design of a sonoelectrochemical degradation process at lab scale in batch configuration can be carried out. All these trials provide a clear idea of the feasibility of the sonoelectrochemical treatment from a technical and environmental point of views. In spite of the fact that not always a systematic analysis has been followed, in the following subsections, we will develop the most popular approaches found in the literature, leaving the specific case of the sonoelectrochemical treatment of organics for Chap. 14.

10.7.2.1 Sonoelectrochemical Treatment by Oxidation

The oxidation process is the most popular route for the degradation of chemical pollutants. Two strategies are normally followed: (1) the transformation of the initial compounds to CO_2, H_2O, and other small molecules and (2) the transformation of the initial compounds to other forms with a higher biodegradability. An example of this is the oxidation of the cyanide anion. Cyanide can be electrochemically degraded by direct oxidation, first giving cyanate, which is further oxidized, or by indirect oxidation using chloride (forming ClO^-) as an oxygen carrier [107]. Both electrochemical procedures have been used in combination with ultrasound [108] analyzing six types of electrodic materials. Cyanide indirect electrochemical degradation under sonication provided better results.

Other applications of sonoelectrochemical oxidation are the improvement of the electrochemical water disinfection. The disinfecting species, mainly hypochlorous acid and hypochlorite, can be electrochemically produced from chloride ions in electrolyzed water, but the deposition of calcareous deposits on the cathode surface is the main drawback of the process. In this case, the combination of ultrasound with this process has been studied by Kraft et al. [109] who showed that the sonotrode efficiently removed calcium carbonate scales from the cathode surface. Thus, electrochemical water disinfection in potable water can be performed on a long timescale, without the necessity for cathode cleaning, through the use of acids or the polarity reversal method.

10.7.2.2 Sonoelectrochemical Treatment by Reduction

The other route in electrochemical processes is the reduction, which is currently the most widely employed method for metal recovery. The low concentration of metals in wastewater is one of the most important problems of electrodeposition, because the mass transport of metallic cations to the cathode, often the rate-controlling step, can be very slow. Therefore, the deposition of metals under the influence of an ultrasonic field has received significant attention [93]. Nevertheless, this work has been focused on the electrodeposits rather than on the recovery processes [110]. Much of this work has been carried out from an industrial standpoint, particularly on metals that are important in the electroplating industry. The results of various studies [110–112] dealing with copper removal show an improvement in the deposition rate with an ultrasonically agitated bath. Farooq et al. [111] used electrodeposition and ultrasound at 35 kHz to remove copper in wastewater. The copper removal efficiency was enhanced from 55.1% to 94.6% in the presence of an ultrasonic field. The removal of zinc [110] and lead [112] from industrial wastewater has also been studied. Hyde et al. [113] concluded that in electrodeposition under the influence of ultrasound, the critical effect is the increase in mass transport, which may be high enough to change from a diffusion-controlled system to a charge-controlled system. In addition, ultrasound also ablates material from the electrode surface, but has no effect on growth through charge transfer from the electrode to the metal ion. All these observations demonstrate that ultrasound can dramatically improve the efficiency of heavy metal wastewater treatment.

Other example of the sonoelectrochemical methodology in metal recovery application is developed in the removal of silver from photographic processing solutions, which consist mainly of sodium thiosulfate, sodium bisulfite, and silver halides. Currently, the most widely employed method for silver removal is electrolysis, where vigorous agitation is required in order to overcome electrode fouling. The reduction of silver cations has been deeply studied by Pollet and coworkers [114–118] at platinum, stainless steel, and carbon stationary or rotating electrodes in the absence and presence of ultrasound. In these studies, low-frequency experiments were performed using either a typical single-compartment voltammetric cell, or using a cell similar in design to that of Compton et al. [119] except that the ultrasound source was placed at the bottom of the electrochemical cell, while high-frequency experiments were performed using a three-compartment cell [114], and bulk electrolysis was carried out in a specially designed cell enabling the use of three ultrasonic probes [116]. Their results indicate that the Ag^{+} cathodic discharge potential does not significantly vary with the ultrasonic frequency [114], but rather with ultrasonic intensity [115] and the position of the ultrasonic probe with respect to the electrode [117]. More results were obtained using a combination of ultrasonic irradiation at 20 kHz and bulk potential–controlled electrolysis with sonoelectrochemical cells adapted to use three ultrasonic probes in "face-on" (probe 2) and "side-on" (probes 1 and 3) geometries (Fig. 10.22 [116]). Their results confirm that the cell geometry is an important parameter in the removal of silver and for most sonoelectrochemical processes. The specially designed cell permitted different ultrasound probe combinations. Operating the one side probe ("side-on") and the bottom ultrasonic probes ("face-on") simultaneously (i.e., probes 1 and 2) led to a higher rate constant for a given ultrasonic power. Other probe combinations (probes $2+3$ or $1+2+3$) were less effective, and the use of the three ultrasonic probes operating at the same power did not significantly improve the rate constant of silver removal compared with that obtained for the "face-on" geometry.

10.7.2.3 Sonoelectrochemical Treatment by Dual Degradation

Finally, it is important to point out the interrelation and mutual influence between anodic and cathodic reactions during the process. In this way and following with the previous processes shown, the anodic reaction for silver removal reaction from photographic "fixer" processing solutions has also been studied [120]. The sonoelectrochemical treatment used in silver removal enhances not only the silver recovery rates, but also the thiosulfate oxidation process. A 50-fold increase was observed compared with silent conditions in the degradation of thiosulfate on platinum. Under the conditions studied, the thiosulfate degradation increased as the ultrasonic power was increased, suggesting that the oxidation of thiosulfate was also limited by diffusion.

In the same way, industrial wastewaters usually contain metal complexes, e.g., copper-ethylenediaminetetraacetic acid (EDTA). During the electrodeposition, organic pollutants can be degraded at the anode while heavy metals are reduced at the cathode. Nevertheless, a very low current efficiency for treating wastewater containing chelated heavy metals is observed, reducing the economical viability of the electrodeposition technique; e.g., a 97.7% recovery rate, but a 3.1% current

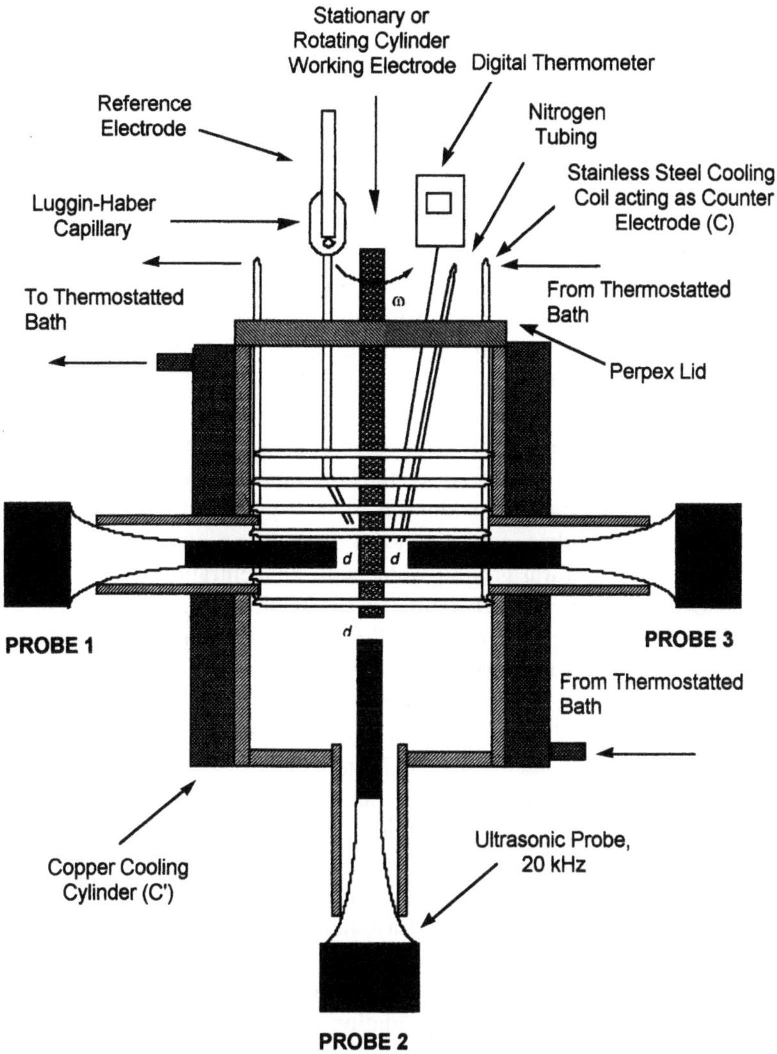

Fig. 10.22 Sonoelectrochemical cell for bulk electrolysis fitted with three ultrasonic probes (Reproduced from Ref. [116]. With kind permission of © Elsevier (2000))

efficiency, for EDTA-Cu system was obtained for the Cu electrodeposition process [121]. Therefore, the influence of ultrasound on electrodeposition applied to EDTA-Cu solutions has been studied [122]. In this case, an ultrasonic transducer with a round tip was placed in the middle of the reactor to produce the ultrasonic wave instead of using an ultrasonic bath. Again, it is concluded that the mass transport mechanism changes to a charge-controlled system under ultrasonic conditions, with the oxidation of EDTA by the combined technique and an 84% COD removal.

10.8 Energetic Costs Analysis

In spite of the increasing number of sonochemical and hybrid ultrasonic applications for environmental remediation, very few processes have actually been scaled-up with industrial purposes due to the lack of standard strategies for the design of large and efficient reactors. As stated by different authors [123, 124], the economical viability of sonochemical and hybrid ultrasonic processes is, without any doubt, the main goal to achieve in order to consider overall ultrasonic processes as competitive as other technologies. This is due to the high energy consumption associated with the application of ultrasound inside the reactors and the nonhomogeneous distribution of cavitation phenomena, both of which make the current sonochemical and hybrid ultrasonic wastewater treatments hardly feasible for the industry from an economic point of view.

Unfortunately, this key issue has received poor attention from most of the scientific community, as only a few studies found in the literature include an estimation of the energetic consumption of the sonochemical treatment. In this sense, Peters and coworkers found that the degradation of highly volatile chlorinated compounds in natural ground water with concentrations in the mg/L range consumed 640 kWh m^{-3} [125], while in the μg/L range 500 kWh m^{-3} [126] was necessary for the removal of the pollutants. In a different study, Shemer and Narkis [127] calculated that 300–1,400 kWh m^{-3} would be necessary to reach a degradation of 90% of the trihalomethanes ($CHCl_3$, $CHBr_3$, and CHI_3, among others) present in a water sample with an initial concentration of 10 mg/L.

For hybrid ultrasonic technologies, Ragaini et al. [69] evaluated the energy consumption of the sonophotocatalytic degradation of 2-chlorophenol in water, comparing the results with other techniques (photocatalysis, photocatalysis with ozone, sonophotocatalysis with ozone, among others). They estimated the energy consumed to obtain a degradation of 70% for all different experimental techniques. They found that the combination of ultrasound and ozone (190 kWh m^{-3}, reaction time of 130 min) was quite competitive compared with the ozone treatment (177.5 kWh m^{-3}, reaction time of 170 min). A similar procedure was carried out by Selli et al. [128] to study the degradation of methyl *tert*-butyl ether in water, where they measured the energy required to reach a degradation of 90% of the initial concentration in the solution (10^{-3} M). The sonophotocatalytic treatment with intermittent stirring was the most efficient (4,900 kWh m^{-3}, reaction time of 148 min), compared to the sonophotocatalytic treatment with continuous stirring (9,550 kWh m^{-3}, reaction time of 280 min) or the photocatalytic treatment (8,750 kWh m^{-3}, reaction time of 322 min). More recently, Rokhina et al. [129] studied the silent and ultrasound-assisted oxidation of phenol over RuI_3 catalyst, estimating that the sonocatalytic oxidation yielded lower energy consumption (501.8 kWh m^{-3}) than the sonochemical treatment in presence of hydrogen peroxide (2,297.8 kWh m^{-3}) or the sonolysis alone (31,265.5 kWh m^{-3}). Analogous conclusions have been made by Mahamuni and Adewuyi regarding the combination of ultrasound and other AOTs for wastewater treatment [130]. In their work, a quite complete review of different hybrid ultrasonic processes is presented, with a focus on the estimation of the economic cost. In their analysis, they concluded that the combination of ultrasound with different AOTs was economically more attractive than the use of ultrasound alone for the degradation of diverse pollutants, specifically for the degradation of reactive azo dyes, where the combination of ultrasound, ultraviolet light and H_2O_2 achieved a competitive cost of $65.17/1,000 gal, which is an order of magnitude less than commercially available processes.

In order to shed more light on the benefits of combining ultrasound with other technologies, we comment on the different degradation strategies followed by Sáez and coworkers's work on the removal of perchloroethylene (PCE) and its degradation by-products from water. While the more optimized electrochemical treatment (ECT) in sodium sulfate aqueous media (sodium sulfate acted as background electrolyte) yield a conversion of PCE of 84% [131], the nonoptimized 20 kHz sonochemical treatment (SCT) reached a 99% [132]. However, poor degradation efficiency was obtained in the removal of C–Cl bonds (30% for SCT vs. 68% for ECT) and a tremendous energetic consumption was observed (900 kWh m^{-3} for SCT vs. 3.8 kWh m^{-3} for ECT). A complete degradation of PCE was observed with the nonoptimized 20 kHz background-electrolyte-assisted sonoelectrochemical treatment (BEA-SECT) [133], although a lower C–Cl degradation efficiency (57%) and higher energy consumption (902 kWh m^{-3}) were observed compared with the ECT process. And more recently, the nonoptimized 20 kHz sonoelectrochemical treatment with no further addition of a background electrolyte (SECT) [134] enabled to reach not only the complete degradation of PCE in the sample, but also to achieve an outstanding C–Cl degradation efficiency of 99%, while reducing the energy consumption (682 kWh m^{-3}) compared with the SCT and the BEA-SECT processes. Yet, the energy consumption of the ECT degradation of PCE remains much lower compared with the SECT process; but we have to keep in mind that, from an environmental point of view, the benefits of the SECT degradation (complete removal of the toxicity inherent to the presence of C–Cl bonds in polluted freshwater) far outweigh the drawback of the energy consumption.

Besides the combination of ultrasound with other technologies as those briefly commented in this chapter, other efforts are being pursued by different research groups in order to increase the energetic efficiency of all kinds of chemical processes involving ultrasound. These include the improvement of high power ultrasonic devices [135], the previously commented use of multiple frequencies, the use of pulsed ultrasound [136], the proper selection of the working frequency near a resonance peak of the system [137] and the design of the reactor taking into account its geometry and materials. For the latter, a proper understanding of the spatial distribution of the acoustic pressure in the reactor is, therefore, mandatory in order to precisely predict the cavitation phenomena inside the reactor and to effectively design large-scale reactors. Regarding this issue, new developments with computer-aided simulations have been recently raised involving linear [138–140] and nonlinear [141–143] acoustics. These developments would eventually lead to the efficient design of ultrasonic systems and devices that would also enable in controlling the extent of transient and stable cavitation phenomena [144], depending on the specifications of the desired application.

10.9 Concluding Remarks

Although an increasing number of studies have been reported, the ultrasound-assisted degradation of organic pollutants is still in its early stage of development, especially with respect to large-scale treatment. The rate of chemical reactions depends on various operational parameters of ultrasound such as frequency, acoustic power, reactor size and design, initial temperature, and mode of operation. The nature of the contaminant, influence of oxidizing agent, mass transport of catalysts, etc., also play a considerable role in the degradation process. During ultrasonic irradiation of a contaminated fluid, many intermediate products are formed that requires a long time for complete mineralization. On the other hand, the coupled method of oxidation can considerably enhance the complete mineralization of an organic compound within a short period of sonication time. Applying double, triple, or multiple acoustic frequencies was found to provide an additive effect on the chemical reactivity generated in most of the studies undertaken. Low-frequency ultrasound is able to distribute acoustic energy more efficiently within a reaction vessel for large-scale reactors than with higher acoustic frequencies. In addition, the development of hybrid technologies involving ultrasound is in progress in order to obtain a practical technology for the degradation of recalcitrant pollutants. Among them, the sonophotocatalysis and sonoelectrochemistry have provided promising results from a technical and environmental point of views while the economics aspects are still being optimized.

References

1. Richards WT, Loomis AL (1927) The chemical effects of high frequency sound waves I. A preliminary survey. J Am Chem Soc 49:3086–3100
2. Wood R, Loomis AL (1927) The physical and biological effects of high-frequency sound-waves of great intensity. Philos Mag Ser 4:417–436
3. Ashokkumar M (2007) Sonochemistry. In: Kirk-Othmer encyclopedia of chemical technology. Wiley, Hoboken
4. Ashokkumar M, Sunartio D, Kentish S, Mawson R, Simons L, Vilkhu K, Versteeg C (2008) Modification of food ingredients by ultrasound to improve functionality: a preliminary study on a model system. Innov Food Sci Emerg Technol 9:155–160
5. Suslick KS (1988) Ultrasound, its chemical, physical and biological effects. VCH, New York
6. Bang JH, Han K, Skrabalak SH, Kim H, Suslick KS (2007) Porous carbon supports prepared by ultrasonic spray pyrolysis for direct methanol fuel cell electrode. J Phys Chem C 111:10959–10964
7. Neppolian B, Celik E, Anpo M, Choi H (2008) Ultrasonic-assisted pH Swing method for the synthesis of highly efficient TiO_2 nano-size photocatalysts. Catal Lett 125:183–191
8. Neppolian B, Haeryong J, Choi H, Lee JH, Kang JW (2002) Sonolytic degradation of methyl tert-butyl ether: the role of coupled Fenton process and persulphate ion. Water Res 36: 4699–4708
9. Neppolian B, Park JS, Choi H (2004) Effect of Fenton-like oxidation on enhanced oxidative degradation of para-chlorobenzoic acid by ultrasonic irradiation. Ultrason Sonochem 11: 273–279

10. Singla R, Grieser F, Ashokkumar M (2004) The mechanism of the sonochemical degradation of benzoic acid in aqueous solutions. Res Chem Intermed 30:723–733
11. Singla R, Grieser F, Ashokkumar M (2009) Kinetics and mechanism for the sonochemical degradation of a nonionic surfactant. J Phys Chem A 113:2865–2872
12. Vecitis CD, Wang Y, Cheng J, Park H, Mader B, Hoffmann MR (2010) Sonochemical degradation of perfluorooctanesulfonate in aqueous film-forming foams. Environ Sci Technol 44:432–438
13. Yang L, Sostaric JZ, Rathman JF, Weavers LK (2008) Effect of ultrasound frequency on pulsed sonolytic degradation of octylbenzene sulfonic acid. J Phys Chem B 112:852–858
14. Yang L, Rathman JF, Weavers LK (2005) Sonochemical degradation of alkylbenzene sulfonate surfactants in aqueous mixtures. J Phys Chem B 110:18385–18391
15. Zhang GM, Zhang PY, Yang JM, Chen YM (2007) Ultrasonic reduction of excess sludge from the activated sludge system. J Hazard Mater 145:515–519
16. Lim M, Son Y, Park B, Khim J (2010) Sonophotochemical degradation of Bisphenol A with solid catalysts. Jpn J Appl Phys 49:07HEO6
17. Gultekin I, Tezcanli-Guyer G, Ince NH (2009) Degradation of 4-n-nonylphenol in water by 20 kHz ultrasound. J Adv Oxid Technol 12:105–110
18. Hamdaoui O, Naffrechoux E (2008) Sonochemical and photosonochemical degradation of 4-chlorophenol in aqueous media. Ultrason Sonochem 15:981–987
19. Guo ZB, Zheng Z, Zheng SR, Hu WY, Feng R (2005) Effect of various sono-oxidation parameters on the removal of aqueous 2,4-dinitrophenol. Ultrason Sonochem 12:461–465
20. Little C, Hepher MJ, El-Sharif M (2002) The sono-degradation of phenanthrene in an aqueous environment. Ultrasonics 40:667–674
21. Kim JK, Martinez F, Metcalfe IS (2007) The beneficial role of use of ultrasound in heterogeneous Fenton-like system over supported copper catalysts for degradation of p-chlorophenol. Catal Today 124:224–231
22. Suh WH, Jang AR, Suh YH, Suslick KS (2006) Porous, hollow, and ball-in-ball metal oxide microspheres: preparation, endocytosis, and cytotoxicity. Adv Mater 18:1832–1837
23. Muthukumaran S, Kentish SE, Stevens GW, Ashokkumar M (2006) Application of ultrasound in membrane separation process: a review. Rev Chem Eng 22:155–194
24. Neppolian B, Doronila A, Ashokkumar M (2010) Sonochemical oxidation of arsenic(III) to arsenic(V) using potassium peroxydisulfate as an oxidizing agent. Water Res 44:3687–3695
25. Neppolian B, Doronila A, Grieser F, Ashokkumar M (2009) Simple and efficient sonochemical method for the oxidation of arsenic (III) to arsenic (V). Environ Sci Technol 43:6793–6798
26. Okitsu K, Ashokkumar M, Grieser F (2005) Sonochemical synthesis of gold nanoparticles: effects of ultrasound frequency. J Phys Chem B 109:20673–20675
27. Singla R, Grieser F, Ashokkumar M (2009) Sonochemical degradation of martius yellow dye in aqueous solution. Ultrason Sonochem 16:28–34
28. Vinodgopal K, He Y, Ashokkumar M, Grieser F (2006) Sonochemically prepared platinum-ruthenium bimetallic nanoparticles. J Phys Chem B 110:3849–3852
29. Ashokkumar M, Hodnett M, Zeqiri GF, Price GJ (2007) Study of the coalescence of acoustic bubbles as a function of frequency, power, and water-soluble additives. J Am Chem Soc 129:2250–2258
30. Neppolian B, Ciceri L, Bianchi CL, Grieser F, Ashokkumar M (2011) Sonophotocatalytic degradation of 4-chlorophenol using $Bi_2O_3/TiZrO_4$ as a visible light responsive photocatalyst. Ultrason Sonochem 18:1832–1837
31. Weavers L, Malmstadt N, Hoffmann MR (2000) Kinetics and mechanism of pentachlorophenol degradation by sonication, ozonation, and sonolytic ozonation. Environ Sci Technol 34:1280–1285
32. Ferguson MA, Hering JG (2006) TiO_2-photocatalyzed As (III) oxidation in a fixed-bed, flow-through reactor. Environ Sci Technol 40:4261–4267
33. Iqbal J, Kim HJ, Yang JS, Baek K, Yang JW (2007) Removal of arsenic from groundwater by micellar-enhanced ultrafiltration (MEUF). Chemosphere 66:970–976

34. Johnston RB, Singer PC (2007) Redox reactions in the Fe-As-O_2 system. Chemosphere 69:517–525
35. Kanel SR, Manning B, Charlet L, Choi HC (2005) Removal of arsenic (III) from groundwater by nanoscale zero-valent iron. Environ Sci Technol 39:1291–1298
36. Kim MJ, Nriagu J (2000) Oxidation of arsenite in groundwater using ozone and oxygen. J Sci Total Environ 247:71–79
37. Lee H, Choi W (2002) Photocatalytic oxidation of arsenite in TiO_2 suspension: kinetics and mechanisms. Environ Sci Technol 36:3872–3878
38. Madhavan J, Kumar S, Anandan A, Grieser F, Ashokkumar M (2010) Sonophotocatalytic degradation of monocrotophos using TiO_2 and Fe^{3+}. J Hazard Mater 177:944–949
39. Madhavan J, Grieser F, Ashokkumar M (2010) Degradation of orange-G by advanced oxidation processes. Ultrason Sonochem 17:338–343
40. Madhavan J, Grieser F, Ashokkumar M (2010) Combined advanced oxidation processes for the synergistic degradation of ibuprofen in aqueous environments. J Hazard Mater 178: 202–208
41. Madhavan J, Grieser F, Ashokkumar M (2010) Degradation of formetanate hydrochloride by combined advanced oxidation processes. Sep Purif Technol 73:409–414
42. Madhavan J, Kumar S, Anandan S, Zhou M, Grieser F, Ashokkumar M (2010) Ultrasound assisted photocatalytic degradation of diclofenac in an aqueous environment. Chemosphere 80:747–752
43. Madhavan J, Kumar S, Anandan S, Grieser F, Ashokkumar M (2010) Degradation of acid red 88 by the combination of sonolysis and photocatalysis. Sep Purif Technol 74:336–341
44. Madhavan M, Grieser F, Ashokkumar M (2009) Kinetics of the sonophotocatalytic degradation of orange G in presence of Fe^{3+}. Water Sci Technol 60:2195–2202
45. Kamat PV (2011) Graphene-based nanoassemblies for energy conversion. J Phys Chem Lett 2:242–251
46. Kamat PV (2010) Graphene-based nanoarchitectures. Anchoring semiconductor and metal nanoparticles on a two-dimensional carbon support. J Phys Chem Lett 1:520–527
47. Lightcap IV, Kosel TH, Kamat PV (2010) Anchoring semiconductor and metal nanoparticles on a 2-dimensional catalyst mat. Storing and shuttling electrons with reduced graphene oxide. Nano Lett 10:577–583
48. Rao CNR, Sood AK, Voggu R, Subrahmanyam KS (2010) Some novel attributes of graphene. J Phys Chem Lett 1:572–580
49. Seger B, Kamat PV (2009) Electrocatalytically active graphene-platinum nanocomposites. Role of 2-D carbon support in PEM fuel cells. J Phys Chem C 113:7990–7995
50. Vinodgopal K, Neppolian B, Lightcap IV, Grieser F, Ashokkumar M, Kamat PV (2010) Sonolytic design of graphene-Au nanocomposites. Simultaneous and sequential reduction of graphene oxide and Au(III). J Phys Chem Lett 1:1987–1993
51. Williams G, Seger B, Kamat PV (2008) TiO_2-Graphene nanocomposites. UV-assisted photocatalytic reduction of graphene oxide. ACS Nano 2:1487–1491
52. Williams G, Kamat PV (2009) Graphene-semiconductor nanocomposites. Excited state interactions between ZnO nanoparticles and graphene oxide. Langmuir 25:13869–13873
53. Zhang H, Lv X, Li Y, Wang Y, Li J (2010) P-25 graphene composite as a high performance photocatalyst. ACS Nano 4:380–386
54. Brotchie A, Ashokkumar M, Grieser F (2007) Effect of water-soluble solutes on sonoluminescence under dual-frequency sonication. J Phys Chem C 111:3066–3070
55. Kanthale PM, Brotchie A, Ashokkumar M, Grieser F (2008) Experimental and theoretical investigations on sonoluminescence under dual frequency conditions. Ultrason Sonochem 15:629–635
56. Brotchie A, Ashokkumar M, Grieser F (2008) Sonochemistry and sonoluminescence under dual-frequency ultrasound irradiation in the presence of water-soluble solutes. J Phys Chem C 112:10247–10250
57. Vijayanand SM (2009) Mechanistic optimization of a dual frequency sonochemical reactor. Chem Eng Sci 64:5255–5267

58. Sivakumar M, Pandit AB (2001) Ultrasound enhanced degradation of Rhodamine B: optimization with power density. Ultrason Sonochem 8:233–240
59. Gogate PR, Sivakumar M, Pandit AB (2004) Destruction of Rhodamine B using novel sonochemical reactor with capacity of 7.5l. Sep Purif Technol 34:13–24
60. Gondrexon N, Renaudin V, Petrier C, Boldo P, Bernis A, Gonthier Y (1999) Degradation of pentachlorophenol aqueous solutions using a continuous flow ultrasonic reactor: experimental performance and modeling. Ultrason Sonochem 5:125–131
61. Entezari MH, Petrier C, Devidal P (2003) Sonochemical degradation of phenol in water: a comparison of classical equipment with a new cylindrical reactor. Ultrason Sonochem 10:103–108
62. Son Y, Lim M, Khim J (2009) Investigation of acoustic cavitation energy in a large-scale sonoreactor. Ultrason Sonochem 16:552–556
63. Vinu R, Madras G (2009) Kinetics of sonophotocatalytic degradation of anionic dyes with nano-TiO_2. Environ Sci Technol 43:473–479
64. Dorathi RPJ, Palanivelu K (2010) Sonochemical degradation of p-chlorophenol in aqueous solution using hypervalent iron. Ind J Chem Technol 17:111–119
65. Bremner DH, Molina R, Martinez F, Melero JA, Segura Y (2009) Degradation of phenolic aqueous solutions by high frequency sono-Fenton systems (US-Fe_2O_3/SBA-15-H_2O_2). Appl Catal B-Environ 90:380–388
66. Luo T, Ai ZH, Zhang LZ (2008) Fe@Fe_2O_3 core-shell nanowires as iron reagent. 4. Sono-Fenton degradation of pentachlorophenol and the mechanism analysis. J Phys Chem C 112:8675–8681
67. Ioan I, Wilson S, Lundanes E, Neculai A (2007) Comparison of Fenton and sono-Fenton bisphenol A degradation. J Hazard Mater 142:559–563
68. Gogate PR, Pandit AB (2004) A review of imperative technologies for wastewater treatment II: hybrid methods. Adv Environ Res 8:553–597
69. Ragaini V, Selli E, Bianchi CL, Pirola C (2001) Sono-photocatalytic degradation of 2-chlorophenol in water: kinetic and energetic comparison with other techniques. Ultrason Sonochem 8:251–258
70. Stock N, Peller J, Vinodgopal K, Kamat PV (2000) Combinative sonolysis and photocatalysis for textile dye degradation. Environ Sci Technol 34:1747–1750
71. Kang JW, Hoffmann MR (1998) Kinetics and mechanism of the sonolytic destruction of methyl *tert*-butyl ether by ultrasonic irradiation in the presence of ozone. Environ Sci Technol 32:3194–3199
72. Hua I, Hoffmann MR (1997) Optimization of ultrasonic irradiation as an advanced oxidation technology. Environ Sci Technol 31:2237–2243
73. Lesko T, Colussi AG, Hoffmann MR (2006) Sonochemical decomposition of phenol: evidence for a synergistic effect of ozone and ultrasound for the elimination of total organic carbon from water. Environ Sci Technol 40:6818–6823
74. Neppolian B, Bruno A, Bianchi CL, Ashokkumar M (2012) Graphene oxide based Pt-TiO_2 photocatalyst: Ultrasound assisted synthesis, characterization and catalytic efficiency. Ultrason Sonochem 19:9–15.
75. Bang JH, Suslick KS (2010) Applications of ultrasound to the synthesis of nanostructured materials. Adv Mater 22:1039–1059.
76. Didenko YT, Suslick KS (2005) Chemical aerosol flow synthesis of semiconductor nanoparticles. J Am Chem Soc 127:12196–12197.
77. Zhou S, Yuan R, Lou S, Wang Y, Yuan H, Zhu G, Liu L, Hao Y, Li N (2011) Sonochemical synthesis and optical properties of amorphous ZnO nanowires. J. Nanoparticle Res 13:4511–4518.
78. Chen Y, Zhu CL, Xiao G (2006) Reduced-temperature ethanol sensing characteristics of flower-like ZnO nanorods synthesized by a sonochemical method. Nanotech 17:4537–4541.
79. Suh WH, Jang AR, Suh YH, Suslick KS (2006) Porous, hollow, and ball-in-ball metal oxide microspheres: reparation, endocytosis, and cytotoxicity. Adv Mater 18:1832–1837.

80. Neppolian B, Wang Q, Jung H, Choi H (2008) Ultrasonic-assisted sol-gel method of preparation of TiO_2 nano-particles: characterization, properties and 4-chlorophenol removal application. Ultrason Sonochem 15:649–658
81. Morigushi N (1934) The effect of supersonic waves on chemical phenomena (III). The effect on the concentration polarization. J Chem Soc Jpn 55:749–750
82. Mason TJ, Lorimer JP, Walton DJ (1990) Sonoelectrochemistry. Ultrasonics 28:333–337
83. Pollet BG, Phull SS (2001) Sonoelectrohemistry-theory, principles and applications. Recent Res Dev Electrochem 4:55–78
84. Brett C (2008) Sonoelectrochemistry. In: Arnau Vives A (ed) Piezoelectric transducer and applications. Springer, Berlin/Heidelberg
85. Walton DJ, Phull SS (1996) Sonoelectrochemistry. Adv Sonochem 4:205–284
86. Compton RG, Eklund JC, Marken F (1997) Sonoelectrochemical processes. A review. Electroanalysis 9:509–522
87. González-García J, Esclapez MD, Bonete P, Vargas-Hernández Y, Gaete-Garretón L, Sáez V (2010) Current topics on sonoelectrochemistry. Ultrasonics 50:318–322
88. Klima J (2011) Application of ultrasound in electrochemistry. An overview of mechanisms and design of experimental arrangement. Ultrasonics 51:202–209
89. González-García J (2011) Sonoelectrochemical synthesis of materials. In: Pankaj, Ashokkumar M (eds) Theoretical and experimental sonochemistry involving inorganic systems. Springer, Dordrecht/Heidelberg/London/New York
90. González-García J, Iniesta J, Aldaz A, Montiel V (1998) Effects of ultrasound on the electrodeposition of lead dioxide on glassy carbon electrodes. New J Chem 22:343–347
91. Hielscher Ultrasonics GmbH. http://www.hielscher.com
92. Sonics & Materials, INC. http://www.sonics.com/
93. Klíma J, Bernard C (1999) Sonoassisted electrooxidative polymerisation of salicylic acid: role of acoustic streaming and microjetting. J Electroanal Chem 462:181–186
94. Klíma J, Bernard C, Degrand C (1994) Sonoelectrochemistry: effects of ultrasound on voltammetric measurements at a solid electrode. J Electroanal Chem 367:297–300
95. Mohapatra SK, Raja KS, Misra M, Mahajan VK, Ahmadian M (2007) Synthesis of self-organized mixed oxide nanotubes by sonoelectrochemical anodization of Ti-8Mn alloy. Electrochim Acta 53:590–597
96. Esclapez MD, Sáez V, Milán-Yáñez D, Tudela I, Louisnard O, González-García J (2010) Sonoelectrochemical treatment of water polluted with trichloroacetic acid: from sonovoltammetry to pre-pilot plant scale. Ultrason Sonochem 17:1010–1020
97. Sáez V, Esclapez MD, Frías-Ferrer AJ, Bonete P, Tudela I, Díez-García MI, González-García J (2011) Lead dioxide film sonoelectrodeposition in acidic media: preparation and performance of stable practical anodes. Ultrason Sonochem 18:873–880
98. Durant A, François H, Reisse J, Kirsch-DeMesmaeker A (1996) Sonoelectrochemistry: the effects of ultrasound on organic electrochemical reduction. Electrochim Acta 41:277–284
99. Compton RG, Hardcastle JL, del Campo J (2003) Sonoelectrochemistry: physical aspects. In: Bard AJ, Stratmann M (eds) Encyclopedia of electrochemistry, vol 3. Wiley-VCH, Weinheim
100. Birkin PR, Offin DG, Joseph PF, Leighton TG (2005) Cavitation, shock waves and the invasive nature of sonoelectrochemistry. J Phys Chem B 109:16997–17005
101. Marken F, Compton RG (1998) Sonoelectrochemically modified electrodes: ultrasound assisted electrode cleaning, conditioning, and product trapping in 1-octanol/water emulsion systems. Electrochim Acta 43:2157–2165
102. Rejňák M, Klíma J, Svodoba J, Ludvik J (2004) Synthesis and electrochemical reduction of methyl 3-halo-1-benzothiophene-2-carboxylates. Collect Czechoslov Chem Commun 69:242–260
103. Zhang H, Coury LA Jr (1993) Effects of high-intensity ultrasound on glassy carbon electrodes. Anal Chem 65:1552–1558
104. Compton RG, Eklund JC, Page SD, Sanders GHW, Booth J (1994) Voltammetry in the presence of ultrasound. Sonovoltammetry and surface effects. J Phys Chem 98:12410–12414

105. Cooper EL, Coury LA Jr (1998) Mass transport in sonovoltammetry with evidence of hydrodynamic modulation from ultrasound. J Electrochem Soc 145:1994–1999
106. Advanced Sonic Processing Systems Company. http://www.advancedsonics.com/Reaction%20Cells.htm
107. Perret A, Haenni W, Skinner N, Tang NM, Gandini D, Comninellis C, Correa B, Foti G (1999) Electrochemical behavior of synthetic diamond thin film electrodes. Diam Relat Mater 8: 820–823
108. Iordache I, Nechita MT, Rosca I, Aelenei N (2004) Ultrasound assisted electrochemical degradation of cyanides: influence of electrode type. Turk J Eng Environ Sci 28:377–380
109. Kraft A, Blaschke M, Kreysig D (2002) Electrochemical water disinfection part III: hypochlorite production from potable water with ultrasound assisted cathode cleaning. J Appl Electrochem 32:597–601
110. Walker R (1997) Ultrasound improves electrolytic recovery of metals. Ultrason Sonochem 4:39–43
111. Farooq R, Wang Y, Lin F, Shaukat SF, Donaldson J, Chouhdary AJ (2002) Effect of ultrasound on the removal of copper from the model solutions for copper electrolysis process. Water Res 36:3165–3169
112. Yaqub A, Ajab H, Khan S, Farooq R (2009) Electrochemical removal of copper and lead from industrial wastewater: mass transport enhancement. Water Qual Res J Can 44:183–188
113. Hyde ME, Compton RG (2002) How ultrasound influences the electrodeposition of metals. J Electroanal Chem 531:19–24
114. Pollet BG, Lorimer JP, Phull SS, Mason TJ, Walton DJ, Hihn JY, Ligier V, Wéry M (1999) The effect of ultrasonic frequency and intensity upon electrode kinetic parameters for the $Ag(S_2O_3)_2^{3-}$/Ag redox couple. J Appl Electrochem 29:1359–1366
115. Lorimer JP, Pollet BG, Phull SS, Mason TJ, Walton DJ (1998) The effect upon limiting currents and potentials of coupling a rotating disc and cylindrical electrode with ultrasound. Electrochim Acta 43:449–455
116. Pollet BG, Lorimer JP, Phull SS, Hihn JY (2000) Sonoelectrochemical recovery of silver from photographic processing solutions. Ultrason Sonochem 7:69–76
117. Pollet BG, Lorimer JP, Phull SS, Mason TJ, Hihn J-Y (2003) A novel angular geometry for the sonochemical silver recovery process at cylinder electrodes. Ultrason Sonochem 10:217–222
118. Pollet BG, Lorimer JP, Hihn JY, Touyeras F, Mason TJ, Walton DJ (2005) Electrochemical study of silver thiosulphate reduction in the absence and presence of ultrasound. Ultrason Sonochem 12:7–11
119. Compton RG, Eklund JC, Marken F, Waller DN (1996) Electrode processes at the surfaces of sonotrodes. Electrochim Acta 41:315–320
120. Pollet BG, Lorimer JP, Hihn J-Y, Phull SS, Mason TJ, Walton DJ (2002) The effect of ultrasound upon the oxidation of thiosulphate on stainless steel and platinum electrodes. Ultrason Sonochem 9:267–274
121. Allen HE, Chen PH (1993) Remediation of metal contaminated soil by EDTA incorporating electrochemical recovery of metal and EDTA. Environ Prog 12:284–293
122. Chang JH, Ellis AV, Yan CT, Tung CH (2009) The electrochemical phenomena and kinetics of EDTA-copper wastewater reclamation by electrodeposition and ultrasound. Sep Purif Technol 68:216–221
123. González-García J, Sáez V, Tudela I, Díez-García MI, Esclapez MD, Louisnard O (2010) Sonochemical treatment of water polluted by chlorinated organocompounds. A review. Water 2:28–74
124. Gogate PR, Sutkar VS, Pandit AB (2011) Sonochemical reactors: important design and scale up considerations with a special emphasis on heterogeneous systems. Chem Eng J 166: 1066–1082
125. Krüger O, Schulze Th-L, Peters D (1999) Sonochemical treatment of natural ground water at different high frequencies: preliminary results. Ultrason Sonochem 6:123–128
126. Peters D (2001) Sonolytic degradation of volatile pollutants in natural ground water: conclusions from a model study. Ultrason Sonochem 8:221–226

127. Shemer H, Narkis N (2005) Sonochemical removal of trihalomethanes from aqueous solutions. Ultrason Sonochem 12:495–499
128. Selli E, Bianchi CL, Pirola C, Bertelli M (2005) Degradation of methyl *tert*-butyl ether in water: effects of the combined use of sonolysis and photocatalysis. Ultrason Sonochem 12:395–400
129. Rokhina EV, Repo E, Virkutyte J (2010) Comparative kinetic analysis of silent and ultrasound-assisted catalytic wet peroxide oxidation of phenol. Ultrason Sonochem 17: 541–546
130. Mahamuni NN, Adewuyi YG (2010) Advanced oxidation processes (AOPs) involving ultrasound for waste water treatment: a review with emphasis on cost estimation. Ultrason Sonochem 17:990–1003
131. Sáez V, Esclapez MD, Tudela I, Bonete P, Louisnard O, González-García J (2010) Electrochemical degradation of perchloroethylene in aqueous media: influence of the electrochemical operational variables in the viability of the process. Ind Eng Chem Res 49:4123–4131
132. Sáez V, Esclapez MD, Bonete P, Walton DJ, Rehorek A, Louisnard O, González-García J (2011) Sonochemical degradation of perchloroethylene: the influence of ultrasonic variables, and the identification of products. Ultrason Sonochem 18:104–113
133. Sáez V, Esclapez MD, Tudela I, Bonete P, Louisnard O, González-García J (2010) 20 kHz sonoelectrochemical degradation of perchloroethylene in sodium sulphate aqueous media: influence of the operational variables in batch mode. J Hazard Mater 183:648–654
134. Sáez V, Tudela I, Esclapez MD, Bonete P, Louisnard O, González-García J (2011) Sonoelectrochemical degradation of perchloroethylene in water: enhancement of the process by the absence of background electrolyte. Chem Eng J 168:649–655
135. Gallego-Juárez JA, Rodriguez G, Acosta V, Riera E (2010) Power ultrasonic transducers with extensive radiators for industrial processing. Ultrason Sonochem 17:953–964
136. Tuziui T, Yasui K, Lee J, Kozuka T, Towata A, Iida Y (2008) Mechanism of enhancement of sonochemical-reaction efficiency by pulsed ultrasound. J Phys Chem A 112:4875–4878
137. Abdelsalam ME, Birkin PR (2002) A study investigating the sonoelectrochemical degradation of an organic compound employing Fenton's reagent. Phys Chem Chem Phys 4:5340–5345
138. Yasui K, Kozuka T, Tuziuti T, Towata A, Iida Y, King J, Macey P (2007) FEM calculation of an acoustic field in a sonochemical reactor. Ultrason Sonochem 14:605–614
139. Louisnard O, González-García J, Tudela I, Klíma J, Sáez V, Vargas-Hernández Y (2009) FEM simulation of a sono-reactor accounting for vibrations of the boundaries. Ultrason Sonochem 16:250–259
140. Tudela I, Sáez V, Esclapez MD, Bonete P, Harzali H, Baillon F, González-García J, Louisnard O (2011) Study of the influence of transducer-electrode and electrode-wall gaps on the acoustic field inside a sonoelectrochemical reactor by FEM simulations. Chem Eng J. doi:10.1016/j.cej.2011.03.064
141. Vanhille C, Campos-Pozuelo C (2011) Nonlinear ultrasonic standing waves: two-dimensional simulations in bubbly liquids. Ultrason Sonochem 18:679–682
142. Louisnard O (2011) A simple model of propagation in a cavitating liquid. Part I: theory, nonlinear attenuation and traveling wave generation. Ultrason Sonochem 19:56–65
143. Louisnard O (2011) A simple model of propagation in a cavitating liquid. Part II: Primary Bjerkness force and bubble structures. Ultrason Sonochem 19:66–76
144. Ashokkumar M, Lee J, Iida Y, Yasui K, Kozuka T, Tuziuti T, Towata A (2009) The detection and control of stable and transient acoustic cavitation bubbles. Phys Chem Chem Phys 11:10118–10121

Chapter 11
Hybrid Sonochemical Treatments of Wastewater: Sonophotochemical and Sonoelectrochemical Approaches. Part II: Sonophotocatalytic and Sonoelectrochemical Degradation of Organic Pollutants

B. Neppolian, M. Ashokkumar, V. Sáez, M.D. Esclapez, and P. Bonete

11.1 Introduction

After discovering the photocatalytic splitting of water using TiO_2 photoelectrode by Fujishima and Honda in 1972, there have been widespread applications of photocatalysis for the degradation of both organic as well as inorganic pollutants in environmental remediation. Numerous research reports have demonstrated the fundamental processes in photocatalysis [1–12]. The major advantages of this technology are as follows: (1) complete degradation of organics into carbon dioxide and water in a relatively short period of time, (2) OH radicals obtained from this method are nonselective oxidants, which can be used to degrade a variety of hazardous compounds in different wastewater streams, (3) the technology can be applied to aqueous and gaseous-phase treatments, (4) the operating conditions for photocatalysis are simple, (5) the efficiency can be improved by doping with transition metals, and (6) visible light photocatalysts can also be used to harvest sunlight.

The common disadvantages of photocatalysis are: (1) the strong adsorption of pollutants on the photocatalyst particles that deactivates the catalyst and (2) mass transfer limitations: catalyst particles settle at the bottom and direct contact between catalyst and pollutants is largely prevented that need continuous stirring at all times.

B. Neppolian
SRM Research Institute, SRM University, Kattankulathur, Chennai 603203, India
e-mail: b_neppolian@yahoo.com

M. Ashokkumar
School of Chemistry, University of Melbourne, Melbourne, Australia
e-mail: masho@unimelb.edu.au

V. Sáez • M.D. Esclapez • P. Bonete (✉)
Grupo de Nuevos Desarrollos Tecnológicos en Electroquímica: Sonoelectroquímica y Bioelectroquímica, Grupo de Fotoquímicay Electroquímica de semiconductores, Departamento de Química Física e Instituto de Electroquímica, Universidad de Alicante, Alicante, Spain
e-mail: veronica.saez@ua.es; desiree_ev@hotmail.com; Pedro.Bonete@ua.es

S.K. Sharma and R. Sanghi (eds.), *Advances in Water Treatment and Pollution Prevention*, DOI 10.1007/978-94-007-4204-8_11,

Recently, sonochemical oxidation has gained much attention as an advanced oxidation process for the degradation of organic contaminants in aqueous environment [13–20]. The chemical reactions that occur from the ultrasonic irradiation of a solution are produced through the phenomenon of cavitation. The process of cavitation refers to the rapid growth and implosive collapse of bubbles in a liquid resulting in an extreme reaction environment within and in the vicinity of bubbles. The temperature produced during the collapse of a cavitation bubble is as high as 4,500 K. During ultrasonication of an aqueous solution, thermolytic cleavage of water molecules occurs, leading to the formation of H and OH radicals. OH radicals are powerful nonselective oxidants that can be used for the complete degradation of organic pollutants in aqueous solutions. Generally, the sonochemical oxidation of organic compounds in aqueous solution occurs by two-reaction pathways: (1) volatile compounds evaporate into the cavity during the expansion cycle and degrade via pyrolytic reaction within the collapsing bubble and (2) it proceeds by the reaction of OH radicals with the solute adsorbed at the bubble interface. The nature of the reaction pathway depends on the volatility, hydrophobicity and surface activity of the compound in question [16]. Advantages of this method are: (1) there is no requirement to use any added chemicals or catalysts for the oxidation of organic pollutants, (2) a simple ultrasonic transducer is sufficient for the complete oxidation of organics in water, and (3) effective mass transfer. Disadvantages of this method are: (1) complete mineralization of pollutants takes relatively longer time and (2) sonochemically formed H_2O_2 is stable in the presence of low levels of pollutants.

Combining photocatalysis and sonochemistry offers a potentially useful way of overcoming the existing problems of the individual methods, i.e., (1) the aggregation of photocatalyst nanoparticles in aqueous solutions is prevented by the physical effects of acoustic cavitation, leading to an increase in the active surface area, (2) the catalyst surface is continuously cleaned due to acoustic microstreaming, which has the ability to further increase the catalytic performance of the photocatalyst, (3) increase in the mass transport of the pollutants to the catalyst surface, (4) sonochemically formed H_2O_2 is stable in the presence of low levels of pollutants due to the facilitated transport by shockwave propagation, and can be cleaved into OH radicals during photolysis [21], (5) enhance the rate of mineralization of organic pollutants, and (6) more free radicals are available for degradation (from photocatalysis as well as sonolysis).

Electrochemical degradation of pollutants in wastewater treatment is not being routinely used because cheaper conventional treatments are available. They are usually composed of a physicochemical step followed by a biological degradation. However, these methods have continuously presented deficiencies, mainly treating some refractory compounds because they are toxic for the microorganisms in the biological plant [22].

From an environmental point of view, one of the most important requirements is the use of effective procedures that avoid, if possible, the use of stoichiometric quantities of nonreusable chemicals. In this sense, in the last decades, ultrasonic irradiation has been used as degradation treatment or in combination with other techniques. Although sonolysis does not require the use of chemicals, it has been observed that the addition of some salts improves the degradation of some pollutants. On the other hand, the electrochemical treatment is a versatile technology with a well-known theoretical basis that has made possible its application not only in synthesis processes but also for the degradation of contaminants in water remediation. The electrochemical treatment requires, a priori, a supporting electrolyte when the solution to electrolyze does not have enough electrical conductivity. This is the only chemical required which increases the salinity of solutions when its use is needed. Polluted freshwaters usually show low electrical conductivity but industrial effluents have a high salinity degree.

Textile dyes are one of the major classifications under the organic dyes, which are commonly used in textile industries and considered to be potentially carcinogenic. During dye production as well as textile manufacturing processes, a large quantity of wastewater containing dyestuffs with intensive color and toxicity is introduced into the aquatic systems. Due to the large degree of organics present in these molecules and stability of modern textile dyes, conventional biological treatment methods are ineffective for their decolourization and degradation [23–32]. This led to the study of other effective methods. Recent studies have demonstrated that photocatalysis and sonochemical methods can be used to degrade dye compounds using semiconductor photocatalysts under ultrasonic irradiation [2, 3, 33–41].

Among various organic contaminants, pharmaceutical products are of significant concern because of their large variety and high consumption over the last few decades. Main sources of pharmaceutical products are emission from production sites due to inadequate treatment of manufacturing effluents, direct disposal of unused medicine and the release of drug-containing waste by human beings and animals. The presence of low levels of pharmaceutical products in water can have potential health effects on humans, through drinking water or consumption of food irrigated by polluted water. Therefore, it is necessary to treat the effluents containing pharmaceutical compounds adequately before discharging them into water streams [42].

A large number of pesticides and herbicides are currently in use worldwide for agricultural and nonagricultural activities. The use of both pesticides and herbicides in the targeted areas is found to contaminate the surrounding aqueous environment. The treatment of wastewater containing pesticides and herbicides is important in order to protect the ecosystem and, hence, has attracted much attention of many researchers in recent years [43–45]. The use of advanced oxidation technologies (AOTs), such as heterogeneous photocatalysis, Fenton and photo-Fenton oxidation, ozonation, sonolysis, and UV/H_2O_2, for the degradation of aqueous organic pollutants have been widely described in the literature [46–50].

Similarly, surfactants have been widely used in various applications to remove contaminants from surfaces of human skin, textiles, and also used in industries such as paper, food, polymers, pharmaceuticals, and oil recovery for various purposes [51]. Every year large quantities of surfactants are produced worldwide for the above purposes to face the demand. For example, the average annual production of detergent is $\sim$8 million tons from the western countries alone [52]. Many types of synthetic surfactants that include alkyl-benzenesulfonate such as dodecylbenzenesulfonate (DBS) have been let into water systems after the use, which constitutes a serious environmental pollution and attracted much environmental concerns [13]. DBS is nonbiodegradable in nature and accumulated in water systems over a period of time [13]. The conventional treatment methods for the total removal and degradation of DBS are still in primitive stage.

The persistent presence of halogenated hydrocarbons in all environmental media is still a serious problem, but especially in aqueous medium. Disposal and spills of chlorinated solvents at industrial sites, military bases, government facilities, or dry cleaner sites have led to extensive groundwater contamination [53, 54]. As a consequence of the poor efficiency of the biological methods for the degradation of chlorinated compounds, a large number of alternative technologies have emerged and been developed in the last decade. These include not only simple procedures such as catalytic hydrogenation [55], metal alloys in dilute aqueous alkaline solutions [56], photochemistry [57], or the techniques mentioned above such as sonochemistry [58] and also electrochemistry [59], but also combined treatments such as reductive dehalogenation and biodegradation [60], which are being considered as possible methods for the degradation of chlorinated organocompounds. On the other hand, several authors have pointed out, not only that the treatment of water with chlorine in the distribution system produces trihalomethanes and haloacetic acids, which prevents its use as drinking water [61], but also the resistance of these compounds to transformation via mainstream biochemical processes [62] used in the water treatment plants, and the formation and fate of haloacetic acid within these plants [63].

The aim of this chapter is to provide an overview of the studies available on the effective degradation of few selected organic pollutants such as (1) organic dyes, (2) pesticides and pharmaceuticals, and (3) surfactants using combined processes. This may provide a broad scope for the researchers to know about the advantageous of these methodologies for environmental remediation. In addition, the application of sonoelectrochemistry to the protection of the environment is an emerging field of research that has been applied successfully for the degradation of a range of important contaminants in water. Sonoelectrochemistry for the degradation of contaminants has several advantages over the conventional methods in that it can be carried out under mild conditions, at room temperature and without the need for additional chemical treatment. The sonoelectrochemistry section summarizes the treatment of major pollutants including textile dyes, phenolic derivatives, nitrocompounds, chlorinated pollutants, and metals that have been studied to date.

11.2 Sonophotocatalytic Degradation of Organic Pollutants

11.2.1 Degradation of Organic Dyes

Information available for the sonophotocatalytic degradation of some organic dyes in recent years is listed in Table 11.1, which provides a clear outline of sonophotocatalytic degradation of various organic compounds, the type of ultrasonicator, catalysts, and name of the organic compound used along with experimental conditions and important observations.

Table 11.1 Overview of the work done in sonophotocatalytic degradation of organic dyes, surfactants and pesticides, and pharmaceuticals

S. No.	References	Experimental details	Key findings
1	Stock et al. [64]	Naphthol blue black (NBB), P-25-TiO_2 (Degussa) 640 kHz ultrasonicator, 240 W	Enhancement observed in the rates of decoloration and mineralization of the azo dye, NBB – the simultaneous process is better than sequential
2	Madhavan et al. [72]	Orange G (OG), P-25-TiO_2 (Degussa) 213 kHz ultrasonicator, continuous mode, 20 W	An additive effect was observed during the sonophotocatalytic oxidation of OG
3	Vinu and Madras [74]	Six different anionic dyes, TiO_2, horn type sonicator, 25 kHz, 36 W, 80 W high pressure mercury UV lamp	Sonophotocatalytic degradation of all dyes was higher compared to the individual processes. P-25 TiO_2 led to lesser mineralization and their prepared TiO_2 showed better conversion
4	Son et al. [75]	Reactive black 5 (RB5), 35 kHz ultrasonicator, 254 nm wavelength UVC lamp	Observed enhanced degradation under acidic conditions and there was no effect of NaCl with high concentration due to physical effect of ultrasound
5	Madhavan et al. [76]	Acid red 88 (AR88), Fe^{3+} is used for homogeneous photocatalysis, P-25-TiO_2 (Degussa), 213 kHz ultrasonicator, continuous mode, 55 mW L^{-1}	Heterogeneous sonophotocatalysis showed 1.3 synergy index whereas, Fe^{3+} assisted homogeneous sonophotocatalysis showed 2.3 synergy index for the dye degradation
6	Ashokkumar et al. [65]	Orange G (OG), Fe^{3+}, 213 kHz ultrasonicator, continuous mode, 20 W	Photolysis was found to be responsible for the increase in the synergy index. Sonophotocatalysis was shown to be more efficient than the individual processes
7	Yuan et al. [66]	TiO_2 nano array, methylene blue (MB), different frequencies from 18 to 30 kHz	Frequency has great effect on the synergy, at 27 kHz, the synergistic effect was as high as 22.1%

(continued)

Table 11.1 (continued)

S. No.	References	Experimental details	Key findings
8	Zhang et al. [77]	Visible light irradiation, Fe-TiO_2-MWCNTs, methylene blue (MB), 28 kHz	The sonophotocatalysis was faster than the respective individual processes due to the increase of the active surface area of Fe/TiO_2-MWCNT under visible light
9	Bejarano-Perez and Suarez-Herrera [71]	Congo red and methylene orange, TiO_2, 4 W lamp, 47 kHz, 81 W	A positive effect of ultrasound not only occurred on oxidation reactions, but also on reduction reactions, a remarkable increase on the rate of combined process
10	Kaur and Singh [67]	Reactive red 198, visible light, dye-sensitized TiO_2, 47 kHz, 130 W	Synergy effect was observed, reaction took place via formation of singlet oxygen, superoxide and hydroxyl radicals
11	Berberidou et al. [68]	Malachite green (MG), TiO_2, 80 kHz, up to 150 W	The parent molecules are easily oxidizable, its degradation products are more stable to total oxidation and synergy index was observed with combined process
12	Wang et al. [69]	Methyl orange, CNTs/TiO_2, 20 kHz, 0–50 W	Results indicated that the photocatalytic efficiency of CNTs/TiO_2 remarkable increased in the presence of ultrasound, compared with P25, CNTs/TiO_2 nanoparticle showed much higher sonophotocatalytic ability
13	González and Martínez [70]	Basic blue 9, TiO_2, 20 kHz, 76 W,15 W UV lamp, 352 nm	Rate of the reaction increased many folds with combined process. Reaction was dependent on the pH solution with the highest at pH 7
14	Wang et al. [73]	Methylene orange (MO), P-25 TiO_2, Ag-TiO_2, 40 kHz, 180 W	Degradation ratio of MO increased with increase of ultrasonic power. MO can be remarkably degraded by the sonophotocatalytic reaction in the presence of photocatalysts under simulated sunlight
15	Madhavan et al. [78]	Ibuprofen (IBP), P-25-TiO_2, Fe^{3+}, 213 kHz frequency, 55 mW mL^{-1}	A significant synergy effect was observed for the sonophotocatalysis in the presence of Fe^{3+}, sonication of IBP led to the formation of its mono and di-hydroxylated intermediates

(continued)

Table 11.1 (continued)

S. No.	References	Experimental details	Key findings
16	Madhavan et al. [79]	Diclofenac (DF), an anti-inflammatory drug, TiO_2, ZnO, and Fe–ZnO, 213 kHz, 55 mW mL^{-1}	Combined process showed a slight synergy effect with TiO_2, an additive effect with Fe-ZnO. However, the mineralization process during the combined process showed a detrimental effect
17	Bahena et al. [80]	Alazine and gesaprim (herbicides), P-25 TiO_2, 20 kHz, 500 W, 15 W UV light emits at 352 nm	90% of the active component in the gesaprim and alazine were completely degraded, photodegradation of these commercial herbicides enhanced by ultrasound, a complete mineralization in both commercial herbicides
18	Ashokkumar et al. [81]	Formetanate hydrochloride (pesticide), TiO_2, Fe^{3+}, 213 kHz, 55 mW mL^{-1}	TiO_2-assisted sonophotocatalysis resulted in a negative effect (synergy index 0.7), whereas with Fe^{3+} showed a synergistic enhancement (synergy index 1.7)
19	Ashokkumar et al. [82]	Monocrotophos (MCP), pesticide, TiO_2, Fe^{3+}, 213 kHz, 55 mW mL^{-1}	The mineralization process was additive for both TiO_2 and Fe^{3+}
20	Anandan and Ashokkumar [84]	Nonylphenol ethoxylate (NPE), surfactant, Au-TiO_2, 358 kHz, 17 W	Photocatalysis was efficient for the degradation of an organic surfactant, synergetic effects could not be observed in the sonophotocatalytic process
21	Neppolian et al. (submitted) [83]	Dodecylbenzene-sulphonate (DBS), surfactant, Pt-graphene oxide (GO)-TiO_2, 215 kHz, 17 W	There was no pH effect on the sonophotocatalytic degradation of DBS, additive effect was observed with combined process. Fast mineralization was observed

One of the major concerns in any advanced oxidation technique is the formation of reactive intermediates during the oxidation of organics. A fast mineralization of any organic compound should be the goal to minimize the reaction time as it takes prolonged time for complete mineralization. A combined sonophotocatalytic approach has led to faster mineralization. TiO_2 and TiO_2-based photocatalysts have been widely used for this combined study for the degradation of organic dyes including textile dyes. Generally, the following observations were reported for the sonophotocatalytic degradation of organic dyes:

1. The rate of sonophotocatalytic degradation of dyes mostly follows pseudo-first-order kinetics.
2. The rate of degradation of organics dyes mainly depends on the initial concentration of the dye. Similarly, the amount of photocatalyst also controls the rate of degradation. The optimization of both amount of catalyst as well as concentration of the dyes is inevitable.
3. The solution pH may be an important factor, depending upon the nature of pollutants.
4. The nature and the structure of the dyes are very important factors for the mineralization of the dyes.
5. The nature of the catalysts plays a considerable role: size, shape, surface areas, and other morphologies of the catalysts need to be considered.
6. Experimental parameters such as nature (UV or visible) and intensity of the light source and the reactor assembly are important factors.
7. Intensity of ultrasound is also taken into consideration as one of the important parameters during the dye degradation.

A few studies have reported on the effects of various experimental parameters on the sonophotocatalytic degradation of textile dyes [64–71]. Kamat et al. in their study on the sonolytic degradation of a textile dye, naphthol blue back (NBB), showed that the rupture of azo group by OH radical attack produced several organic intermediates, which needed to be further degraded to achieve complete mineralization [64]. It was mentioned that the reaction intermediates compete with the parent compound for OH radicals and, hence, slow down the overall degradation process. The enhancement observed in the combined method with TiO_2 photocatalysts was an additive effect. About 68% mineralization in 12 h was obtained by the photocatalytic degradation of NBB solution. Only 35% mineralization was noted in the sonolysis experiment. Nearly 80% in 12 h was obtained when they performed simultaneous photocatalysis and sonolysis treatments. Two types of combinations were attempted: (1) simultaneous and (2) sequential. The simultaneous sonolysis and photocatalysis experiment was found to be better than the sequential combination. It has been suggested that the sonication prevents the catalyst from aggregation and keeps them as individual nanoparticles. This is almost a general observation for most of the textile dyes degradation by this combined [64–77].

It is generally seen that neutral solution pH is good for sonophotocatalytic degradation of textiles dyes. The TiO_2 surface charge plays an important role on the kinetics of photo-oxidation of ionic dyes. It has been reported that TiO_2 has a superficial pH of ~6.3, which coincides with its zero point charge (zpc). Therefore, the attractive interaction with cations is stronger at higher pH values (pH > pH zpc), but at lower pH the interactions forces are repulsive; the anions behave in the opposite way. For example, Congo red dye molecules have two sulfonic groups that are charged negative at alkaline pH, and it prevents the adsorption of the dye molecule to the negative surface of TiO_2. On the other hand, at acidic pH, the adsorption of Congo red seems to be high and does not allow the adsorption of O_2 on TiO_2 particles, making the photo-oxidation process very difficult. Experiments performed at different pH values showed that the optimum pH value for the photo-oxidation process of Congo red was near to the isoelectric point of TiO_2, as reported by Bejarano-Pérez et al. [71]. Madhavan et al. also reported similar results during the degradation of textile dye orange G. They have explained the results on the basis of pKa of the dye [72, 73].

The effect of dissolved gases was studied by Vinu and Madras [74]. The sonophotocatalytic degradation of textile dyes in the presence of different gases follows the order, $O_2 > N_2 > Ar >$ dissolved natural air. The observed behavior could be explained with the solubility of the gases in water. Ar is highly soluble in water compared to N_2 and O_2 and, hence, it exhibited the slowest degradation. Among N_2 and O_2, although N_2 is less soluble in water, it exhibited a lower activity, which can be attributed to the scavenging of hydroxyl radicals by N_2 to yield nitrogen and nitrous oxides. Moreover, O_2 chemically participates in the reaction pathway of photocatalysis and the formation of oxygen radicals in sonocatalysis, thereby leading to complete degradation. But, Ar and N_2 are inert gases and, hence, their relative contribution to the UV and ultrasound pathways is diminished. Among the different gases, presence of oxygen enhances the rate of reaction substantially. After carrying out a complete investigation on the sonophotocatalytic degradation of 6 different textile dyes, Vinu and Madras proposed that UV light played a significant role in the degradation of dyes in the initial period, whereas ultrasound had a more influential effect at longer exposure hours in TOC removal [74].

The effect of additives such as sodium chloride on the sonophotocatalytic degradation of dyes was studied by Son et al. [75]. Interestingly, there was no significant difference in the removal trends of the dye under various NaCl concentrations ranging from 50 to 5,000 mgL^{-1} (0.86–86 mM). When salt was present in the solution, salt could be an inhibitor in AOTs as NaCl acted as OH radical scavenger and generated by-product radicals, which are considered as weak oxidants. When the concentration of a pollutant was high, NaCl acted as an inhibitor, but there was no significant change in removal rate at various NaCl concentrations. On the contrary, the presence of NaCl largely increased removal rate at low pollutant concentrations. Owing to a relatively high initial concentration of the pollutant, it appeared that salt concentration did not affect ultrasonic degradation in this study. The physical forces of ultrasound are able to clean the surface of TiO_2 continuously and, hence, the problem associated with NaCl was largely prevented during this combined method.

The effect of ultrasound power during sonophotocatalysis was reported by many researchers [21, 73, 74, 76]. Recently, Madhavan et al. studied the effect of ultrasound power on the sonophotocatalytic degradation of textile dye acid red 88 [76]. It was reported that the degradation rate increased with an increase in the ultrasound power. For example, a higher degradation rate of 36.9×10^{-7} M min^{-1} was obtained at 55 mW mL^{-1}, whereas only about 0.77×10^{-7} M min^{-1} of dye degradation was observed at 16 mW mL^{-1}. Vinu and Madras also observed a similar effect on the degradation of textile dyes. This is due to the formation of more cavitation bubbles and their transient behavior in solution [74], which may lead to their increase in average size. Generally, the transient cavitation bubbles contain vapor, which implode violently, as there is not much gas to cushion their collapse. This contributes to higher local temperature and pressure and, hence, the generation of more hydroxyl radicals. It is worthwhile to note that the rate increased with an increase in ultrasound intensity, signifying that increasing the ultrasound power amplitude imparts a cleaning effect on the catalyst particles. This leads to desorption of the dye from TiO_2 active sites.

Generally, UV irradiation is harmful to human beings. Many researchers have been using solar light as a source of irradiation, which is freely available with abundant source of energy. Zhang et al. studied the effect of visible light on the degradation of textile dyes using different types of photocatalysts [77]. Methylene blue (MB) degradation was monitored by means of simultaneous visible light and ultrasound in the present of pure TiO_2, Fe/TiO_2 and Fe/TiO_2-multi-walled carbon nanotube (MWCNT) catalyst. It is reported that the sonophotocatalytic degradation of MB dye by Fe/TiO_2-MWCNT catalyst was better than that of pure TiO_2 and Fe/TiO_2 catalyst under visible light irradiation.

Homogeneous photocatalysis using Fenton (Fe^{2+}/H_2O_2) and modified Fenton reagents (Fe^{3+}/H_2O_2, ferrioxalate, Fe^{3+}/oxone) have also been reported as a useful method for the degradation of aqueous organic pollutants [76]. Madhavan et al. [76] have developed a method for the degradation of a mono-azo textile dye, acid-red 88 (AR88) using sonophotocatalysis. During this homogeneous sonophotocatalytic oxidation method, the synergy index was observed more than two times. In addition, Fe^{3+} can also be regenerated during the combined method as described in Part 1 in Chap.10. The simultaneous operation of sono-Fenton and photo-Fenton reactions is likely to be the underlying reason for the observed synergy in the presence of Fe^{3+}-assisted sonophotocatalysis. This approach is relatively simple and an efficient way for the degradation of organics.

The above-mentioned studies clearly suggest that sonophotocatalysis is an effective method for the degradation of textile dyes in terms of both fast degradation as well as fast mineralization, i.e., an additive effect can be achieved by this method. The general conclusion is that sonolysis is effective for inducing faster degradation of the parent dye, while TiO_2 photocatalysis is effective for promoting mineralization.

11.2.2 Degradation of Pesticides and Pharmaceuticals

Madhavan et al. described the sonophotocatalytic degradation of two commonly used pharmaceuticals (ibuprofen and diclofenac) with TiO_2, Fe-ZnO, and Fe^{3+}-mediated homogeneous sonophotocatalysis (Table 11.1) [78, 79]. The results discussed are almost similar to those results obtained for the degradation of textile dyes. The rate of the degradation of pharmaceuticals followed first-order degradation. Ibuprofen (commonly used as analgesic and anti-inflammatory drug) can be easily degraded by TiO_2 and Fe^{3+} homogenous sonophotocatalysis. Both catalysts showed a slight synergy in the degradation of ibuprofen whereas an additive effect was observed with mineralization of ibuprofen. On the other hand, a synergistic effect was observed with homogeneous photocatalysis in the presence of Fe^{3+} ion. This is due to the complex formation of photoactive complexes between Fe^{3+} and carboxylic acid intermediates. The degradation of diclofenac (anti-inflammatory drug) using TiO_2 and Fe–ZnO photocatalysts showed an additive effect. However, mineralization studies revealed that TiO_2 exhibited a detrimental effect but Fe–ZnO attributed an additive effect. The rate of degradation of pharmaceuticals also depends on the nature of TiO_2 and organic compounds. Many intermediate products were identified during the degradation processes. Most of the intermediates were more highly toxic than their parent compounds. Hence, complete mineralization of pharmaceuticals is a very important aspect in this method. Most of the pharmaceuticals are nonvolatile in nature; hence, there was less chance for them to undergo pyrolytic degradation during sonolysis. Most of the reactions are induced by OH radicals. All other parameters discussed for the dye degradation have shown almost similar trend for the degradation of pharmaceuticals.

Details of some herbicides and pesticides degradation studies with sonophotocatalysis performed in recent years are provided in Table 14.1. Bahena et al. and Ashokkumar et al. described the sonophotocatalytic degradation of few herbicides and pesticides using TiO_2 and Fe^{3+} photocatalysts [80–82]. Bahena et al. attempted to degrade alazine and gesaprim, commercial herbicides, in aqueous TiO_2 suspensions under UV light (15 W, 352 nm) and using an ultrasound source of 20 kHz [80]. Alazine and esaprim are herbicides used for the control of broadleaf weeds and some grassy weeds. Alazine contains alachlor and atrazine as active compounds, and esaprim comprises atrazine as active compound. HPLC was used to monitor the parent compounds as well as active compounds during the reaction. The amount of catalysts and concentration of herbicides were optimized before proceeding for further experiments. TOC results clearly indicated that both the active compounds present in alazine and esaprim were completely degraded by the combined method of degradation in 2 h. According to TOC results, the degradation of gesaprim preceded this order: sonophotocatalysis > photocatalysis > sonolysis > photolysis.

Madhavan et al. have studied the sonophotocatalytic degradation of commonly used insecticides such as formetanate hydrochloride and monocrotophos (MCP) [81, 82], which have been found as pollutants in aqueous environments. The reaction was carried out in the presence of homogeneous (Fe^{3+}) and heterogeneous (TiO_2) photocatalysts with high frequency ultrasound 213 kHz and high intensity UV lamp (400 W). The combination of sonolysis with TiO_2 photocatalysis showed a negative effect on the degradation of formetanate hydrochloride with a synergy index of 0.7. The degradation rates obtained for US + Fe^{3+}, UV + Fe^{3+}, and US + UV + Fe^{3+} were 14.1, 10.1, and 40.9×10^{-7} M min^{-1}, respectively. Fe^{3+} addition during the sonophotocatalysis process was found to be beneficial with a synergy index of 1.6. The total organic carbon analysis showed that the mineralization process was synergistic by combining the two processes in the presence of Fe^{3+}. On the other hand, the rate of photocatalytic degradation of monocrotophos using TiO_2 was found to be lower than that of sonolysis alone due to the interference of phosphate ions formed as an intermediate product. A 15-fold enhancement in the degradation rate was achieved when photolysis was carried out in the presence of Fe^{3+} compared to the rate observed with photolysis alone. The combination of sonolysis and photocatalysis (using either TiO_2 or Fe^{3+}) showed a detrimental effect. Synergy indices of 0.62 and 0.87 were noted for the sonophotocatalytic degradation of MCP in the presence of TiO_2 and Fe^{3+}. The mineralization process was additive for both TiO_2 and Fe^{3+} sonophotocatalysis according to TOC results. It is clearly seen that again the complete mineralization of organics strongly depends on the nature of pollutants. Dimethyl phosphate, dimethylphosphonate, 3-hydroxy 2-buteneamide, and N-methyl 3-oxobutanamide were identified as intermediates during the degradation of monocrotophos and all of them degraded effectively under this process.

11.2.3 Degradation of Surfactants

Ashokkumar and coworkers studied the sonophotocatalytic degradation of commonly used surfactants dodecylbenzenesulphonate (DBS) and nonylphenol ethoxylate Teric GN9 (GN9) in aqueous solution using TiO_2 based photocatalysts (Table 11.1) [submitted 83, 84]. In their study, it was found that the initial pH of the solution played an important role during the photocatalytic degradation of DBS; whereas, there was no effect of pH during both sonolytic as well as sonophotocatalytic degradation using Pt-graphene oxide (GO)-TiO_2 nanoparticles. It is an added advantage of coupling both methods especially for surfactants degradation at any pH condition. Moreover, in the presence of GO, an enhanced rate of DBS oxidation was observed. When doped with platinum, mineralization of DBS was further enhanced. Many intermediate products were identified during the DBS degradation within 10 min of reaction using electrospray mass spectrometry, mostly by the removal of CH_2-groups from the chain and also subsequent addition of very few CH_2- and OH-groups onto DBS. In 1-h reaction time, almost all the intermediate products were completely decomposed with a small amount of parent compound remaining, as a result more than 75% mineralization was observed (0.2 mM DBS concentration). Ag doped TiO_2 was used as a photocatalysts for the degradation of nonylphenol ethoxylate under the visible region of light. Ag-TiO_2 was able to degrade GN9 within 3 h of reaction. There was no synergetic effect on the rate of this combined method for the degradation of both surfactants. This is due to the formation of many intermediates during the degradation, which took long time for the complete mineralization of surfactants.

11.3 Sonoelectrochemical Degradation of Organic Pollutants

11.3.1 Degradation of Textile Dyes

Lorimer et al. studied the decolorization of several dye effluents [85, 86] such as Sandolan Yellow and Maxilion Blue 5G using sonolysis (SCT), electrolysis

(ECT), and background electrolyte-assisted sonoelectrolysis (BEA-SECT). Sandolan Yellow solutions were not decolorized under ultrasound (20 and 40 kHz). Decolorization could be achieved using electro-oxidation and the decolorization process was further enhanced under simultaneous sonication (40 kHz) [85]. This work showed that the decolorization process was dependent on the type of electrode used, the current density applied, and the nature and concentration of the electrolyte. Only alkali metal chloride solutions promoted the decolorization of dye solutions due to the production of hypochlorite during electrolysis. Furthermore, platinum electrodes led to better efficiencies than carbon electrodes when galvanostatic electrolysis was carried out. Similar results were observed using a basic dye such as Maxilion Blue 5G in high-frequency (510 kHz) sonoelectrochemical oxidation [86].

Sonovoltammetric oxidation studies of Procion Blue on a boron-doped diamond (BDD) electrode were carried out in buffered aqueous solutions [87] as a function of pH, dye concentration, and ultrasound treatment. The oxidation of the aqueous solution of Procion Blue was most easily achieved in acidic solutions and at low dye concentrations; and electrode surface fouling occurred in more alkaline solutions and at very high dye concentrations. Upon increasing the dye concentration, a proportional increase in the limiting current was observed. It was concluded that a resistive layer was formed on the electrode surface as a result of the adsorption and desorption of the large dye molecules at the BDD electrode prior and after oxidation.

Other dyes including Lissamine Green B [88], Trupocor Red [88], Reactive Black 5 [88], Acid Black [88], Methyl Orange [88, 89], Rhodamine B [89], Methylene Blue [89], Reactive Brilliant X-3B [89], and Reactive Blue 19 [90] have been used as reactive dyes for acoustic cavitation coupled with electrochemical treatment. Hydrodynamic cavitation has also been used to enhance the electrochemical treatment of Brilliant Red X-3B [91]. For all these studies, various electrode materials such as Ti-IrO_2 [91], graphite [88], platinum [89], PbO_2 [90], and three-dimensional electrodes [92] were employed and the effect of several parameters were studied including the ultrasonic power/intensity, the cell design, come along the pH and initial dye concentrations.

It can be concluded that the use of electro-oxidation combined with ultrasound increases the efficiency compared to the individual treatments, and in most of the studied cases synergic effects were observed.

11.3.2 Degradation of Aromatic and Phenolic Derivatives

Trabelsi et al. [93] reported the BEA-SECT of aqueous phenol solutions under low (20 kHz) and high-frequency (540 kHz) ultrasound irradiation. Electrochemical oxidation of phenol was carried out using 0.1 $g\,L^{-1}$ phenol solutions in 0.5 $g\,L^{-1}$ NaCl as supporting electrolyte in galvanostatic mode (6.8 mA cm^{-2}) with a cylindrical nickel foam as cathode and a cylindrical expanded platinized titanium as anode. When 20 kHz ultrasonic waves were applied, conversion of 75% of initial

Table 11.2 Sonoelectrochemical degradation of phenol at 20 kHz and 540 kHz

Time/minute	Phenol/mg L^{-1}	*p*-Quinone/mg L^{-1}	Acetic acid/mg L^{-1}	Phenol conversion/%
20 kHz				
0	100	0	0	0
5	54	7.9	2.4	46
10	24	12.6	3.1	76
20	12.8	11.1	4.6	87
45	7.8	7.6	4.8	92
60	3	5.5	6.2	97
540 kHz				
0	100	–	0	0
5	26.4	–	2.7	73.6
10	5	–	3.5	95
45	–	–	11.5	100
60	–	–	7.4	100

Reproduced from Ref. [93]. With kind permission of © Elsevier (1996)

phenol was achieved within 10 min of treatment. However, a toxic intermediate, *p*-quinone, was formed. At 540 kHz, a conversion of 95% of the initial phenol concentration was obtained at the same treatment time, and final products of degradation were acetic and chloroacrylic acids. Main by-products of the sonoelectrochemistry treatment of phenol are shown in Table 11.2.

Zhao et al. analyzed the electrochemical oxidation of phenol on BDD and Pt electrodes under silent and ultrasonic conditions (30 kHz, 50 W), from both fundamental and applied aspects point of view [94]. Although both anode materials were highly fouled, a much stronger electrode passivation took place on Pt. It was found that the ultrasonic irradiation improved the phenol degradation leading to high current efficiencies on both electrodes; although much higher when using BDD electrodes. Mass transport and adsorption phenomena, as well as electrode reactions, can explain the different behaviors of these electrode materials. It was observed that, for the BDD electrode, (1) reaction rates and current efficiencies were increased by 300% and 100%, respectively, under sonication and (2) lower yields of intermediates compared to silent conditions. Figure 11.1 shows the main by-products for both treatments. Although in the presence of ultrasound the variety of intermediates do not change for both electrodes used, it was found that the production and degradation rates of intermediates were especially enhanced by ultrasound on BDD electrodes. In addition, the energy consumption of the electro-oxidation of phenol decreased when ultrasound was used, due to the fact that the treatment time was significantly reduced.

The potentiostatic sonoelectrochemical oxidation of 2,4-dihydroxybenzoic acid (0.1–0.3 $g\,L^{-1}$) was also performed using both 20 and 500 kHz ultrasonic irradiations [95]. The geometry and position of the electrodes were optimized in order to maximize the acoustic effects. The platinized titanium grid anode (cylinder or disc at high-frequency or low-frequency ultrasonic experiments, respectively)

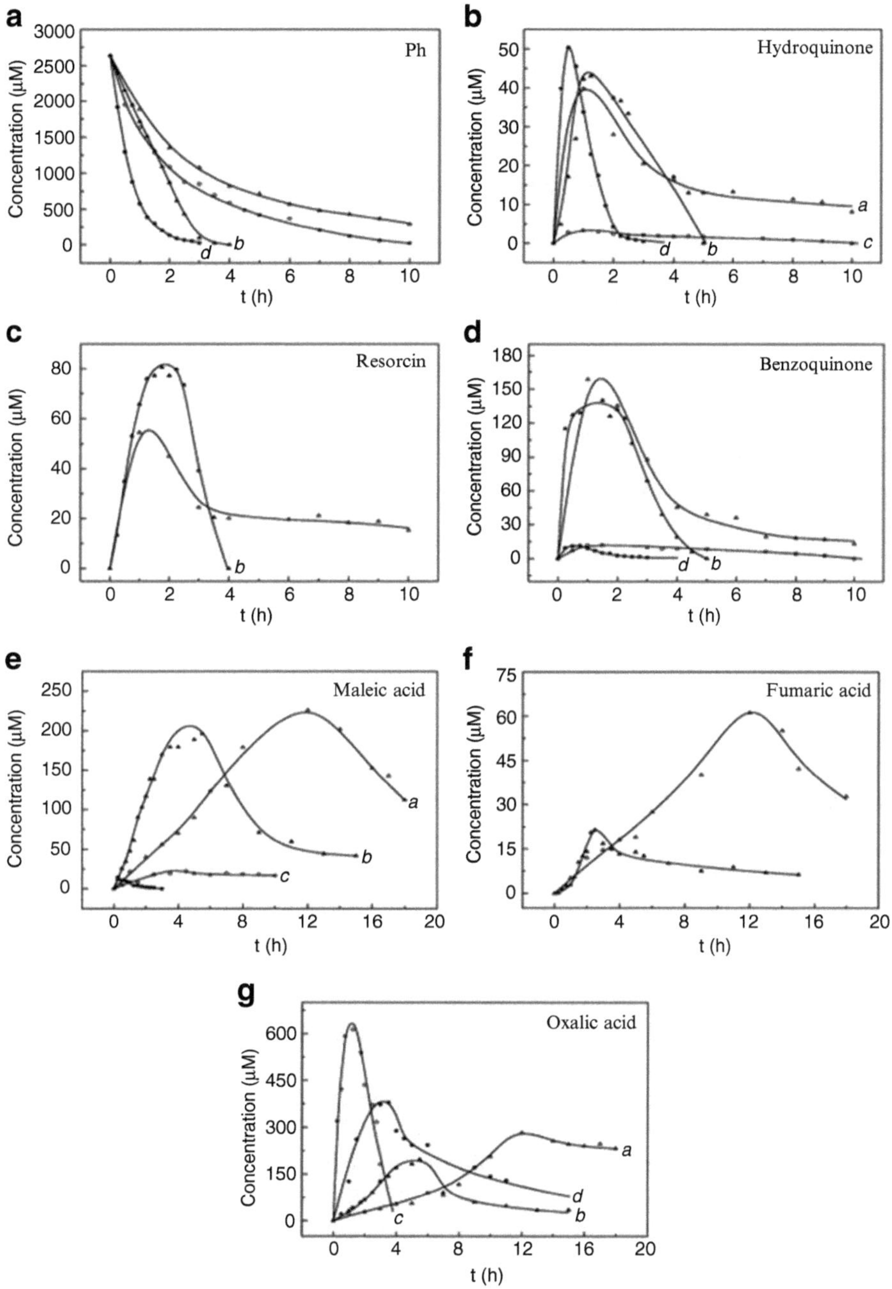

Fig. 11.1 Variation of the concentration of phenol and by-products with the degradation time: (**a**) ECT and (**b**) BEA-SECT on platinum; (**c**) ECT and (**d**) BEA-SECT on BDD (Reproduced from Ref. [94]. With kind permission of © Elsevier (2008))

was located either in the region of highest ultrasonic intensity (situated near the air–liquid interface) or at the transducer surface for low-frequency and high-frequency ultrasonic experiments, respectively.

11.3.3 Degradation of Nitrocompounds

The treatment of nitroaromatic compounds using sonoelectrochemistry methods has been analyzed sonovoltammetrically [96] and galvanostatically [97]. Aqueous solution of nitrobenzene (pH = 13) was reduced on glassy carbon and gold electrodes [96] under both silent and ultrasonic conditions (20 kHz and up to 40 W cm^{-2}). A reduction mechanism was proposed. First, a chemically reversible one electron process took place followed by an irreversible three electron reduction and, furthermore, an overall four electron process occurred at sufficiently negative potentials, leading to the production of phenylhydroxylamine. In these experiments, the glassy carbon electrode surface was damaged for very short electrode–horn distances and the gold electrode led to more complicated mechanisms due to surface reaction pathways.

The effect of ultrasonic irradiation on galvanostatic reduction of 1,3-dinitrobenzene and 2,4-dinitrotoluene in acid media have also been studied [97]. Ultrasound irradiation enhanced the electrochemical reduction rate.

11.3.4 Degradation of Chlorinated Compounds

2,4-Dichlorophenoxyacetic acid is one of the most widely used herbicides worldwide. It is a poorly biodegradable pollutant with mild toxicity, which can be converted into highly toxic chlorinated compounds such as 2,4-dichlorophenol. The degradation of 2,4-dichlorophenol has been studied sonoelectrochemically [98] using platinum electrodes in an ultrasonic bath (44 kHz). Cyclic sonovoltrammograms showed an increase in currents under insonation leading to an overall overpotential decrease together with no deactivation of the electrode. Bulk electrolyses using 1.0 M Na_2SO_4 as supporting electrolyte were carried out galvanostatically at various current densities. High degradation rates were obtained when the current density was increased to 70 mA cm^{-2}, leading to a reduction of 78% in total organic carbon value. The mass balance based on chloride ions indicated the formation of unanalyzed low-molecular–weight chlorinated compounds.

Other chlorinated compounds such as perchloroethylene (PCE) and trichloroacetic acid [99, 100] have been treated using the sonoelectrochemistry method. Alicante group has extensively studied the degradation of PCE using electrochemical (ECT), sonochemical (SCT), and sonoelectrochemical treatments [101–107]. The following section shows the main results obtained in the degradation of this pollutant.

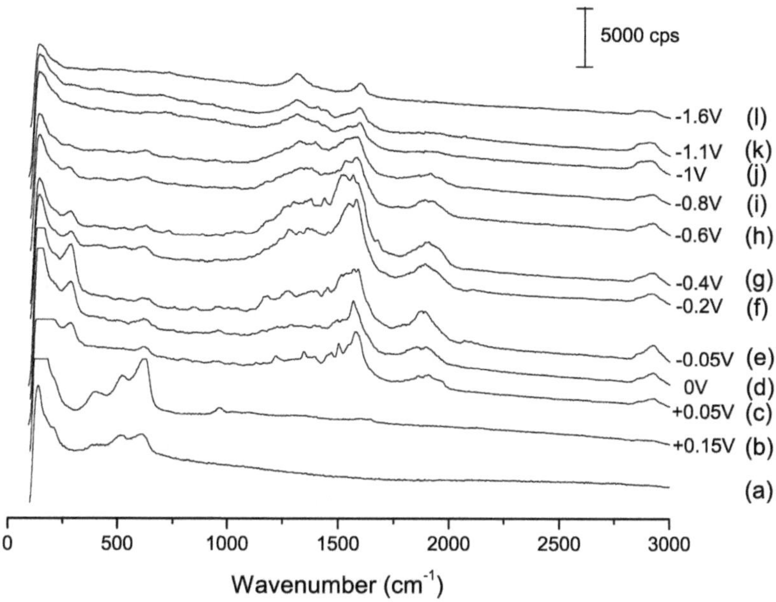

Fig. 11.2 Surface-enhanced Raman spectra of (**a**) bared copper, (**b**) 0.05 M Na_2SO_4 and (**c–l**) 10 ppm PCE + 0.05 M Na_2SO_4 adsorbed on SERS-active copper electrode at different potentials showed in the figure (Reproduced from Ref. [101]. With kind permission of © Elsevier (2008))

11.3.4.1 Fundamental Studies

The study of application of ECT started with the study of the voltammetric behavior of the aqueous supporting electrolyte/electrode system, followed by the use of surface-enhanced Raman spectroscopy (SERS) analysis. It continues with rotating disc electrode and bulk electrolyses experiments to give whole information about optimal variable parameters and reactor configuration. SERS, HPLC, GC, and ion chromatography analyses were used for the identification of degradation by-products, their quantification and mechanistic details.

Taking into account the industrial application, only a few electrode materials were used in this study. Copper showed a good performance on SERS experiments and exceptional catalytic properties and was used by Sáez et al. [101] on the preliminary voltammetric and SERS study of aqueous PCE electroreduction, using Na_2SO_4 as supporting electrolyte. Authors found that as the electrode potential decreases, the reduction of PCE is favored. The potential-dependent formation of PCE derivatives on a copper surface reveals that PCE undergoes reduction at $E \leq -0.3$ V vs. Ag/AgCl/KCl (3 M) electrode (all potential data on the PCE degradation study are quoted with reference to this electrode). This is confirmed by the emergence of intense Cu–Cl stretching (290 cm^{-1}), C–H stretching (contribution to the band at 2,920 cm^{-1}) band on SERS analysis (Fig. 11.2). The presence of Cl^- and dichloroethylene (DCE) in the catholyte was observed during electrolysis at −0.3 V detected by IC, HPLC, and GC.

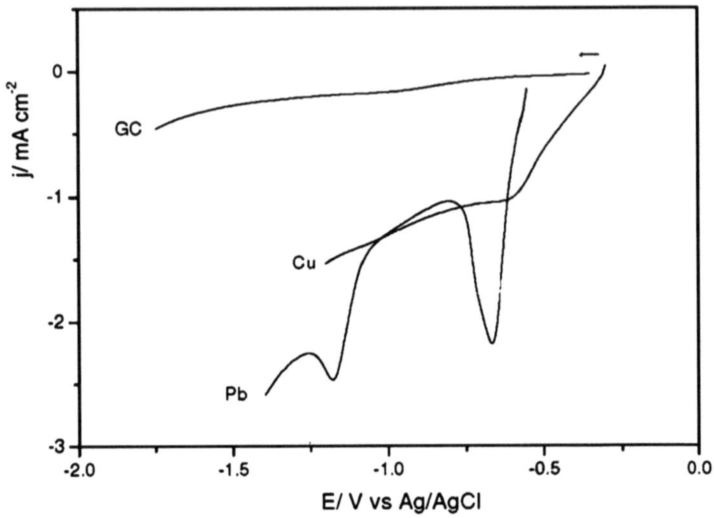

Fig. 11.3 Linear voltammograms at electrode material in 50 ppm of PCE + 0.05 M Na_2SO_4 aqueous solution, 0.5 V s^{-1}, first scan (Reproduced from Ref. [102]. With kind permission of © J New Mater Electrochem Syst (2008))

A band centered at 1,900 cm^{-1} (C–C structures) was detected in the 0 to −0.9 V potential region. The presence of carbon on the surface of the copper electrode is supported by the appearance of two bands at 1,330 and 1,590 cm^{-1} at potentials lower than −1.0 V, and by the XPS analysis of the electrode surface after preparative electrolysis carried out at −1.0 V. During the preparative electrolysis, a C–Cl multibond cleavage process occurs as it suggests by the appearance of DCE and the absence of trichloroethylene (TCE). The direct generation of DCE in front of a step process via TCE is favored.

11.3.4.2 Electrodic Material Study

Other electrodic materials such as glassy carbon and lead have been studied on the ECT of PCE by Sáez et al. [102]. No peaks can be observed on the cyclic voltammetry of an aqueous PCE/Na_2SO_4 solution on glassy carbon; and, therefore, slow kinetics for PCE reduction could be expected or its electroreduction could take places at potentials close to the supporting electrolyte decomposition. Lead electrode presents a stronger interaction with PCE and a peak at −0.7 V and a shoulder at −1.2 V can be observed on cyclic voltammograms. The peak at −0.7 V is associated to the lead oxide/lead sulfate surface reduction process and the reduction of PCE at this potential. On copper electrodes this process is associated with the shoulder appearing at −0.55 V. Figure 11.3 summarizes the voltammetric behavior of the PCE electroreduction with the three electrode materials.

11.3.4.3 Bulk Electrolyses

The controlled potential bulk electrolyses (divided glass cell, Nafion 450 cationic membrane) confirm the electroreduction of PCE in aqueous medium at the three electrodes (at −1.0 and −1.2 V for glassy carbon, at −0.4, −0.55, −0.7, −1.0, and −1.2 V for lead; and −0.3, −0.55 and −1.0 V for copper electrodes), but showing different electrocatalytic behavior. The lead electrode gave the best results and higher chlorinated or C3 reaction products have not been detected on any electrode. The reduction mechanism does not seem to follow a soluble radical route, and authors suggest that reduction seems to take place in a step-wise process giving products that are soluble and can be easily detected. A molar ratio TCE/DCE close to 2 obtained in the preparative electrolyses is in agreement with a predominant step-wise electroreduction pathway. Chemical reduction trials discarded the reduction of PCE by the zero-valent metal.

For all the electrode materials studied, TCE, DCE, and Cl^- anions were clearly detected as major intermediates and final products. The major part of the chlorinated compounds was detected and quantified, but some nonchlorinated organic compounds remain unidentified according to the mass balances calculations. The current efficiencies were low but they increased as the working electrode potential decreases.

Galvanostatic electrolyses are preferable for industrial purposes because of the cheapest investment and easiest handling, and were carried out in the next stage. Sáez et al. [103] studied not only the influence the initial concentration of PCE, but also several glass cell configurations with different cathodic (divided cell, electroreduction), anodic (divided cell, electrooxidation) and both simultaneously (undivided cell, dual degradation) treatments. Aqueous and gas phases were analyzed and the mass balance of the processes was checked. Lead was chosen as cathode material in this study because of its high overpotential for the reduction of water in spite of the theoretical drawbacks, from an environmental point of view. An anode of lead dioxide was used (Fig. 11.4).

The highest degradation (~50%) was obtained using the undivided configuration. PCE can be nearly degraded (<7% of the initial product remaining) and Cl^-, DCE, and TCE were found as major reactions products. The mass balance on total chlorine is closed. In the anodic treatment of PCE chloroacetic anions have been clearly detected along with the presence of $CHCl_3$ and CCl_4 compounds suggesting that the anodic degradation follows a slow and complex multipath route.

Following the study of ECT of PCE in water a scale-up to a laboratory scale electrochemical filter-press reactor and an extensive study of the influence of the operational variables was carried out by Sáez et al. [104]. The influences of the initial concentration, volumetric flow and electrode geometry, as well as the nature on the kinetics, degradation efficiency and mechanism were analyzed by determining the values of process performance parameters such as fractional conversion (FC), degradation efficiency (DE), current efficiency (CE), mass balance error based on chlorine (Cl-MBE) and selectivity (S_{Cl}).

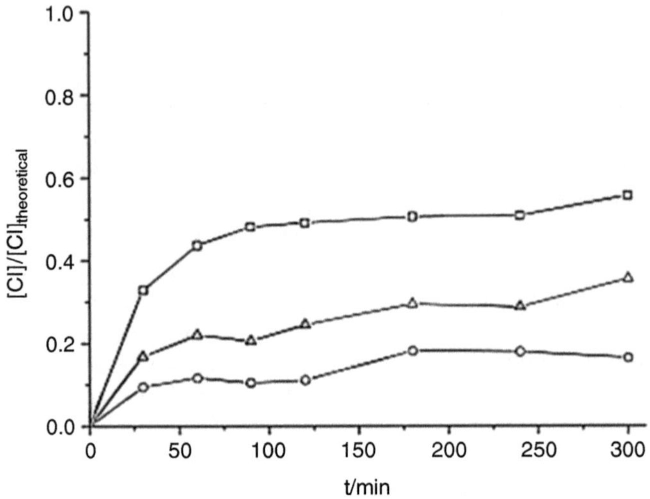

Fig. 11.4 Degradation of PCE for the three ECTs: (□) dual, (Δ) cathodic, and (O) anodic (Reproduced from Ref. [103]. With kind permission of © Elsevier (2009))

$$FC = \frac{[PCE]_{t=0} - [PCE]_{t=t_f}}{[PCE]_{t=0}} \times 100 \tag{11.1}$$

$$DE = \frac{[Cl^-]_{t=t_f} - [Cl^-]_{t=0}}{4 \times [PCE]_{t=0}} \times 100 \tag{11.2}$$

$$CE = \frac{[Cl^-]_{t=t_f} \times V \times 2 \times F}{\text{Passed charge}} \times 100 \tag{11.3}$$

$$Cl - MBE = \left(1 - \frac{\sum^{P=\text{compound}} (\text{mg of Clin P})_{t=t_f}}{(\text{mg of Clin PCE})_{t=0}}\right) \times 100 \tag{11.4}$$

$$S_{Cl} = \frac{\text{mol of chloride ion}}{\sum \text{mol of detected chlorinated compounds} + \text{mol of chloride ion}}, \tag{11.5}$$

where $[PCE]_{t=0}$ and $[PCE]_{t=t_f}$ are the initial and the final PCE concentration, $[Cl^-]_{t=0}$ and $[Cl^-]_{t=t_f}$ are the initial and the final chloride anion concentration, respectively; V represents the total volume of the solution and F the Faraday constant.

Several electrode materials and electrode geometries were tested. Three-dimensional carbon electrodes did not provide a more competitive option. The best results on the scale-up were an FC higher than 85%, DE of 65%, S_{Cl} close to 0.8 and energetic consumption around 3 kWh m^{-3} obtained with two-dimensional electrodes. In summary, the ECT provides a high DE for PCE with low energetic consumption; but toxic chlorinated by-products are not fully degraded (Table 11.3, Fig. 11.5).

Table 11.3 Main results for ECT, SCT, BEA-SECT, and SECT at batch scale

	FC/%	DE/%	Cl-MBE/%	S_{Cl}/%	CE/%	Time/min
ECT						
Q_V /L h^{-1}						
100	84	51	4	0.75	2	300
150	87	65	2	0.87	2.5	300
200	86	42	5	0.71	1	300
250	84	68	13	0.84	1	300
SCT						
I_a/Wcm^{-2}–Π/Wcm^{-3}						
1.84–0.065	86	24	53	0.79	–	300
3.39–0.120	98	29	59	0.88	–	300
5.09–0.180	99	30	62	0.92	–	300
6.36–0.225	98	25	64	0.90	–	300
7.64–0.270	99	21	72	0.91	–	300
BEA-SECT						
I_a/Wcm^{-2}–Π/Wcm^{-3}						
1.84–0.065	95	30	52	0.83	2.3	150
	97	37	38	0.86	1.2	300
3.39–0.120	96	46	45	0.97	12	150
	97	46	47	0.96	5	300
5.09–0.180	96	50	50	1.00	11	150
	100	57	39	0.98	6	300
6.36–0.225	97	49	48	0.99	13	150
	100	53	43	0.99	7	300
7.64–0.270	97	50	46	0.98	14	150
	100	56	41	0.99	8	300
SECT						
I_a/Wcm^{-2}–Π/Wcm^{-3}						
1.84–0.065	82	36	23	0.78	19	150
	96	50	40	0.95	9	300
3.39–0.120	96	49	46	0.98	21	150
	100	57	34	0.93	14	300
5.09–0.180	95	65	20	0.98	20	150
	96	74	29	0.98	13	300
6.36–0.225	97	77	16	0.99	12	150
	100	84	13	0.99	10	300
7.64–0.270	100	99	1	1.00	13	150
	100	100	0	0.99	10	300

Reproduced from Refs. [104, 107]. With kind permission of © Elsevier (2010) and © The American Chemical Society (2011)

11.3.4.4 Sonochemical Treatment

SCT has routinely been used for the degradation of chlorinated compounds in order to check water remediation applications and it has been reported in the literature of some papers about the SCT of PCE [108–112]. A deep analysis of the reaction

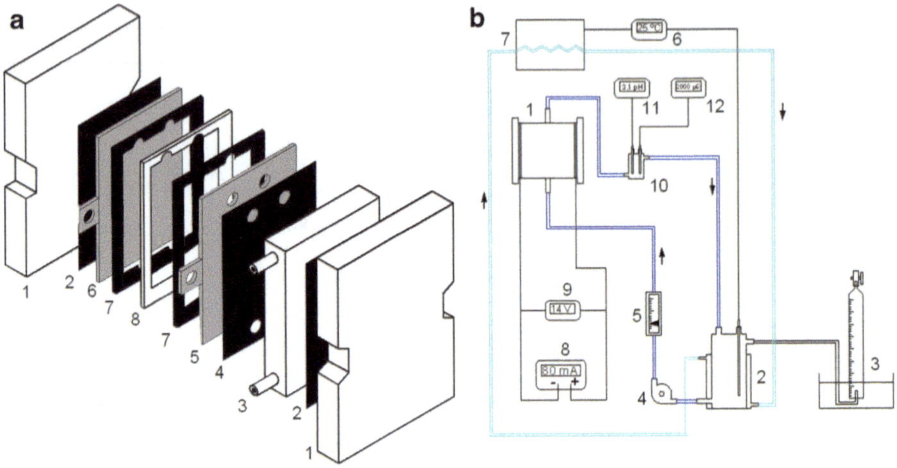

Fig. 11.5 (**a**) Electrochemical filter-press reactor: (*1*) backplate, (*2*) backplate joint, (*3*) polypropylene block with flow channels, (*4*) electrode joint, (*5*) cathode, (*6*) anode, (*7*) compartment joint, (*8*) compartment. (**b**) Experimental setup: (*1*) electrochemical filter-press reactor, (*2*) electrolyte reservoir, (*3*) gasmeter, (*4*) pump, (*5*) flowmeter, (*6*) temperature control, (*7*) cooling system, (*8*) power supply, (*9*) voltmeter, (*10*) probes holder, (*11*) pH meter, (*12*) conductimeter (Reprinted from Ref. [104]. With kind permission of © The American Chemical Society (2010))

by-products, the effect of ultrasonic operational variables upon PCE degradation (frequency, power, and system geometry as well as substrate concentration) and a mass balance were carried out by Sáez et al. [107]. They studied these parameters and noted Cl^- and CO_2/CO as major reaction products, and also TCE, and DCE at ppm concentrations as reported earlier in the literature. The formation of very low (ppb) concentrations of small halocompounds ($CHCl_3$, CCl_4), and also of higher mass species, such as pentachloropropene, hexachloroethane, is noteworthy. They also found significant quantities of chloroacetate derivatives at ppm concentrations.

The best results for PCE degradation (in terms of DE and Cl-MBE) were obtained at 580 and 850 kHz. Taken into account the hydrophobic and volatile nature of PCE, the authors suggest that the first step of the mechanism is the degradation of PCE into other chloroethenes in the interior of cavitation bubbles by pyrolysis or in the interfacial region of a cavitation bubble, rather than in the aqueous solution. Then, two pathways can interfere: pyrolysis of the remaining compounds and oxidation by the OH radicals. At high frequencies, the pyrolytic pathway is prevalent, taking part in the degradation of the initial PCE and the formation of residual (at ppb level) 1C, 3C, and 4C chlorinated compounds. At low frequency the oxidation by the OH radicals is predominant and highly soluble oxygenated compounds such as chloroacetates can be formed because the reaction time inside the cavitating bubbles is longer whereas at high frequency OH radicals are rapidly ejected to the solution.

Sonolysis was very effective for PCE degradation and Sáez et al. obtained FC $>$ 97%, and a mass balance close to 100% in the same conditions. From an application point of view, it should be noted that chlorinated reaction products are of some intrinsic toxicity, so the overall environmental benefit of insonation to destroy pollutants of this type remains debatable. In summary, the SCT of PCE proceeds with high DE values but with very high energetic cost and produces a higher by-products speciation than the ECT (Table 11.3).

11.3.4.5 Sonoelectrochemical Treatment

The use of hybrid technologies such as sonoelectrochemical treatment has been proposed by the authors in order to achieve a good level of water remediation. The 20 kHz BEA-SECT of PCE in aqueous Na_2SO_4 using controlled current density sonoelectrolyses in batch mode gave an important improvement in the viability of the SCT process when the electrochemistry is implemented [105]. On the BEA-SECT, a fractional conversion near 100% and degradation efficiency around 55% are obtained independent of the ultrasound power used (Fig. 11.6 and Table 11.3). Although the main volatile compounds produced, TCE and DCE, are not only totally degraded, similar concentration levels are achieved at shorter reaction time than in the SCT or ECT. The current efficiency is enhanced compared to the ECT and a higher speciation is also detected.

The combination of ultrasound and electrical fields provides a reaction environment, which greatly improves the SCT and ECT of PCE solutions, providing an acceptable procedure from a technical point of view. Moreover, the treatment time is significantly reduced, and then, the energetic consumption with BEA-SECT is lower than that presented by SCT.

The treatment of PCE solutions simulating real wastewater samples completes this study of technical viability [106]. The authors studied the sonoelectrolysis, without the addition of the supporting electrolyte (SECT) (20 kHz, Pb/PbO_2 as cathode/anode), of low conductivity PCE solutions. A controlled-voltage mode was selected for the electrolysis (a maximum applied voltage between anode and cathode of 14.3 V) and the maximum permitted current density was fixed at 3.5 mA cm^{-2}. The absence of supporting electrolyte implies that in the early stages of the sonoelectrolysis experiments the current is quite low because of the low conductivity of the solution. As the process proceeds, a background electrolyte is generated due to the fact that the electrochemical process provides, at least, chloride, chlorate and hydrogen ions.

Controlled-voltage PCE solutions sonoelectrolysis experiments at different ultrasound intensities (I_a) showed that when I_a was increased, an increase in the rate of disappearance of PCE was observed. However, the CE was not influenced by this parameter. The higher values of CE (>50% for chlorinated compounds) were maintained during the first hour of treatment (Fig. 11.7) in comparison with those obtained during the ECT (<8% in the best situation) or during the BEA-SECT. PCE, TCE and DCE were totally degraded in the first 2.5 h of the process (Fig. 11.8 and Table 11.3) in a similar way to those of BEA-SECT and in contrast to the ECT, for which even at high volumetric flows, a steady-state was observed in the remaining

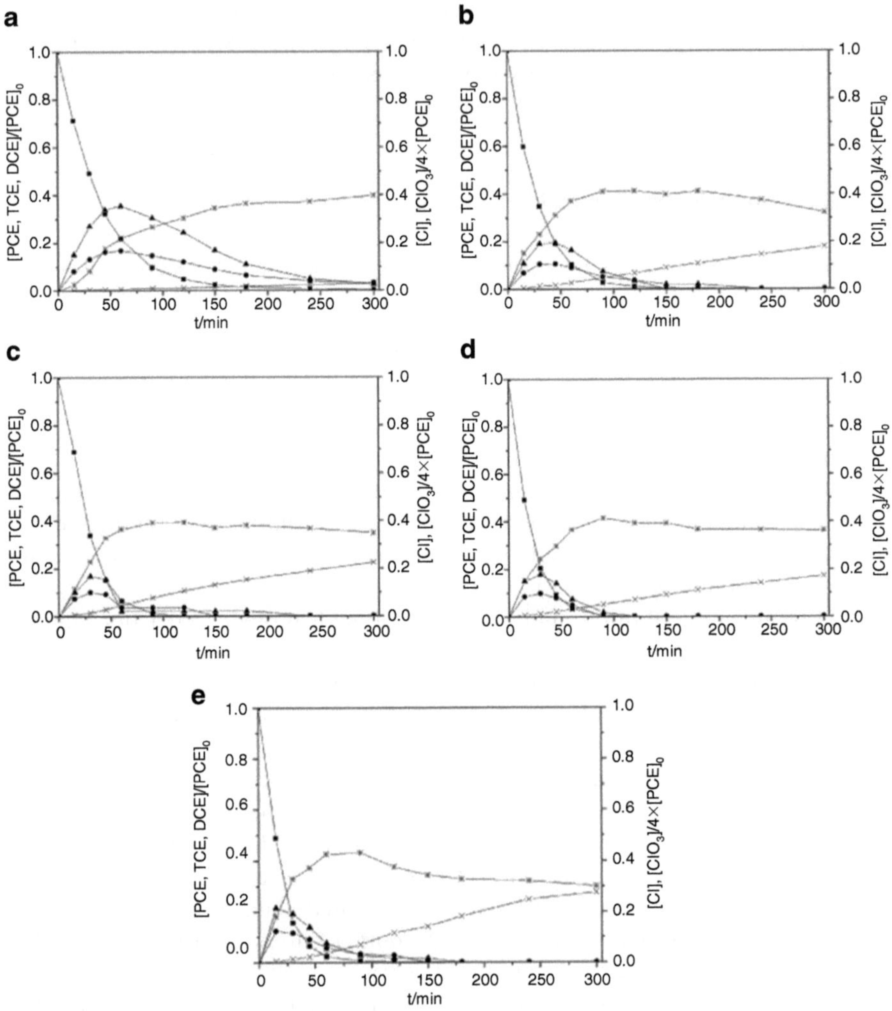

Fig. 11.6 Normalized concentration evolution vs time: (■) PCE, (▲) TCE, (●) DCE, (x) Cl^-, (*) ClO_3^- for the BEA-SECT at different ultrasonic intensities (**a**) 1.84; (**b**) 3.39; (**c**) 5.09; (**d**) 6.36 and (**e**) 7.64 W cm^{-2}, f = 20 kHz, 3.5 mA cm^{-2}, 20°C (Reproduced from Ref. [105]. With kind permission of © Elsevier (2010))

concentration of TCE and DCE. The results obtained in the SECT suggest that the mechanism is not fully controlled by the mass transport.

Therefore, the implementation of an ultrasound field during the electrochemical treatment of polluted water with low electrical conductivity appears to be a promising option enabling the electrochemical degradation of halogenated organic compounds in polluted water, employing an electrical current as the unique reactant,

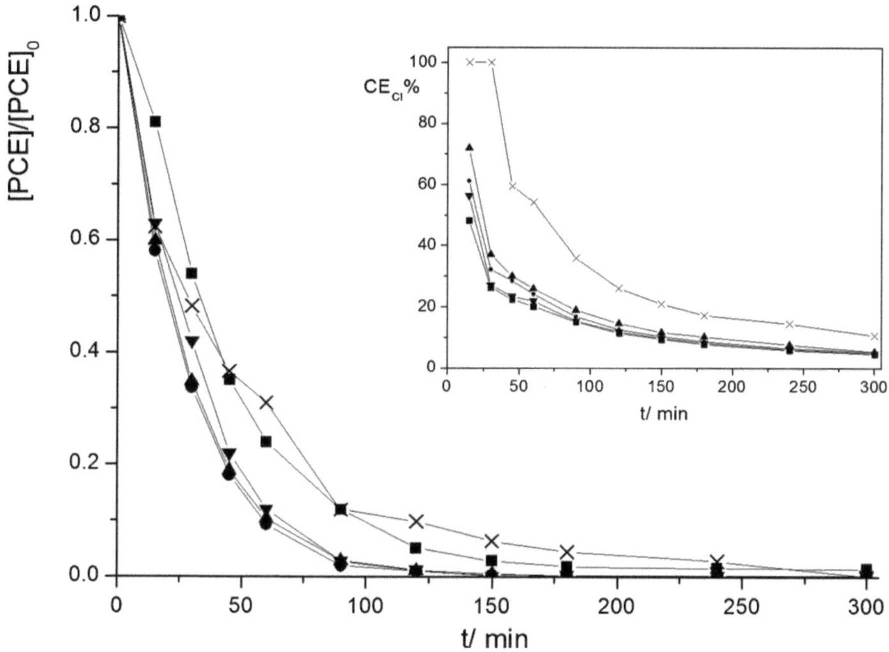

Fig. 11.7 Normalized degradation of PCE by sonoelectrochemical treatment and in the inset the CE for different concentrations (g L^{-1}) of Na_2SO_4 (mM ionic strength in brackets): (x) 0, (■) 0.5 (3.5), (●) 2.5 (17.6), (▲) 7.1 (50), (▼) 12 (84.5); f = 20 kHz, 3.5 mA cm^{-2}, 3.39 W cm^{-2}, 20°C (Reproduced from Ref. [106]. With kind permission of © Elsevier (2011))

and avoiding the undesirable addition of any chemicals. The high energetic cost of this treatment is the main drawback.

11.3.5 Hybrid Sonoelectrochemical Methods

Presently, further efforts in the development of the emerging AOTs, going deep into the possibility of achieving high synergic effects when some of those processes are applied simultaneously, are subject of interest. In this sense, combinations of sonoelectrochemical treatments with ozonation, photocatalysis, and Fenton techniques have been addressed.

As previously stated, photocatalytic degradation of water contaminants has been successfully used alone or together with the irradiation of an ultrasonic field. Zhang and coworkers were able to combine properly the advantages of photocatalysis, electrocatalysis, and sonocatalysis achieving a synergistic effect [113]. The authors raised the azodye methyl orange degradation at the 1.0×10^{-5}–2.0×10^{-4} M concentration range through different oxidative strategies in a comparative study.

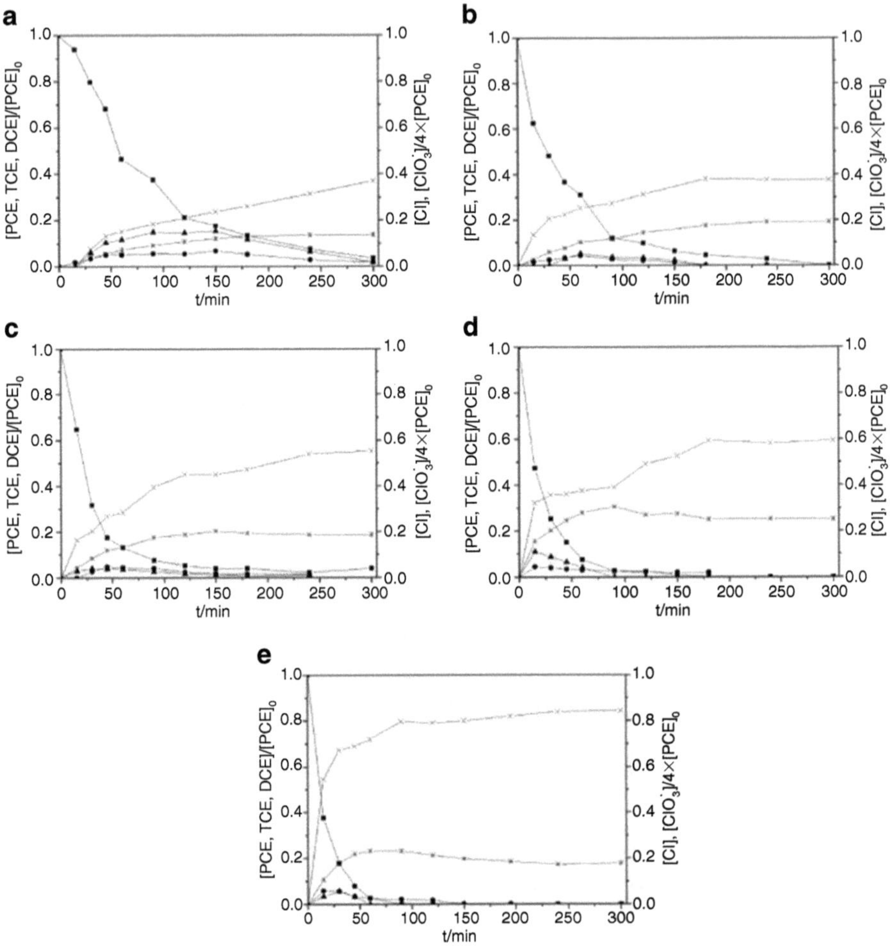

Fig. 11.8 Normalized concentration vs time: (■) PCE, (▲) TCE, (●) DCE, (*) Cl^-, (x) ClO_3^- for the PCE SECT at different ultrasonic intensities (**a**) 1.84; (**b**) 3.39; (**c**) 5.09; (**d**) 6.36 and (**e**) 7.64 $W\,cm^{-2}$, f = 20 kHz, V_{A-C} = 14.3 V, 20°C, 0 $g\,L^{-1}$ Na_2SO_4 (Reproduced from Ref. [106]. With kind permission of © Elsevier (2011))

Different batch photocatalytic, sonophotocatalytic, photoelectrocatalytic, and sonophotoeletrocatalytic experiences were carried out in the same multipurpose reactor. A photoreactor immersed in 40 kHz ultrasonic cleaning bath operating at several acoustic powers ranging from 60 to 150 W was used. The working electrode was a homemade TiO_2 nanotube anode, the supporting electrolyte was 0.5 M Na_2SO_4 aqueous solution, and electrode potential applied was 0.6 V vs. SCE for the ECT. The reactor used had a bottom-attached 11 W UV lamp with central wavelength of 253.7 nm (0.1 Hz chopped UV irradiation was used).

The degradation processes tested followed pseudo-first-order kinetics in all case and their rate constants increased as follows: photocatalytic (0.0035 min^{-1}) < sonophotocatalytic (0.0073 min^{-1}) < photoelectrocatalytic (0.0523 min^{-1}) < sonophotoelectrocatalytic (0.0732 min^{-1}). Results showed that the hybrid photoelectrocatalytic and sonophotoelectrocatalytic processes could efficiently enhance the degradation efficiency of methyl orange dye achieving a degradation of 88% of the initial pollutant after 60 min of treatment ($Co = 5 \times 10^{-5}$ M, ultrasound power 150 W).

The possibility of the combination of a BEA-SECT with ozonation has also been explored by Abramov et al. in their work on the destruction of 1,3-dinitrobenzene and 2,4-dinitrotoluene [97]. In that case, a titanium ultrasonic horn (5 cm^2) was used as a cathode immersed in a 100 cm^3 catholyte containing 100 mg L^{-1} of 1,3-dinitrobenzene and 2,4-dinitrotoluene. The use of ultrasound (5 W cm^{-2}, calorimetrically measured) enhanced the rate of electrochemical reduction (current: 50 mA), increasing the current efficiency from ~15% to 55–57%, but the overall reaction rate remained still slow. On the other hand, the compounds were shown to be stable to reaction with ozone, even under ultrasonic activation. However, the simultaneous application of ultrasound and ozonation (flow rate 0.1–3 L^{-1} h^{-1}, O_3 concentration 3–10%) to the electrochemical reaction allowed complete destruction of the compounds in short times.

In this case, the effect was attributed to the ultrasonic enhancement of the electrochemical process giving intermediates that are susceptible to ozone oxidation. Ozone is a very efficient oxidizing agent able to interact with electron donors or be reduced in the cathode to O_3^-. This highly oxidant anion rapidly forms the anion radical $O\bullet^-$ in acidic media, or alternatively can generate OH radical, a very active oxidant. It is assumed that this process is improved by the mechanical effects provided by cavitation such as enhanced mixing, diffusion, and mass transfer.

On the other hand, different groups have tried to improve the performance of the Fenton process applied to environmental remediation by the application of sonoelectrochemistry. Fenton reaction is based on the H_2O_2 decomposition to OH radicals in acid aqueous medium, catalyzed by the addition of a small amount of Fe^{2+} [114]. It involves further radical reactions among the organic compounds and the OH radicals formed.

$$H_2O_2 + Fe^{2+} \rightarrow Fe^{3+} + HO^{\bullet} + HO^- \quad (11.6)$$

The potentiality of the application of electrochemical processes coupled with Fenton treatment has revealed itself as a useful technology that improves significantly the original approach [115]. One of the main advantages of this combination is the possibility to generate H_2O_2 electrochemically in a safer process compared with H_2O_2 addition to the polluted effluent. Moreover, the possibility of Fe^{2+} regeneration in the cathode makes the process more environmentally friendly, since less amount of iron is needed initially. As in other electrochemical processes, some authors have taken advantage of the outstanding beneficial effects of ultrasound in sonochemical processes.

The first study found in literature dealt with the sonoelectro-Fenton degradation (SEF) of the organic dye molecule meldola blue (MDB) [116]. The authors used a sonoelectrochemical reactor containing 100 cm^3 of solution (1×10^{-4} M MDB, 0.050 mM Na_2SO_4, 0.010 mM H_2SO_4, 0.5 mM $FeSO_4$) with a reticulated carbon felt cathode (15 mm × 10 mm × 5 mm) working at −0.7 V vs Ag. The solution was exposed to 124 kHz ultrasound field. Assuming pseudo-first-order kinetics, it was found that the rate constant for the dye degradation reached for the different tested treatments in the following order: ECT (negligible) < SCT (0.0044 min^{-1}) < sono-Fenton (0.0081 min^{-1}) < BEA-SECT (0.0126 min^{-1}) < electro-Fenton (0.0142 min^{-1}) < SEF (0.0237 min^{-1}). In the ECT, SCT, and BEA-SECT treatments exposed, no $FeSO_4$ was added to the reaction solution.

Recently, the effect of low-frequency ultrasonic irradiation on the SEF degradation has been tested again for decomposition of cationic red X-GRL [117]. It has been found that both total organic carbon removal efficiency and mineralization current efficiency were efficiently promoted with the electro-Fenton process.

Furthermore, the SEF treatment is a helpful technology for the degradation of organochlorinated compounds in wastewater [118]. Outran et al. found a synergistic effect while using an undivided electrolytic cell (250 mL) with a three-dimensional carbon-felt cathode (15 cm × 8 cm) applying constant current (200 mA) for the destruction of the herbicides, 4,6-dinitro-o-cresol (DNOC), and 2,4-dichlorophenoxyacetic acid (2,4-D) [119]. A 28 kHz ultrasound field operating at 20 W was used for the SEF processes. The apparent rate constants for the electro-Fenton and SEF processes were 0.025 min^{-1} and 0.046 min^{-1} for 2,4-D and 0.023 min^{-1} and 0.051 min^{-1} for DNOC, respectively, and followed pseudo-first-order kinetics in all cases.

The enhancement observed in the SEF process is partially ascribed to the enhancement in the mass transfer of both reactants (Fe^{3+} and O_2) toward the cathode for the electrochemical generation of Fenton's reagent ($Fe^{2+} + H_2O_2$) and its subsequent transfer. González-García et al. observed an enhancement of the electrochemically produced H_2O_2 in aqueous solution by means of ultrasound application [120]. Furthermore, the amount of OH radical in solution is expected to be higher in irradiated systems than in silent conditions due to its formation inside cavitation bubbles.

11.4 Conclusion

It has been shown that sonophotocatalysis seems to be a promising method for the degradation of organic pollutants in aqueous solutions. The degradation process is mostly governed by free radical reactions because all the organics discussed here are nonvolatile in nature. The extent of degradation depends on many parameters such as nature of catalysts, concentration of pollutant, amount of catalyst, intensity of light, ultrasonic power, solution temperature, pH, and dissolved gas. Although the initial rate of sonochemical degradation is relatively fast, complete mineralization is achieved along with TiO_2 photocatalysts in most cases.

This chapter also has summarized the applications found in the literature on the use of sonoelectrochemical treatment of organic pollutants in aqueous solutions. Greater effort is needed to fully understand the degradation process that occurs when combining both methods. This will bring improvements in existing and future applications. Given that the contamination of water effluents is a raising problem, an increase on the research into the new possibilities of hybrid sonoelectrochemical technologies is also expected in the near future.

References

1. Hoffmann MR, Martin ST, Choi W, Bahnemann DW (1995) Environmental applications semiconductor photocatalysis. Chem Rev 95:69–96
2. Neppolian B, Sakthivel S, Palanichamy M, Arabindoo B, Murugesan V (1998) Photocatalytic degradation of textile dye commonly used in cotton fabrics. Stud Surf Sci Catal 113:329–335
3. Neppolian B, Sakthivel S, Palanichamy M, Arabindoo B, Murugesan V (1999) Degradation of textile dye using TiO_2 and ZnO photocatalysts. J Environ Sci Health A 43:1829–1838
4. Serpone N, Pelizzetti E (1989) Photocatalysis: fundamentals and applications. Wiley, New York
5. Herrmann JM (1995) Heterogeneous photocatalysis – an emerging discipline involving multiphase systems. Catal Today 24:157–164
6. Vinodgopal K, Wynkoop DE, Kamat PV (1996) Environmental photochemistry on semiconductor surfaces: photosensitized degradation of a textile azo dye, acid orange 7, on TiO_2 particles using visible light. Environ Sci Technol 30:1660–1666
7. Das S, Kamat PV, Padmaja S, Madison SA (1999) Free radical induced oxidation of the azo dye Acid Yellow 9. J Chem Soc Perkin Trans 2:1219–1223
8. Stafford U, Gary KA, Kamat PV (1997) Photocatalytic degradation of 4-chlorophenol: the effects of varying TiO_2 concentration and light wavelength. J Catal 167:25–32
9. Davis RJ, Gainer JL, Neal GO, Wenwu I (1994) Photocatalytic decolourization of wastewater dyes. Water Environ Res 66:50–53
10. Bahnemann D, Boule P (eds) (1999) Photocatalytic degradation of polluted waters, The handbook of environmental chemistry, vol 2, Part L: Environmental photochemistry. Springer, Berlin
11. Singla R, Grieser F, Ashokkumar M (2009) Sonochemical degradation of martius yellow dye in aqueous solution. Ultrason Sonochem 16:28–34
12. Choi W (2006) Pure and modified TiO_2 photocatalysts and their environmental applications. Catal Surv Asia 10:16–28
13. Yang L, Rathman LK, Weavers LK (2005) Sonochemical degradation of alkylbenzene sulfonate surfactants in aqueous mixtures. J Phys Chem B 110:18385–18391
14. Neppolian B, Jung H, Choi H, Lee JH, Kang JW (2002) Sonolytic degradation of methyl tert-butyl ether: the role of coupled Fenton process and persulphate ion. Water Res 36:4699–4708
15. Singla R, Ashokkumar M, Grieser F (2004) The mechanism of the sonochemical degradation of benzoic acid in aqueous solutions. Res Chem Intermed 30:723–733
16. Singla R, Grieser F, Ashokkumar M (2009) Kinetics and mechanism for the sonochemical degradation of a nonionic surfactant. J Phys Chem A 113:2865–2872
17. Peller J, Wiest O, Kamat PV (2001) Sonolysis of 2, 4-dichlorophenoxyacetic acid in aqueous solutions. Evidence for OH-radical-mediated degradation. J Phys Chem 105:3176–3181

18. Yang L, Sostaric JZ, Rathman JF, Weavers LK (2008) Effect of ultrasound frequency on pulsed sonolytic degradation of octylbenzene sulfonic acid. J Phys Chem B 112:852–858
19. Naldoni A, Schibuola A, Bianchi CL, Bremner DH (2011) Mineralization of surfactants using ultrasound and the advanced Fenton process. Water Air Soil Pollut 215:487–495
20. Neppolian B, Park JS, Choi H (2004) Effect of Fenton-like oxidation on enhanced oxidative degradation of para–chlorobenzoic acid by ultrasonic irradiation. Ultrason Sonochem 11: 273–279
21. Neppolian B, Ciceri L, Bianchi CL, Grieser F, Ashokkumar M (2011) Sonophotocatalytic degradation of 4-chlorophenol using $Bi_2O_3/TiZrO_4$ as a visible light responsive photocatalyst. Ultrason Sonochem 18:135–139
22. Tarr M (2003) Chemical degradation methods for wastes and pollutants. Environmental and industrial applications. Marcel-Dekker, New York
23. Souther RH, Alspaugh TA (1957) Textile wastes- recovery and treatment. J Water Pollut Control Fed 29:804–810
24. Hamza A, Hamoda MF (1980) Multiprocess treatment of textile wastewater. In: Proceedings of the 35th Purdue industrial waste congress, West Lafayette, IN, USA
25. Sheng HL, Chi ML (1993) Treatment of textile waste effluents by ozonation and chemical coagulation. Water Res 27:1743–1748
26. Arslan I, Balcioglu IA, Tuhkanen T, Bahnemann D (2000) H_2O_2/UV-C and Fe^{2+}/H_2O_2/UV-C versus TiO_2/UV-A treatment for reactive dye wastewater. J Environ Eng 126:903–911
27. Chaudhuri SK, Sur B (2000) Oxidative decolorization of reactive dye solution using fly ash as catalyst. J Environ Eng 126:583–594
28. Neppolian B, Sakthivel S et al (1999) Photoassisted degradation of textile dye using ZnO catalyst. Bull Catal Soc India 81:164–171
29. Neppolian B, Choi H, Sakthivel S, Arabindoo B, Murugesan V (2002) The influence of solar light induced and TiO_2 assisted degradation of textile dye reactive blue 4. Chemosphere 46:1173–1181
30. Yang YQ, Wyatt DT, Bahorsky M (1998) Decolorization of dyes using UV/H_2O_2 photochemical oxidation textile dye. Chem Color 30:27–35
31. Balcioglu IA, Arslan I (1997) Treatment of textile waste water by heterogeneous photocatalytic oxidation processes. Environ Technol Lond 18:1053–1059
32. Neppolian B, Shankar MV, Murugesan V (2002) Semiconductor assisted photodegradation of textile dye, reactive red 2 by ZnO. J Sci Indus Res 61:224–230
33. Neppolian B, Kanel SR, Choi H, Shankar MV, Arabindoo B, Murugesan V (2003) Photocatalytic degradation of reactive yellow 17 in aqueous solution in the presence of cement. Int J Photoenergy 5:45–49
34. Hua I, Hochemer RH, Hoffmann MR (1995) Sonochemical degradation of p-nitrophenol in a parallel-plate near-field acoustic processor. Environ Sci Technol 29:2790–2796
35. Kang JW, Hung HM, Lin A, Hoffmann MR (1999) Sonolytic destruction of methyl tert-butyl ether by ultrasonic irradiation: the role of O_3, H_2O_2, frequency, and power supply. Environ Sci Technol 33:3199–3205
36. Kotronarou A, Mills G, Hoffmann MR (1992) Decomposition of parathion in aqueous solution by ultrasonic irradiation. Environ Sci Technol 26:1460–1475
37. Hua I, Hoffmann MR (1996) Kinetics and mechanism of sonolytic degradation of CCl_4: intermediates and byproducts. Environ Sci Technol 30:864–871
38. Hua I, Hoffmann MR (1997) Optimization of ultrasonic irradiation as an advanced oxidation technology. Environ Sci Technol 31:2237–2243
39. Lin JH, Ma YS (2000) Oxidation of 2-chlorophenol in water by ultrasound/Fenton method. J Environ Eng 126:130–137
40. Lin JH, Chang CN, Wu JR (1996) Decomposition of 2-chlorophenol in aqueous solution by ultrasound/H_2O_2 process. Water Sci Technol 33:75–81
41. Kang JW, Hoffmann MR (1998) Kinetics and mechanism of the sonolytic destruction of methyl tert-butyl ether by ultrasonic irradiation in the presence of ozone. Environ Sci Technol 32:3194–3199

42. Magureanu M, Piroi D, Mandache N, David V, Medvedovici A, Parvulescu V (2010) Degradation of pharmaceutical compound pentoxifylline in water by non-thermal plasma treatment. Water Res 44:3445–3453
43. Pérez MH, Peñuela G, Maldonado MI, Malato O, Fernández-Ibáñez P, Oller I, Gernjak W, Malato S (2006) Degradation of pesticides in water using solar advanced oxidation processes. Appl Catal B Environ 64:272–281
44. Shankar MV, Nelieu S, Kerhoas L, Einhorn J (2008) Natural sunlight NO3-/NO2–induced photo-degradation of phenylurea herbicides in water. Chemosphere 71:1461–1468
45. Santos TCR, Rocha JC, Alonso RM, Martínez E, Ibañez C, Barceló D (1998) Rapid degradation of propanil in rice crop fields. Environ Sci Technol 32:3479–3484
46. Lee DJ, Senseman SA, Sciumbato AS, Jung SC, Krutz LJ (2003) The effect of titanium dioxide alumina beads on the photocatalytic degradation of picloram in water. J Agric Food Chem 51:2659–2664
47. Tamimi M, Qourzal S, Barka N, Assabbane A, Ait-Ichou Y (2008) Methomyl degradation in aqueous solutions by Fenton's reagent and the photo-Fenton system. Sep Purif Technol 61:103–108
48. Chelme-Ayala P, El-Din MG, Smith DW (2010) Kinetics and mechanism of the degradation of two pesticides in aqueous solutions by ozonation. Chemosphere 78:557–562
49. Rajeswari R, Kanmani S (2009) A study on synergistic effect of photocatalytic ozonation for carbaryl degradation. Desalination 242:277–285
50. Acero JL, Real FJ, Benitez FJ, González A (2008) Oxidation of chlorfenvinphos in ultrapure and natural waters by ozonation and photochemical processes. Water Res 42:3198–3206
51. Bandala ER, Pelaez MA, Salgado MJ, Torres L (2008) Degradation of sodium dodecyl sulphate in water using solar driven Fenton-like advanced oxidation processes. J Hazard Mater 151:578–584
52. Lea J, Adesina AA (1998) The photo-oxidative degradation of sodium dodecyl sulphate in aerated aqueous TiO_2 suspension. J Photochem Photobiol A: Chem 118:111–122
53. Rao PSC, Lee LS, Pinal R (1990) Cosolvency and sorption of hydrophobic organic chemicals. Environ Sci Technol 24:647–654
54. Lee LS, Rao PSC, Nikedi-Kizza P, Delfino JJ (1990) Influence of solvent and sorbent characteristics on distribution of pentachlorophenol in octanol-water and soil-water systems. Environ Sci Technol 24:654–661
55. Song H, Carraway ER (2008) Catalytic hydrodechlorination of chlorinated ethenes by nanoscale zero-valent iron. Appl Catal B: Environ 78:53–60
56. Lee CC, Doong RA (2008) Dechlorination of tetrachloroethylene in aqueous solutions using metal-modified zerovalent silicon. Environ Sci Technol 42:4752–4757
57. Satuf ML, Brandi RJ, Cassano AE, Alfano OM (2008) Photocatalytic degradation of 4-chorophenol: a kinetic study. Appl Catal B: Environ 82:37–49
58. González-García J, Sáez V, Tudela I, Díez-García MI, Esclapez MD, Louisnard O (2010) Sonochemical treatment of water polluted by chlorinated organocompounds. A review. Water 2:28–74
59. Isse AA, Gottardello S, Durante D, Gennaro A (2008) Dissociative electron transfer to organic chlorides: electrocatalysis at metal cathodes. Phys Chem Chem Phys 10:2409–2416
60. Kastanek F, Maleterova Y, Kastanek P (2007) Combination of advanced oxidation and/or reductive dehalogenation and biodegradation for the decontamination of waters contaminated with chlorinated organic compounds. Sep Sci Technol 42:1613–1625
61. Villanueva CM, Kogevinas M, Grimalt JO (2003) Haloacetic acids and trihalomethanes in finished drinking waters from heterogeneous sources. Water Res 37:953–958
62. McRae BM, Lapara TM, Hozalski RM (2004) Biodegradation of haloacetic acids by bacterial enrichment cultures. Chemosphere 55:915–925
63. Rodriguez MJ, Serodes J, Roy D (2007) Formation and fate of haloacetic acids (HAAs) within the water treatment plant. Water Res 41:4222–4232
64. Stock N, Peller J, Vinodgopal K, Kamat PV (2003) Combinative sonolysis and photocatalysis for textile dye degradation. Environ Sci Technol 34:1747–1750

65. Madhavan J, Grieser F, Ashokkumar M (2009) Kinetics of the sonophotocatalytic degradation of orange G in presence of Fe^{3+}. Water Sci Technol 60:2195–2202
66. Yuan S, Yu L et al (2009) Highly ordered TiO_2 nanotube array as recyclable catalyst for the sonophotocatalytic degradation of methylene blue. Catal Commun 10:1188–1191
67. Kaur S, Singh V (2007) Visible light induced sonophotocatalytic degradation of Reactive Red dye 198 using dye sensitized TiO_2. Ultrason Sonochem 14:531–537
68. Berberidou C, Poulios I, Xekoukoulotakis NP, Mantzavinos D (2007) Sonolytic, photocatalytic and sonophotocatalytic degradation of malachite green in aqueous solutions. Appl Catal B Environ 74:63–72
69. Wang S, Gong Q, Liang J (2009) Sonophotocatalytic degradation of methyl orange by carbon nanotube/TiO_2 in aqueous solutions. Ultrason Sonochem 16:205–208
70. González A, Martínez S (2008) Study of the sonophotocatalytic degradation of basic blue 9 industrial textile dye over slurry titanium dioxide and influencing factors. Ultrason Sonochem 15:1038–1042
71. Bejarano-Perez NJ, Suarez-Herrera MF (2007) Sonophotocatalytic degradation of congo red and methyl orange in the presence of TiO_2 as a catalyst. Ultrason Sonochem 14:589–595
72. Madhavan J, Grieser F, Ashokkumar M (2010) Degradation of orange-G by advanced oxidation processes. Ultrason Sonochem 17:338–343
73. Wang H, Niu J, Long X, He Y (2008) Sonophotocatalytic degradation of methyl orange by nano-sized Ag/TiO_2 particles in aqueous solutions. Ultrason Sonochem 15:386–392
74. Vinu R, Madras G (2009) Kinetics of sonophotocatalytic degradation of anionic dyes with nano-TiO_2. Environ Sci Technol 43:473–479
75. Son Y, Cho E, Lim M, Khim J (2010) Effects of salt and pH on sonophotocatalytic degradation of azo dye reactive black 5. Jpn J Appl Phys 49:07HE05–3–07HE05–1
76. Madhavan J, Kumar P, Anandan S, Grieser F, Ashokkumar M (2010) Degradation of acid red 88 by the combination of sonolysis and photocatalysis. Sep Purif Technol 74:336–341
77. Zhang K, Oh WC (2010) Kinetic study of the visible light-induced sonophotocatalytic degradation of MB solution in the presence of Fe/TiO_2-MWCNT catalyst. Bull Korean Chem Soc 31:1589–1595
78. Madhavan J, Grieser F, Ashokkumar M (2010) Combined advanced oxidation processes for the synergistic degradation of ibuprofen in aqueous environments. J Hazard Mater 178: 202–208
79. Madhavan J, Kumar P, Anandan S, Zhou M, Grieser F, Ashokkumar M (2010) Ultrasound assisted photocatalytic degradation of diclofenac in an aqueous environment. Chemosphere 80:747–752
80. Bahena C, Martínez S, Guzman D et al (2008) Sonophotocatalytic degradation of alazine and gesaprimcommercial herbicides in TiO_2 slurry. Chemosphere 71:982–989
81. Madhavan J, Grieser F, Ashokkumar M (2010) Degradation of formetanate hydrochloride by combined advanced oxidation processes. Sep Purif Technol 73:409–414
82. Madhavan J, Kumar P, Anandan S, Grieser F, Ashokkumar M (2010) Sonophotocatalytic degradation of monocrotophos using TiO_2 and Fe^{3+}. J Hazard Mater 177:944–949
83. Neppolian B, Bruno A, Bianchi CL, Ashokkumar M (2012) Graphene oxide based TiO_2 photocatalyst: synthesis, characterization and catalytic efficiency. Ultrason Sonochem 19:9–15
84. Anandan S, Ashokkumar M (2009) Sonochemical synthesis of Au–TiO_2 nanoparticles for the sonophotocatalytic degradation of organic pollutants in aqueous environment. Ultrason Sonochem 16:316–320
85. Lorimer JP, Mason TJ, Plattes M, Phull SS (2000) Dye effluent decolourisation using ultrasonically assisted electro-oxidation. Ultrason Sonochem 7:237–242
86. Lorimer JP, Mason TJ, Plattes M, Phull SS, Walton DJ (2001) Degradation of dye effluent. Pure Appl Chem 73:1957–1968
87. Foord JS, Holt KB, Compton RG, Marken F, Kim DH (2003) Mechanistic aspects of the sonoelectrochemical degradation of the reactive dye Procion Blue at boron-doped diamond electrodes. Diamond Relat Mater 10:662–666

88. Rivera M, Pazos M, Sanromán MA (2009) Improvement of dye electrochemical treatment by combination with ultrasound technique. J Chem Technol Biotechnol 84:1118–1124
89. Ai Z, Li J, Zhang L, Lee S (2010) Rapid decolorization of azo dyes in aqueous solution by an ultrasound-assisted electrocatalytic oxidation process. Ultrason Sonochem 17:370–375
90. Siddique M, Farooq R, Khan ZM, Khan Z, Shaukat SF (2011) Enhanced decomposition of reactive blue 19 dye in ultrasound assisted electrochemical reactor. Ultrason Sonochem 18:190–196
91. Wang X, Jia J, Wang Y (2010) Electrochemical degradation of reactive dye in the presence of water jet cavitation. Ultrason Sonochem 17:515–520
92. He P, Wang L, Xue J, Cao Z (2010) Electrolytic treatment of methyl orange in aqueous solution using three-dimensional electrode reactor coupling ultrasonics. Environ Technol 31:417–422
93. Trabelsi F, Ait-lyazidi H, Ratsimba B, Wilhelm AM, Delmas H, Fabre PL, Berlan J (1996) Oxidation of phenol in wastewater by sonoelectrochemistry. Chem Eng Sci 51:1857–1865
94. Zhao G, Shen S, Li M, Wu M, Cao T, Li D (2008) The mechanism and kinetics of ultrasound-enhanced electrochemical oxidation of phenol on boron-doped diamond and Pt electrodes. Chemosphere 73:1407–1413
95. Lima Leite RH, Cognet P, Wilhelm AM, Delmas H (2002) Anodic oxidation of 2,4-dihydroxybenzoic acid for wastewater treatment: study of ultrasound activation. Chem Eng Sci 57:767–778
96. Marken F, Kumbhat S, Sanders GHW, Compton RG (1996) Voltammetry in the presence of ultrasound: surface and solution processes in the sonovoltammetric reduction of nitrobenzene at glassy carbon and gold electrodes. J Electroanal Chem 414:95–105
97. Abramov VO, Abramov OV, Gekhman AE, Kuznetsov VM, Price GJ (2006) Ultrasonic intensification of ozone and electrochemical destruction of 1,3-dinitrobenzene and 2,4-dinitrotoluene. Ultrason Sonochem 13:303–307
98. Padilla KR, Blanco LM, Orozco S, Jáuregui F (2004) Degradación oxidativa del 2,4-diclorofenol por vía sonoelectroquímica. Ciencia UANL VII:51–59
99. Esclapez MD, Sáez V, Milán-Yáñez D, Tudela I, Louisnard O, González-García J (2010) Sonoelectrochemical treatment of water polluted with trichloroacetic acid: from sonovoltammetry to pre-pilot plant scale. Ultrason Sonochem 17:1010–1020
100. Esclapez MD, Díez-García MI, Sáez V, Tudela I, Pérez JM, González-García J, Bonete P (2011) Spectroelectrochemical study of trichloroacetic acid reduction at copper electrodes in an aqueous sodium sulfate medium. Electrochim Acta 56:8138–8146
101. Sáez V, Esclapez MD, Bonete P, González-García J, Pérez JM (2008) Spectroelectrochemical study of perchloroethylene reduction at copper electrodes in neutral aqueous medium. Electrochim Acta 53:3210–3217
102. Sáez V, Esclapez MD, Frías-Ferrer A, Bonete P, González-García J (2008) Electrochemical reduction of perchloroethylene in aqueous media: influence of the electrode material. J New Mater Electrochem Syst 11:287–295
103. Sáez V, Esclapez MD, Frías-Ferrer A, Bonete P, González-García J (2009) Electrochemical degradation of perchloroethylene in aqueous media: an approach to different strategies. Water Res 43:2169–2178
104. Sáez V, Esclapez MD, Tudela I, Bonete P, González-García J (2010) Electrochemical degradation of perchloroethylene in aqueous media: influence of the electrochemical operational variables in the viability of the process. Ind Eng Chem Res 49:4123–4131
105. Sáez V, Esclapez MD, Tudela I, Bonete P, Louisnard O, González-García J (2010) 20 kHz sonoelectrochemical degradation of perchloroethylene in sodium sulfate aqueous media: influence of the operational variables in batch mode. J Hazard Mater 183:648–654
106. Sáez V, Esclapez MD, Tudela I, Bonete P, Louisnard O, González-García J (2011) Sonoelectrochemical degradation of perchloroethylene in water: enhancement of the process by the absence of background electrolyte. Chem Eng J 168:649–655
107. Sáez V, Esclapez MD, Bonete P, Walton DJ, Rehorek A, Louisnard O, González-García J (2011) Sonochemical degradation of perchloroethylene: the influence of ultrasonic variables, and the identification of products. Ultrason Sonochem 18:104–113

108. Colussi AJ, Hung HM, Hoffmann MR (1999) Sonochemical degradation rates of volatile solutes. J Phys Chem A 103:2696–2699
109. Bhatnagar A, Cheung HM (1994) Sonochemical destruction of chlorinated C1 and C2 volatile organic compounds in dilute aqueous solution. Environ Sci Technol 28:1481–1486
110. Inazu K, Nagata Y, Maeda Y (1993) Decomposition of chlorinated hydrocarbons in aqueous solutions by ultrasonic irradiation. Chem Lett 22:57–60
111. Clark CJ II, Annable MD, Rao PS (2000) Evaluation of sonochemical destruction of PCE in situ flushing waste fluids. J Environ Eng 126:1033–1038
112. Krüger O, Schulze ThL, Peters D (1999) Sonochemical treatment of natural ground water at different high frequencies: preliminary results. Ultrason Sonochem 6:123–128
113. Zhang Z, Yuan Y, Liang L, Fang Y, Cheng Y, Ding H, Shi G, Jin L (2008) Sonophotoelectrocatalytic degradation of azo dye on TiO_2 nanotube electrode. Ultrason Sonochem 15:370–375
114. Lipczynska-Kochany E (1991) Degradation of aqueous nitrophenols and nitrobenzene by means of the Fenton reaction. Chemosphere 22:529–536
115. Brillas E, Sirés I, Oturan MA (2009) Electro-Fenton process and related electrochemical technologies based on Fenton's reaction chemistry. Chem Rev 109:6570–6631
116. Abdelsalam ME, Birkin PR (2002) A study investigating the sonoelectrochemical degradation of an organic compound employing Fenton's reagent. Phys Chem Chem Phys 4:5340–5345
117. Li H, Lei H, Yu Q, Li Z, Feng X, Yang B (2010) Effect of low frequency ultrasonic irradiation on the sonoelectro-Fenton degradation of cationic red X-GRL. Chem Eng J 160:417–422
118. Yasman Y, Bulatov V, Gridin VV, Agur S, Galil N, Armon R, Schechter I (2004) A new sonoelectrochemical method for enhanced detoxification of hydrophilic chloroorganic pollutants in water. Ultrason Sonochem 11:365–372
119. Oturan MA, Sirés I, Oturan N, Pérocheau S, Laborde JL, Trévin S (2008) Sonoelectro-Fenton process: a novel hybrid technique for the destruction of organic pollutants in water. J Electroanal Chem 624:329–332
120. González-García J, Banks CE, Sljukic B, Compton RG (2007) Electrochemical synthesis of hydrogen peroxide assisted by ultrasound. Ultrason Sonochem 13:405–412

Chapter 12
Nature Is the Answer: Water and Wastewater Treatment by New Natural-Based Agents

Jesús Sánchez-Martín and Jesús Beltrán-Heredia

12.1 Introduction

It is well known that water is the most relevant and affecting factor in life development [1]. The immediate need of water in adequate conditions (quality and quantity) is a crucial factor for allowing a dignified life, and it is more than obvious that no economical development is possible without water [2]. Drinkable water scarcity is directly linked to high morbi-mortality levels and to low economical development; therefore, several vicious and virtuous circles are established from and to water situation (Fig. 12.1).

As can be seen, water is not only one of the first necessities in the Maslow's pyramid [3], but also the promoting agent in the economic and social development. Bad water makes this growth difficult (even impossible). These links are not theoretical; the real situation confirms the importance of water in the global scenario. Figure 12.2 shows three world maps that were modified according to a specific study variable.

The first two are very similar while they are almost opposed to the third one. Poverty is directly proportional to water scarcity (first and second world maps) and inversely to industrial development (the third map).

Consequently, water must be considered a valuable and scarce good, not only regarding environmental concerns, but above all bearing in mind the impact it has on the human wellness. If it is adequately managed, treated, and cared for, water can be an equity factor in the fight against poverty, since an appropriate and sustainable development model that allows dignified life conditions to humankind is needed.

J. Sánchez-Martín (✉) • J. Beltrán-Heredia
Department of Chemical Engineering and Physical Chemistry, University of Extremadura, Badajoz, Spain
e-mail: jsanmar@unex.es; jbelther@unex.es

S.K. Sharma and R. Sanghi (eds.), *Advances in Water Treatment and Pollution Prevention*, DOI 10.1007/978-94-007-4204-8_12,

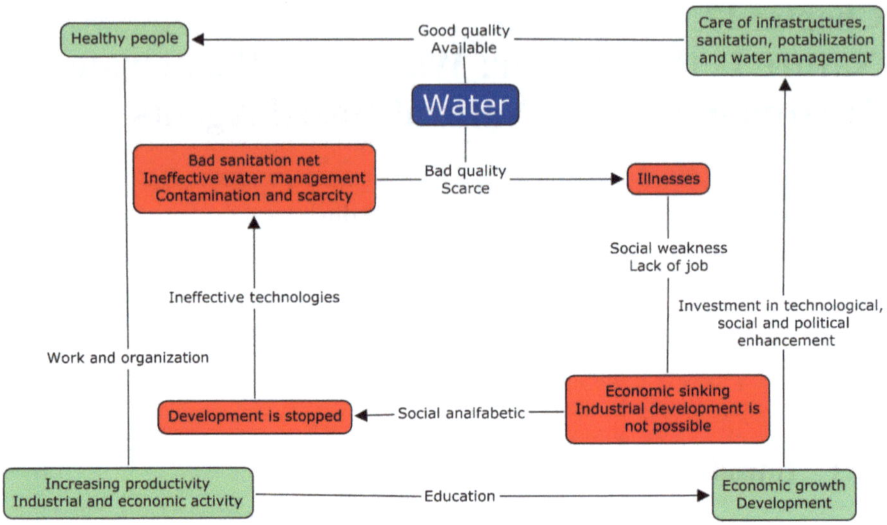

Fig. 12.1 Relationship between water, poverty and human development

In this sense, research is needed with the clear focus on people's development, according to the current developing model [4]; i.e., the promotion of people's chances in agreement with sustainability, governability, democracy, and participation criteria. This kind of investigation is supported by several international actors [5] and must also be implemented inside the university context.

The current chapter is within this general scenario. Natural resources are evidently widespread, cheap and easy-to-handle materials. Their possibilities are almost unlimited and many of them are still unknown. Our researching groups have been working on natural-based solutions to water pollution problems, so the technical solutions we are presenting are also easy to implement in developing areas.

This chapter will explore the possibilities of two main groups of natural resources in water and wastewater treatment: *Moringa oleifera* seed extract as a coagulant and tannin-derived products. The latter ones can be used under the form of cationic coagulants (for the removal of anionic colloidal pollutants) or anionic tannin rigid resins (*tannin gels*) for the removal of cationic contaminants.

12.2 Present Pollution Problem: Refractory Contaminants

The water scarcity is a global situation that reaches unacceptable levels: 40% of the total earth population lack proper sanitation infrastructures and problems with water quality are the fifth cause of death (above AIDS, tuberculosis, or malaria). Nowadays, more than one in six does not have access to safe freshwater [6]. This fact is even more acute if we take into account that the shortage of safe water has a concomitant effect with living without basic sanitation, which is the case

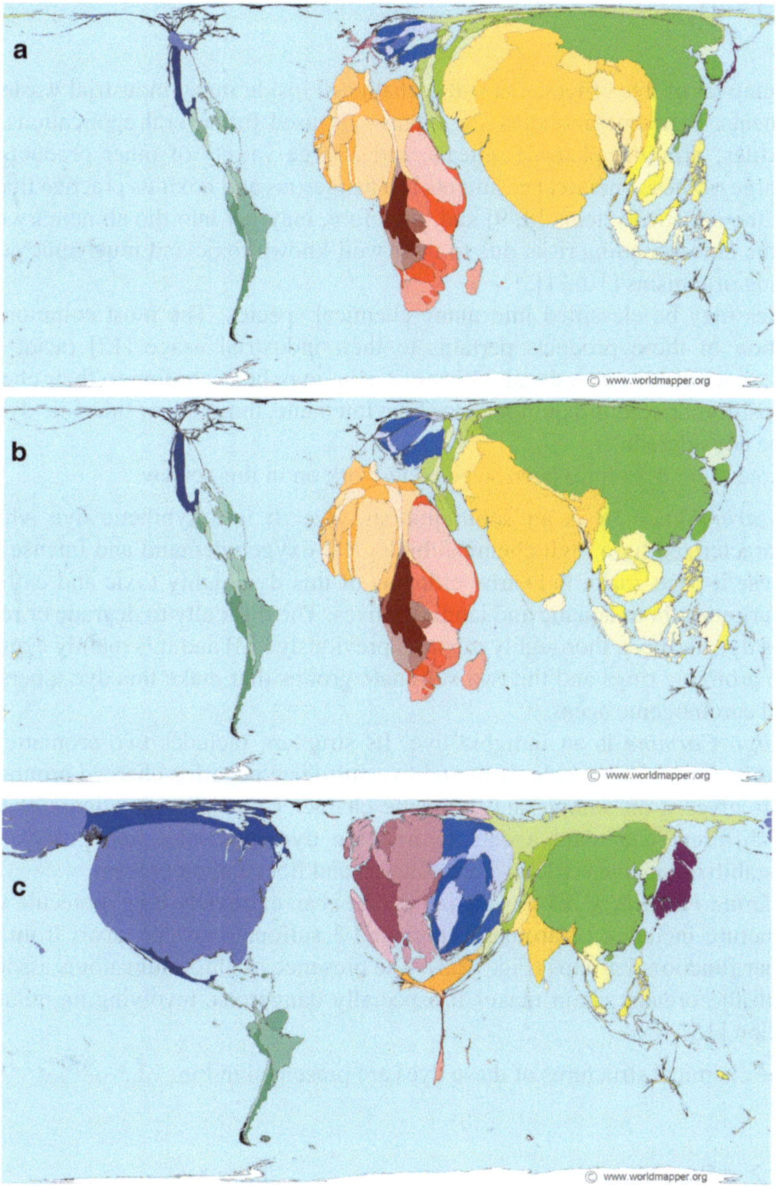

Fig. 12.2 Three world maps. The areas are proportionally deformed to (**a**) Human poverty index, (**b**) Safe water scarcity and (**c**) Industrial water consumption (Source: www.worldmapper.org)

of 2.5 billion people [7]. In addition, emerging economies like India or China are currently facing a pollution problem with special contaminants that are not easy to remove from water effluents. These are called *refractory contaminants* and the most important groups are the dyes, surfactants, and heavy metals.

12.2.1 Dyes

The removal of dye effluents is still a challenge inside many industrial wastewater treatments. Pigments and dyes are extensively used for several applications, such as textiles, printing, pharmaceuticals, and a large variety of other products. The discharge of these chemical pollutants is a dangerous and noxious practice that may affect the aquatic systems [8, 9] and, therefore, may get into the alimentary chain, with the corresponding risks due to their well-known toxic and mutagenic activity to living organisms [10, 11].

Dyes may be classified into many chemical species. The most common classification of these products pertains to their industrial usage [12] (acid, basic, direct, reactive, disperse dyes). Other criterion may be according to their chemical composition: azo, anthraquinonic, triphenylmethane, indigoid, or thiazinic dyes can also be considered.

Three main dyes have been selected to work on in this review.

- *Alizarin Violet 3R* is an anthraquinonic dye. It is a synthetic dye which is characterized by a high chemical/biological oxygen demand and intense color. These aspects make industrial effluents of this dye highly toxic and extremely injurious to both aquatic and landform lives. The difficulty to degrade or remove this dye has been thoroughly reported previously [13] and it is mainly caused by the aromatic rings and the two sulfonate groups that make this dye a persistent and carcinogenic agent.
- *Indigo Carmine* is an indigoid dye. Its structure includes two aromatic rings with a double link inside them and two sulfonate negative-charged groups [14]. This gives it an anionic and aromatic character that allows cationic character of different coagulant agents to link the dye molecules and provoke their destabilization and settle in a coagulation and flocculation process.
- *Palatine Fast Black WAN* is an azo dye. It is an extremely long molecule whose structure includes 12 aromatic rings and 3 sulfonate groups, apart from many other functional groups (Fig. 12.3). The presence of chromium atoms associated with the organic chain makes it especially dangerous, involving its mutagenic action [15].

The chemical structures of these dyes are presented in Fig. 12.3.

12.2.2 Surfactants

Surfactants have become a very important group of compounds in modern life. They are present in a large variety of usual and normal products, like soaps, detergents, pharmaceuticals, personal care products, etc.; they are also used in the chemical industry, "high-tech" devices, paints, and leather [16]. As can be appreciated, surfactants have achieved a main position in human activity in many fields. Analyzing the last statistical data, more than 12 M tons per year are used [17], so surfactants can be considered as a first important chemical group. Figure 12.4 presents some of the most important anionic surfactants.

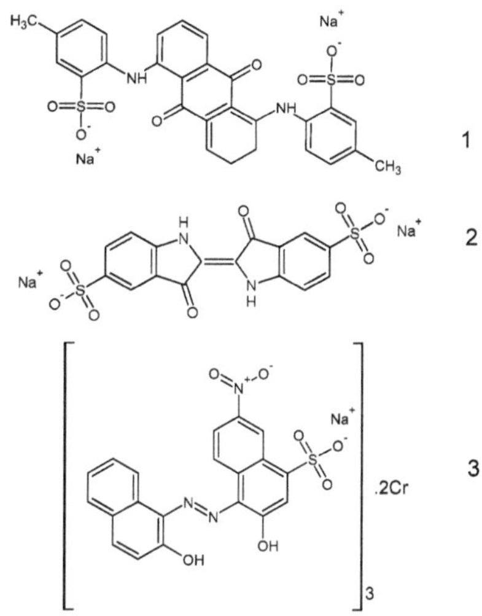

Fig. 12.3 Chemical structures of different dyes considered in this review. (*1*) *Alizarin Violet* 3R; (*2*) *Indigo Carmine*; (*3*) *Palatine Fast Black WAN*

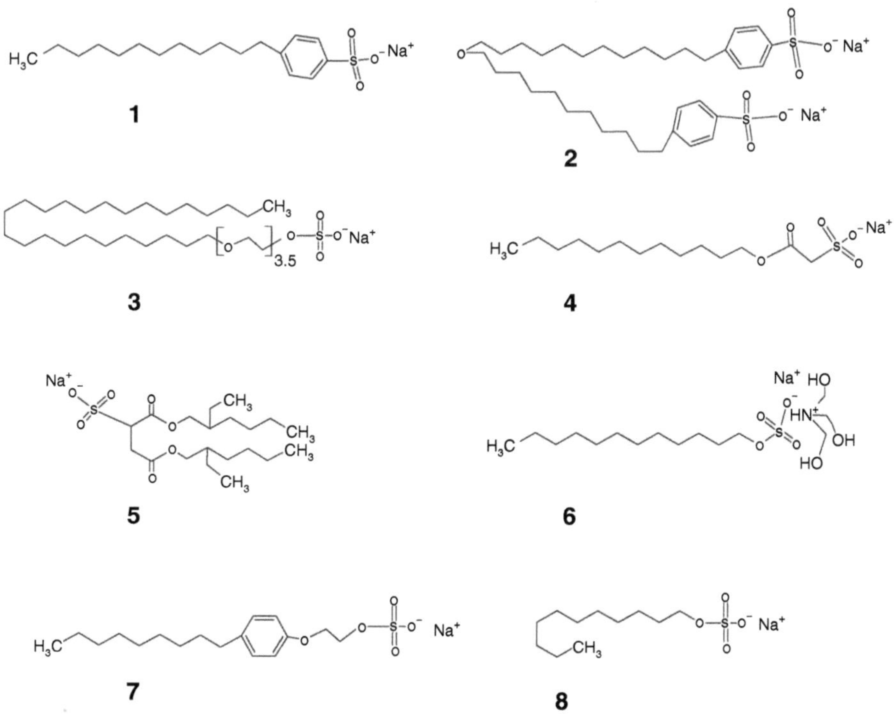

Fig. 12.4 Chemical structures of different anionic surfactants. (*1*) Sodium dodecyl benzene sulfonate (SDBS); (*2*) Sodium dodecyl diphenyl ether disulfonate (SDDED), (*3*) POE (3.5) Sodium lauryl ether sulfate (SLES); (*4*) Sodium lauryl sulfoacetate (SLSA); (*5*) Sodium dioctyl sulfosuccinate (SDSS); (*6*) Sodium triethanolamine lauryl sulfate (TEA-LS); (*7*) POE sodium nonylphenol sulfate (SNS) and (*8*) Sodium lauryl sulfate (SLS)

Surfactants dumping into the environment represent a harmful and noxious practice. They may be useful and needed compounds, but they are also considered dangerous and undesirable substances because of their impact on water animal and vegetal life. The main routes by which surfactants can alter environmental equilibrium are contamination of groundwater and lakes [18], association with pharmaceuticals (thereby, considerably enhancing their contaminant potential), biopersistence and toxicity for animals including humans, etc. [19, 20].

Effluents with high content in surfactants usually presents high COD levels (up to 50 $g\,L^{-1}$) [21], which is a cause of problems in normal water treatment plants, so new and cheaper methods for treating these kinds of wastewaters are still needed.

Nowadays, surfactants can be removed by several mechanisms; most of them imply adsorption on activated carbon [19, 22], chemical oxidation [23] or electrochemical removal [24]. However, new removal methods should be researched on because the impact of surfactants and tensioactives is quite high.

Due to these reasons, removing surfactants from water flows has become a priority for a large number of researchers. As it is known, there are several types of surfactants depending on its ionic character: anionic, cationic, amphoteric, nonionic, etc. Cationic surfactants enter aquatic ecosystems together with polluted waters because they are widely used in many industries, including petroleum, oil refining, petrochemical, and gas industries. They are also used in pest control in aquaculture for combating pathogenic organisms in fish [25].

12.2.3 Heavy Metals

Finally, the pollution with heavy metals is well known. Effluents from industrial facilities, mainly linked to minery, are highly charged with Cu^{2+}, Hg^{2+}, Pb^{2+}, Zn^{2+}, Ni^{2+}, or Cd^{2+}. These wastewaters represent cumulative pollution in the food chain [26], so the elimination of such harmful compounds has been the subject of research by scientists for a long time and is achieved by different ways: chemical precipitation [27] coagulation and flocculation [28], or nanofiltration [29], among others.

12.3 Natural Products: A Feasible Chance of Becoming Clean

One of the possible paths of minimizing the multiple concerns that have recently arisen on water and refractory pollutants is the implementation of new treatments based on natural products. Both adsorption and coagulation are well-known processes for eliminating these kinds of contaminants; yet, new agents must be developed.

Today there is a global challenge regarding water management: Water resources may be optimized in order to guarantee an adequate availability for the large majority of the people. This is the first and main motivation to continue research on these kinds of resources.

12.3.1 Appropriate Technology and the Needs of the Present World

Undoubtedly, the water shortage hits in an unequal way, depending on the developing situation of countries. Israel can produce shrimp inside the desert by using advanced techniques of water extraction from beneath the ground [30], while almost 40% of total Nicaraguan population has no home access to tap water [31]. That fact reveals that water subject is not linked to global technological advances, but to particular and adequate systems of solving specific situations [32].

Among the most influencing parameters in this situation is the fact that the pollution sources are widespread and often nonlocalized. Contaminants such as surfactants, dyes, or heavy metals get into the alimentary chain and appear almost everywhere, and usually there is not a specific procedure for their removal, mostly in places where water treatment is almost a luxuriant process. Because of this, it is necessary for the development of new materials and methods to fit into the large variety of economical situations, both in the First and the so-called Third World. Natural products are perhaps one of the most interesting options of making water treatment a universal chance. Becoming clean is then a cheaper and affordable possibility.

Technology is one of the mechanisms human beings can apply to fight against poverty [33], but as it is a cultural construct, it is not free from social implications [34]. The concept of *appropriate technology* (Fig. 12.5) arose in the mid-1970s and included some aspects that may be especially cared for in order to guarantee social and environmental feasibility of a technical proposal. The target is to develop a new technological paradigm suitable mostly in developing countries (but not only) that can be executed on the following principles [35]:

- Environmental-friendly (sustainability): It does not endanger subsequent generations' resources
- Cultural suitability: The solutions to given problems does not interfere with social manners or modals
- Technological transfer, not external dependency of mechanisms, apparatuses, or equipment

These statements are summarized in one: Technology must match both the user and the need in complexity and scale [34]. Relatively, new research concerns are recently open in order to provide enough first-line knowledge in this sense [36] for applying new discoveries to the Third World.

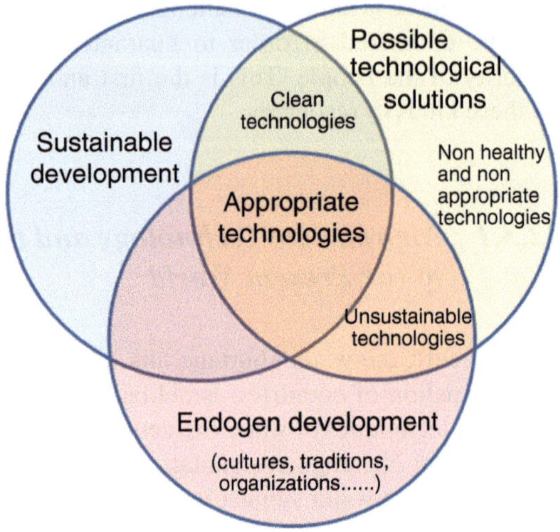

Fig. 12.5 Appropriate technologies inside the general scenario of possible technologies (Source: Own elaboration from [113])

Bearing in mind all the above, this chapter will deal with two main natural sources: *Moringa oleifera* and tannin derivatives. Both seem to present rather interesting properties in order to become important agent of water treatment in removing contaminants according to appropriate technology principles.

12.3.2 Moringa oleifera

Moringa oleifera is a tropical tree that comes from the sub-Himalayan valleys. It is a multipurpose tree and is generally known in the developing world as a vegetable, a medicine plant and a source of vegetal oil. Its multiple properties have been known for a long time [37] and several authors have referred to the importance of some *Moringa oleifera* aspects: human [38] and animal nutrition [39, 40], pharmacology [41, 42], cosmetics [43], etc.

The leaves and young seeds of *Moringa oleifera* are rich in calcium, iron, and vitamin C, which serve as nutritious source for communities. The fruits are called "pods" and roots of the tree are used as vegetables. The fruits range usually from 20 to 30 cm long. Each fruit contains more or less 20 seeds. Seeds are globular, weighing on an average between 0.3 and 0.4 g, 0.6–1 cm long, and 0.7–1.2 cm wide (Fig. 12.6).

Moringa oleifera is a widely spread, readily available, and easy to store product, especially in the developing countries. It can be a social-change factor, as it allows water treatment without external dependence [32].

According to Goh [44], the cultivation cost for producing 1 kg of *Moringa oleifera* is approximately US$2. Although the cost of *Moringa oleifera* seems more expensive than alum, it is more beneficiary to communities in terms of health and economy. Community could gain income from the sale of the seeds to companies or institutions processing them to produce coagulant or oil [45].

Fig. 12.6 Seeds of *Moringa oleifera*

Moringa oleifera has a big advantage considering its high added value, i.e., expensive oil is obtained from it [46], which can be used for many scopes. As Bhuptawat et al. [47] pointed out, the simple aqueous extract that can be obtained from the press cake residue remaining after oil extraction may be used for waste and drinking water treatment, being a real cheap and easy way of getting profit from this main usage of *Moringa oleifera*. The interest on *Moringa oleifera* in this aspect has been pointed by institutions such as Food and Agricultural Organization (FAO) [48].

This chapter is focused on the removal of surfactants by means of natural products. *Moringa oleifera* seed extract is one of these agents, and it acts as a coagulant. Particularly, this coagulant presents some characteristics that make it affordable and technologically friendly: It is obtained directly from the seeds with a little help of reasonably low dosages of sodium chloride. The complete extraction process involves the following steps. First, the seeds are reduced into powder by a domestic mill. A 1 M sodium chloride solution is prepared and 5 g of powder is added to 100 mL of it. The sodium chloride solution with powder is stirred for 30 min at room temperature (around 25°C). No pH modification is needed, as natural pH 7 is achieved. Then, the extract is filtered twice, once through commercial filter paper on Büchner funnel and again through a fine filtering Millipore system (0.45 μm glass fiber). The result is a clear, milk-like liquid. This raw material presents the physicochemical characteristics showed in Table 12.1.

12.3.3 *Tannin Derivatives*

The term *tannin* covers many families of chemical compounds. Traditionally, they have been used for tanning animal skins, hence their name, but one also finds several of them used as natural-derived coagulants and adsorbents. Their natural origin is as

Table 12.1 Chemical characteristics of *Moringa oleifera* seed extract

Parameter	Value
Dry residue (sodium chloride excluded)	3.24 g L^{-1}
Ammonium	0.06 g L^{-1} (N)
Nitrate	0.55 g L^{-1} (N)
Nitrite	0 g L^{-1} (N)
$KMnO_4$ oxidability	1.08 g L^{-1} (O_2)
Phosphate	0.05 g L^{-1} (P)
Total phosphorus	0.07 g L^{-1} (P)

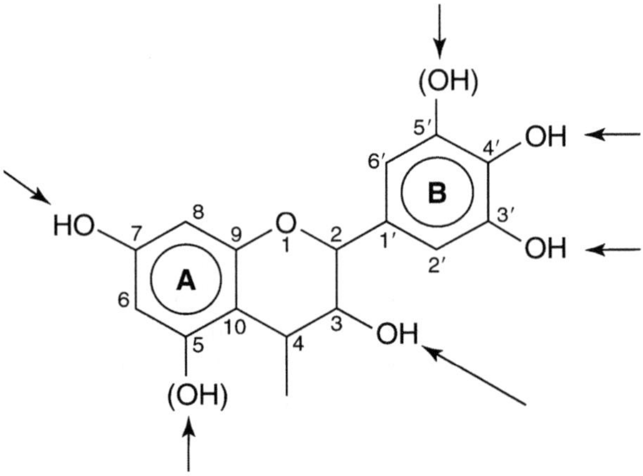

Fig. 12.7 Chemical structure of a standard tannin (Reprinted from Ref. [51]. With kind permission of © The American Chemical Society (2009))

secondary metabolites of plants [49], occurring in the bark, fruit, leaves, etc. While *Acacia* and *Schinopsis* bark constitute the principal source of tannins for the leather industry, the bark of other nontropical trees such as *Quercus ilex*, *Quercus suber*, *Quercus robur*, *Castanea*, and *Pinus* can also be tannin-rich.

Tannins are mostly water-soluble polyphenolic compounds of molecular weight ranging between 500 and some thousands of Daltons [50]. The chemical complexity of tannins and the fact that they are usually taken from natural matrix without a very thorough purification make it a very difficult task to understand their structure. Some approaches are found in literature review and one of them is shown in Fig. 12.7. A full study about tannins, its chemical structure, and properties can be found in Pizzi [52].

Tannin structure involves multiple aromatic rings that provide a useful matrix in which active centers can be introduced by means of an adequate polymerization process [52], either for coagulant or adsorbent production. Two classes of chemical compounds of mainly phenolic nature are included as vegetable tannins: condensed and hydrolyzable tannins [53]. Inside the first group, one can present tannins from *Acacia mearnsii* de Wild, *Schinopsis balansae*, or *Pinus pinaster*. Regarding the hydrolyzable tannins, this group includes barks from *Castanea sativa* or *Caesalpinia spinosa*.

Cationization of tannins is a chemical procedure that confers cationic character to the organic tannin matrix, so the main characteristics (such as solubility, stability at different pH levels or heavy metals quelating activity) are retained while others are added. These new abilities that appear have to do with the coagulating potential, as positive charged agents may destabilize anionic colloids if mixed in aqueous solution.

Destabilization and subsequent settlement provoke the removal of a wide variety of anionic substances such as dyes, surfactants, or organic matter. The chemical procedure for making tannins become cationic is known to follow a Mannich reaction path, and different variations have been reported under several patents [54–57]. Namely, tannins undergo Mannich aminomethylation by reaction with an aldehyde and an amine [58]. The resulting tannin Mannich polymer possesses a higher molecular weight due to formaldehyde and Mannich base cross-linking, and also possesses ampholytic character due to the presence of both cationic amines and anionic phenols on the polymer. Briefly, Mannich reaction is described as the introduction of a quaternary nitrogen into the tannin complex structure [59].

Two main ways of obtaining Mannich bases are reported in scientific literature—one involving the use of ammonium chloride and another involving other types of nitrogen compounds, such as mono- or diethanolamine. The reaction is completed in both cases by the controlled addition of formaldehyde in a lower proportion than in gelation. Higher doses of formaldehyde would lead to tanning gelation and, therefore, makes the product insoluble [60].

Although these kinds of coagulants are quite well known, very few examples of investigations have been found about them. Regarding the main scope of water treatment, according to these prescriptions, they found several previous papers that pointed out the use of tannins as a coagulant aid [61] or other cationic compounds [62]. We point out the use of tannin-based coagulants that has been researched by Graham et al. [63, 64] as well as in our own previous papers.

Commercial tannin-based coagulants are available mainly as a way of getting profit from tannery wastes. These are the situations for *Acacia mearnsii* de Wild (black wattle), which is the feedstock for TANFLOC and for ACQUAPOL, and *Schinopsis balansae* (quebracho colorado), which gives SILVAFLOC. Although the production processes of these coagulants are under patent law, following the rules of Mannich reactions we can surely make an approach of their synthesis, e.g., SILVAFLOC may be the reaction product of three reagents: monoethanolamine (MEA), formaldehyde, and the tannin extract mixture. First, MEA and formaldehyde may react in the way described in Eq. 12.1:

$$H_2N - CH_2 - CH_2OH + HCHO \rightarrow CH_2 = N - CH_2 - CH_2OH + H_2O \quad (12.1)$$

which yields the reactive species called *imine*. This may react with aqueous tannin extract mixture to form SILVAFLOC, whose chemical formula may respond to Eq. 12.2:

$$\text{tannin} - (CH_2 - NH - CH_2 - CH_2OH)_n \quad (12.2)$$

In the case of TANFLOC, it is said to be a vegetal water-extract tannin, mainly constituted of flavonoid structures with an average molecular weight of 1.7 kDa. It is presented as powder. More groups as hydrocolloids gums and other soluble salts are included in TANFLOC structure. The industrial production of TANFLOC involves the reaction between formalin, ammonium chloride and commercial hydrochloric acid. The mixture is stirred and heated and tannin extract is added. The reaction continues for several hours and then a viscous mixture with 40% of solids content is achieved. Allowing it to evaporate, TANFLOC in its powder form is produced [65].

On the other hand, tannin gelation is a chemical procedure that immobilizes tannins in an insoluble matrix [60] so that their properties of interest, e.g., metal chelation, are then available in an efficient adsorbent agent. In addition, the material resulting from their gelation (sometimes called *tannin rigid resin*) presents interesting properties in terms of resistance, nonflammability and mechanical undeformability [66, 67].

Gelation of tannins has been widely described in the scientific literature and in patents. The experimental conditions for gelation may involve the use of formaldehyde or other aldehyde in a basic or acid medium. One may find examples of basic gelation in the scientific literature [68–70] and in patents such as US patent 5,158,711 and acid gelation is described by other workers [71, 72].

The chemical basis of the tannin gelification are widely reported [52]. Formaldehyde and other aldehydes react with tannins to induce polymerization through methylene bridge linkages at reactive positions on the tannin flavonoid molecules. The reactive positions of the rings depend on the type of tannin; but mainly involve the upper terminal flavonoid units. For example, the A-rings (Fig. 12.7) of *Acacia mearnsii* and *Quebracho* tannins show reactivity toward formaldehyde compared to that of resorcinol.

However, aspects such as size and shape make the tannin molecules lose mobility or flexibility at relatively low level of condensation, so that the available reactive sites are too far apart for further methylene bridge formation. The result may be incomplete polymerization and, therefore, poor material properties. There is a need to search for constitutional differences between several tannin extracts in order to predict the gelation product. For example, among condensed tannins from mimosa bark (*Acacia mearnsii*), the main polyphenolic pattern is represented by flavonoid analogues based on resorcinol A-rings and pyrogallol B-rings. This is similar in the case of *Quebracho* bark extract, but no phloroglucinol A-ring pattern exists in the first or in the second type. The A-rings of pine tannin possess only the phloroglucinol type of structure, much more reactive toward formaldehyde than resorcinol-type counterpart. Elemental probable reaction mechanism in two steps between generic condensed tannin and formaldehyde is shown in Fig. 12.8 [73].

Therefore, it is very difficult to state the exact and real gelation mechanism for each type of tannin extract. Presumably, it depends on the real structure and composition of each one, which not only has to do with the tannin source (the vegetal bark), but also with the extraction process. In addition, researchers are not interested in working with pure tannin extracts but with natural wood materials, the chemical

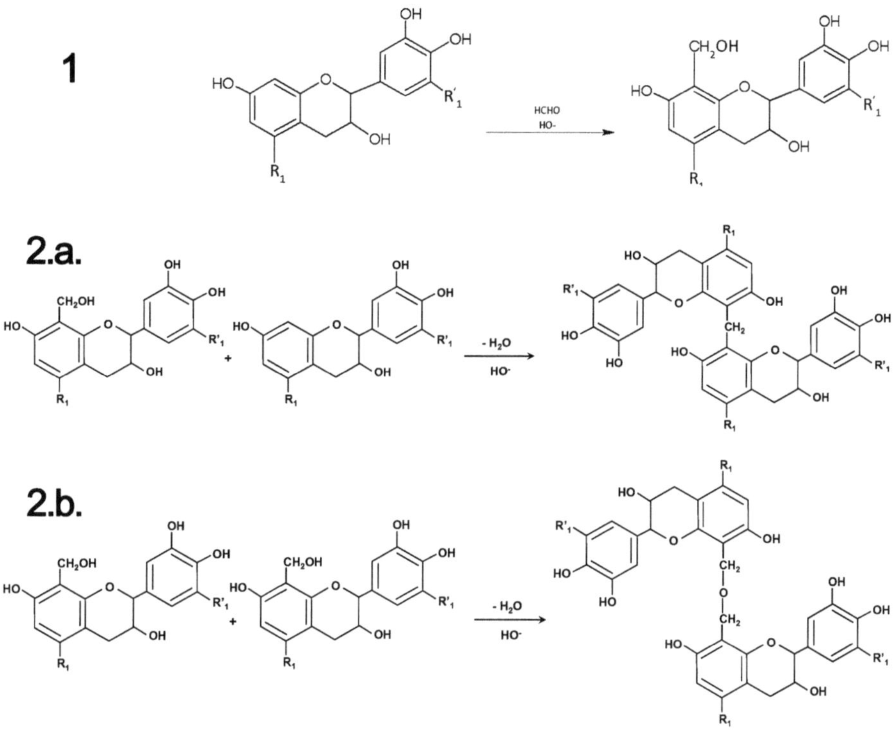

Fig. 12.8 Probable generic gelation mechanism between condensed tannin and formaldehyde. (*1*) First step: hydroxy-methylation; (*2a*) Second step: Methylene bridge immovilization; (*2b*) Methylene ether bridge immovilization (Reprinted from Ref. [73]. With kind permission of © Elsevier (2011))

characterization of gelified tannins is rather complex. However, researchers who have worked with these types of adsorbents performed several characterization procedures. This is the case of our own work [74], where a FTIR analysis was performed, or the work of Yurtsever and Sengil [75] where SEM was presented.

12.4 Natural Coagulants

When talking about natural coagulants, one must present on the one hand the seed extract from *Moringa oleifera*; and on the other hand, tannin-derived coagulants can be commercial ones (SILVAFLOC, ACQUAPOL or TANFLOC mainly) or lab-synthesized ones (from *Acacia mearnsii* de Wild and diethanolamine, ammonium chloride or glycidyltrimethyl ammonium chloride).

The first product is the seed extract from *Moringa oleifera*. It can be easily obtained by a well-known procedure. *Moringa oleifera* has been found to present a

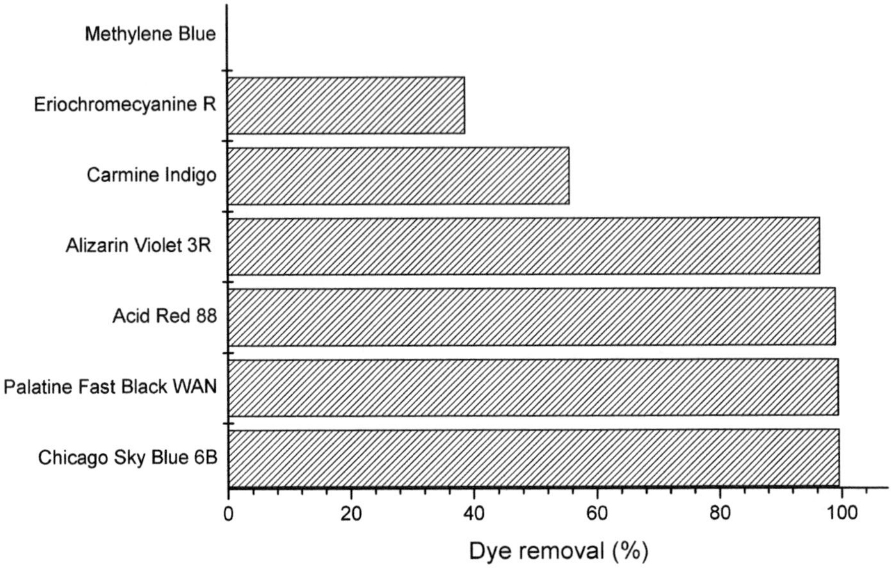

Fig. 12.9 Preliminary screening for dyes-*Moringa oleifera* seed extract interaction

protein with coagulating properties [76, 77]. The usual way of extracting it from the whole kernel has been thoroughly described for the first time by Ndabigengesere et al. [78] and has been later improved with different modifications. The most important of these changes are referred by Okuda et al. [79, 80] and it includes addition of salt (sodium or potassium chloride) to the extraction solution. Following this last method, *Moringa oleifera* seed extract is usually obtained in this way: A certain amount of dry shelled or nonshelled seeds are crushed in a domestic miller. Then, the resulting mass is added to a sodium chloride solution and the active protein principle is extracted for a given period with magnetic stirring. A clear, milk-like liquid is then separated by filtration.

This seed extract is usually carried out in the following conditions: 100 mL of distilled water, pH 7, 20°C, and 5 g of crushed shelled seed. In addition, a sodium chloride concentration of 0.1 M can enhance the extraction process.

This protein extract has been proved to remove dyes. *Moringa oleifera* seed extract has been found effective in removing at least four types of anionic dyes: triphenylmethane, anthraquinonic, indigoid, and azo dyes.

The ability of *Moringa oleifera* in removing all of these kinds of dyes has been thoroughly tested in our previous cited works. Figure 12.9 presents a preliminary screening with a fixed dose of *Moringa oleifera* seed extract (around 150 mg L^{-1}) and a fixed initial concentration of each dye (100 mg L^{-1}). As can be appreciated, *Moringa oleifera* presents a very high capability of removing azo dyes. For the three studied azo dyes (*Chicago Sky Blue 6B, Acid Red 88* and *Palatine Fast Black WAN*), the percentage of dye removal was higher than 98.8%.

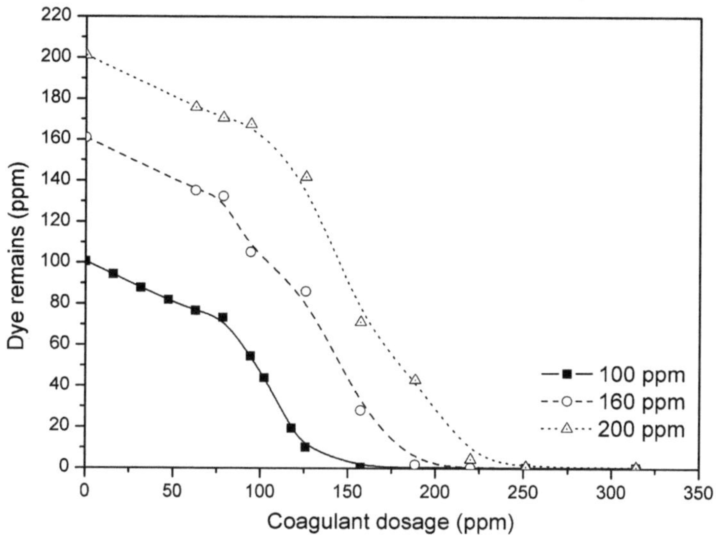

Fig. 12.10 *Chicago Sky Blue 6B* reduction by coagulation with *Moringa oleifera* (Reprinted from Ref. [81]. With kind permission of © John Wiley and Sons (2008))

Consequently, we have studied the removal of azo dyes thoroughly [81]. The interaction between *Chicago Sky Blue 6B* and the coagulation principle was evidenced and the suitability of *Moringa oleifera* for removing color from dye solutions was confirmed. Figure 12.10 shows the rapid depletion of dye concentration as coagulant dosage is increased.

Anthraquinonic dyes (such as *Alizarin Violet 3R*) also were demonstrated to be easily removed by *Moringa oleifera* [82], so were the indigoid dyes [83]. Figure 12.11 depicts the fast elimination of almost 100% of dye (initial dye concentration [IDC] equal to 100 mg L^{-1}) with an average coagulant dosage equal to 150 mg L^{-1}. Coagulant–dye interaction may be predicted by a theoretical model; in this case, by Langmuir hypothesis.

A significant difference in dye removal is shown in the case of *Eriochromecyanine R*. Although almost 40% of dye removal is achieved, this lack of effectiveness may be due to nonlinear structure that may cause steric difficulties, and the coagulation process may prefer linear molecules rather than other space distributions.

In the case of *Methylene Blue*, no removal was achieved by *Moringa oleifera* seed extract. Its cationic nature implies no cationic polyelectrolytes, including proteinic flocculants such as *Moringa*, which may cause its destabilization.

Kinetics of dye removal has been shown to be very fast. In the first 10 min, the coagulation and adsorption process are quite complete, so up to 95% of total dye removal is achieved in this first stage.

Moringa oleifera has been confirmed as a fully working coagulant agent even in a wide range of pH and IDC, although the different nature of several dyes can affect in the efficiency of the process [18]. Figure 12.12 reflects this fact within

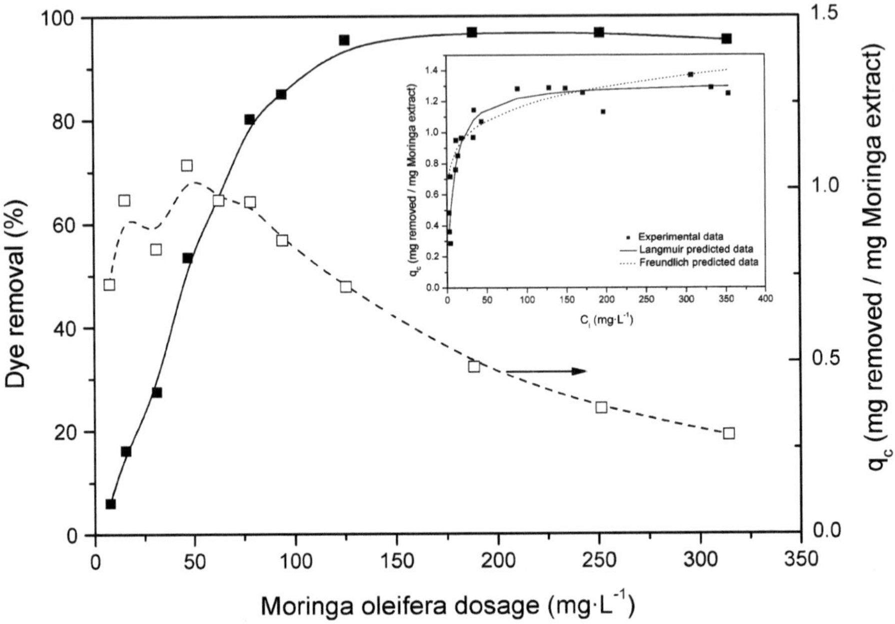

Fig. 12.11 *Alizarin Violet 3R* reduction by coagulation with *Moringa oleifera*. On the *upper corner*: Theoretical modelization of coagulant extract-dye interaction

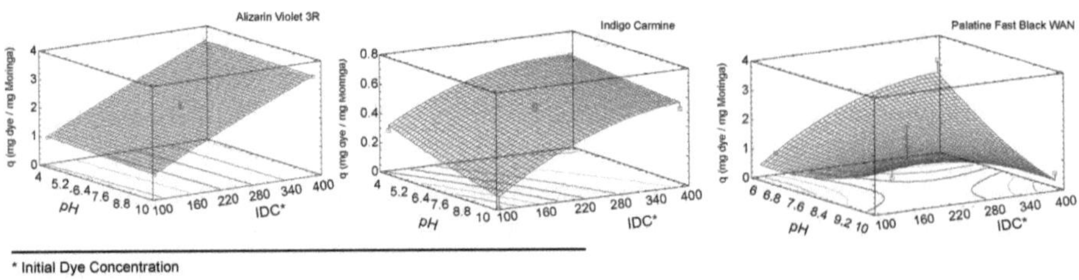

Fig. 12.12 *Moringa oleifera* seed extract coagulation in removing three types of dyes. Analysis through response surface methodology (Reprinted from Ref. [81]. With kind permission of © John Wiley and Sons (2009))

an analysis with response surface methodology (RSM). Coagulation process can be appreciated taking into account the shape of the surface: As the curve becomes convex the interaction *Moringa*-dye is more affected by pH and IDC and, therefore, the coagulation process is less efficient.

Extensive studies on the removal of *Alizarin Violet 3R* [82] and on *Chicago Sky Blue 6B* [81] have concluded that temperature is not a significant variable in the treatment efficiency. This is an advantage in order to implement a scale-up system, as the flexibility in working ranges allows in treating several kind of industrial effluents.

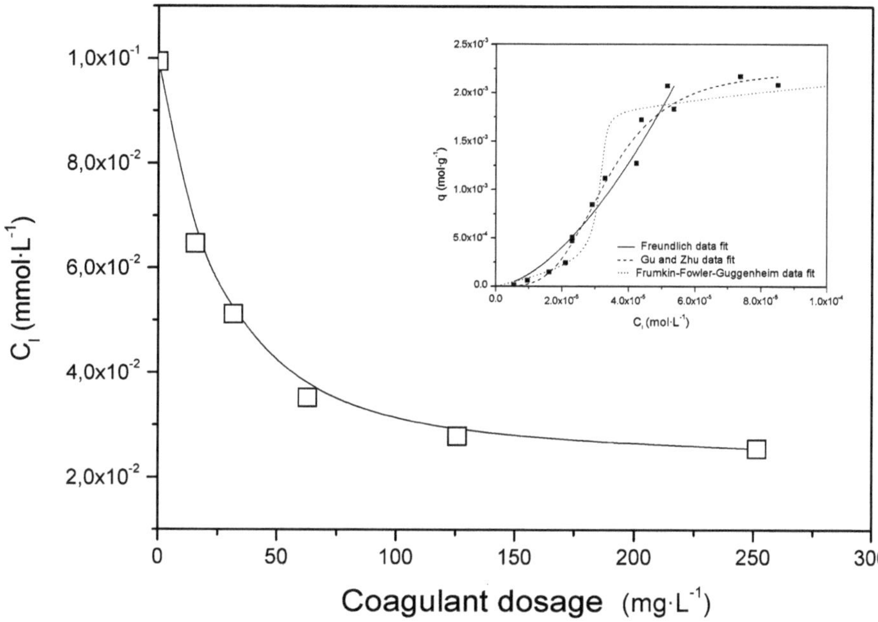

Fig. 12.13 Sodium lauryl sulfate removal by coagulation with *Moringa oleifera* seed extract. On the *upper corner*: Theoretical modelization of coagulant extract-surfactant interaction

Apart from dyes, *Moringa oleifera* seed extract fully works in the removal of several anionic surfactants. It has been a long time since the interaction among natural polymers and surfactants is well-known [84]. The interest of testing *Moringa oleifera* in removing surfactants from water lies on the fact that they are dangerous and are widely spread pollutants [18]. Because of the cationic character of *Moringa oleifera* seed extract, it is expected to be effective against anionic surfactants. Our study [85] evidenced that the coagulation and flocculation processes fully work for the removal of sodium lauryl sulfate (Fig. 12.13) and this interaction *Moringa*-surfactant obeys the adsorption-like phenomenon rule postulated by the Gu and Zhu theoretical model [86]. Other models, such as Freundlich or Frunkim-Fowler-Guggenheim can also be used to predict this behavior. Some recent studies have shown that *Moringa oleifera* actually interacts with anionic surfactants and also with cationic ones, although in a slight manner.

The pH level does not have a very acute influence on surfactant removal. *Moringa oleifera* presents a high efficiency in removing sodium lauryl sulfate from pH 5 to pH 9. This fact may be due to the anionic character of the surfactant that should not be dramatically reduced by lowering the pH; whereas the proteinic nature of the vegetal extract makes this agent to be cationic. Therefore, its cationic form would be higher at acidic pH. Electrostatic attraction between *Moringa oleifera* cationic proteins and negative-charge surfactant active centers is reinforced. Both effects should explain this behavior.

Temperature does not present a significant influence on surfactant elimination. Experiments at 10°C, 20°C, 30°C and 40°C revealed removal efficiencies in the range of 67–69%. In a general way, this variable does not seem to be significantly important if it is fixed at room temperature or higher, as the only different data in surfactant appear at 10°C.

Finally, by increasing the initial surfactant concentration, the removal efficiency tends to be higher up to a maximum, then it appears to decrease (around 10 mg L^{-1}). It may be due to the fact that we are reaching the critical micellar concentration (so properties may change dramatically) and to the denaturation of proteins in the presence of surfactants [87].

Models can be applied to surfactant coagulation and adsorption. According to the theoretical modelization for this particular system (surfactant-*Moringa*), q_{max} appears around 2.1×10^{-3} mol g^{-1} (500 mg g^{-1}, approximately), which is a very high removal capacity.

The second type of natural products is tannin-based coagulants and adsorbents. Tannin derivatives can be commercial or noncommercial. Regarding the first group, as said before, the more representative are TANFLOC, SILVAFLOC, and ACQUAPOL. They differ from each other either in the tannin feedstock source (*Acacia* or *Quebracho*) or in the polymerization agent (*ammonium chloride* or *monoethanolamine*).

Commercial tannin coagulants show a very interesting ability in removing heavy metals (not presented in the case of *Moringa oleifera*). The chelating activity of tannins is supposed to enhance dissolved heavy metals removal from water [88].

We have researched on the ability of TANFLOC in removing heavy metals from polluted wastewater [89] on two basic hypothesis: the effective chelating activity and the adsorption process between metal ions and suspended matter (such as humic or fulvic matters) that may be measured as turbidity. The result is that heavy metals such as Zn^{2+}, Ni^{2+}, or Cu^{2+} are easily removed by dosages between 50 and 150 mg L^{-1}.

The pH dependency of the metal ions in aqueous solution makes this removing process to be very pH-sensitive. As pH is an important factor for metal hydroxide precipitation, it is needed to adjust the pH in order to avoid this effect and evaluate metal removal caused only by coagulation and flocculation process. An orientative pH working value for each metal is given by the solubility-product constant values at each pH, which shows an idea of what quantity of metal is dissolved and what quantity is under solid form. Solubility-product constants of Ni^{2+}, Zn^{2+} and Cu^{2+} hydroxides are well known and their values are 5.54×10^{-16}, 1.2×10^{-17}, and 1.6×10^{-19}, respectively [90]. It should be difficult to calculate the solubility of these species accurately, as the working water in the referred article is not distilled, but river water (so it has a lot of impurities that may interfere in the equilibrium process), and these values give an idea of how Cu^{2+} will be insoluble at lower pH than Ni^{2+} and Zn^{2+}.

In fact, pH precipitation is a common method for metal removal. Contaminated water is pH-adjusted till a basic value (pH 10–11) is attained. At this pH, metal ions become into a hydroxide form according to Eq. 12.3.

$$M^{2+} + 2(OH)^- \rightarrow M(OH)_2 \downarrow \quad (12.3)$$

As pH is supposed to be favorable to metal removal, it is very important to adjust its value and evaluate the amount of metal that is removed by TANFLOC effect and the metal ion loss due to natural precipitation. In this paper, natural removal by hydroxide precipitation is excluded from metal removal results, as the initial metal concentration was considered after the first Jar-test assay, in which no flocculant was added.

Taking this fact into account, optimum pH levels were found for each metal and equilibrium metal concentration was reached after standard Jar test. These pH values were 7, 8, and 9 for Cu^{2+}, Zn^{2+}, and Ni^{2+}, respectively. Figure 12.14 shows the main insights this investigation reached. Up to 85% of metal removal (from an initial metal concentration equal to 20 mg L^{-1}) was obtained in every case.

The interaction with anionic dyes was also fully proved. Bearing in mind the increasing importance of dyes in industries all over the world [91], it is easy to understand that these compounds mean a very significative pollutant, as it is known over 50,000 tons of dyes are discharged into environmental effluent annually, with the severe consequences to fauna and flora equilibrium [92].

Coagulation has been one of the most popular ways of removing dyes from aqueous systems because of its simplicity and high efficiency [93]. The utilization of these new tannin-based coagulants in removing anionic dyes has been found in previous scientific literature in some works where they were used in a preliminary screening for dye removal [81, 82], but there are at least two papers where they are the main agents: Graham et al. [64] and Beltrán-Heredia et al. [83]. The first of them works on already known TANFLOC and its use in colored humic water (which is not properly dye removal) and the second one develops the study on a similar commercial tannin coagulant called ACQUAPOL-C1, which is also derived from *Acacia mearnsii* de Wild tannin.

ACQUAPOL-C1 has demonstrated to be a very effective coagulant agent in removing at least two kinds of anionic dyes: anthraquinonic and azo dyes. Its ability in removing triphenylmethane (such as *Eriochromecyanine R*) or indigoid dyes (such as *Indigo Carmine*) is much lower (Fig. 12.15). In the case of ACQUAPOL-C1 with *Alizarin Violet 3R* (anthraquinonic dye), the efficiency of the coagulant is quite high [82]. An average q_{max} was found around 0.5 mg mg^{-1}, which is a very interesting value, and the process seems to follow a Langmuir adsorption phenomenon (Fig. 12.16).

The feasibility of removing surfactants with these kinds of commercial tannin-based coagulants was also confirmed. As said before, surfactants are a very important chemical group in modern life. Due to their biopersistance and implications on environmental equilibrium [18], their removal from aqueous systems has become an important task for wastewater treatment. Tannin-derived coagulants have been used with this scope in recent publications.

The ability of TANFLOC and SILVAFLOC in removing anionic surfactants has been thoroughly tested by Beltrán-Heredia et al. [51, 94].

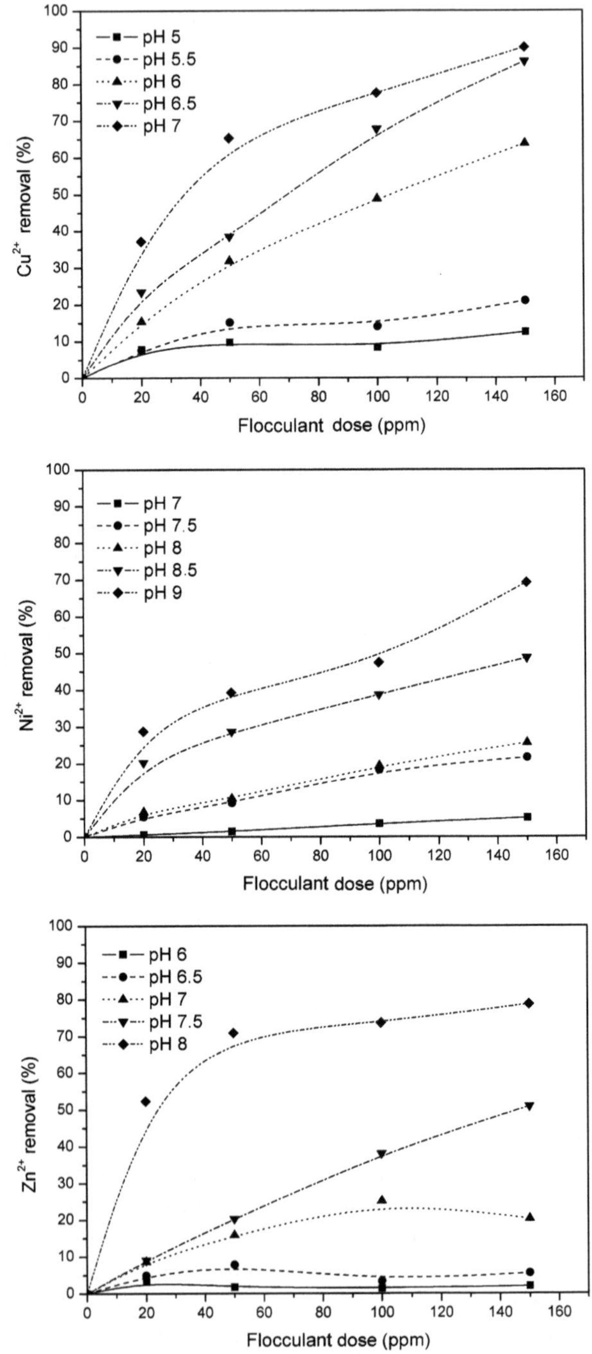

Fig. 12.14 Heavy metal removal by coagulation with TANFLOC (Reprinted from Ref. [89]. With kind permission of © Elsevier (2009))

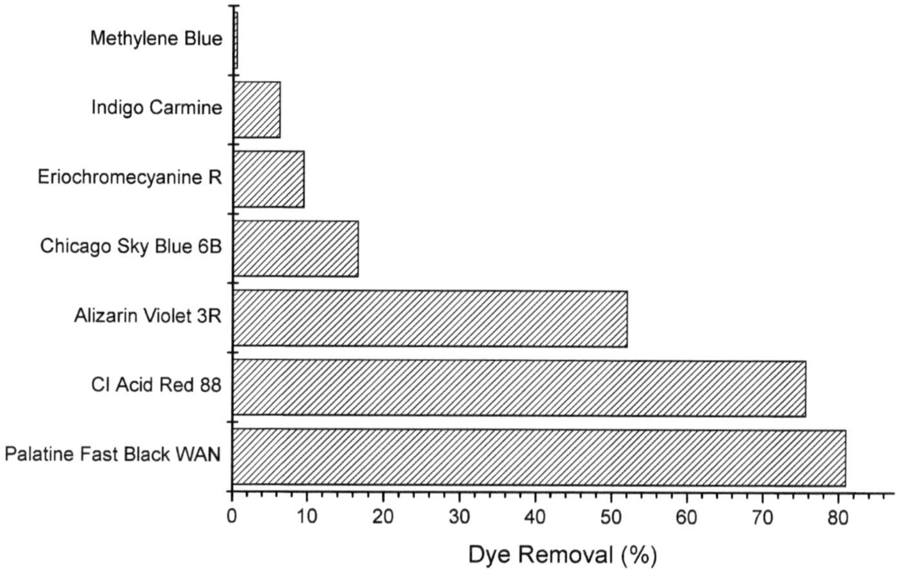

Fig. 12.15 Removal of different dyes by ACQUAPOL-C1. Initial dye concentration: 100 mg L^{-1}. Coagulant dosage: 100 mg L^{-1} (Reprinted from Ref. [82]. With kind permission of © Elsevier (2009))

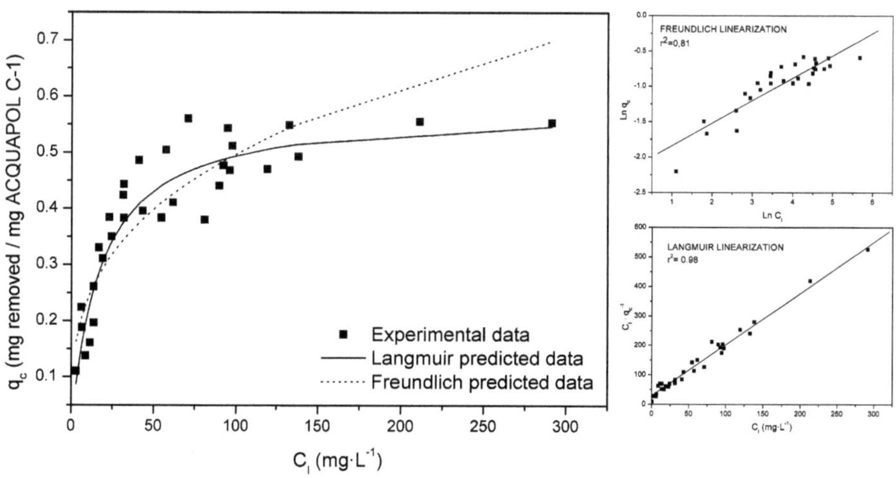

Fig. 12.16 Equilibrium data and models adjustment for *Alizarin Violet 3R* removal by ACQUAPOL-C1 (Reprinted from Ref. [82]. With kind permission of © Elsevier (2009))

Figure 12.17 depicts the efficient decrease in sodium dodecyl benzene sulfonate (SDBS, anionic surfactant) concentration when treated with these coagulants. As can be appreciated, in both cases, near 90% of initial surfactant concentration (50 mg L^{-1}) is achieved with a reasonably coagulant dosage.

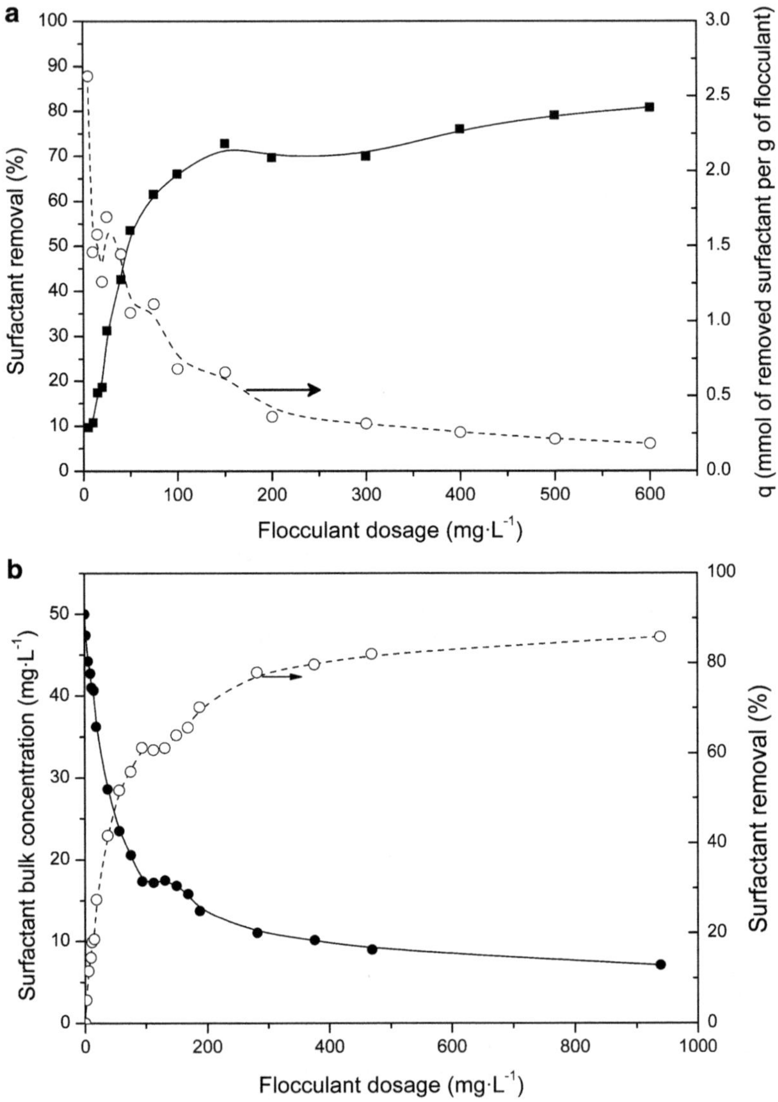

Fig. 12.17 Sodium Dodecyl Benzene Sulfonate (SDBS, anionic surfactant) removal by (**a**) TANFLOC and (**b**) SILVAFLOC

The pH seems to be a rather important variable in surfactant removal by these tannin-based coagulants [51, 94]. In the same way, the initial surfactant concentration also refers as a reliable influence in the efficiency of the process. Bearing in mind these two variables, Beltrán-Heredia et al. [95] developed a multistep analysis by means of a Design of Experiments in order to obtain the maximum efficiency by varying both these parameters. The result was analyzed

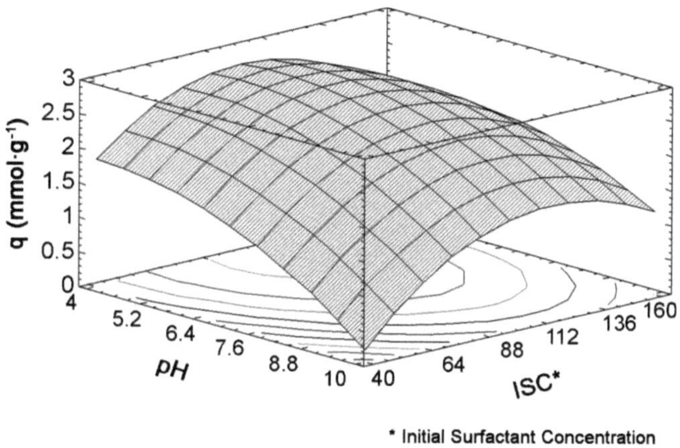

Fig. 12.18 Response surface of a design of experiments for obtaining the optimum q in SDBS removal by TANFLOC (Reprinted from Ref. [51]. With kind permission of © The American Chemical Society (2009d))

according to RSM and this optimum was estimated to be 0.96 mg mg^{-1} for q in a pH level of 4.9 and initial surfactant concentration equal to 103.2 mg L^{-1}. This is shown in Fig. 12.18.

Regarding the special surface characteristics of anionic surfactants [96] and their adsorption affinity onto polymers [97], removal phenomenon has been treated as coagulation followed by adsorption, so the theoretical basis of this last operation can be applied in order to explain how the surfactant is removed from the aqueous matrix. In the case of SDBS and TANFLOC, three theoretical models have been tested–Freundlich, Frumkin-Fowler-Guggenheim and Gu-Zhu.

As can be appreciated in Fig. 12.19, a very well data fit is obtained with the three models. Beltrán-Heredia et al. [94] carried out the linear and nonlinear data adjustments and concluded that the best model (regarding r^2 values) was the Gu-Zhu hypothesis, which is properly defined for surfactant–polymer interactions [86].

Lastly, several attempts were carried out for optimizing the synthesis of tannin-based coagulants. Some of them were performed following a categorical optimization process, others were focused following an orthogonal rotatable cubic centered design (CCD).

The reagents involved in the cationization process were:

- Tannin extracts from *Schinopsis balansae* and from *Acacia mearnsii* de Wild. Commercial trademarks are Quebracho ATO (Q) for the first one and *Clarotan* (C) and Weibull black (W) for the second one. The three tannin extracts were supplied by TANAC Inc. (Brazil).
- Reagents involved in the coagulant synthesis are *Ammonium Chloride* (Cl), *Diethanolamine* (D), *Glycidyltrimethylammonium Chloride* (G) and *Formaldehyde* (F). The four products were supplied by SIGMA in commercial purity grade.

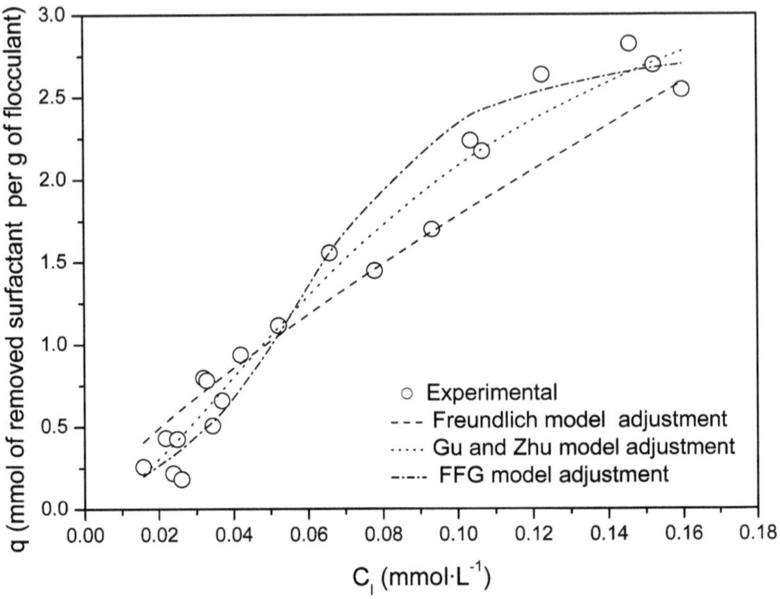

Fig. 12.19 Equilibrium data and models adjustment for SDBS removal by TANFLOC (Reprinted from Ref. [51]. With kind permission of © The American Chemical Society (2009d))

Table 12.2 Production processes of different tannin coagulants

Amine	Tannin (g)	Amine (g)	Formaldehyde (g)	Reference
Ammonium chloride (Cl)	2.5	2.5	1.5	[65]
Diethanolamine (D)	2.5	10.8	1.5	[98]
Glycidyltrimethylammonium chloride (G)	2.5	2.5	0.06	[99]

Table 12.2 presents the specific amounts of each reagent. The cationization processes were conducted as follows: certain fixed amount of tannin extract was diluted in distilled water at room temperature. Then the sample was thermostated at the reaction temperature (30°C). Then, certain amount of amine was added to the mix: diethanolamine, ammonium chloride or glycidyltrimethylammonium chloride. Finally, always under thermal control, formaldehyde was added to the reaction mixture. A peristaltic pump was used in this step, so that it lasted for 90 min at least.

The product so obtained was kept under agitation and at the same temperature for 24 h. The final product was put in a 50 mL-flask and filled up to the mark with distilled water.

Figure 12.20 shows the interaction graphics of the involved variables according to the factorial model design. The trials with each nitrogenant agents are represented by the different lines, tannin extracts are placed along the X-axis. *Palatine Fast Black WAN* was the model compound for simulating textile wastewater. Tests were

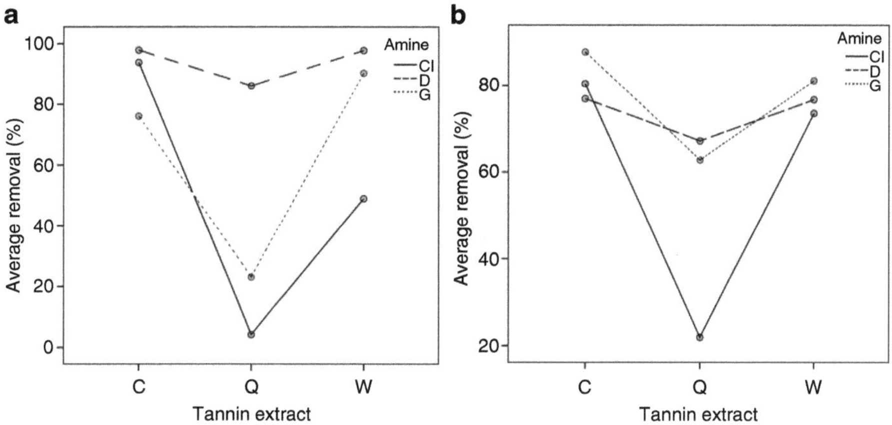

Fig. 12.20 Interaction graphics in qualitative optimizing. (**a**) Simulated textile wastewater with *Palatine Fast Black WAN*; (**b**) Simulated laundry wastewater with SDBS

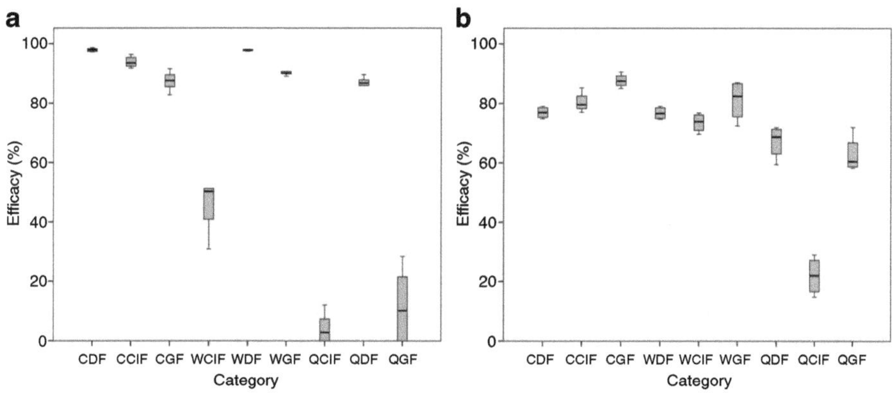

Fig. 12.21 Box plot for the nine products in coagulation of two types of water. (**a**) Simulated textile wastewater with *Palatine Fast Black WAN*; (**b**) Simulated laundry wastewater with SDBS

carried out with an IDC of 100 mg L^{-1} or SDBS initial concentration of 50 mg L^{-1}. Coagulant dosage was 500 mg L^{-1} in the case of textile and laundry simulated wastewater.

For wastewater samples, the interaction is clear, not only from the graphical analyses, but also because of the ANOVA report, which gives a *p*-value for interaction lower than 0.05 in each case. The complexity of the different pollutants should interfere in the destabilization of each one [100] and the model is not so simple in this case.

Because of that fact, the nine combinations can be compared with every objective variable–dye removal and surfactant elimination. A box-and-whisker plot, which shows the results of the tests, is presented in Fig. 12.21. This is a multicategorical analysis in which each product is tested and compared to the rest of them.

For evaluating the system coagulant-SDBS or coagulant-*Palatine Fast Black WAN*, we have to study ANOVA test, which gives us significativity data about these results. In a first approach, we can see that there is a set of coagulants that work significantly better than the rest of them; they are those derived from *Acacia mearnsii* de Wild, *Clarotan* and *Weibull black*. Combinations with DEA are especially effective, either in terms of pollutant removal or low dispersion. This tendency is presented more clearly in the removal of *Palatine Fast Black WAN*, while a wider dispersion of results is presented in the case of SDBS.

Tukey's test for multiple comparison did not give significative differences above 75% in both dye and surfactant removal. Three subsets are arranged in the case of dye removal, while up to five were established regarding SDBS removal. This has to do with the increase in dispersion of the last data presented, so more groups can be identified. It is also important to point out the fact that no overlapping is observed among the subsets that are related to dye removal, so clear differences can be appreciated in this case. This again may be explained by the more affinity the dye presents towards the coagulant.

On the contrary, the five subsets that are obtained for SDBS elimination presented several cases that are not distinguishable and could be inserted in various groups.

Quebracho-derived coagulants seem to be less effective, according not only to the graphical representation, but also to Tukey's multiple comparisons (data not shown). Probably the differences between coagulants may be found in the fact that tannin extracts are not exactly equal in their chemical composition; so the cationization and polymerization are affected not only by the chemical nitrogenant agent, but also by the specific tannin structure [101, 102].

In view of these results, an intermediate solution was selected. *Clarotan* was chosen as tannin source for a standard cationization. So-called CDF (*Clarotan* with DEA and formaldehyde) coagulant was used in dye removal and surfactant elimination.

The way this new coagulant forms coagules and flocs suggests it follows a bridging procedure. This is caused by a flocculent clarification of the surface water and it is characteristically slow, without the typical sedimentation zones [100, 103].

Regarding the coagulant dosage on dye removal (Fig. 12.22), it is shown that dye concentration undergoes a rapid and dramatic decrease from the first coagulant dosage [104]. The efficacy of the coagulant is very high, therefore, 100 mg L^{-1} can reduce up to 80% the initial dye concentration. However, a flocculent sedimentation is presented regarding the value of sludge production (ca. 200 ml L^{-1} in Imhoff's cone test). This yields to a Sludge Volume Index (SVI) equal to 317 ml L^{-1}. This relatively high value recommends the usage of a flocculant agent, beyond the coagulant effect of the tannin-derived product [105, 106].

Other experimental series were done in order to determine the influence of variables on the removal of SDBS. A fixed dose of 0.10 mmol L^{-1} of surfactant was evaluated to be removed with different doses of coagulant. As is shown in the same figure, Fig. 12.22, final surfactant concentration tends to decrease as CDF dose increases. However, it is observed that the efficiency of the process arrives

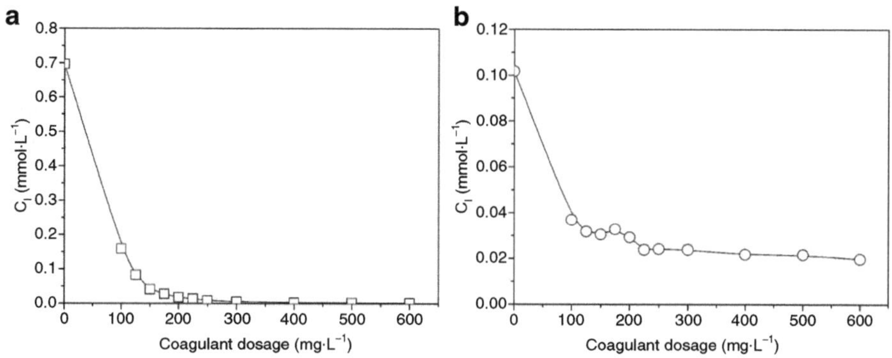

Fig. 12.22 Wastewater treatment with optimum CDF coagulant. (**a**) Simulated textile wastewater with *Palatine Fast Black WAN*; (**b**) Simulated laundry wastewater with SDBS (Reprinted from Ref. [104]. With kind permission of Elsevier (2010))

at a maximum, and a higher dose of coagulant does not achieve lower surfactant concentrations. There is a residual surfactant concentration that is no possible to remove through this flocculation process, and it seems to be about 0.02 mmol L^{-1}. This can be due to the existence of an "equilibrium surfactant concentration" which is highly difficult to remove, as reported previously [107].

12.5 Natural Adsorbents

Tannin gelation is a feasible way of obtaining natural adsorbents that can easily removed these kinds of pollutants. For this purpose, a complete study on types of tannin gels has been carried out by Sánchez-Martín et al. [108].

Commercially available tannin extracts, such as *Acacia mearnsii* de Wild (Weibull black), *Schinopsis balansae* (Quebracho colorado), *Castanea sativa* (Chestnut), and *Caesalpinia spinosa* (Tara) were kindly supplied by TANAC (Brazil). They were products involved in the leather treatment and they are presented as powder.

Lab-extracted tannins such as *Pinus pinaster* (Pine) and *Cupressus sempervivens* (Cypress) are extracted according to the following procedure [109]. A quantity of 100 g of bark was milled in a cutting mill and placed in 600 mL of tap water. Then 5 g of sodium hydroxide (PANREAC) was added and the mixture was stirred in magnetic stirrer at 90°C for 1 h. Solids were separated by filtration and liquid fraction was dried in an oven (65°C) overnight. The resultant was considered the tannin extract.

Tannin rigid resins are useful for the removal of cationic pollutants, not anionic ones. For this reason, different pollutants were used in the current investigation. *Methylene blue* (MB) was used as dye model compound, although others were included in preliminary screening. It was provided by Sigma-Aldrich.

Regarding heavy metal, Zn^{2+} was selected as model compound and it was supplied by Panreac. Finally, Cetyltrimethylammonium bromide (CTAB) was the model compound for cationic surfactants. It was supplied by Sigma-Aldrich.

Tannin gels were prepared according to the basis of Nakano [69]. Five grams of tannin extract was dissolved in 32 mL of sodium hydroxide (PANREAC) 0.125 mol L^{-1} and 30 mL of distilled water at 80°C. When the mixture was homogeneous, 2 mL of formaldehyde (commercial purity grade) was added and the reaction was kept at the same temperature for 8 h until polymerization was considered as being completed. Then, the apparent gummy product was led to complete evaporation of remaining water and dried in an oven (65°C) overnight.

After gelation, tannin rigid foams were crushed and sieved to produce 38–53 μm diameter particles. They were washed successively with distilled water and nitric acid 0.01 mol L^{-1} (PANREAC) to remove unreacted sodium hydroxide. Finally, the adsorbent was dried again in the oven. Differences are found between this preparation and the description made by Yurtsever and Sengil [75], mainly concerning the amount of formaldehyde.

Chemical polymerization of tannin extracts by means of formaldehyde is a well-known process for producing *tannin gels*. Our previous work on adsorbents derived from *Schinopsis balansae* (red quebracho) presented some interesting data about the removal of MB from aqueous solutions. So the obtained adsorbent is called QTG.

As a first approach, the tannin gel was synthesized strictly according to referred Nakano et al. [69] even with the formaldehyde proportions (2 mL each 5 g of tannin extract). The first physical characterization of this preliminary product was made on the basis of FTIR. The spectra of both samples are shown in Fig. 12.23. Wide bands in the range of 3,600–3,100 cm^{-1} correspond to –OH bridging groups in all systems and are attributed to O–H stretching (phenolic or alcoholic group) and to water molecules hydrogen bonded with –OH groups. The small peaks in the region of 2,950–2,850 cm^{-1} are associated with the methylene ($-CH_2-$) bridges. Also, stretching vibrations of C–H groups in the aromatic rings give absorption bands in this region. The absorption bands between 1,620 and 1,450 cm^{-1} are characteristic of the elongation of the aromatic –C=C– bonds. The deformation vibration of the C–C bonds in the phenolic group absorbs in the region of 1,500–1,400 cm^{-1}. The peak at 1,390–1,370 cm^{-1} is associated with the O–H deformation vibration of phenolic or alcoholic group. The peaks in the region 1,280–1,210 cm^{-1} are associated with the –CO stretchings of the aromatic ring and the methylene ether bridges formed by reaction with formaldehyde. The peaks at 1,160–975 cm^{-1} are due to asymmetrical C–O–C stretching and C–H deformation. The deformation vibrations of the C–H bond in the aromatic rings give absorption bands in the range of 835–650 cm^{-1}.

For confirming the validity of equilibrium periods (up to 15 days), a kinetic study was carried out with *Methylene Blue* (MB). A series of trials was performed with a fixed initial dye concentration (IDC) (100 mg L^{-1}) and with different proportions of QTG and MB (mmol g^{-1} of adsorbent). Figure 12.24 reports the

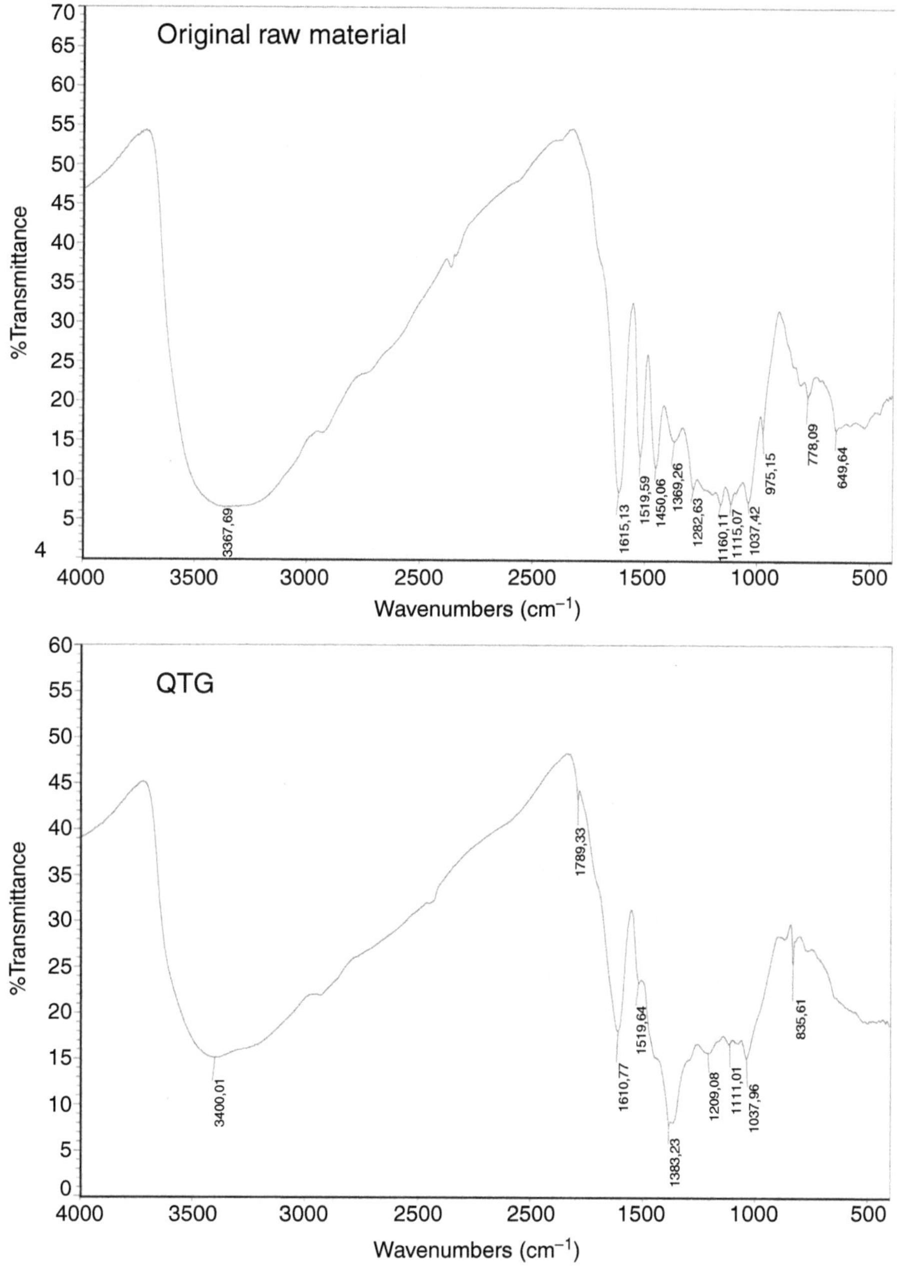

Fig. 12.23 FTIR spectra of raw and gelified Quebracho tannin (Reprinted from Ref. [74]. With kind permission of © Elsevier (2010))

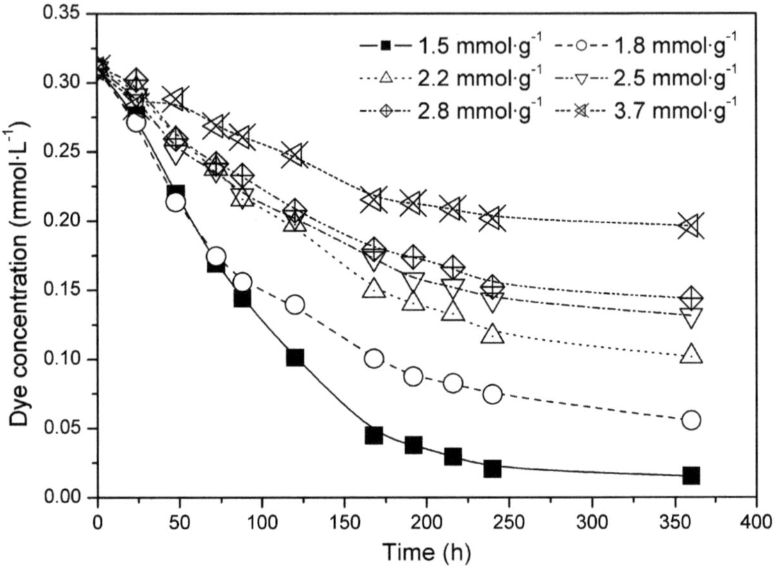

Fig. 12.24 Kinetics of dye removal with classical tannin gel from quebracho (Reprinted from Ref. [74]. With kind permission of © Elsevier (2010))

decreasing concentration of dye in six experiments with 1.5, 1.8, 2.2, 2.5, 2.8 and 3.7 mmol g^{-1}. A rather rapid dye removal is achieved in the first 150 h, although complete equilibrium dye concentration is reached at 350 h.

The application of the three classical models in kinetic studies (Lagergren, Ho and Elovich) has been carried out by a nonlinear adjustment and results are presented in Fig. 12.25. As can be appreciated, the three of them fit reasonably well to the experimental situations, so regression coefficient r^2 may be considered in order to discriminate the goodness of each data fit. All of them presented r^2 levels above 0.95. According to this, the three models explain rather well the adsorption process. With little differences, Lagergren model gives a 0.98 regression coefficient. Due to its simplicity and to the goodness of the linear correlation, not only the nonlinear regression, this hypothesis may be assumed as the best theoretical model in this adsorption case. Similar phenomena have been reported in *Methylene Blue* adsorption on other natural products [110].

Finally, equilibrium studies on MB and CTAB adsorption onto tannin gels were carried out. The adsorption of MB onto classical QTG is presented in Fig. 12.26 and follows the Langmuir hypothesis as presented in Eq. 12.4 (Fig. 12.26).

$$q = k_{l1} \frac{C_l}{1 + k_{l2} C_l} \tag{12.4}$$

where k_{l1} is the first Langmuir adsorption constant (L [g of adsorbent]$^{-1}$), and k_{l2} is the second Langmuir adsorption constant (L [mmol of pollutant]$^{-1}$).

Fig. 12.25 Kinetic of dye removal with classical tannin gel from quebracho data adjustment in the removal of *Methylene Blue* (Reprinted from Ref. [74]. With kind permission of © Elsevier (2010))

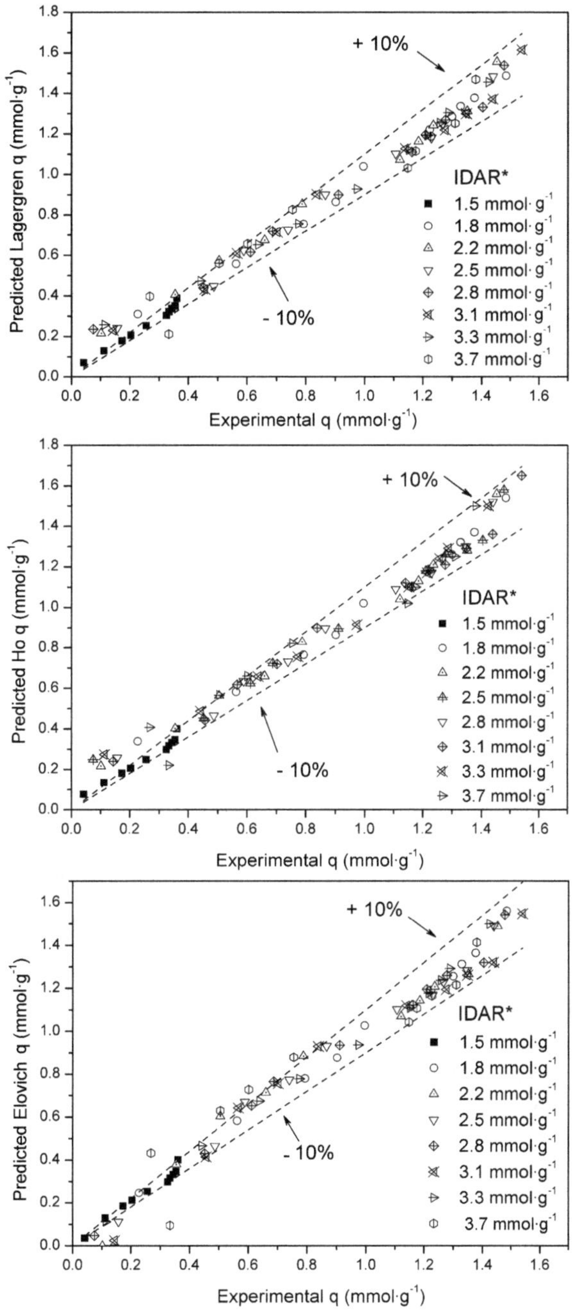

* Initial Dye-Adsorbent Ratio

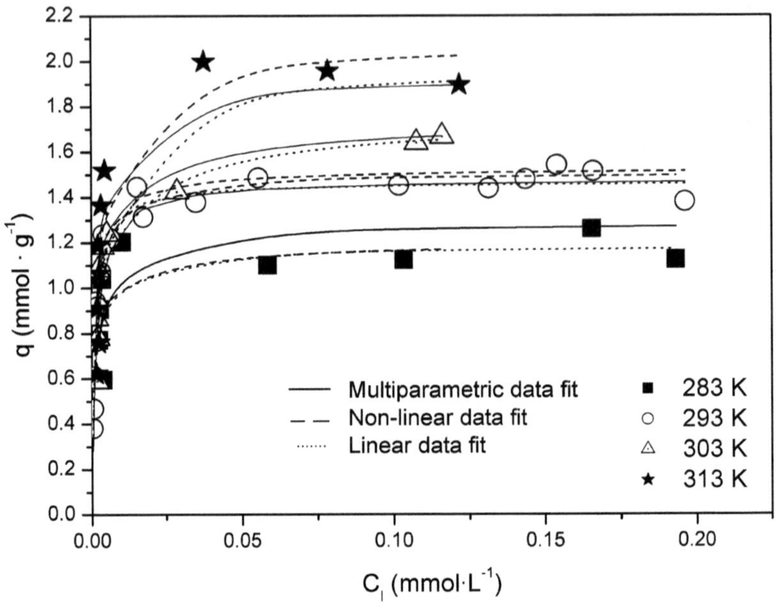

Fig. 12.26 Data fit of multiparameter, nonlinear and linear adjustments of Langmuir equation (Reprinted from Ref. [74]. With kind permission of © Elsevier (2010))

In addition, temperature can be included in a generalized expression of Langmuir hypothesis by considering Langmuir constants according to Arrhenius correlation. That assumes that k_{l1} and k_{l2} may have the following form (12.5):

$$k = k_0 \exp\left(-\frac{E}{RT}\right) \tag{12.5}$$

where k_0 is the basic constant, whose units are equal to k, E is the activation energy (J mol^{-1}), R is the universal constant for perfect gases (8.314 J mol^{-1} K^{-1}), and T is the temperature of the adsorption process (K).

The inclusion of definition (12.5) into Langmuir expression (12.4) leads to the expression (12.6):

$$q = \frac{k_{o1} \exp\left(-\frac{E_1}{RT}\right) C_l}{1 + k_{o2} \exp\left(-\frac{E_2}{RT}\right) C_l} \tag{12.6}$$

The specific r^2 is 0.86 in this case. It stands near to the average value of r^2 in the case of nonlinear individual adjustments for each temperature (0.84). Both the nonlinear procedures give a more accurate idea of the goodness of the model, while linear regression checks the adequacy of Langmuir's hypothesis in this adsorption process. In addition, multiparametric adjustment gives us two more

Table 12.3 Fitting parameters for equilibrium adsorption process. Units in text

Model	Parameters	r^2
Nonlinear Langmuir		
10°C	$k_{11} = 0.69$; $k_{12} = 19.4$	0.52
20°C	$k_{11} = 0.87$; $k_{12} = 34.4$	0.52
30°C	$k_{11} = 0.77$; $k_{12} = 105.2$	0.77
40°C	$k_{11} = 0.81$; $k_{12} = 94.4$	0.77
Linear Langmuir		
10°C	$k_{11} = 0.67$; $k_{12} = 21.8$	0.84
20°C	$k_{11} = 0.80$; $k_{12} = 27.4$	0.95
30°C	$k_{11} = 0.82$; $k_{12} = 48.2$	0.97
40°C	$k_{11} = 0.80$; $k_{12} = 95.3$	0.99
Multiparameter Langmuir		
	$k_{01} = 1.59$; $k_{02} = 1.27$ $E_1 = 0.18\ 10^4$, $E_2 = 5.37\ 10^4$	0.77

data–activation energies. According to the mathematical results, the first of these parameters (E_1), which corresponds to adsorption energy (while E_2 corresponds to desorption energy) is equal to 9,973.4 J mol^{-1}. E_2 is equal to zero, so desorption process is not temperature-dependent. In every case, the values of each k belongs to similar magnitude order, as can be appreciated according to the adjustment of experimental to predicted data.

Regarding CTAB adsorption onto *Pinus pinaster* classical *tannin gel*, the same three fitting procedures were performed: A linear and a nonlinear fit for each temperature, and a multiparameter fit for the entire temperature range. Table 12.3 lists the values of the parameters and the correlation coefficients r^2.

The nonlinear Langmuir fits for the specific temperatures gave lower correlation coefficients (mean of 0.65) because the error bars are wider. However, previous studies have shown this method to be more accurate than linear fits [111, 112].

Finally, the goodness of the multiparameter fit is reflected in the reasonably high value of the r^2 coefficient (0.77). In addition, this model yields two further parameters: the activation energies. According to the mathematical results, one of these parameters (E_{02}), which corresponds to the equilibrium adsorption energy, is equal to 5.37×10^4 J mol^{-1}, while the other one is equal to 0.18×10^4 J mol^{-1}. Since the difference between them is positive, the adsorption process reaches equilibrium. The fact that its value is very low has to do with the enhancement of the adsorption process due to the thermal effect on the pores, i.e., to thermal activation. In each case of the linear and nonlinear fits, the values of the parameters are similar in order of magnitude (Fig. 12.27).

The energies involved in the adsorption process are high enough for it to be considered as chemisorption (the links between adsorbate and adsorbent are stable and irreversible). The SEM images also led us to think that this chemisorption must be governed by the external diffusion stage, since they showed the material to have little porosity so that the adsorption process must take place on the surface at regularly distributed active centers.

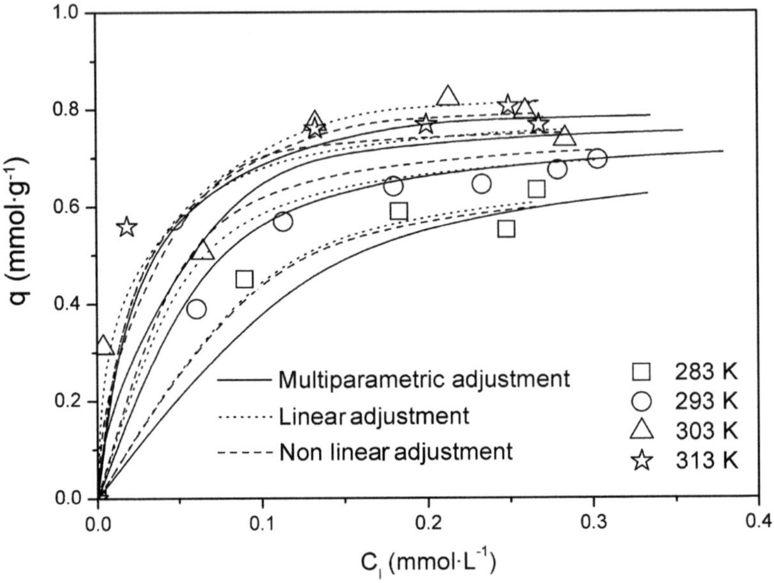

Fig. 12.27 Multiparameter, nonlinear, and linear fits of the Langmuir equation to the CTAB adsorption onto *Pinus pinaster* classical tannin gel data

Pinus tannin gel presents a reasonably high level of maximum *q* as shown by the nonlinear Langmuir model. Of the relatively few materials referred to in the scientific literature for CTAB removal, none belongs to the category of the so-called *waste* or *low-cost* materials. One observes that most of the adsorbents are synthetic in origin, such as perlite or modified silica gel. *Pinus tannin gel* presents an intermediate situation in terms of CTAB removal, so its capacity is comparable to that presented by other materials.

12.6 Conclusion

This chapter aimed to present some novel natural products as a possible response to concomitant and urgent pollution process. It is more than evident that environmental concerns are more significant every day and human beings are forced to face this challenge if we want to keep the current level of development. Moreover, the appropriate implementation of these and others technical resources should be properly developed for promoting economical and social development in emerging countries and in the so-called Third World.

We presented two main sources for controlling water pollution: *Moringa oleifera* seed extract and tannin derivatives. The last ones can be considered either as coagulants and adsorbents. The performance of every agent is promising and further investigations must be carried out, but it is clearly shown that *Moringa* seed extract and tannin coagulants (both commercial and lab-synthesized) are fully efficient in the removal of anionic dyes and surfactants. On the other hand, tannin adsorbents can easily remove cationic dyes, heavy metals, and cationic surfactants from wastewater samples.

References

1. UNESCO (2009) World water assessment programme. The United Nations world water development report 3: water in a changing world. EarthScan, London
2. PEP (2006) Linking poverty reduction and water management. Poverty-Environment Partnership and World Health Organization, Ginebra
3. Maslow A (1987) Motivation and personality, Brandeis University, Late
4. Unceta K (2007) La cooperación al desarrollo en las universidades españolas/Development cooperation in the Spanish universities. Spanish Agency for International Cooperation, Madrid
5. SPAIN (2009) Plan Director de Cooperación para el Desarrollo 2009–2010/Development Cooperation Director Plan 2009–2010, Madrid
6. WHO, UNICEF (2009) Joint programme on water supply and sanitation. WHO Library Cataloguing-in-Publication Data, Geneva
7. WHO (2008) United Nation-global annual assessment of sanitation and drinking water. WHO Library Cataloguing-in-Publication Data, Geneva
8. Lee JW, Choi SP, Thiruvenkatachari R, Shim WG, Moon H (2006) Evaluation of the performance of adsorption and coagulation processes for the maximum removal of reactive dyes. Dye Pigment 69:196–203
9. Shaul GM, Holdsworth TJ, Dempsey CR, Dostal KA (1991) Fate of water soluble azo dyes in the activated sludge process. Chemosphere 22:107–119
10. Guibal E, Roussy J (2007) Coagulation and flocculation of dye-containing solutions using a biopolymer (Chitosan). React Funct Polym 67:33–42
11. Shi B, Li G, Wang D, Feng C, Tang H (2007) Removal of direct dyes by coagulation: the performance of preformed polymeric aluminum species. J Hazard Mater 143:567–574
12. O'Neill C, Hawkes FR, Hawkes DL, Lourenco ND, Pinheiro HM, Deleé W (1999) Colour in textile effluents-sources, measurement, discharge consent and simulation: a review. J Chem Technol Biotechnol 74:1009–1018
13. Zollinger H (1987) Colour chemistry-synthesis. Properties and application of organic dyes and pigments. VCH Publishers, New York
14. Gaini LE, Lakraimi M, Sebbar E, Meghea A, Bakasse M (2009) Removal of indigo carmine dye from water to Mg–Al–CO3-calcined layered double hydroxides. J Hazard Mater 161:627–632
15. Chung KT, Chen SC, Claxton LD (2006) Review of the *Salmonella typhimurium* mutagenicity of benzidine, benzidine analogues, and benzidine-based dyes. Mutat Res/Rev Mutat Res 612:58–76
16. Myers D (2006) Surfactant science and technology. Wiley, New Jersey
17. Edser C (2008) Status of global surfactant markets. Focus Surfactant 11:1–2
18. Cserháti T, Forgács E, Oros G (2002) Biological activity and environmental impact of anionic surfactants. Environ Int 28:337–348
19. Basar CA, Karagunduz A, Cakici A, Keskinler B (2004) Removal of surfactant by powdered activated carbon and microfiltration. Water Res 38:2117–2124
20. Clara M, Scharf S, Scheffknecht C, Gans O (2007) Occurrence of selected surfactants in untreated and treated sewage. Water Res 41:4339–4348

21. Patterson DA, Metcalfe IS, Xiong F, Livingston AG (2006) Wet air oxidation of linear alkylbenzene sulfonate: 1. Effect of temperature and pressure. Ind Eng Chem Res 40:5507–5516
22. Pasieczna-Patkowska S, Czech B, Ryczkowski J, Patkowski J (2010) Removal of recalcitrant pollutants from wastewater. Appl Surf Sci 256:5434–5438
23. Rosu M, Marlina A, Kaya A, Schumpe A (2008) Surfactant adsorption onto activated carbon and its effect on absorption with chemical reaction. Chem Eng Sci 62:7336–7343
24. Önder E, Koparal AS, Ögütveren ÜB (2007) An alternative method for the removal of surfactants from water: electrochemical coagulation. Sep Purif Technol 52:525–532
25. Ostroumov SA (2006) Biological effects of surfactants. CRC Press, Boca Raton
26. Kurniawan TA, Chan GYS, Lo W-H, Babel S (2006) Physicochemical treatment techniques for wastewater laden with heavy metals. Chem Eng J 118:83–98
27. Tünai O, Kabdasly NI (1994) Hydroxide precipitation of complexed metals. Water Res 28:2117–2124
28. Charerntanyarak L (1999) Heavy metals removal by chemical coagulation and precipitation. Water Sci Technol 39:135–138
29. Yoon J, Amy G, Chung J, Sohn J, Yoon Y (2009) Removal of toxic ions (chromate, arsenate, and perchlorate) using reverse osmosis, nanofiltration, and ultrafiltration membranes. Chemosphere 77:228–235
30. Appelbaum S, Garada J, Mishra JK (2002) Growth and survival of the white leg shrimp (*Litopenaeus vannamei*) reared intensively in the brackish water of the Israeli Negev desert. Isr J Aquac–Bamidgeh 54:41–48
31. WHO and UNICEF (2007) Meeting the MDG drinking water and sanitation target: the urban and rural challenge of the decade. WHO Library Cataloguing-in-Publication Data, Geneva
32. Wilderer PA (2004) Applying sustainable water management concepts in rural and urban areas: some thoughts about reasons, means and needs. Water Sci Technol 49:8–16
33. WHO (1985) Appropriate technology for the treatment of wastewaters for small rural communities. WHO Library Cataloguing-in-Publication Data, Geneva
34. Dorf RC (2001) Sustainable and appropriate technologies, technology, humans, and society. Academic, San Diego
35. Bowonder B (1979) Appropriate technology for developing countries: some issues. Technol Forecast Soc Change 15:55–67
36. Miller SM, Fugate EJ, Craver VO, Smith JA, Zimmerman JB (2008) Towards understanding the efficacy and mechanism of *Opuntia* spp. as a natural coagulant for potential application in water treatment. Environ Sci Technol 42:4274–4279
37. Fuglie LJ (2001) The miracle tree. The multiple attributes of *Moringa*. Technical Centre for Agricultural and Rural Cooperation, Wageningen
38. Makkar HPS, Becker K (1996) Nutritional value and antinutritional components of whole and ethanol extracted *Moringa oleifera* leaves. Anim Feed Sci Technol 63:211–228
39. Aregheore EM (2002) Intake and digestibility of *Moringa oleifera*-Batiki grass mixtures by growing goats. Small Rumin Res 46:23–28
40. Richter N, Siddhuraju P, Becker K (2003) Evaluation of nutritional quality of Moringa (*Moringa oleifera* Lam.) leaves as alternative protein source for Nile Tilapia (*Oreochromis niloticus* L.). Aquaculture 217:599–611
41. Cáceres A, Cabrera O, Morales O, Mollinedo P, Mendia P (1991) Pharmacological properties of *Moringa oleifera*. 1: preliminary screening for antimicrobial activity. J Ethnopharmacol 33:213–216
42. Cáceres A, Saravia A, Rizzo S, Zabala L, De León E, Nave F (1992) Pharmacologic properties of *Moringa oleifera*. 2: Screening for antispasmodic, antiinflammatory and diuretic activity. J Ethnopharmacol 36:233–237
43. Armand-Stussi I, Basocak V, Pauly G, McCaulley J (2003) *Moringa oleifera*: an interesting source of active ingredients for skin and hair care. SOFW-J 129:45–52
44. Goh CW (2009) Effect of room temperature on coagulation performance of *Moringa oleifera* seeds. B. Sc. dissertation, Faculty of Engineering, University of Putra, Malaysia

45. Katayon S, Noor MM, Asma M, Thamer AM, Liew AG, Idris A, Suleyman AM, Aminuddin MB, Khor BC (2004) Effects of storage duration and temperature of *Moringa oleifera.* Stock solution on its performance in coagulation. Int J Eng Technol 1:146–151
46. Abdulkarim SM, Long K, Lai OM, Muhammad SKS, Ghazali HM (2005) Some physico-chemical properties of *Moringa oleifera* seed oil extracted using solvent and aqueous enzymatic methods. Food Chem 93:253–263
47. Bhuptawat H, Folkard GK, Sanjeev C (2007) Innovative physico-chemical treatment of wastewater incorporating *Moringa oleifera* seed coagulant. J Hazard Mater 145:120–126
48. Jahn SA, Musnad HA, Burgstalle H (1986) The tree that purifies water: cultivating multipurpose *Moringaceae* in Sudan. Unasylva 38:23–28
49. Schofield P, Mbugua DM, Pell AN (2001) Analysis of condensed tannins: a review. Anim Feed Sci Technol 91:21–40
50. Haslam E (1989) Plant polyphenols-vegetables and tannins revisited. Cambridge University Press, Cambridge
51. Beltrán-Heredia J, Sánchez-Martín J, Solera-Hernández C (2009) Anionic surfactants removal by natural coagulant/flocculant products. Ind Eng Chem Res 48:5085–5092
52. Pizzi A (2008) Tannins: major sources, properties and applications. In: Belgacem MN, Galdini A (eds) Monomers, polymers and composites from renewable sources. Elsevier, Amsterdam
53. Hagerman A (1995) Tannin analysis. Miami University, Ohio
54. Mitchel DB, Minnis RL, Curran TP, Deboo SM, Kelly JA, Patwardhan R, Tai W-T (1998) Treatment of aqueous systems using a chemically modified tannin. US Patent 5,843,337
55. Quamme JE, Kemp AH (1985) Stable tannin based polymer compound. US Patent 4,558,080
56. Reed PE, Finck MR (1997) Modified tannin Mannich polymers. US Patent 5,659,002
57. Vasconcellos SR, Boyce PD, Smith LP (1993) Methods for the flocculation of coal fines and insoluble metals in coal mine waters. US Patent 4,183,575
58. Roux DG, Ferreira D, Hundt HL, Malan E (1975) Structure, stereochemistry, and reactivity of natural condensed tannins as basis for their extended industrial application. Appl Polym Symp 1:335–353
59. Tramontini M, Angiolini L (1994) Mannich bases. Chemistry and uses. CRC Press, Boca Raton
60. Pizzi A (1994) Advanced wood adhesives technology. Marcel Dekker, New York
61. Roussy J, Chastellan P, Van Vooren M, Guibal E (2005) Treatment of ink-containing wastewater by coagulation/flocculation using biopolymer. Water SA 31:369–376
62. Polasek P, Mutl S (2002) Cationic polymers in water treatment. Part 1: Treatability of water with cationic polymers. Water SA 28:69–82
63. Graham N, Gang F, Fowler G, Watts M (2008) Characterisation and coagulation performance of a tannin-based cationic polymer: a preliminary assessment. Coll Surf A 327:9–16
64. Graham N, Gang F, Fowler G, Watts M, Camm R (2008) Evaluation of a tannin-based cationic polymer as a coagulant for coloured humic water. J Water Supply Res Technol AQUA 58:75–84
65. Lamb LH, Decusati OG (2002) Manufacturing process for quaternary ammonium tannate, a vegetable coagulating and flocculating agent. US Patent 6,478,986 B1
66. Tondi G, Zhao W, Pizzi A, Du G, Fierro V, Celzard A (2009) Tannin-based rigid foams: a survey of chemical and physical properties. Bioresour Technol 100:5162–5169
67. Zhao W, Fierro V, Pizzi A, Du G, Celzard A (2010) Effect of composition and processing parameters on the characteristics of tannin-based rigid foams. Part II: Physical properties. Mater Phys Chem 123:210–217
68. Kim Y-H, Nakano Y (2005) Adsorption mechanism of palladium by redox within condensed-tannin gel. Water Res 39:1324–1330
69. Nakano Y, Takeshita K, Tsutsumi T (2001) Adsorption mechanism of hexavalent chromium by redox within condensed-tannin gel. Water Res 35:496–500
70. Tondi G, Oo CW, Pizzi A, Thevenon MF (2008) Metal absorption of tannin-based rigid foams. Ind Crop Prod 29:336–340

71. Vázquez G, Antorrena J, González J, Doval MD (1994) Adsorption of heavy metal ions by chemically modified *Pinus pinaster* bark. Bioresour Technol 48:251–255
72. Vázquez G, González-Álvarez J, Freire S, López-Lorenzo ML, Antorrena G (2002) Removal of cadmium and mercury ions from aqueous solution by sorption on treated *Pinus pinaster* bark: kinetics and isotherms. Bioresour Technol 82:247–251
73. Sánchez-Martín J, Beltrán-Heredia J, Carmona-Murillo C (2011) Adsorbents from *Schinopsis balansae*: optimisation of significant variables. Ind Crop Prod 33:409–417
74. Sánchez-Martín J, González-Velasco M, Beltrán-Heredia J, Gragera-Carvajal J, Salguero-Fernández J (2010) Novel tannin-based adsorbent in removing cationic dye (Methylene Blue) from aqueous solution. Kinetic and equilibrium studies. J Hazard Mater 174:9–16
75. Yurtsever M, Sengil IA (2009) Biosorption of Pb(II) ions by modified quebracho tannin resin. J Hazard Mater 63:58–64
76. Broin M, Santaella C, Cuine S, Kokou K, Peltier G, Joët T (2002) Flocculent activity of a recombinant protein from *Moringa oleifera* Lam. seeds. App Microbiol Biotechnol 60:114–119
77. Kwaambwa HM, Maikokera R (2007) A fluorescence spectroscopic study of a coagulating protein extracted from *Moringa oleifera* seeds. Coll Surf B 60:213–220
78. Ndabigengesere A, Narasiah KS, Talbot BG (1995) Active agents and mechanism of coagulation of turbid waters using *Moringa oleifera*. Water Res 29:703–710
79. Okuda T, Baes AU, Nishijima W, Okada M (2001) Coagulation mechanism of salt solution-extracted active component in *Moringa oleifera* seeds. Water Res 35:830–834
80. Okuda T, Baes AU, Nishijima W, Okada M (2001) Isolation and characterization of coagulant extracted from *Moringa oleifera* seed by salt solution. Water Res 35:405–410
81. Beltrán-Heredia J, Sánchez-Martín J (2008) Azo dye removal by *Moringa oleifera* seed extract coagulation. Color Technol 124:310–317
82. Beltrán-Heredia J, Sánchez-Martín J, Delgado-Regalado A, Jurado-Bustos C (2009) Natural coagulants in removing Alizarin Violet 3R anthraquinonic dye. J Hazard Mater 170:43–50
83. Beltrán-Heredia J, Sánchez-Martín J, Delgado-Regalado A (2009) Removal of Carmine Indigo dye with *Moringa oleifera* seed extract. Ind Eng Chem Res 48:6512–6520
84. Merta J, Stenius P (1995) Interactions between cationic starch and anionic surfactants. Coll Polym Sci 273:974–983
85. Beltrán-Heredia J, Sánchez-Martín J (2009) Removal of sodium lauryl sulfate by coagulation/flocculation with *Moringa oleifera* seed extract. J Hazard Mater 164:713–719
86. Gu T, Zhu B-Y (1990) The S-type isotherm equation for adsorption of nonionic surfactants at the silica gel-water interface. Coll Surf 44:81–87
87. Holmberg K, Jönsson B, Kronberg B, Lindman B (2003) Surfactants and polymers in aqueous solution. Wiley, Chichester
88. Palma G, Freer J, Baeza J (2003) Removal of metal ions by modified *Pinus radiata* bark and tannins from water solutions. Water Res 37:4974–4980
89. Beltrán-Heredia J, Sánchez-Martín J (2008) Removing heavy metals from polluted surface water with a tannin-based flocculant agent. J Hazard Mater 165:1215–1218
90. Frankenthal RPC (1963) Handbook of analytical chemistry. McGraw-Hill, New York
91. Sikka P (1991) Strategies for technology development in India. Technovation 11:445–452
92. Brown D (1987) Effects of colorants in the aquatic environment. Ecotoxicol Environ Saf 13:139–147
93. Allegre C, Maisseu M, Charbit F, Moulin P (2004) Coagulation, flocculation and decantation of dye house effluents: concentrated effluents. J Hazard Mater 116:57–64
94. Beltrán-Heredia J, Sánchez-Martín J, Frutos-Blanco G (2009) *Schinopsis balansae* tannin-based flocculant in removing sodium dodecylbenzene sulfonate. Sep Purif Technol 67:292–303
95. Beltrán-Heredia J, Sánchez-Martín J, Solera-Hernández C (2009) Removal of sodium dodecylbenzene sulfonate from wáter by means of a new tannin-based coagulant: optimisation studies through design of experiments. Chem Eng J 153:56–61
96. Rosen MJ (2004) Surfactants and interfacial phenomena. Wiley, New Jersey

97. Esumi K, Ueno M (2003) Structure-performance relationships in surfactants. Marcel Dekker, New York
98. Lindert AT, Wolpert SM (1990) Tannin Mannich adducts for improving corrosion resistance of metals. US Patent 4,944,812
99. Hemingway RW, Laks PE (1992) Preparation and performance of tannin based flocculants. In: Biogenesis, chemical properties and significance. Second North American tannin conference on plant polyphenols. Plenum Press, New York
100. Kim YH (1995) Coagulants and flocculants. Theory and practice. Tall Oak Publishing, Littleton
101. Streit W, Fengel D (1994) Purified tannins from *Quebracho colorado*. Phytochemistry 34:481–484
102. Vivas N, Nonier MF, Vivas de Gaulejac N, Absalon C, Bertrand A, Mirabel M (2004) Differentiation of proanthocyanidin tannins from seeds, skins and stem of grapes (*Vitis vinifera*) and heartwood of quebracho (*Schinopsis balansae*) by matrix-assisted laser desorption/ionization time-of-flight mass spectrometry and thiacidolysis/liquid chromatography/electrospray ionization mass spectrometry. Anal Chim Acta 513:247–256
103. Nazaroff WW, Álvarez-Cohen L (2001) Environmental engineering science. Wiley, New York
104. Beltrán-Heredia J, Sánchez-Martín J, Gómez-Muñoz MC (2010) New coagulant agents from tannin extracts: preliminary optimisation studies. Chem Eng J 162:1019–1025
105. Özacar M, Sengil A (2001) Effectiveness of tannins obtained from *Valonia* as a coagulant aid for dewatering of sludge. Water Res 34:1407–1412
106. Özacar M, Sengil A (2003) Evaluation of tannin biopolymer as a coagulant aid for coagulation of colloidal particles. Coll Surf A 229:85–96
107. Ayranci E, Duman O (2007) Removal of anionic surfactants from aqueous solutions by adsorption onto high area activated carbon cloth studied by in situ UV spectroscopy. J Hazard Mater 148:75–82
108. Sánchez-Martín J, Beltrán-Heredia J, Gibello-Pérez P (2011) Adsorbent biopolymers from tannin extracts for water treatment. Chem Eng J 168:1241–1247
109. Vázquez G, González-Álvarez J, Freire S, López-Suevos F, Antorrena G (2001) Characteristics of *Pinus pinaster* bark extracts obtained under various extraction conditions. Eur J Wood Wood Prod 59:451–456
110. Dogan M, Abak H, Alkan M (2009) Adsorption of methylene blue onto hazelnut shell: kinetics, mechanisms and activation parameters. J Hazard Mater 164:172–181
111. Kumar KV, Porkodi K, Rocha F (2008) Isotherms and thermodynamics by linear and non-linear regression analysis for the sorption of methylene blue onto activated carbon: comparison of various error functions. J Hazard Mater 151:794–804
112. Kumar KV, Sivanesan S (2006) Pseudo second order kinetics and pseudo isotherms for malachite green onto activated carbon: comparison of linear and non-linear regression methods. J Hazard Mater 136:721–726
113. Boni A, Ferrero G (1997) Introduction to development cooperation. Politechnical University of Valencia, Valencia, Spain

Chapter 13
Polysaccharide-Based Macromolecular Materials for Decolorization of Textile Effluents

Vandana Singh, Tulika Malviya, and Rashmi Sanghi

13.1 Introduction

Dyes are highly dispersible aesthetic pollutants contributing to aquatic toxicity. Reminiscent or unspent dye materials are mainly responsible for the colored effluents from industries, and its removal is therefore a necessary and inevitable step for the recycling and disposal of the textile industrial effluents prior to it being discharged into sewers. As most of the dyes are highly stable molecules which resist degradation by light, chemical, biological, and other treatment methods, the textile waste containing dye pollutant are difficult to treat and pose a great challenge to the existing conventional wastewater treatment techniques. The high cost and toxic nature of the synthetic chemicals used in most of the conventional treatments is a major drawback whereby the demand for the development of alternative environment-friendly treatment methods is on the rise. In recent years, usage of natural polymeric materials to treat dye effluents has drawn considerable attention because they are renewable, biodegradable, nontoxic, and potentially environmentally friendly. Several polysaccharides and their derivatives have been utilized in conventional water treatment techniques like adsorption and coagulation–flocculation. Water-soluble polysaccharides are used for coagulation–flocculation, while insoluble polysaccharide-/polysaccharide-derived materials such as cross-linked polysaccharides, polysaccharide composites, and chemically modified polysaccharides are used as adsorbents through which passive dye sequestration is feasible with many advantages over conventional techniques. To overcome the requirement of large

V. Singh (✉) • T. Malviya
University of Allahabad, Allahabad 211002, India
e-mail: singhvandanasingh@rediffmail.com; tulika.au@gmail.com

R. Sanghi
Indian Institute of Technology, Kanpur, UP 208016, India
e-mail: rsanghi@gmail.com; rsanghi@iitk.ac.in

S.K. Sharma and R. Sanghi (eds.), *Advances in Water Treatment and Pollution Prevention*, DOI 10.1007/978-94-007-4204-8_13,

dosages of polysaccharide flocculants and to improve the stability and strength of flocs obtained with polysaccharide flocculants, attempts have been made to suitably tailor or modify the properties of the polysaccharides in obtaining macromolecular materials having high shear stability and improved shelf life. Such shear stable and controlled biodegradable flocculants are the most sought flocculants for treating the industrial effluents and mineral-processing wastes.

13.2 Synthetic Dyes

Natural dyes are plant-sourced colored substances which are isolated from fruits, berries, bark, leaves, root, and wood of several plants. As the isolation and processing processes of natural dyes are time demanding and often difficult, the use of synthetic dyes is more popular in the industry. Though synthetic dyes are toxic, they have lower cost as compared to their natural counterparts and are able to impart stable and bright colors to the dyed substrates. Dyes are mostly applied to a substrate in an aqueous solution. Many synthetic dyes are now known which have almost replaced the traditional natural dyes because of their quite low cost, vast color range, and good color stability.

Dyes can be classified according to the procedures used in the dying processes [1] and on the basis of the chromophoric groups they have, such as acridine dyes, anthraquinone dyes, arylmethane dyes, azo dyes, nitro dyes, nitroso dyes, phthalocyanine dyes, quinone-imine dyes, indamines, xanthine, and fluorene dyes, and their toxicity is related with their structure. Structure and dyeing details of some popular dyes are summarized in Table 13.1 and Fig. 13.1a,b.

13.3 Dye Removal from Waste Water

Colored effluents from dye-using industries contain unspent toxic synthetic dyes which are frequently used by industries due to their broad color range and cost-effectiveness. Improper disposal of such effluents can pose major environmental and health threat; thus, decolorization of such effluents becomes an integral step of the waste treatment process and is performed before such effluents can be discharged into the receiving water body. Dyestuffs are mostly highly structured polymers with low or no biodegradability [12, 13], and the color removal mostly involves the breaking of the conjugated bond in dye molecules [14] for which many chemical treatment processes such as membrane separation [11], oxidation or ozonation [13, 15], electrocoagulation, chemical precipitation, adsorption by activated carbon [16], and chemical [17, 18] and photocatalytic oxidation [19] are commonly employed. Although effective, these methods are expensive and involve secondary pollution effects.

Table 13.1 Some important dye types and their applications

S.No.	Type of dye	Structure	Dying substrates	Dye bath	Reference
1.	Acid dyes	Sodium or ammonium salt of a sulfonic, carboxylic, or phenol organic acid	For coloring amphoteric fibers such as silk, wool, nylon, and modified acrylic fibers	Neutral to acid	[2]
2.	Basic dyes	A colored cation combined with a colorless anion	Acrylic fibers and paper. Occasionally for dying wool and silk fibers	Acid	[3]
3.	Direct dyes	Planar, highly conjugated molecular structures that also contain one or more anionic sulfonate group	Hot water dyes for dying cotton, paper, leather, wool, silk, and nylon	Sodium chloride or sodium sulfate is added to the dye baths	[4]
4.	Mordant dyes	As azo dyes and anthraquinone	Cotton and wool	Tannic acid, oleic and stearic acids, and mordants, e.g., soluble salts of chromium, aluminum, iron, copper, and tin	[5]
5.	Vat dyes	Anthraquinonoid and thio-indigoid class of dyes	Dye cotton wool, as well as other fibers	Alkaline liquor	[6]
6.	Reactive dyes	Having substituents with chromophoric group, e.g., haloheterocycle or an activated double bond	Cotton and other cellulose fibers	–	[7]
7.	Disperse dyes	Small, planar, and nonionic, with polar functional groups like $-NO_2$ and –CN	Polyester, to some extent for nylon, cellulose triacetate, and acrylic fibers	–	[8]
8.	Azo dyes	Diazo or triazo groups			[9, 10]
9.	Sulfur dyes		Cotton	Sodium sulfide or sodium hydrosulfite	[11]

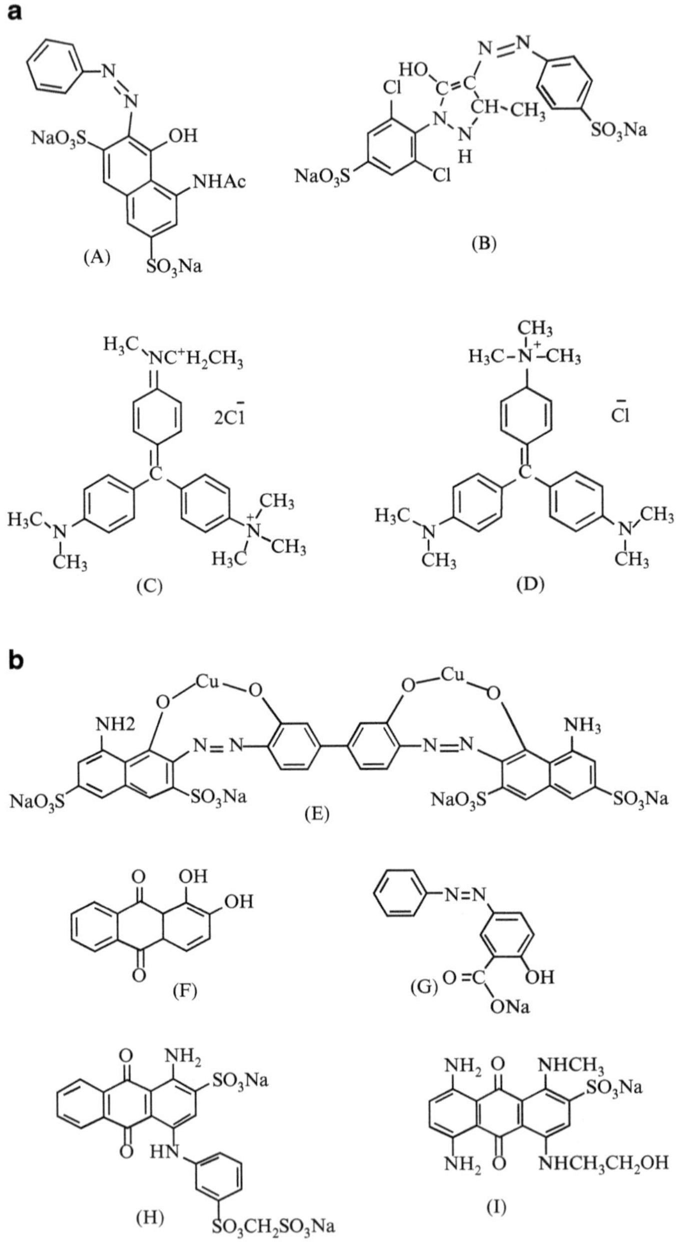

Fig. 13.1 (**a**) Structures of some of the acid and basic dyes: (*A*) Acid Red G; (*B*) Acid Light Yellow 2G; (*C*) Methyl Green; (*D*) Methyl Violet; (**b**) Chemical structure of (*E*) Direct Blue 218; (*F*) Mordant Red 11; (*G*) Mordant Yellow; (*H*) Reactive Blue; (*I*) Disperse Blue

As a step toward utilization of environment-friendly and cost-effective techniques, naturally occurring polysaccharides and their derivatives have been utilized in conventional water treatment techniques like adsorption and coagulation–flocculation. The wastewater may be considered safe to human health when the colored impurities are removed from it which can be done either by coagulation or adsorption. Natural seed gums are good alternatives to the conventional chemical coagulants as their use has lesser environmental impact. The polysaccharides may or may not be water soluble and based on their solubility, the method of choice could be either adsorption or coagulation. Water-soluble polysaccharide solutions are used for coagulation–flocculation, while insoluble polysaccharide-/polysaccharide-derived materials such as cross-linked polysaccharides, polysaccharide composites, and chemically modified polysaccharides find application as adsorbents.

13.3.1 Coagulation–Flocculation

In wastewater treatment process, the coagulation and flocculation are used to separate suspended solids from water. Coagulation is the destabilization of colloids by neutralizing the forces that keep them apart, while flocculation is the action of polymers to form bridges between the flocs and bind the particles into large agglomerates or clumps [20]. For the primary treatment of wastewater, coagulation–flocculation [18] chemical method is usually employed where alum [21] and polyaluminium chloride (PAC) [22] are commonly used as coagulants though many other organic and inorganic flocculants are known for treatment of colored effluents. The effectiveness of these coagulants is highly pH dependent, and the finished water thus obtained has a contamination of residual aluminum. Moreover, the process results into a significant amount of sludge which further complicates the handling and disposal procedures. The cost of importing alum and other required chemicals for conventional treatment processes may often be detrimental and sometimes prohibitive for developing countries. Currently, polymeric flocculants, synthetic as well as natural, have become more acceptable in industrial effluent treatment processes because of their inertness to pH changes, handling comfort and effectiveness.

Synthetic polymers behave as effective flocculants due to their versatile tailorability, but they normally have poor shear stability. Economically viable alternative of these polymeric flocculants may be the natural polymers [23, 24] which are locally grown and harvested [25, 26]. These are low-cost, nontoxic, and highly biodegradable materials endowed with sufficient shear stability. Since the time immemorial, the use of natural coagulants for water clarification is known, e.g., the seeds of the Nirmali [27] and seeds of *Moringa oleifera* [28] tree are known to clarify turbid surface water. Seed extracts from Nirmali (*Strychnos potatorum*), tamarind (*Tamarindius indica*), guar (*Cyamopsis psoraloides*), red sorella (*Hibiscus sabdariffa*), fenugreek (*Trigonella foenum*), and lentils (*Lens esculenta*) behave as effective coagulant aids at high raw water turbidities and can reduce required alum doses by 40–50% [27]. *Cassia tora* and *Cassia angustifolia* polysaccharides which are structurally related to guar gum have also been utilized as coagulant aids [23, 29]. These alternative naturally occurring coagulants have several advantages over those synthesized conventionally like alum. Their use considerably reduces the sludge volume in comparison to what is produced with alum; additionally, the natural alkalinity is not consumed during the treatment process.

Natural coagulants extracted from plants or animals [30] are thus workable alternatives to synthetic polyelectrolytes because of their biodegradability and biocompatibility. Their use is safe to human health, and they have a wider effective dosage range for flocculating various colloidal suspensions. They are also cost-effective as they are isolated and harvested from locally grown plants. It has also been possible to develop efficient, shear stable, and less biodegradable flocculants by modifying the synthetic polymers by grafting, e.g., grafting of poly(acrylamide) onto the rigid backbone of the natural polysaccharides [31].

13.3.2 Adsorption

In removing contaminants from the effluent of dye-using industries, adsorption method is very effective as it can completely remove dyes even from diluted solution. Insoluble materials can be used as adsorbents for dyes and other pollutants under aqueous conditions. The removal is either physisorption or chemisorption depending upon the adsorbent type and structure. Besides conventional sorbents such as fly ash, carbon, and silica gels, many biosorbents have been developed for the removal of dyes from the effluents. Commercially available sorbents are expensive, besides their regeneration produce small additional effluent. This warranted the utilization of several biosorbents for dye removal. Biosorption is a rapid phenomenon of passive dye sequestration by the adsorbents developed from biomass. Numerous cost-effective biosorbents have been worked out [32] as alternatives of conventional adsorbents.

Removal of dyes through these biosorbents has many advantages compared with conventional techniques [33]. Among other, polysaccharide-derived sorbents are very effective in dye removal from wastewater. Binding sites in these sorbents may be either the appropriate functional groups, e.g., grafted polysaccharides or the porosity, e.g., polysaccharide clay or silica composites. Various polysaccharides have been converted to insoluble materials which can be used under aqueous conditions as sorbents or certain water-soluble polysaccharides may find such application as such, e.g., chitosan (insoluble at neutral pH), cellulose, etc. Cross-linking of the polysaccharide macromolecules is the one of the most common way to turn them insoluble for such applications. This can also be achieved by introducing synthetic polymeric graft chains whereby solubility of a polysaccharide may be tailored for the targeted adsorption application. Traditional dye adsorbents are bauxite and Fuller's earth.

This chapter summarizes the use of polysaccharides and polysaccharide-derived macromolecular materials in dye removal from industrial wastes along with a detailed discussion on the advantages of using these materials for the remediation of industrial wastes over conventional polymeric materials. The discussion has been subdivided polysaccharide-wise for easy understanding and clarity. The following section presents a brief description of polysaccharides.

13.4 Polysaccharides

Polysaccharides are stereoregular, abundant natural polymers of sugars with unique biological and chemical properties. They are biocompatible, nontoxic, biodegradable biopolymers, possessing high chemical reactivity, polyfunctionality, chirality, and chelating properties. The polysaccharide materials can be exploited for the fabrication of excellent adsorbents [34] as they possess (1) hydroxyl groups which makes them hydrophilic, (2) a large number of reactive functional groups (acetamido, primary amino, and/or hydroxyl groups) depending upon the polysaccharide type, and (3) flexible structure of the polysaccharide chain.

Among the known polysaccharides, starch is the most abundant. It is known to exist in nature as energy reserve for living plants as the mixtures of two polyglucans, amylopectin and amylose; both of them are homoglycans as they contain only a single type of carbohydrate, glucose. Other important and abundant polysaccharides are the chitin, chitosan, and cellulose. Both starch and chitin are safe for humans and possess several characteristics and advantages that are suitable for their exploitation in deriving excellent materials for industrial use [35]. The type of derivatization is chosen depending upon the desired application of the final material. In deriving a flocculant for wastewater or a derivative which can be used as additive in paper manufacturing, a random conversion of hydroxyl groups of starch to aminopropyl [36], hydroxyalkyl [37], or betaine [38] is done, while for the adsorbent resins a rigid macromolecular structure is necessary.

Like chitosan and starch, guar gum and other nonconventional galactomannans have also been utilized for water remediation application such as metal ion and dye removal from wastewater. Galactomannans are water-soluble hydrophilic heterobiopolymers of galactose and mannose. The ratio of galactose to mannose in guar gum is 1:2 while this ratio varies in other nonconventional seed gums isolated from *Cassia* and *Ipomoea* plants. For particular genera, this ratio varies from species to species and also with climatic distribution of the source plant. In the recent past many of the *Cassia* polysaccharides [23, 31] have been evaluated for dye removal from wastewater in their natural as well as in derived form.

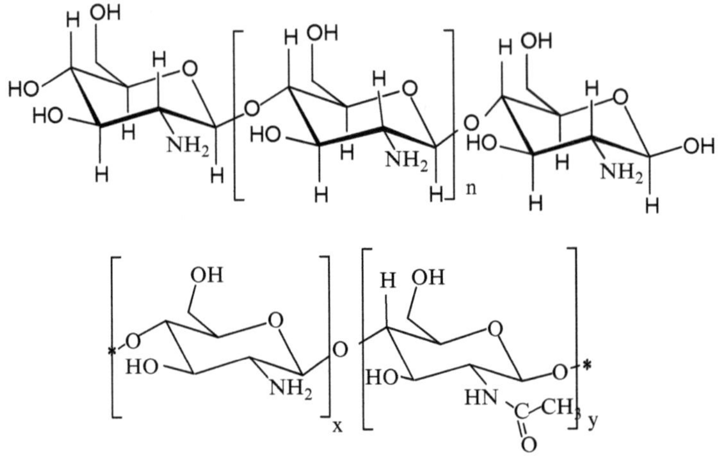

Fig. 13.2 Chemical structure of chitosan and chitin

13.4.1 Chitin and Chitosan

Chitin (Fig. 13.2) is the exoskeletons material of molluscs and insects and constitutes the cell wall of some fungi. It is the main component in the shells of crustaceans, such as shrimp, crab, and lobster, and is the characteristic polysaccharide of several important phyla, e.g., Arthropoda, Annelida, Mollusca, and Coelenterata and of many fungi such as ascomycetes, zygomycetes, basidiomycetes, and deuteromycetes. The polymer contains 2-acetamido-2-deoxy-β-D-glucose linked through a $\beta(1 \rightarrow 4)$ linkage. It is structurally related to cellulose and differs from cellulose in having acetamido group at position C-2 in place of hydroxyl groups present in cellulose. Chitin may now be produced commercially at low cost since such wastes (shrimp, lobster, crab shells) are abundantly available in nature.

Much attention has been paid to chitosan as a potential polysaccharide resource [39] for deriving macromolecular materials effective in wastewater treatment. There have been tremendous efforts for preparing functional derivatives of chitosan through chemical modifications [40–43]; however, only few among these are known to dissolve in conventional organic solvents. In column studies, powder form or flakes of chitin and chitosan is mostly used in which, after certain time, there is a significant drop in the column pressure [44], and to avoid this drop, the use of chitin and chitosan beads is preferred.

Chitosan (Fig. 13.2) is derived from chitin through N-deacetylation, and there has been recent attraction for this material in deriving macromolecular materials of interest in water remediation. Though presence of amino group makes this polysaccharide very versatile for chemical modifications and cross-linking, amino groups restrict water solubility of the biopolymer at neutral pH. It is only soluble in aqueous solutions of some acids and through some selective N-acylations [45].

13.4.1.1 Chitosan-Derived Materials

Chitosan and chitin are among the most used polysaccharides for water remediation application. These polysaccharides in their native or derived form have high affinity for many classes of dyes [46]. These bioadsorbents being versatile materials have been used in different forms, from flake types to gels, bead types, or fibers. Though several disadvantages are associated with the use of natural chitin and chitosan, their use for water remediation is very well reported. The adsorption properties of these natural polysaccharides depend on the chitin source from which chitosan is derived, its degree of N-acetylation, molecular weight, and solution properties, and also vary with crystallinity, affinity for water, percent deacetylation, and amino group content [47]. All these parameters, as determined by the selection of the preparation conditions, control of the swelling and diffusion properties of the biopolymer that finally influence the characteristics of the biomaterials. Performance is dependent on the type of material used, and the efficiency of adsorption depends on the accessibility of sorption sites. The dye uptake by chitin and chitosan is strongly pH dependent. Complicated structures of dyes are the deciding factor for uptake of the dyes by these polysaccharide materials [48]. The dye uptake mechanism is mostly controlled by intraparticle diffusion [49] though reports on ion-exchange mechanisms are also proposed. The major adsorption site of chitosan is a primary amine group which is easily protonated to form $-NH_3^+$ in acidic solutions. The strong electrostatic interaction between the NH_3^+ groups and dye anions can be used to explain the sorption mechanism [50]. The degree of adsorption by chitosan is different for different dyes depending upon their chemical structures [48]. In general, chitosan-based materials have been established as promising biosorbents for adsorption processes.

Chitosan can remove dissolved dyes from aqueous solutions in single systems of dyes [48, 51, 52], especially in the removal of acid [48] and reactive dyes [49]. Maximum adsorption capacities of chitosan [48] for acid orange 12, acid orange 10, acid red 73, and acid red 18 dyes have been evaluated to be 973.3, 922.9, 728.2, and 693.2 mg/g, respectively. Mahmoodi et al. [53] studied dye removal from colored textile wastewater using chitosan in single as well as binary dye systems. The removal of combination of Direct Red 23 and Acid Green 25 has been studied [53] by evaluating chitosan as an effective biosorbent for the removal of anionic dyes. The adsorption kinetics of dyes followed a pseudo-second-order kinetic model at various pH values. The biopolymer was suitable for decolorizing acidic-colored textile wastewater having low concentration of the dyes.

Though both chitosan beads and flakes are effective in adsorbing dyes from aqueous solution, chitosan beads show better results and faster kinetics [49, 54]. Performance of the chitosan also depends on the source of fishery wastes from which it has been derived. The maximum adsorption capacity for Reactive Red 222 by chitosan flakes and beads have been showed to be 293 and 1103 mg/g, respectively. Owing to high performance of chitosan in bead form, chitosan beads have been extensively used for the removal of dyes [49, 54–57]. The high surface area of chitosan beads as compared to flakes explains the better performance of the beads. In general, both batch-contacting and column processes have been undertaken for the dye removal using chitosan [58]. Chitosan nanoparticles (particle size = 180 nm; degree of deacetylation = 74%) have been used for the removal [51] of Acid Green 27 (AG27), from an aqueous solution. The dye concentration at equilibrium (Q_e, mg/g) was calculated using the weight of the nanoparticles in the mixed solution (Q_{es}) and the weight of chitosan in the nanoparticles (Q_{ep}). The Langmuir monolayer adsorption capacity (Q_0) with Q_{es} and Q_{ep} was found to be 1051.8 and 2103.6 mg/g, respectively, which were much ahead of micron-sized chitosan.

13.4.1.2 Chitin Gels

Chitin gels obtained by N-acetylation of chitosan in water–alcohol mixtures [59] have also been used in the dye removal, e.g., as Acid Blue 74, Reactive Violet 5, or Direct Red 28. Three kinds of transport mechanisms were investigated: (1) the sorption of solutes interacting with chitin, (2) the desorption of solutes without significant interaction with the chitin, and (3) osmosis phenomena. The sorption of Acid Blue 74 and Reactive Violet 5 depended on the charge density of the polymer network which was governed by the degree of association, pH, and the dielectric constant of the media. The sorption of Direct Red 28 has been attributed to the hydrophobic interactions and H-bonding at the extreme surface of the gel.

13.4.1.3 Chitosan-Derived Materials

Several derivatives of chitosan are known for effective removal of dyes from wastewater. Chitosan biomolecule is usually modified to obtain improved solubility in acidic and alkaline media in addition to solubility in some of the organic solvents. Many such modifications usually target for increase in the temperature and pH dependence of the dye adsorption. Most frequent and common method for chitosan modification is its cross-linking that enhances the performance of this biopolymer while maintaining its properties and original characteristics. Chemical modification in general improves the performance of chitosan besides increasing sorption selectivity, diffusion properties, and sensitivity to environmental conditions. Some of the common chitosan derivatives that have been utilized for dye removal are being summarized below.

Cross-Linked Chitosan

As chitosan is soluble in acidic media (due to its amino groups), its use as adsorbent under aqueous acidic conditions is not feasible. To change its solubility, many

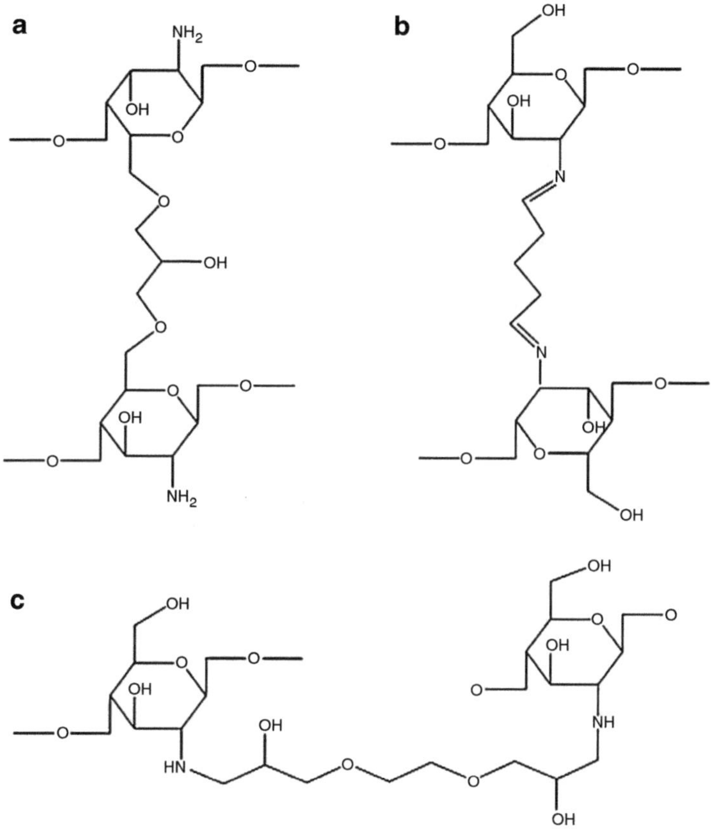

Fig. 13.3 Schematic representation of cross-linked chitosan beads: Chitosan-EPI, Chitosan-GLA, Chitosan EDGE (Reprinted from Crini et al. [60]. With kind permission of © Elsevier)

physical and chemical methods have been used, and among the various possible routes for modifying chitosan properties for the target application, cross-linking is the best option. Modification of the chitosan through cross-linking is known to be an inexpensive and easy route for fabricating chitosan-based biosorbents. On cross-linking, chitosan turns insoluble in aqueous acidic or alkaline media and most of the organic solvents. Such cross-linked derivatives are more resistant to temperature and low-pH conditions as compared to the native chitosan. It has been reported that cross-linked chitosan macromolecular materials perform well in dye removal as they are least affected by the environmental conditions [34]. The biomolecule has been cross-linked in the bead form using three common cross-linking agents, e.g., glutaraldehyde, epichlorohydrin, or ethylene glycol diglycidyl ether [50, 61, 62] (Fig.13.3). Among these three cross-linked chitosan beads, the chitosan–epichlorohydrin beads [62] show best performance for dye uptake. Cross-linked chitosan beads show outstanding removal capabilities for direct dyes and exhibit excellent performance for adsorption of anionic dyes when compared to activated carbon (~3–15 times better) at the same pH [50], while cross-linked chitosan show low affinity for cationic (basic) dyes.

A novel biosorbent, more hydrophilic than synthetic resins, has been developed where cyclodextrins and chitosan have been combined with several spacer arms without affecting the selectivity of both biopolymers. This biosorbent has been used for the decontamination of effluents containing the textile dyes [63]. The chitosan–cyclodextrin beads were characterized by a rate of sorption, and their efficiency was superior to that of the parent polysaccharide chitosan bead without CD and of the cross-linking cyclodextrin–epichlorohydrin gels. The maximum adsorption capacities of cross-linked cyclodextrin and chitosan, cyclodextrin mixed sorbents for Acid Blue 25 were 249, 88, and 77.4 mg/g, respectively.

Chemically Modified Chitosan

Though cross-linked chitosan has tremendous use in adsorbing anionic dyes, it is not of much help in the removal of cationic dyes. Useful adsorbents for cationic dyes have been derived by modification of chitosan functional groups such as enzymatic grafting of carboxyl groups [64]. The presence of carboxylic acid functional groups at grafted chitosan beads offer more surface polarity and sorption site density to improve the sorption selectivity of the basic dyes.

In general, grafting increases the efficiency of chitosan macromolecule (CTS) toward dye uptake especially the cationic dyes [64]. Four kinds of phenol derivatives: 4-hydroxybenzoic acid (BA), 3,4-dihydroxybenzoic acid (DBA), 3,4-dihydroxyphenyl-acetic acid (PA), and hydrocaffeic acid (CA) have been grafted at chitosan using tyrosinase enzyme. The grafting was followed by taking FTIR spectra of the graft copolymers. The carboxyl groups content in the grafted chitosan were measured to be 46.36 for BA, 70.32 for DBA, 106.44 for PA, and 113.15 for CA. In a batch adsorption experiment, these modified chitosans proved efficient in removing cationic dyes such as crystal violet (CV) at pH 7 and Bismarck brown Y (BB) at pH 9 at 30°C. Unilayer adsorption was established for the adsorption where the maximum adsorption capacities for both the dyes were in the following order: CTS-CA > CTS-PA > CTS-DBA > CTS-BA.

Vinyl modification of chitosan (Fig. 13.4) also proved very useful for deriving bioadsorbents from chitosan. Poly(methyl methacrylate)- [65] and poly(acrylamide)- [66] grafted chitosan have been evaluated for removing dyes. Chitosan-graft-poly(methyl methacrylate) (Ch-g-PMMA) proved efficient in adsorbing anionic azo dyes such as Procion Yellow MX, Remazol Brilliant Violet, and Reactive Blue H5G. The copolymer showed better performance than native chitosan in terms of uptake of anionic dyes from synthetic dye solutions and real textile wastewater. Unlike chitosan, which worked best at pH 4, Ch-g-PMMA worked well in a broad pH range (pH 4–10). The concentration of the dyes and temperature of the adsorption experiment also had a marked effect on

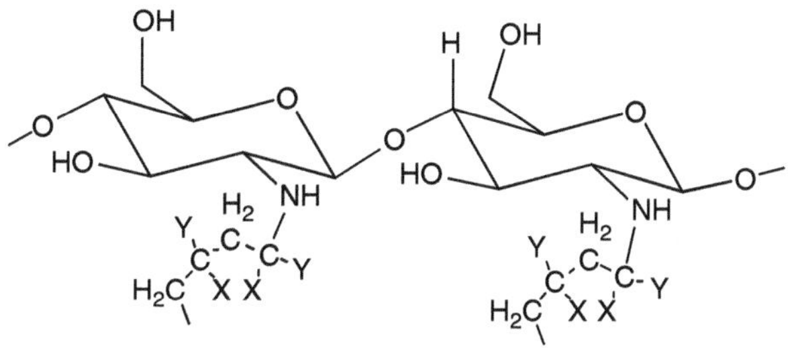

(A) X=$CONH_2$Y = H Polyacrylamide grafts

(B) X=COOMe;Y = Me Polymethylmethacrylate grafts

Fig. 13.4 Vinyl-grafted chitosan. (*A*) Chitosan-graft-poly(acrylamide); (*B*) Chitosan-graft-poly(methyl methacrylate)

the dye uptake. On increasing %G in the copolymer, uptake of the dyes could be significantly enhanced which confirmed the role of the graft chains in the removal mechanism. The dye uptake by the copolymer was fast and could reach the equilibrium in just ~3 h. The adsorption equilibrium data for the dye adsorption best fitted to pseudo-second-order kinetic model indicating chemisorption was taking place. The adsorbent affinity toward the dyes was in the order violet > yellow > blue for both chitosan and Ch-g-PMMA.

Another grafted chitosan, chitosan-graft-poly(acrylamide) (Ch-g-PAM) [66] has also proved equally effective in the removal of anionic dyes from real textile waste and synthetic dye solution. This modification also broadened the pH range of adsorption (pH 2–8), but adsorption by this copolymer had slightly higher equilibrium time (6 h) than that by Ch-g-PMMA. Chemisorption governed the adsorption, and the copolymers had reusability for almost ten consecutive cycles. Thus, overall, it may be concluded that by introducing vinyl grafts, chitosan may be converted to useful and effective biosorbent for anionic dyes.

Chitosan Composites

A more recent approach to utilize chitosan for dye removal involves the preparation of chitosan composites [67]. Many chitosan composites have been used for dye removal, e.g., chitosan clay composites [67–72], chitosan/polyurethane composites [73], chitosan/oil palm ash composites [74], chitosan/silica [75–77], and chitosan/titania composites [78].

Chitosan–Clay Composites

Montmorillonite (a natural phyllosilicate)–chitosan composite showed efficient removal of Congo red dye [70]. The adsorption property of the composite was decided by the optimum molar ratio of chitosan to montmorillonite in the composite. X-ray analysis of the composite revealed intercalation of chitosan macromolecules in between the montmorillonite interlayers which disrupt the crystalline structure of montmorillonite. Electrostatic attraction and chemisorption mechanisms were found to be responsible for the sorption [70]. At low pH values, electrostatic interaction is possible between the protonated amino groups at chitosan and anionic dyes; however, at alkaline pH, the excessive amount of hydroxyl ions compete with the anionic dyes for the active sites at the composite and reduce the dye uptake. Chemisorption was suggested to be the sorption phenomenon as the adsorption kinetic data followed pseudo-second-order kinetics. The monolayer adsorption suggested as the adsorption equilibrium data could be successfully modeled using Langmuir isotherm with an $Q_{\max}$ value of 54.52 mg/g.

Chitosan/activated clay composite beads [68] have been fabricated, and it was found that the addition of activated clay can enhance the hardness of the chitosan beads which agglomerate easily to facilitate the separation of the adsorbents from the solution without swelling. The adsorption studies of Methylene Blue and reactive dye (RR22) onto chitosan composite revealed that the composites had a comparable adsorption to chitosan beads.

Cross-linked chitosan/bentonite composites [67] have been used to adsorb tartrazine azo dye. Bentonite contains a high proportion of swelling clays consisting of SiO_2, Al_2O_3, CaO, MgO, Fe_2O_3, Na_2O, and K_2O [79, 80] and is a 2:1 type aluminosilicate [81]. The chitosan composites, cross-linked with epichlorohydrin, had improved performance as adsorbent over native chitosan composite [34]. A cross-linking agent stabilizes chitosan in acid solutions to make it insoluble. It has been established [67] that the pH of the tartrazine played a crucial role in the adsorption process. As pH was increased, the surface of the adsorbent became negatively charged to increase repulsion between tartrazine and the adsorbent surface which decreased the adsorption capacity of the composite.

Using kaolinite (1:1 aluminosilicate, consisting of SiO_2, Al_2O_3, and H_2O) [69] and Fe_2O_3 [72] a composite bead was prepared. Scanning electron microscope (SEM) and tunneling electron microscope (TEM) images revealed many pores and pleats on the surface of the composites that were attributed to be the active sites for dye entrapment. The composites exhibited good adsorption ability for methyl orange at pH 6. Addition of anions depressed the adsorption of methyl orange that was due to preferred adsorption of anions over methyl orange.

Chitosan/Polyurethane Composites

Chitosan/polyurethane composite [73] has been used for the removal of Violet 48 dye. Chitosan can be easily immobilized into the polyurethane matrix foam at 0.25 wt.% of glutaraldehyde concentration [82]. SEM analysis revealed that the composite had an open structure [83, 84] which can be responsible for its accessibility to acid dyes [82]. It was observed that the adsorption capacity of neat polyurethane was relatively lower than the composite indicating that the amine groups of the neat polyurethane are unable to specifically react to an active site of the dye.

Chitosan/Oil Palm Ash Composites

Cross-linked chitosan/oil palm composite beads [74] have been used to remove Reactive Blue 19 by adsorption. The dye uptake was dependent on pH, at lower pH the protonated amino groups of chitosan are electrostatically attracted by the negatively charged dye molecules while at pH >10, the adsorption capacity of chitosan beads declines because of the decrease in porosity whereby the pore diameter makes transfer of dye molecules difficult.

Chitosan/SiO_2 Composite

Chitosan/SiO_2 and chitin–silica composites have been developed for the removal of toxic metals and dyes [75–77]. Acid-catalyzed polymerization of tetraethoxysilane in presence of chitosan furnished the sol–gel-derived hybrid composites [76]. The composite was tested for the removal of dyes having different chemical properties, toxicity, and industrial applications like the Remazol Black B, textile and food dye (Erythrosine B), biological dye (Gentian Violet), and biological and chemical dye (Neutral Red). The polysaccharides in the form of silica hybrid hydrogels had complete compatibility with the ethanol that was released during the hydrolysis of tetraethoxysilane. In both matrices, chitosan and chitin retain their capability of dye removal inside the SiO_2 network while the composites at the same time enjoy the improved mechanical properties in comparison to native polysaccharides. The adsorption was found to be pH dependent which indicated a spontaneous charge-associated interaction mechanism of dye adsorption. The composite had both polysaccharide and silica network affinity for the dyes.

Chitosan/TiO_2

TiO_2/chitosan microporous composites [78] have been used for adsorption and degradation of two reactive dyes, Methylene Blue (MB) and Benzopurpurin (BP), from aqueous solutions. An irreversible adsorption that decreases with the increase of temperature was indicated for these dyes.

Since the traditional methods of extraction of chitin and its conversion to chitosan generate large quantities of waste, it is difficult to develop chitosan-based materials as adsorbents at an industrial scale. Though a huge volume of published articles is available on using chitin and chitosan for dye removal, there has been little understanding on the sorption mechanisms since these articles mostly focus on the evaluation of sorption performances. Mainly ion-exchange interactions, hydrophobic attraction, and physical adsorption are said to be operating simultaneously in most of these sorption processes. Wide ranges of chemical structures, pH, salt concentrations, and presence of ligands often add to the complication.

13.4.1.4 Starch and Cyclodextrins

Starch is a low-cost abundantly available renewable biopolymer which has a wide range of applications in its native and modified form. In its crude form, starch is a mixture of amylose and amylopectin. Amylose is a linear polymer of 1,4-linked α-D-glucopyranosyl units [85], while amylopectin is a highly branched polymer of D-glucopyranosyl residues linked together mainly by $1 \rightarrow 4$ linkages with $1 \rightarrow 6$ bonds at the branch points [86]. Adsorptions by starch based materials are reported to take place by physical adsorption, complexation, and ion-exchange interactions [87, 88]. Cyclodextrins are the cyclic derivatives of starch; three smallest cyclodextrins, alpha-cyclodextrin, beta-cyclodextrin, and gamma-cyclodextrin, having six, seven, and eight α-1,4-linked D-glucopyranose units, respectively, are well known and are commercially used. Use of cyclodextrins as low-cost sorbents is well studied [87].

Cyclodextrin hydroxyl groups are utilized to cross-link with coupling agents like epichlorohydrin to form water-insoluble cross-linked networks. As the cross-linking units are hydrophilic in nature, the cross-linked cyclodextrins have remarkably high swelling capacity in water, which allow the swelling of their network for a fast diffusion of the pollutants. They also contain hydrophobic sites for trapping nonpolar dyes efficiently [87]. Cross-linked starch polymers [88] having amine groups have also been reported (Fig. 13.5).

The cross-linked polymers have been prepared in one step by reticulation of starch-enriched flour using epichlorohydrin with NH_4OH. Using this biosorbent, in a column experiment, several hundred ppm of dyes could be effectively removed from water. The control of the cross-linking reaction offers control on the material's properties in terms of the sorption. The adsorption by this material was highly dependent on pH as protonation of the amine groups on the surface of the sorbent is possible only at acidic pH.

Amphoteric starch has also been used for dye removal [89–91]. Such starch derivatives being amphoteric (containing cationic or anionic groups) in nature can quickly remove pollutants. Starch has been modified to amphoteric starch derivatives [92] having quaternary ammonium and carboxymethyl groups by semidry reaction. These derivatives have been used for cationic and anionic dye uptake and are suitable for both acid and basic dyes. Cross-linked starch was first etherified by reacting with 65% 3-chloro-2-hydroxypropyl trimethylammonium chloride for 2 h at 80°C in a cylindrically shaped reactor with a mechanical stirrer. The resulting product after rinsing and neutralization with 80% ethanol solution containing hydrochloric acid was carboxymethylated using a fixed dose of chloroacetic acid with 30% water in the whole reaction system at 45°C. Nitrogen content and the quaternary ammonium groups of the adsorbent have been used for interacting with acid dyes, while basic dyes interact with carboxymethyl groups. It was found that the low temperature facilitates the adsorption of acid dyes while the basic dyes have the highest adsorption capacity at 303 K. Both Langmuir isotherm and Freundlich isotherm models were well applied for the adsorption, and the kinetic study showed the pseudo-second-order model for the adsorption, indicating chemisorption. Until now, there are only limited data on the adsorption of dyes onto amphoteric starch, and this area of investigation need further investigations in terms of their use for the removal of dyes.

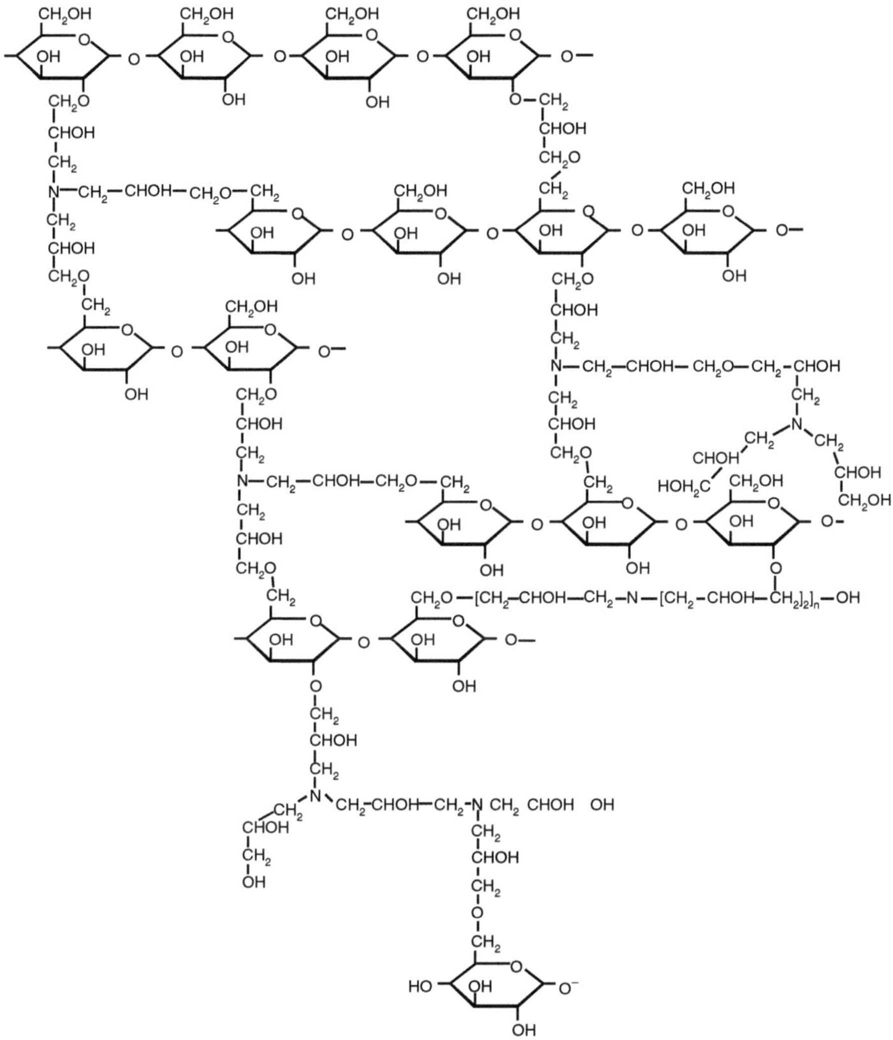

Fig. 13.5 A possible structure of cross-linked starch containing tertiary amine group (Reprinted from Crini et al. [60]. With kind permission of © Elsevier)

In spite of varied characteristics and properties of starch-based derivatives, a limited number of dye adsorption studies using this polysaccharide material have been carried.

13.4.1.5 Seed Gums as Coagulants

Using natural seed gums as a replacement coagulant for proprietary coagulants can meet the need for water and wastewater technology in developing countries which is simple to use, robust, and cheap to both install and maintain. Natural coagulants have been used for centuries in traditional water treatment practices throughout certain areas of the developing world. The utilization of *Cyamopsis tetragonolobus* (family Leguminosae) seeds for this purpose is an event which has resulted in the emergence of one of the most important plant-based industries of India and has contributed in its economic development. This galactomannan, known as guar gum, has found varied uses in modern industries throughout the world. Guar has US Public Health Service approval in potable water treatment as coagulant aid together with alum, ferric sulfate, and lime.

Polymeric flocculants, synthetic as well as natural, because of their natural inertness to pH changes, low dosage, and easy handling, are becoming popular in industrial effluent treatment. Among natural and synthetic polymers, synthetic polymers are very effective due to their versatile tailorability, but they have poor shear stability. Natural polymers are of low cost, shear stable, nontoxic, and highly biodegradable. It has been possible to develop efficient, shear stable, and less biodegradable flocculants by grafting synthetic polymers like polyacrylamide branches onto the rigid backbone of the natural polysaccharides. The dangling grafted chains have easy approachability to contaminants in effluents.

Tamarind Gum

Tamarind trees are known to be rich in mucilaginous content. The native tamarind gum and its graft copolymers are widely used as flocculant in color removal [93] of industrial effluents. Poly(acrylamide) grafting increases the shelf life of tamarind seed mucilage though the graft copolymer (Tam-g-PAM) still remains biodegradable. Compared to native tamarind gum, Tam-g-PAM showed better flocculation efficiency in the removal of dyes. Tamarind gum-derived flocculants show better performance for azo and reactive dyes than basic dyes and are attractive because of its low cost and biodegradability as compared to other synthetic flocculants.

Psyllium Seed Gum

Psyllium (*Plantago ovata*) [94] is a profusely branched medicinal plant whose seed husk [95] has been used as traditional and herbal medicine. The psyllium seeds are coated with mucilage (10–30%), a gelatinous material which swells upon exposure to moisture. Chemically, *Psyllium* seed husk is xylan which occurs in association with cellulose. The seed husk [96] has been evaluated for the decolorized three different synthetic dyes: direct – Kahi Green (DKG), acidic – Sandolan Red RSNI (ASR), and reactive – Procion Brilliant Blue RS (PBB) dyes at laboratory scale. Its dose and pH were found two important factors [97] that governed the coagulation mechanism. Psyllium husk proved very effective for the decolorization of anionic dyes when used with polyaluminium chloride.

Removal of color from dye solutions generally takes place by physiochemical mechanisms of coagulation and/or chelation-complexation-type reactions [98] which were found to be mainly of physiochemical nature. The structure of the dyes is such that they may result in a chelation–complex formation with chemical coagulants that leads to the formation of insoluble complexes between metal and dye. The complex may either precipitate from solution or may be removed by adsorption onto metal hydroxy species [98]. The color removal was suggested by aggregation/precipitation and adsorption of coloring substances onto the polynuclear coagulant species [99] and onto hydrated flocs [100]. DKG was easily decolorized by psyllium as coagulant aids whereas the decolorization of ASR was found to be moderately effective. The reactive dye, which is the most difficult to remove by conventional methods, could also be decolorized to a large extent using the psyllium seed gum. Psyllium gum though much less efficient than PAC when used as a coagulant aid with a very low dose of PAC becomes much effective for the decolorization of the dyes.

Guar Gum

Traditional water treatment utilizes the natural coagulants [101] for centuries. The galactomannan from seed of *Cyamopsis tetragonolobus* N.O. Leguminosae is popular as coagulant, the guar gum increases the size of the floc initially formed by the coagulant, thereby increasing the rate of setting of solid impurities, reducing solid carry over to the filters, and increasing periods between backwashes. In industrial waters, guar flocculates clays, carbonates, hydroxides, and silica when used alone or in conjunction with inorganic coagulants. The gum helps in increasing the size of the flocs which is initially formed by the coagulant. In industrial waters, guar has been used to flocculate clays, carbonates, hydroxides, and silica individually or in conjunction with inorganic coagulants.

Ipomoea Seed Gums

The seed galactomannans of *Ipomoea* genera N.O. Convolvulaceae have been used as coagulants and for the decolorizing of textile dye effluents [29]. Physicochemical and chemical studies showed that the protein-free *Ipomoea* seed gum exhibits similar behavior to the guar galactomannan. These seed gums on chemical modifications produce materials which can be modified to fine products [25, 102] having a broader spectrum of properties for various industrial applications. The decolorization [103] of three different classes of synthetic dye was undertaken on laboratory scale with *Ipomoea* seed gums and their poly(acrylamide) derivatives [104]. The *Ipomoea* seed gums from *Ipomoea turpethum* (IT) [105], *Ipomoea dasysperm* [24], and *Ipomoea quamocalit* (IQ) in comparison to guar gum (GG) [103] have been used for the coagulation studies. Structure of the repeating units of some of seed gums is shown in Fig. 13.6.

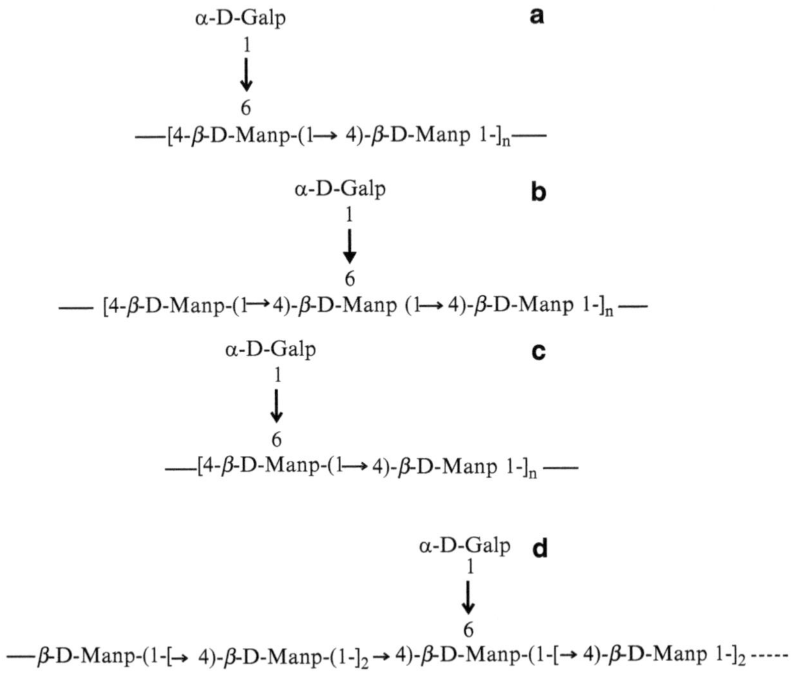

Fig. 13.6 Structures of the repeating units of seed gums. (**a**) *Ipomoea turpethum*, (**b**) *Ipomoea quamoclit*, (**c**) Guar gum, (**d**) *Ipomoea turpethum*

Coagulation experiments were done with taking conventional coagulant PAC as reference [22]. Among the varied coagulation parameters, coagulation dose and coagulation pH were found to be the most important factors which influenced the types of hydrolysis products and therefore the coagulation mechanism. These seed gums when used as coagulant aid with a very small dose of PAC could effectively remove each class of dye as compared to when PAC was used alone.

Ipomoea dasysperma seed gum is a galactomannan polysaccharide having galactose to mannose ratio in 1:6 ratio [24]. The seed gum has a branched structure having a linear chain of $\beta(1 \rightarrow 4)$-linked mannopyranosyl units to which D-galactose side chains are attached by $\alpha(1 \rightarrow 6)$ linkages. The decolorization of dye solutions by coagulation with ID seed gum [24] is dependent largely on the type of dye, pH, and coagulant dosages, pH 9.5 being the optimum. The seed gum showed the most efficient removal of the direct dyes, while when used with a very low dose of PAC, the utility of the biosorbent was extended for acid as well as Procion dye. The natural galactomannans from guar gum and ID gum offer a sustainable method of effluent treatment, as these gums are low cost and are readily available.

Cassia Seed Gums

Cassia is an abundantly available plant of tropical countries including India. These plants are a good source of mucilages, flavonoids, anthraquinones, and polysaccharides. *C. javahikai* seed gum was the first seed gum from *Cassia* plant that has been investigated as a coagulant in textile wastewater treatment. Its seed endosperm contains a nonionic, water-soluble galactomannan having 1:2 galactose to mannose ratio [106]. Like guar gum, the seed gum has a branched structure consisting of a linear chain of $\beta(1 \rightarrow 4)$-linked mannopyranosyl units having side branches of $\alpha(1 \rightarrow 6)$-linked galactose units. *C. javahikai* (CJ) seed gum [31] and its poly(acrylamide)-grafted (CJG) derivative proved to be efficient coagulants for the decolorization of direct dyes. In combination with very low dose of PAC, both CJ and CJG were found to be efficient coagulant aids for acid as well as Procion dye. The performance of CJG over CJ was enhanced more so for direct dyes.

13.4.1.6 Polysaccharide Hydrogels

Superabsorbent hydrogel (SH) based on polysaccharides may be used in an alternative procedure for removal of some aqueous solutes. The SH are formed of three-dimensional cross-linked polymer networks of flexible chains that are able to absorb and retain water and solute molecules. The high water content and porous network structure of the hydrogels allow solute diffusion [107] into it. These can be prepared by the copolymerization of polysaccharides with hydrophilic monomers and polyfunctional monomers, with cross-linkers like bisacrylamide, glutaraldehyde, etc., in the presence of redox initiators. On water immersion, hydrophilic polymer chains of the supra-adsorbent hydrogels are not dissolved into the aqueous phase as they are cross-linked [108]. Ionic groups at the SH hydrogels can absorb and trap ionic dyes from wastewater, such as Methylene Blue from an aqueous environment. A reasonably priced and biodegradable SH based on arabic gum (AG) has been [109] obtained by chemically modified AG with acrylamide and sodium acrylate by a copolymerization reaction. To obtain supra-hydrogel, AG 5% (m/v) was modified with glycidyl methacrylate. Modified AG, sodium acrylate, and acrylamide were stirred for 15 min in presence of 10 mmol sodium persulfate for 20 min at 70°C. The hydrogels were dried using air flow. The maximum MB absorptions were 48 mg g^{-1} (98%) for the (1.5–0.5–0.5) hydrogels and 39 mg g^{-1} (80%) for the (1.0–0.5–0.5) SH. The maximum desorption was 46 mg MB per g (1.5–0.5–0.5) of the hydrogel.

13.5 Conclusion

Dyes are the part and participle of many industries, and residual dye contents in the industrial effluents are becoming a threat to mankind as there has been high increase in the production and utilization of the dyes in past few decades. For removing dyes from industrial wastes, various techniques are in practice, but still, there is a requirement for the low-cost methodology that can be used universally for all types of dyes. Adsorption is a method of choice for removing dyes from wastewater in combination with other procedures. High cost of the conventional adsorbents (e.g., activated carbon) and expensive regeneration procedures is becoming detrimental for the commercialization of this method of dye remediation. Due to increasing awareness of environmental protection, the effectiveness and cost of treatment processes for dyes is important, and to ensure this, the use of polysaccharides has now emerged as important and fast-growing field. Being low cost and renewable, their use as adsorbent has been encouraged in native or modified forms. Available literature revealed that a wide range of nonconventional as well as commercial polysaccharides can be efficiently exploited as low-cost adsorbents. Among the conventional polysaccharides, the chitosan has been most widely studied for such purpose and chitosan-based sorbent demonstrated outstanding removal capabilities for certain dyes in comparison to activated carbon. This area of research however needs more attention in terms of better understanding of the sorption mechanisms and scaling up of the experiments which have been successful on laboratory scale and to test the established polysaccharide-based adsorbents for real industrial effluents. There is a need of time to develop an adsorbent based on these biorenewable polysaccharides which can be universally applied to all the dye types under amiable conditions, like pH, temperature, and in presence of other contaminants which are normally found in the effluents.

References

1. Singh R (2002) Synthetic dyes. Mittal Publication, New Delhi, pp 1–241
2. Janoš Pavel, Coskun S, Pilařová V, Rejnek J (2009) Removal of basic (methylene blue) and acid (egacid orange) dyes from waters by sorption on chemically treated wood shavings. Bioresour Technol 100:1450–1453
3. Clarke EA, Anliker R (1980) Organic dyes and pigments. In: Handbook of environmental chemistry, anthropogenic compounds, part A, vol 3. Springer, New York, pp 181–215
4. Bae JS, Freeman HS (2007) Aquatic toxicity evaluation of new direct dyes to the *Daphnia magna*. Dye Pigment 73:81–85
5. Akgerman A, Guzel B (2000) Mordant dyeing of wool by supercritical processing. J Supercrit Fluid 18:247–252
6. Štěpánková M, Wiener J, Rusinová K (2011) Decolourization of vat dyes on cotton fabric with infrared laser light. Cellulose 18:469–478
7. Yuksel E, Gurbulak E, Eyvaz M (2011) Decolorization of a reactive dye solution and treatment of a textile wastewater by electrocoagulation and chemical coagulation: techno-economic comparison. Environ Prog Sustain Energy (in press)
8. Reife A, Freeman HS (2007) Dyes for polyester: disperse dyes. Chim Oggi 25:38–41
9. Saratale GD, Chang JS, Govindwar SP (2011) Bacterial decolorization and degradation of azo dyes: a review. J Taiwan Inst Chem Eng 42:138–157
10. Ozyurt M, Ataçag H (2003) Biodegradation of azo dyes: a review. Fresenius Environ Bull 12:1294–1302
11. Chattopadhyay SN, Pan NC, Roy AK, Khan A (2009) Coloration of jute fabric using sulphur dyes. Int Dye 194:35–39

12. Banat IM, Nigam SP, Marchant R (1996) Microbial decolorization of textile dye containing effluents: a review. Bioresour Technol 58:217–227
13. Ganesh R, Boardman GD, Michelsen D (1994) Fate of azo dyes in sludges. Water Res 28:1367–1376
14. Solozhenko EG, Soboleva NM, Goncharuck VV (1995) Decolorization of azo dyes solutions by Fentons oxidation. Water Res 29:2206–2210
15. Ibrahim NA, El-Gamal AR, Mahrous F (2008) Improving the environmental aspects of sulphur dyeing of cotton knitted fabrics. J Nat Fiber 5:238–250
16. Robinson T, McMullan G, Marchant R, Nigam P (2001) Remediation of dyes in textiles effluent: a critical review on current treatment technologies with a proposed alternative. Bioresour Technol 77:247–255
17. Kang SF, Liao CH, Chen MC (2002) Pre-oxidation and coagulation of textile wastewater by the Fenton process. Chemosphere 46:923–928
18. Lin SH, Peng FC (1996) Continuous treatment of textile wastewater by combined coagulation, electrochemical oxidation and activated sludge. Water Res 30:587–592
19. Arslan I, Akmehmet BI, Tuhkanen T (1999) Oxidative treatment of simulated dyehouse effluent by UV and near-UV light assisted Fenton's reagent. Chemosphere 39:2767–2783
20. Zonoozi MH, Moghaddam MR, Arami M (2009) Coagulation/flocculation of dye-containing solutions using polyaluminium chloride and alum. Water Sci Technol 59:1343–1351
21. Edzwald JK (1993) Coagulation in drinking water treatment: particles, organics and coagulants. Water Sci Technol 27:21–35
22. Viraraghavan T, Wimmer CH (1988) Polyaluminium chloride as an alternative to alum coagulation: a case study. In: Proceedings of Canadian society of civil engineers annual conference, pp 480–498
23. Sanghi R, Bhattacharya B, Singh V (2002) *Cassia angustifolia* seed gum as an effective natural coagulant for decolorisation of dye solutions. Green Chem 4:252–254
24. Sanghi R, Bhattacharya B, Dixit A, Singh V (2006) *Ipomoea dasysperma* seed gum: an effective natural coagulant for the decolorization of textile dye solutions. J Environ Manage 81:36–41
25. Singh RP, Karmakar GP, Rath SK, Pandey SR, Triphaty T, Panda J, Kanan K, Jain SK, Lan NT (2000) Biodegradable drag reducing agents and flocculants based on polysaccharides: materials and applications. Polym Eng Sci 40:46–60
26. Singh RP, Tripathy T, Karmakar GP, Rath SK, Karmakar NC, Pandey SR, Kannan K, Jain SK, Lan NT (2000) Novel biodegradable flocculants based on polysaccharides. Curr Sci 78:798–803
27. Shultz CR, Okun DA (1983) Treating surface waters for communities in developing countries. J Am Water Work Assoc 75:212–219
28. Jahn SAA (1988) Using Moringa seeds as coagulants in developing countries. J Am Water Work Assoc 6:43–50
29. Joshi VA, Nanoti MV (1999) Lab studies on Tarota as coagulant aid in water treatment. Ind J Environ Prot 19:451–455
30. Christman RF (1967) Performance of chitosan as polyelectrolyte. Food, Chemical and Research Corporation, report for Kypro Co., Bellvina, Washington
31. Sanghi R, Bhattacharya B, Singh V (2006) Use of *Cassia javahikai* seed gum and gum-g-polyacrylamide as coagulant aid for the decolorization of textile dye solutions. Bioresour Technol 97:1259–1264
32. Velmurugan P, Kumar RV, Dhinakaran G (2011) Dye removal from aqueous solution using low cost adsorbent. Int J Environ Sci 1:1492–1503
33. Volesky B (1999) Biosorption for the next century. In: Ballester A, Amils R (eds) Biohydrometallurgy and the environment toward the mining of the 21st century. International biohydrometallurgy symposium proceedings, volume B. Elsevier Sciences, Amsterdam, the Netherlands, pp 161–170

34. Crini G, Badot PM (2008) Application of chitosan, a natural aminopolysaccharide, for dye removal from aqueous solutions by adsorption processes using batch studies: a review of 821 recent literature. Prog Polym Sci 39:399–447
35. Chandra R, Rastogi R (1998) Biodegradable polymers. Prog Polym Sci 23:1273–1335
36. Gonera A, Goclik V, Baum M, Mischnick P (2002) Preparation and structural characterization of O-aminopropyl starch and amylose. Carbohydr Res 337:2263–2272
37. Wesslen BK, Wesslen B (2002) Synthesis of amphiphilic amylose and starch derivatives. Carbohydr Polym 47:303–311
38. Rachel A-V, Rinaudo M (2003) Synthesis of starch derivatives with labile cationic groups. Int J Biol Macromol 31:123–129
39. Le TC, Lacroix M, Ispas-Szabo P, Mateescu MA (2003) N-acylated chitosan: hydrophobic matrices for controlled drug release. J Control Release 93:1–13
40. Kim TH, Jiang HL, Jere D, Park IK, Cho MH, Nah JW, Choi YJ, Akaike T, Cho CS (2007) Chemical modification of chitosan as a gene carrier in vitro and in vivo. Prog Polym Sci 32:726–753
41. Li F, Chen W, Tang C, Zhang S (2009) Development of hydrogen peroxide biosensor based on in situ covalent immobilization of horseradish peroxidase by one-pot polysaccharide-incorporated sol-gel process. Talanta 77:1304–1308
42. Mourya VK, Inamdar NN (2008) Chitosan-modifications and applications: opportunities galore. React Funct Polym 68:1013–1051
43. Wang J, Jin X, Chang D (2009) Chemical modification of chitosan under high-intensity ultrasound and properties of chitosan derivatives. Carbohydr Polym 78:175–177
44. Varma AJ, Deshpande SV, Kennedy JF (2004) Metal complexation by chitosan and its derivatives: a review. Carbohydr Polym 55:77–93
45. Hudson SM, Smith C (1998) Polysaccharide: chitin and chitosan: chemistry and technology of their use as structural materials. In: Kaplan DL (ed) Biopolymers from renewable resources. Springer, New York, pp 96–118
46. Gupta VK, Suhas (2009) Application of low-cost adsorbents for dye removal – a review. J Environ Manage 90:2313–2342
47. Guibal E (2004) Interactions of metal ions with chitosan-based sorbents: a review. Sep Purif Technol 38:43–74
48. Wong YC, Szeto YS, Cheung WH, McKay G (2004) Adsorption of acid dyes on chitosan-equilibrium isotherm analyses. Proc Biochem 39:693–702
49. Wu FC, Tseng RL, Juang RS (2000) Comparative adsorption of metal and dye on flake- and bead-types of chitosan prepared from fishery wastes. J Hazard Mater B73:63–75
50. Chiou MS, Ho PY, Li HY (2004) Adsorption of anionic dyes in acid solutions using chemically cross-linked chitosan beads. Dye Pigment 60:69–84
51. Hu ZG, Zhang J, Chan WL, Szeto YS (2006) The sorption of acid dye onto chitosan nanoparticles. Polymer 47:5838–5842
52. Piccin JS, Vieira MLG, Gonçalves JO, Dotto GL, Pintoz LAA (2009) Adsorption of FD&C Red No. 40 by chitosan: isotherms analysis. J Food Eng 95:16–20
53. Mahmoodi NM, Salehi R, Arami M, Bahrami H (2011) Dye removal from colored textile wastewater using chitosan in binary systems. Desalination 267:64–72
54. Wu FC, Tseng RL, Juang RS (2001) Enhanced abilities of highly swollen chitosan beads for color removal and tyrosinase immobilization. J Hazard Mater B81:167–177
55. Chatterjee S, Chatterjee S, Chatterjee BP, Das AR, Guha AK (2005) Adsorption of a model anionic dye, eosin Y, from aqueous solution by chitosan hydrobeads. J Colloid Interface Sci 288:30–35
56. Chatterjee S, Chatterjee S, Chatterjee BP, Guha AK (2007) Adsorptive removal of congo red, a carcinogenic textile dye by chitosan hydrobeads: binding mechanism, equilibrium and kinetics. Colloid Surf A 299:146–152
57. Morais WA, Fernandes ALP, Dantas TNC, Pereira MR, Fonseca JLC (2007) Sorption studies of a model anionic dye on cross-linked chitosan. Colloid Surf A310:20–31

58. McKay G, Blair HS, Gardner JR (1989) The adsorption of dyes onto chitin in fixed bed column and batch adsorbers. J Appl Polym Sci 28:1499–1544
59. Vachoud L, Zydowicz N, Domard A (2001) Sorption and desorption studies on chitin gels. Int J Biol Macromol 28:93–101
60. Crini G et al (2005) Recent developments in polysaccharide-based materials used as adsorbents in wastewater treatment. Prog Polym Sci 30:38–70
61. Chiou MS, Li HY (2002) Equilibrium and kinetic modelling of adsorption of reactive dye on cross-linked chitosan beads. J Hazard Mater B93:233–248
62. Chiou MS, Li HY (2003) Adsorption behaviour of reactive dye in aqueous solution on chemical cross-linked chitosan beads. Chemosphere 50:1095–1105
63. Martel B, Devassine M, Crini G, Weltrowski M, Bourdonneau M, Morcellet M (2001) Preparation and sorption properties of a beta-cyclodextrin-linked chitosan derivative. J Polym Sci Part A: Polym Chem 39:169–176
64. Chao AC, Shyu SS, Lin YC, Mi FL (2004) Enzymatic grafting of carboxyl groups on to chitosan—to confer on chitosan the property of a cationic dye adsorbent. Bioresour Technol 91:157–162
65. Singh V, Tripathi DN, Tiwari A, Sanghi R (2006) Microwave synthesized Chitosan graft-poly(methylmethacrylate): an efficient Zn^{+2} ion binder. Carbohydr Polym 65:35–41
66. Singh V, Sharma AK, Sanghi R (2009) Poly(acrylamide) functionalized chitosan: an efficient adsorbent for azo dyes from aqueous solutions. J Hazard Mater 166:327–335
67. WanNgah WS, Ariff NFM, Hanafiah MAKM (2010) Preparation, characterization, and environmental application of cross-linked chitosan-coated bentonite for tartrazine adsorption from aqueous solutions. Water Air Soil Pollut 206:225–236
68. Chang MY, Juang RS (2004) Adsorption of tannic acid, humic acid and dyes from water using the composite of chitosan and activated clay. J Colloid Interface Sci 278:18–25
69. Nandi BK, Goswami A, Purkait MK (2009) Adsorption characteristics of brilliant green dye on kaolin. J Hazard Mater 161:387–395
70. Wang L, Wang A (2007) Adsorption characteristics of congo red onto the chitosan/montmorillonite nanocomposite. J Hazard Mater 147:979–985
71. Wang CC, Juang LC, Hsu TC, Lee CK, Lee JF, Huang FC (2004) Adsorption of basic dyes onto montmorillonite. J Colloid Interface Sci 273:80–86
72. Zhu HY, Jiang R, Xiao L (2010) Adsorption of an anionic dye by chitosan/kaolin/_-Fe2O3 composites. Appl Clay Sci 48:522–526
73. Won SL, Lee HC, Jeong YG, Min BG, Lee SC (2009) Preparation and acid dye adsorption behavior of polyurethane/chitosan composite foams. Fiber Polym 10:636–642
74. Hameed BH, Hasan M, Ahmad AL (2008) Adsorption of reactive dye onto cross-linked chitosan/oil palm ash composite beads. Chem Eng J 136:164–172
75. Cestari AR, Vieira EF, Pinto AA, Lopes EC (2005) Multistep adsorption of anionic dyes on silica/chitosan hybrid. 1. Comparative kinetic data from liquid- and solid-phase models. J Colloid Interface Sci 292:363–372
76. Copello GJ, Meberta AM, Raineria M, Pesentia MP, Diaza LE (2011) Removal of dyes from water using chitosan hydrogel/SiO_2 and chitin hydrogel/SiO2 hybrid materials obtained by the sol–gel method. J Hazard Mater 186:932–939
77. Li F, Li J, Zhang S (2008) Molecularly imprinted polymer grafted on polysaccharide microsphere surface by the sol–gel process for protein recognition. Talanta 74:1247–1255
78. Zubieta CE, Messina PV, Luengo C, Dennehy M, Pieroni O, Schulz PC (2008) Reactive dyes remotion by porous TiO_2–chitosan materials. J Hazard Mater 152:765–777
79. Holzer L, Münch B, Rizzi M, Wepf R, Marschall P, Graule T (2010) 3D microstructure analysis of hydrated bentonite with cryo-stabilized pore water. Appl Clay Sci 47:330–342
80. Li Q, Yue QY, Sun HJ, Su Y, Gao BY (2010) A comparative study on the properties, mechanism and process designs for the adsorption of non-ionic or anionic dyes onto cationic-polymer/bentonite. J Environ Manage 91:1601–1611
81. Wei JM, Zhu RL, Zhu JX, Ge F, Yuan P, He HP et al (2009) Simultaneous sorption of crystal violet and 2-naphthol to bentonite with different CECs. J Hazard Mater 166:195–199

82. Lyoo WS, Lee HC, Jeong YG, Min BG, Lee SC (2009) Preparation and acid dyes adsorption behavior of polyurethane-chitosan composite foams. Fiber Polym 10:636–642
83. Jang SH, Min BG, Jeong YG, Lyoo WS, Lee SC (2008) Removal of lead ions in aqueous solution by hydroxyapatite/polyurethane composites foams. J Hazard Mater 152:1285–1292
84. Moisés LP, João P, Ana PC, Manuela BC, João CB (2006) Synthesis and regeneration of polyurethane/adsorbent composites and their characterization by adsorption methods. Micropor Mesopor Mat 89:260–269
85. Buleon A, Colonna P, Planchot V, Ball S (1998) Starch granules: structure and biosynthesis. Int J Biol Macromol 23:85–112
86. Tester RF, Karkalas J, Qi X (2004) Starch-composition, fine structure and architecture. J Cereal Sci 39:151–165
87. Crini G (2003) Studies of adsorption of dyes on beta-cyclodextrin polymer. Bioresour Technol 90:193–198
88. Delval F, Crini G, Vebrel J, Knorr M, Sauvin G, Conte E (2003) Starch-modified filters used for the removal of dyes from waste water. Macromol Symp 203:165–171
89. Lu W, Yang JZ, Cui CX (2003) Poly(sulfobetaine)s and corresponding cationic polymers. XI. Synthesis and aqueous solution properties of a cationic poly(methyl iodide quaternized ethyl vinyl ether/N, N-dimethylaminopropyl maleamidic acid) copolymer. J Appl Polym Sci 89:263–267
90. Xu SM, Feng S, Yue F, Wang JD (2004) Adsorption of Cu(II) ions from an aqueous solution by cross-linked amphoteric starch. J Appl Polym Sci 92:728–732
91. Xu SM, Feng S, Peng G et al (2005) Removal of Pb(II) by cross-linked amphoteric starch containing the carboxymethyl group. Carbohydr Polym 60:301–305
92. Wang JL, Xu SM, Wu RL et al (2006) Adsorption behaviors of acid and basic dyes on cross-linked amphoteric starch. Chem Eng J 117:161–167
93. Mishra A, Bajpai M, Pal S, Agrawal M, Pandey S (2006) *Tamarindus indica* mucilage and its acrylamide-grafted copolymer as flocculants for removal of dyes. Colloid Polym Sci 285:161–168
94. Sandhu JS, Hudson GJ, Kennedy JF (1981) The gel nature and structure of the carbohydrate of isapghula husk ex *Plantago ovata* Forsk. Carbohydr Res 93:247–259
95. Samuelsen AB, Lund I, Djahromi JM, Paulsen BS, Wold LK, Knutsen SH (1999) Structural features and anticomplementary activity of some heteroxylan polysaccharide fractions from the seeds of *Plantago major* L. Carbohydr Polym 38:133–143
96. Sanghi R, Bhattacharya B (2005) Psyllium and chitosan as coagulant aids for decolourization of dye solutions. Water Qual Res J Can 40:97–101
97. Li G, Gregory J (1991) Flocculation and sedimentation of high-turbidity waters. Water Res 25:1137–1143
98. Karthikeyan J (1990) A study on colour removal from textile dye waste by chemical treatment. Dissertation, Indian Institute of Technology, Kanpur, India
99. Kace JS, Linford HB (1975) Reduced cost flocculation of a textile dyeing wastewater. J Water Pollut Control Fed 47:1971–1980
100. Rebhun M, Weimberg A, Narkis N (1970) Treatment of wastewater from cotton. Dyeing and finishing works for reuse. In: Proceedings of the 25th industrial waste conference, Purdue University, West Layfayette, Indiana
101. Sutherland JP, Folkard GK, Mtawali MA, Grant WD (1994) *Moringa oleifera* as a natural coagulant, affordable water supply and sanitation. In: 20th WEDC conference, Colombo, Sri Lanka, pp 297–299
102. Tripathy T, Bhagat RP, Singh RP (2001) The flocculation performance of grafted sodium alginate and other polymeric flocculants in relation to iron ore slime suspension. Eur Polym J 37:125–130
103. Sanghi R, Bhattacharya B, Singh V (2007) Seed gum polysaccharides and their grafted co-polymers for the effective coagulation of textile dye solutions. React Funct Polym 67:495–502
104. Bajpai UDN, Jain A, Rai S (1990) Grafting of polyacrylamide on to guar gum using $K_2S_2O_8$ ascorbic acid redox system. J Appl Polym Sci 39:2187–2204

105. Singh V, Srivastava V, Pandey M, Sethi R, Sanghi R (2003) *Ipomea turpethum* seeds: a potential source of commercial gum. Carbohydr Polym 51:357–359
106. Singh V, Srivastava A, Tiwari A (2009) Structural elucidation, modification and characterization of seed gum from Cassia javahikai: a nontraditional source of industrial gum. Int J Biol Macromol 45:293–297
107. Amsden B (1998) Solute diffusion within hydrogels. Mechanisms and models. Macromolecules 31:8382–8395
108. Sannino A, Pappadà S, Madaghiele M, Maffezzoli A, Ambrosio L, Nicolais L (2005) Cross linking of cellulose derivatives and hyaluronic acid with water soluble carbodiimide. Polymer 46:11206–11212
109. Paulino AT, Guilherme MR, Reis AV, Campese GM, Muniz EC, Nozaki J (2006) Removal of methylene blue dye from an aqueous media using superabsorbent hydrogel supported on modified polysaccharide. J Colloid Interface Sci 301:55–62

Chapter 14
Wastewater Treatment with Concomitant Bioenergy Production Using Microbial Fuel Cells

Liping Huang, Shaoan Cheng, Daniel J. Hassett, and Tingyue Gu

14.1 Introduction

Wastewater treatment currently requires a significant energy input. Current technology does not recover the chemical energy of organic matter contained in wastewaters even though the energy stored in such wastewaters is not trivial. For example, new estimates indicate that domestic wastewater contains 7.6 kJ/L [1]. Roughly 2.2–4.4 × 10^{15} kJ of energy per year is available in wastewaters worldwide. This amount of energy is equivalent to burning 62–124 million tonnes of oil in a modern power station [1]. This estimate would escalate considerably if industrial and agricultural wastewaters were also included. If a significant amount of energy is recovered from wastewaters during treatment, the energy cost of the wastewater industry can be greatly reduced. It is even possible to have a net energy gain.

L. Huang (✉)
Key Laboratory of Industrial Ecology and Environmental Engineering,
Ministry of Education (MOE), School of Environmental Science and Technology,
Dalian University of Technology, Dalian 116024, China
e-mail: lphuang2008@gmail.com

S. Cheng
State Key Laboratory of Clean Energy Utilization, Department of Energy Engineering,
Zhejiang University, Hangzhou 310027, China
e-mail: shaoancheng@zju.edu.cn

D.J. Hassett
Department of Molecular Genetics, Biochemistry and Microbiology, University of Cincinnati
College of Medicine, Cincinnati, OH 45267, USA
e-mail: daniel.j.hassett@gmail.com

T. Gu
Department of Chemical and Biomolecular Engineering, Ohio University, Athens,
OH 45701, USA
e-mail: gu@ohio.edu

S.K. Sharma and R. Sanghi (eds.), *Advances in Water Treatment and Pollution Prevention*, DOI 10.1007/978-94-007-4204-8_14,

Although energy from wastewater treatment will unlikely contribute to the grid power due to the relatively low energy density of wastewaters compared to fuels such as ethanol and hydrogen (H_2), contribution to a local power supply is possible, especially in remote villages, and also military forward operating bases (FOBs) where fuel costs can be ten times more expensive due to trucking risks in a war zone.

Microorganisms can utilize biodegradable solid wastes and wastewaters to produce energy. For example, methane (CH_4) digesters can convert organic wastes to CH_4. An alternative to methane digesters is microbial fuel cells (MFCs), which can convert the energy in the organic wastes directly into electricity. A number of reports have been published on the use of MFCs for treating a variety of wastewaters including municipal, swine, and food processing wastewaters; agricultural and sludge wastes; as well as any biodegradable industrial wastewaters [2]. Recently, there is a growing trend in the literature for the use of biocathodes to replace traditional chemical cathodes. Researchers have attempted optimization of various MFC parameters including electrode material, electrode potential, substrate type, microorganism metabolism, pH, temperature, ionic strength, as well as reactor architecture. The electron-transfer mechanisms from *Geobacter sulfurreducens* and *Shewanella oneidensis* have been described in great detail in MFC studies. MFCs have also been operated with an externally imposed voltage to produce bioproducts such as H_2 and CH_4. This chapter critically reviews the new developments in three aspects: (1) bioanodes, (2) chemical and biological cathodes for wastes and wastewater treatment, and (3) value-added bioproducts through MFCs and the broadened MFCs, and a potential "game-changer" in MFC technology using genetically engineered "superbug" bacteria as electrogens.

14.2 Bioanode MFCs for Waste and Wastewater Treatment

14.2.1 Wastes and Wastewaters Treated at the Anode

All bacteria must have some source of carbon-containing molecules for the building blocks of cellular components including DNA, RNA, protein, and lipid. Autotrophic bacteria can utilize inorganic carbon from carbonate or CO_2. Phototrophic bacteria harvest their energy from light. A chemotroph derives energy from either organic or inorganic compounds. However, the most common and metabolically versatile organisms are termed heterotrophs. Thus, heterotrophic bacteria would arguably be the best candidates for the "engines" in MFCs for the treatment of a variety of "practical" wastewaters, such as sewage, brewery, and rendering wastes at the anode in an MFC.

Earlier MFC investigations almost exclusively focused on anodic fuels because oxidation of organic carbons occurs in the anodic chamber. Pant et al. [2] presented a comprehensive review on the oxidation of these substrates. While synthetic wastewater was often used as fuel for MFCs in lab tests due to its ease of controlling loading strength, pH, and conductivity, wastewaters such as municipal [3, 4],

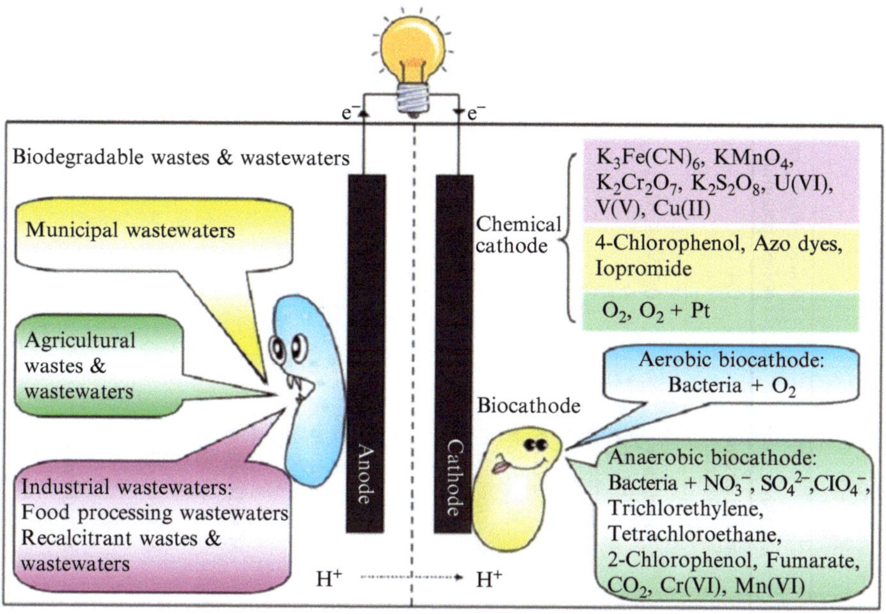

Fig. 14.1 Configuration of diverse wastes and wastewaters treated in MFCs

swine [5], starch processing [6], and brewery wastewaters [7] were tested as fuels in MFCs for power generation during treatment (Fig. 14.1). Landfill leachates were also used as a fuel for power generation from MFCs [8–10]. Recently, researchers have used sewage sludge to produce 9 W/m^3 of power and enriched the hydrophilic fraction of the extracellular biological organic matter by 48–65%, while reducing the hydrophobic acid fraction by 32–15%. They also removed 37% of the dissolved organic carbon and decreased the sludge aromaticity by 66% [11–13]. Lignocellulosic biomass is widely available as agriculture wastes, wood chips, and toilet paper in municipal wastewater. It is a recalcitrant biopolymer, but the success of renewable bioenergy from biomass depends on the successful utilization of lignocellulosic biomass. With the presence of *Clostridium cellulolyticum* and *Clostridium thermocellum*, cellulose was able to fuel MFCs via the mediation of H_2 produced from cellulose [14]. More recent investigations proved that cellulose and algae can be directly used as a fuel [15–22] or via the decomposition by cellulase enzymes [23] (Table 14.1) in MFCs for electricity production using exoelectrogenic bacteria. Further research indicated that more complex lignocellulosic biomass such as corn stover, wheat straw, and *Canna indica* can also be degraded at bioanodes, although only 22–61% degradation was achieved and the power output was quite low [26, 27] (Table 14.1). Pretreatment by dilute acid can improve the biodegradability of these wastes for power production in MFCs [24, 26]. Alternatively, integration of anaerobic digestion with MFCs can successfully achieve COD removal of grass silage with simultaneous power production [28]. The products of lignocellulosic

Table 14.1 Cellulose and lignocellulosic biomass degraded in MFCs

Type of substrate	Type of pretreatment	Reactor, electrode, and microorganisms	Microorganisms	Power (W/m^3) or poised potential	References
Chitin	–	Two-chambered, graphite fiber anode and carbon cloth cathode, Pt-catalyzed O_2 as an electron acceptor	Mixed culture	0.4–0.6	[20]
Cellulose	–	Two-chambered, carbon felt anode and carbon rod cathode, Pt-catalyzed O_2 as an electron acceptor	*Clostridium cellulolyticum*, *C. thermocellum*	0.2 V vs. SHE[a]	[14]
	–	Two-chambered, carbon plate anode and carbon paper cathode, $K_3Fe(CN)_6$ as an electron acceptor	*C. cellulolyticum*, *Geobacter sulfurreducens*	0.9	[17]
	–	Two-chambered, carbon cloth anode and graphite fiber cathode, $K_3Fe(CN)_6$ as an electron acceptor	*Enterobacter cloacae*	0.06	[19]
	–	Two-chambered, carbon plate both anode and cathode, $K_3Fe(CN)_6$ as an electron acceptor	Mixed culture	1.2	[21]
	–	Two-chambered, graphite fiber anode and carbon paper cathode, Pt-catalyzed O_2 as an electron acceptor	Mixed culture	0.06	[23]
	Enzymatic hydrolysis	Two-chambered, graphite fiber anode and carbon paper cathode, Pt-catalyzed O_2 as an electron acceptor	Mixed culture	0.6	[23]
	–	One-chambered, graphite fiber anode and carbon cloth cathode, Pt-catalyzed O_2 as an electron acceptor, batch operation	Mixed culture	0.3	[15]
	–	One-chambered, graphite fiber anode and carbon cloth cathode, Pt-catalyzed O_2 as an electron acceptor, continuous operation	Mixed culture	5.9	[16]

Wheat straw	Hydrothermal treatment, liquid part	Two-chambered, carbon paper both anode and cathode, $K_3Fe(CN)_6$ as an electron acceptor	Mixed culture	2.1	[24]
Corn stover	Steam exploded (neutral pH), liquid part	One-chambered, carbon cloth both anode and cathode, Pt-catalyzed O_2 as an electron acceptor	Mixed culture	1.6	[25]
	Steam exploded (acidic pH), liquid part	One-chambered, carbon cloth both anode and cathode, Pt-catalyzed O_2 as an electron acceptor	Mixed culture	1.3	[25]
	–	One-chambered, carbon paper anode and carbon cloth cathode, Pt-catalyzed O_2 as an electron acceptor	Mixed culture	0.5	[26]
	Steam exploded (neutral pH), solid part	One-chambered, carbon paper anode and carbon cloth cathode, Pt-catalyzed O_2 as electron acceptor	Mixed culture	0.7	[26]
Phytoplankton	–	One-chambered, graphite fiber anode and carbon cloth cathode, Pt-catalyzed O_2 as an electron acceptor	Mixed culture	277	[22]
Macrophyte	–	One-chambered, graphite fiber anode and carbon cloth cathode, Pt-catalyzed O_2 as an electron acceptor	Mixed culture	215	[22]
Aquatic plant	–	One-chambered, carbon paper both anode and cathode, Pt-catalyzed O_2 as an electron acceptor	Mixed culture	0.4	[27]

[a] *SHE* standard hydrogen electrode, – none pretreated

biomass hydrolysis, glucose and xylose, as well as polyalcohols can fuel MFCs with a maximum power generation of 44 W/m^3, Coulombic efficiencies (CEs) of 13–28%, and COD removal of 71–92% [29–31, 234].

Furan derivatives and phenolic compounds except 5-hydroxymethyl furfural can only be co-metabolized in the presence of glucose [32]. Both the type and size of the substrates derived from lignocellulosic biomass affect the treatment efficiency and power generation in MFCs. For example, the size of chitin particles affected electricity generation and the composition of the anodic microbial consortia [20]. The inoculum type also affects system performance for lignocellulosic biomass utilization. An MFC inoculated with activated sludge of municipal wastewater efficiently degraded furfural in the absence of glucose, while an MFC inoculated with a previously acclimated anolyte of a MFC fed with glucose did not degrade furfural [33]. In another study, however, an inoculum from an acclimated microbial electrolysis cell used to degrade cellulose efficiently degraded cellulose and provided more electricity than directly degrading paper recycling wastewater [15, 16, 34].

Apart from these, MFCs also have the potential to treat other recalcitrant wastes [227]. These include pyridine [35], indole [36], quinoline [37], and its derivatives, as well as coking wastewater [38] can be efficiently removed or their levels significantly reduced by MFCs. Apart from the organic wastes, ammonia removal is a major issue for animal wastewater [39, 40]. Ammonia was removed mainly by a physicochemical process, although some losses may occur through biological nitrification [40]. Kim et al. [40] did not detect anaerobic ammonia oxidizing bacteria in their MFC and concluded that ammonia oxidation did not contribute to electricity generation in their MFC fed with swine wastewater. However, He et al. [39] observed electricity generation from ammonia oxidation although the CEs were relatively low in their rotating-cathode MFCs. Nonetheless, this points to the possibility of using ammonia oxidation for power generation.

14.2.2 Factors Influencing System Performance

Various operating parameters such as pH, temperature, external resistance, influent COD, and ionic strength can influence system performances. Electrode characteristics such as electrode material, surface roughness, surface area, and potential, as well as microbial metabolism, also affect MFC performance. Studies on the impact of these parameters on practical MFC applications and basic bioelectrochemistry in MFCs have been performed.

14.2.2.1 Substrate and Microbial Metabolism

Microbial metabolism can be divided into fermentation and respiration. Cellular fermentation is normally composed of internal oxidation and reduction in which closely linked and balanced electron transfer was involved. During this process, microorganisms usually convert substrate to multiple by-products. Microbial respiration, however, requires external electron acceptors to main electroneutrality. The anode-based electron acceptors have been recently shown to overcome the constraints and stoichiometrically converted glycerol into ethanol with a metal-reducing bacterium *S. oneidensis* [41]. Even for the eukaryotic *Saccharomyces cerevisiae*, with the mediation of self-secreted NADH/NAD$^+$ and FADH/FAD$^+$, an efficient glucose metabolism and electricity generation were revealed at a loading rate ranging from 0.91 to 2.86 kg COD/m^3/day [42], indicating a potential method of using yeast based MFCs for treatment of high strength wastewaters with power generation. However, in another study with *Clostridium acetobutylicum* at a high glucose concentration of 50 g/L and a comparatively high external resistor of 10 kΩ, the MFC exhibited a bimodal metabolism, resulting in two output voltage stages in which the first stage was correlated with acidogenic metabolism (acetate and butyrate production) and the second a solventogenic metabolism (acetone and butanol production) [43]. Besides the bacteria, experimental conditions also significantly impact microbial metabolism. For example, when the MFCs were independently operated at different circuit loads, the anodic bacterial community exhibited significant differences when fed with short chain fatty acids [44], and CEs and biomass growth was lower [45–47] in larger circuit loads. These results indicated that increasing external resistance favored fermentative metabolism over anaerobic respiration. It was proposed that fermentative and electrogenic microorganisms competed from substrates (electron donors), a higher resistance correlated with decreased utilization of the substrates by the electrogenic microorganisms and less electricity generation [45, 48]. These results can be also used as a potential platform for controlling electrogenesis, methanogenesis, and anode microbial community [48].

While external circuit load can orient bacterial metabolism, the type of substrate also affects microbial consortia metabolism and system performance. Based on the analysis of electron sink with current, biomass, residual organic compounds, H_2 and CH_4, MFCs fed with non-fermentable (acetate) and fermentable (glucose) electron donors exhibited different electron sink contribution by electrical current (acetate 71%, glucose 49%), biomass (acetate 15%, glucose 26%), and residual organic compounds (acetate 11%, glucose 18%), respectively, while methane gas (3.7%) was found only in the glucose-fed MFC [49]. Jung and Regan [48] also obtained similar results. With a mixed culture, microorganisms did not directly use glucose as a fuel for electricity generation but used H_2 and acetate of the fermentation products. In this process, methanogens could decrease the electricity generation by consuming H_2 but also increase fermentation rates [48, 50]. This electrical current-driven microbial metabolism has led to a series of applications such as bioremediation,monitoring of glucose metabolism, and the production of value-added bioproducts from wastewaters (see Sect. 14.5).

The presence of oxygen in the bioanode also affected bacteria by altering the electron delivery rate and metabolic pathway utilized. It was initially found that the presence of oxygen selected for bacteria with differential expression of proteins resulting in decreasing electron transfer and electricity generation [51–53].

However, Biffinger et al. [54] later reported that the power output of MFCs using *S. onedensis* DSP10 with glucose as substrate could actually be promoted by oxygen. Based on this observation with *S. oneidensis*, Li et al. [55] conducted a series of experimental comparisons using *S. decolorationis* NTOU1 as a model bacterium and concluded that current increased by oxygen was related to the increase in the overall aerobic metabolic rate, rather than an increase in cell number or the concentration of the self-secreted endogenous mediator.

14.2.2.2 Electrode Materials

Electrode materials can substantially affect power generation of MFCs. An efficient anodic electrode requires high conductivity, chemical stability, biocompatibility, and catalytic activity. In addition, a highly porous structure should exhibit internal colonization and strong interaction between the electrode surface and the microbial biofilm, allowing a better mechanical contact and higher electrical conductivity [56]. Various carbon-based materials like carbon foam [57], carbon paper [4], reticulated vitrified carbon (RVC, [58]), carbon cloth [59, 60], graphite brush [61], graphite granule [62], and molybdenum carbide [63] have been extensively investigated for anodes. Among these materials, graphite brush and graphite granules both exhibited a high power production due to their high specific surface areas [61, 62]. The maximum power of MFCs was proportional to the logarithm of the surface area of the anode [64]. The COD removal rates measured for both open circuit and closed circuit operations of MFCs were found to be independent of the various carbon anodes used such as graphite, sponge, paper, cloth, felt, fiber, foam, and RVC. However, MFC peak powers varied from a low of 1.3 for RVC to a high of 568 mW/m^2 for graphite [65, 66]. The low power output of RVC was mainly due to the small anodic surface area and high concentration polarizations caused by the RVC morphology [65, 66]. Ammonium treatment, electrochemical oxidation, and 4(N,N-dimethylamino) benzene diazonium tetrafluoroborate modification can improve the electrocatalytic activity of the anode due to the formation of functional groups of amine, carboxyl, and nitrogen on the anode material surface [67, 229]. Bonding materials to the anode can also increase power by facilitating electron transfer and improving the adhesion of bacteria [68, 69]. For example, the bonded plain G10 graphite with surface-confined anthraquinone-1,6-disulfonic acid (AQDS) or an incorporated Sb(V) complex, oxidized graphite, and oxidized graphite bonded with AQDS can exhibit 1.9–218 times greater kinetic activities than plain G10 graphite without modification [68].

Alternatively, anodes modified with pure β-molybdenum carbide ((Mo_2C)) [63] or polymers or multiwall carbon nanotubes (CNTs) [70] can boost the interaction between the anodic surface and its microbial biofilm and therefore increases the MFC performance. The promoting effect of CNTs was mainly due to the transformation of the irreversible electrochemical behavior of cell surface cytochromes, where electrooxidation was inhibited [71]. A new composite by combining CNT with polyaniline (PAN) also showed enhanced electronic properties of anode for MFC applications [72, 73]. Carbon nanotube-textile (CNT-textile) composite performed biocompatible, highly conductive, and open three-dimensional space for efficient substrate transport and internal colonization by a diverse microflora and can increase maximum current density by 157%, maximum power density by 68%, and energy recovery by 141% compared with the use of the traditional carbon cloth anode [56]. In yet another study, a nanostructured fibrillar polypyrrole was found to perform better than granular polypyrrole, due to the fact that more electrons can be harvested directly from photosynthetic cultures by a higher redox current and a lower interface electron-transfer resistance [74]. The cost-effective energy generated from MFCs can also supply the system of photocatalytic oxidation mediated by TiO_2, showing as a synergetic effect on *p*-nitrophenol degradation [75].

14.2.2.3 Electrode Potential

In the anodic chamber, electrons from substrate oxidation flow from the biofilm to the anode. Microorganisms can harvest some of the energy from substrate oxidation for their own survival and reproduction [76]. In view of this, an electrode potential can produce a selective pressure for bacterial evolution [48, 77, 78]. For example, a strain of *G. sulfurreducens* obtained under a selective pressure of anodic potential of −0.2 V can produce current densities five times greater than the wild-type strain, providing a strategy for the selection of more efficient strains [78]. Additionally, anode potential also influences the evolution time to maximal power production, which is related to bacterial growth. However, power production at different set anode potentials resulted in the same value at the end of the experimental period [77, 79], indicating that the set anode potential mainly regulated the activity and growth of bacteria and not power generation. For example, at different anodic potentials, *G. sulfurreducens* altered the way they exchanged electrons with an electrode in response to an applied electrode potential change [80]. Based on the comparison of biomass evolution at various set potentials and the quantitative analysis between biomass and possible metabolic energy gain (defined by anode potential), a proper set potential can promote the growth of this strain and therefore accelerating MFC activation while lowering its internal resistance. These positive enhancements can also be observed during steady-state MFC operations [80]. However, anodic potentials were highly variable on start-up or current density when different cultures and reactor conditions were used [230]. Therefore, it is problematic to conclude that positive results are related to the higher or lower potentials due to the various culture conditions, electrode materials, and bacterial inoculum in the different studies [81, 82]. Much more work is needed to investigate the effects of various potentials on system performance at various conditions in order to understand the behavior of microbial communities in different redox environments for wastewater treatment [230].

14.2.2.4 pH, Ionic Strength, and Temperature

Equation 14.1 describes electrode potentials as a function of pH at 298 K according to the Nernst equation [83]. The net change of potential with a pH change thus follows Eq. 14.2:

$$E = E_0 - 0.0592\text{pH} \tag{14.1}$$

$$\Delta E = 0.0592\Delta\text{pH} \tag{14.2}$$

Equation 14.2 shows that a decrease of one pH unit can lower the potential by 0.0592 V. In the case of complex bioanodic processes, the medium pH can affect multiple surface properties including cell morphology during cell division in planktonic and biofilm culture, cell surface hydrophobicity, and net surface electrostatic charge as well as biofilm structure [84, 85]. Considering the various chemical reactions on the anode and cathode, the oxidation reactions in the anodic chamber generate protons, and reduction reactions in the cathodic chamber generate hydroxide ion. There is often a pH imbalance because of the resistance by the PEM and, thus, can result in potential and power losses. In the case of the air-cathode system, one-chambered MFC with a Pt catalyst, the higher current generation was achieved at the optimal initial pH of 8–10, not at lower or higher pH [86]. In another reported air-cathode, one-chambered MFCs without Pt catalyst, *S. cerevisiae* exhibited higher power production and substrate degradation at an initial pH of 6.0 than at pH 5.0 or 7.0, mainly due to higher metabolic activities of this microorganism at pH 6.0 [42]. Considering the pH imbalance and the effect of pH, phosphate buffer electrolytes are extensively used to decrease potential losses and obtain an efficient system performance [7, 15, 30, 52, 59–61, 87]. However, the addition of phosphate buffer is not cost-effective and sustainable for practical wastewater treatment. Phosphate buffer electrolytes can be replaced by CO_2/carbonate or CO_2/bicarbonate buffered catholyte systems which are cost-effective [88–90]. The pH can be also balanced by flow through the anode to the cathode or a recycling loop between anode and cathode chambers [91–94].

Temperature also has a significant impact on bacterial activities [7, 52, 65, 66, 95–97], although most studies have been conducted under mesophilic conditions (e.g., 30°C). Liu et al. [52] reported that when the temperature was decreased from 32°C to 20°C, power output was reduced by only 9% in one-chambered, air-cathode MFCs. In contrast to the above observation, another air-cathode, one-chambered MFC acclimated at 30°C exhibited a linearly increasing power production with temperature, from 10.6 at 4°C to 31.5 W/m^3 at 30°C while CEs dropped from 31% at 4°C to 17% at 30°C [96]. Additionally, MFCs initially operated at 15°C or higher achieved stable power generations, but the time required to achieve it increased from 50 h at 30°C to 210 h at 15°C whereas MFCs did not produce appreciable power at starting temperatures below 15°C [96]. While the effect of temperature has been extensively investigated in artificial wastewater [52, 96, 97], it has also been evaluated in practical wastewater-fueled MFCs [7, 65, 66, 95]. Increasing the temperature from 20°C to 30°C increased power from 4.3 to 5.1 W/m^3 while COD removal rates and CEs decreased only slightly using brewery wastewater as fuel in single-chambered, air-cathode MFCs run in batch mode (net working volume 28 mL) [7]. Similarly, Ahn and Logan [95] examined the effects of temperature on power generation with domestic wastewater at 23°C and 30°C using batch and continuous operation modes in one-chambered, air-cathode MFCs. Under the batch operation mode, a higher power of 10.2 W/m^3 was obtained at 30°C while COD removal was greater than 88% at both temperatures. Under the continuous operation mode, a COD removal of 25.8% and a higher power density of 12.8 W/m^3 at an organic loading rate of 54 g COD/L/d were obtained at 30°C, implying the importance of not only temperature but also running mode on practical wastewater treatment efficiency and power generation. In another report [65, 66], both single-chambered and two-chambered, air-cathode MFCs fueled by barley processing wastewater from a brewery diluted in domestic wastewater and run in batch mode exhibited a substantially increased COD removal and power production with an increase of temperature from 4°C to 35°C (net working volume of two-chambered, 200 mL, and one-chambered, 100 mL).

14.2.2.5 Reactor Architecture

Considerable efforts have been devoted to the improvement of reactor architecture in the pursuit of greater MFC performance [59, 60, 87, 98]. Single- and two-chambered MFCs are common reactor architectures and have been extensively reviewed by Du et al. [99]. In comparison with two-chambered MFCs that are separated by a PEM, a single-chambered MFC (normally using air/O_2 as an electron acceptor) inherently possesses multiple advantages. These include a simple structure, easy installation, appropriate electrode space, elimination of PEM, and therefore low cost, convective flow, low internal resistance, higher power generation, and free electron acceptor with high redox potential [15, 16, 30, 59, 60, 62, 231]. In view of practical application, scale-up reactor architectures should satisfy the requirement of large amount wastewater treatment with simultaneous power generation. Four different one-chambered MFCs with various working volumes ranging from 28 mL to 1.6 L are shown in Fig. 14.2. The high costs of Pt catalyst and Nafion binders have limited their use in MFC scale-up. Efforts have been made to reduce costs by using cathode catalysts such as metal tetramethoxyphenylporphyrin (TMPP) and metal phthalocyanine, and substite binders such as Nafion/PTFE mixed binders [101–103]. Practical large-scale MFCs will likely be membrane-less because membranes are expensive and easily fouled and they exert a large internal resistance. This will be discussed in Sect. 14.4.

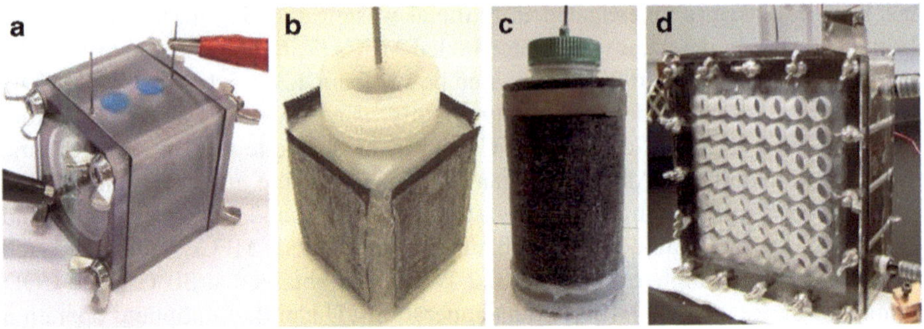

Fig. 14.2 Single-chambered air-cathode MFCs with different working volumes of 28 mL (**a**), 250 mL (**b**), 1 L (**c**), and 1.6 L (**d**) (Reprinted from Cheng and Logan [100] with kind permission of © Elsevier)

14.2.3 Electron-Transfer Mechanisms

MFCs are capable of directly transforming the chemical energy stored in organic materials to electricity via electrochemical reactions associated with biofilm bacteria [104, 105]. These are generated by various microbial consortia, including Gram-negative and Gram-positive bacteria [106] from many microbially rich sources such as sewage sludge, activated sludge, and industrial and domestic effluents. Even the eukaryote *S. cerevisiae* can perform electron transfer from substrate fuels to the anode [42]. A pure culture of *Shewanella* species (typically MR-1) is frequently used as a model bioanode catalyst due to its multiple electron-transfer mechanisms including outer membrane-bound cytochromes [51], conductive pili for direct electron transfer [107], and locally secreted (endogenous) or added (exogenous) mediators such as riboflavin and quinones [108–111] (Fig. 14.3 and Table 14.2). By using optically transparent nanoelectrodes and in situ optical imaging, it had been found that *S. oneidensis* MR-1 performed electron transfer mainly by mediated mechanism, based on the measured currents uncorrelated with the number of bacteria on the electrodes [12, 13]. *S. decolorationis* NTOU1 analyzed by the combination of electrochemical and instrumental analysis such as cyclic voltammetry and high performance liquid chromatography also exhibited similar characters [55].

Members of the *Geobacter* genus can directly reduce insoluble oxidants including minerals and electrodes in the absence of soluble extracellular electron-transfer mediators. Using cyclic voltammetry, it was defined that the *c*-type cytochromes Z, B, S, E (a conductive network of bound electron-transfer mediators) and type IV pili (surface-dependent electron conduits) of *G. sulfurreducens* strains in biofilms possess the capacity for homogeneous (biofilm/bulk solution) electron transfer and/or heterogeneous (biofilm/electrode interface) electron transfer (Table 14.2) [123]. Recently, Pd (0) nanoparticles biologically transformed from Pd(II) cations by sulfate-reducing bacterium *Desulfovibrio desulfuricans* and bound to the cell

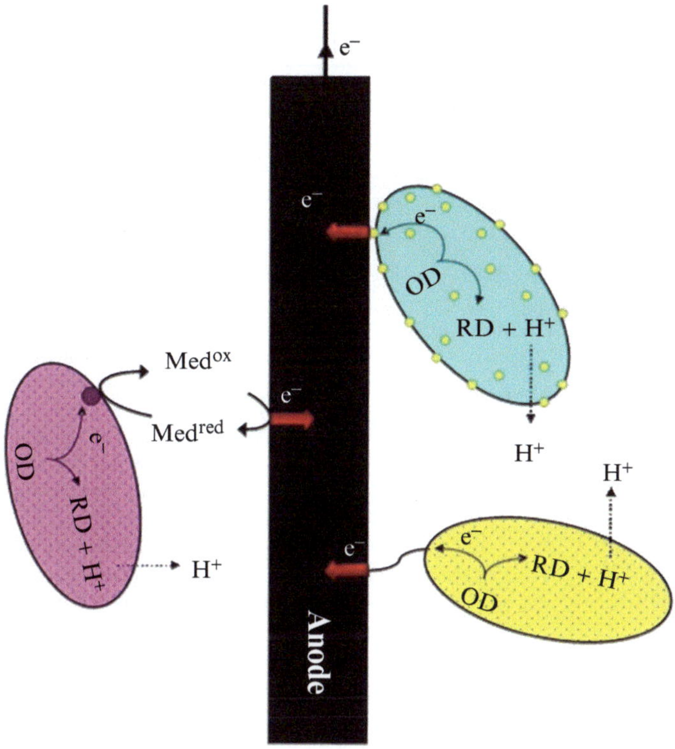

Fig. 14.3 Schematic of electron-transfer mechanisms at the anode via membrane-bound cytochromes ( ), conductive pili ( ), and secreted or added mediators ( ). *OD* oxidized electron donor, *RD* reduced electron donor

membranes were illustrated to facilitate direct electron transfer from *D. desulfuricans* to the anode [114]. These results provided the first direct electrochemical evidence for the role of bound electron-transfer mediators as well as bio-derived metal nanoparticles bound to the cell membrane in anode-reducing microorganisms and, therefore, are useful to advance strategies to increase the power density of MFCs.

Except for the electron-transfer mechanisms of outer membrane-bound cytochromes and secreted or exogenously added redox mediators frequently reported in the literature, the bacterial nanowire electron-transfer mechanism is not thoroughly investigated, except those involving limited observations using scanning/transmission electron microscope. Some studies indicated that the formation of microbial pilus nanowires on anodes was required for maximum power production [117], which is associated with enhanced current production in both pure and mixed culture reactors (Table 14.2) [78, 120]. However, in comparison with the model wild-type strain of *G. sulfurreducens*, *pilA*-deficient *G. sulfurreducens* MA

Table 14.2 Anodic electron transfer mechanisms and methods for debottlenecking

Mechanism category	Molecules or structures	Comments
Direct electron transfer (DET)	c-type cytochromes and heme proteins [76]	Only monolayer sessile cells donate electrons to the anode because cells are poor electron conductors
		Shewanella spp. not only have outer membrane cytochromes for direct electron transfer by contact but also extrude electrically conductive nanowires as well as produce flavins as electron shuttles [107, 111–113]
	Pili (nanowires)	The microbial palladium nanoparticles facilitate the direct electron transfer between *Desulfovibrio desulfuricans* and the anode [114]
		Sessile cells can be linked together in a network by pili. Multiple layers of sessile cells may be able to denote electrons [115]
		Shewanella oneidensis [107], *Geobacter sulfurreducens* and *pilA*-deficient *G. sulfurreducens* MA [78, 116–120]. The *pilT* mutant [121] has more pili to conduct electrons
		Genetically engineered superbugs with hyperpiliation will be a major breakthrough to alleviate the electron-transfer bottleneck
Mediated electron transfer (MET)	Exogenous mediators: neutral red, thionin, and methyl viologen [99]	Cytotoxicity, cost, and disposal problems
	Endogenous mediators: pyocyanin, 2-amino-3-carboxy-1,4-naphthoquinone, riboflavin, flavins, pyocyanin, ACNQ [108–111, 122]	Synergistic microbes secrete them to benefit the entire biofilm community. Genetically engineered superbugs will secrete them at sufficiently large local concentrations

exhibited abundant filaments similar to type IV pili in transmission electron micrographs (Table 14.2). This observation indicated that different microbes can produce filaments that appear similar but have a different function(s) [116]. While PilR, an enhancer binding protein, has been recently illustrated to regulate production and assembly of pili and flagella in *G. sulfurreducens* [124], direct experimental evidences of electron transfer along the line of bacterial nanowires were recently obtained by using two techniques: (1) a nanofabricated electrode patterned on top of an individual nanowire and (2) using a conductive atomic force microscopy (AFM) tip with a dissimilatory model metal-reducing bacterium of *S. oneidensis* MR-1 touching various points along a nanowire linked to a metallic electrode [125].

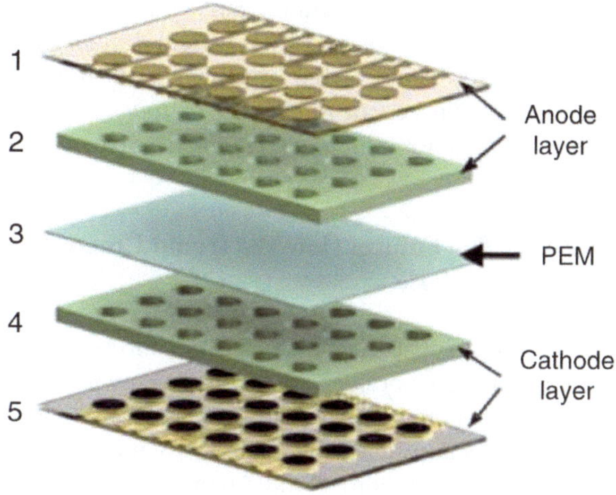

Fig. 14.4 Schematic diagram of a 24-well microbial fuel cell (MFC) array as shown in Hou et al. [126]. Composed of an anode layer (*1* anode electrode layer, *2* anode well layer), a proton exchange membrane (*3* PEM), and a cathode layer (*4* cathode well layer, *5* cathode electrode layer)

In MFCs, electron transfer from bacteria to the anode is often a rate-limiting step in current production. Additional work is needed to investigate the composition of these putative nanowires and their roles in bioanode systems.

14.2.4 Use of "Superbugs" to Resolve Electron-Transfer Bottlenecks

Since the early 1970s, the advent of true cloning techniques, mutagenesis via chemical, transposon, allelic exchange, etc., genome sequencing, and polymerase chain reaction technology have allowed the burgeoning of publications regarding important information about a myriad of microbiological processes, metabolic prowess, competitive tendencies, and overall fitness. However, only until the last 3–4 years has the MFC field gained significant ground to realize the potentially remarkable impact of genetic tools to generate bacteria with significantly greater electrogenic properties as well as the production of biogases and useful by-products of wastewater metabolism.

The major limitations thus far in the generation of "superbugs" is the current "love affair" of many MFC scientists with the known electrogenic organisms such as *Geobacter* and *Shewanella* species. For example, if one plugs in the words "*Shewanella*," "*Geobacter*," and "microbial fuel cell" into the academic search engine PUBMED, there are 97 hits as of March 2011. Thus, despite the fairly large number of known electrogenic bacteria, the majority of studies involve studies on MFCs powered by these organisms. A recent paper by Hou et al. [126] showed elegantly in a high throughput, 24-well format (Fig. 14.4) that the overall power densities of the top electrogenic bacteria varied only modestly ($\pm 10\%$). Although not at all to be considered a major negative, the metabolic properties of *Shewanella* and *Geobacter* species certainly pale with known, highly competitive soil-dwelling organisms such as species of the genera *Actinomyces*, *Bacillus*, *Burkholderia*, and *Pseudomonas*. Many of the organisms with high electrogenic properties include members of the genera *Bacillus*, *Enterobacter*, *Arthrobacter*, *Stenotrophomonas*, *Aeromonas*, *Shewanella*, *Paenibacillus*, and *Pseudomonas*.

14.2.4.1 The Future: How We Would Create "Superbug" Electrogenic Bacteria

First thing's first: enhance the biofilm forming capacity of electrogenic bacteria. Bacteria in the laboratory are typically grown in rich or minimal media and grow as what are commonly referred to as planktonic or free-swimming organisms. However, because electrons cannot "swim" in an aqueous solution, for optimal MFC power generation, it would behoove scientists to optimize bacterial structures on the anodic surface where organisms can develop as highly differentiated, three-dimensional structures known as biofilms, a monumental global problem. Biofilms have been studied for more than three decades when the term "biofilm" was actually coined by the "father" of biofilm research, Bill Costerton of the University of Southern California in 1978. There are at least five known stages of biofilm development. These include (1) the initial attachment phase, (2) irreversible attachment, (3) maturation stage I, (4) maturation stage II, and finally (5) dispersion. A number of genes of the soil-dwelling opportunistic pathogen *Pseudomonas aeruginosa* (PA) are involved in biofilm formation and other features involved in the process of biofilm development, some of which are listed in Table 14.3. Shown in Fig. 14.5 is a cartoon of the types of biofilms that would be optimized with strategically engineered genetic manipulations described in detail below.

14.2.4.2 Mutations to Enhance Electrogenesis in MFCs

Mutations that we would generate using hypothesis-driven science to enhance electrogenesis in MFCs using the metabolically voracious and genetically tractable facultative bacterium, *P. aeruginosa*:

1. *Increase the number of type IV pili on the cell surface.* In the following sections, we will use the paradigm of a metabolically robust, biofilm-forming, mediator-generating bacterium, *P. aeruginosa*, to suggest potential genes that could influence the overall electrogenic properties in an MFC system. Another means by which bacteria can enhance the conductivity of metabolically generated electrons via normal metabolism is via the synthesis of tiny, hollow, and retractable surface appendages termed type 4 pili that emanate from the surface of many bacteria. Relating to the overall efficiency of MFCs, both *Shewanella* and *Geobacter* species have been shown to conduct an electrical current, in part, through their pili [117, 141]. Both of these bacteria possess 3-start helical

Table 14.3 *Pseudomonas aeruginosa* genes known to be involved in biofilm formation, maturation or structure (part of this table from Hassett et al. [127])

Gene	PA#	Function	Biofilm phenotype	References
bdlA	1423	Putative chemotaxis regulator	Inability to disperse	[128]
crc	5332	Catabolite repressor control protein	Dispersed monolayer	[129]
cupA1–5		Cup fimbriae	Early stages of biofilm formation	[130]
fimL		Twitching motility	Abnormal biofilm development	[131]
fimX	4959	Diguanylate cyclase GGDEF domain	Poor biofilm formation	[132]
fliC	1092	Flagellin type B	Little or no biofilm	[133]
lasI	1432	Autoinducer synthase	Flat, undifferentiated	[134]
morA	4601	Motility regulator	Poor biofilm formation	[135]
PA0169	0169	Diguanylate cyclase GGDEF domain	Poor biofilm formation	[132]
PA5487	5487	Diguanylate cyclase GGDEF domain	Poor biofilm formation	[132]
pilA	4525	Type IV pilin	Little or no biofilm	[133]
phoQ	1180	Two-component senor	Poor biofilm formation	[136]
pslAB	2231–2	Exopolysaccharide synthesis	Adherence defects	[137, 138]
sadB	5346	Biofilm formation/swarming	Reversible/irreversible attachment	[139]
tpbB	1120	Diguanylate cyclase GGDEF domain	Poor biofilm formation	[140]

assemblages of tiny, type 4 pili that have high homology with the N-terminal region of the classical type 4 pili of *P. aeruginosa.* Pili can bind directly with a myriad of inanimate or animate surfaces. Once bound, they can undergo a form of organized movement known as twitching motility (TM). The end of the pilus is very much akin to a grappling hook that binds to a surface allowing for TM to take place, release of the pilus, and repetition of this cycle. Chiang and Burrows [121] have noted a lack of twitching motility and hyperpiliation of *pilT* mutant bacteria. They also noted that the formation of a robust biofilm is promoted by attachment and cell-to-cell adhesion rather than twitching motility. Thus, reducing twitching motility not only enhances the formation of *P. aeruginosa* biofilms but also will likely increase the power density of MFCs.

2. *Prevent dispersion of PA biofilms.* To date, it is now well recognized that the last stage of biofilm development and maturation is that involving a chemotactic response by viable cells within the biofilm. This process is called biofilm dispersion, involving a physical dispersal of bacteria from the biofilm matrix. Biofilm dispersion is now a field of intense interest of late due to the rapidly identification of emerging dispersion-related gene products that could one day become viable drug targets against this important organism. Many genetic

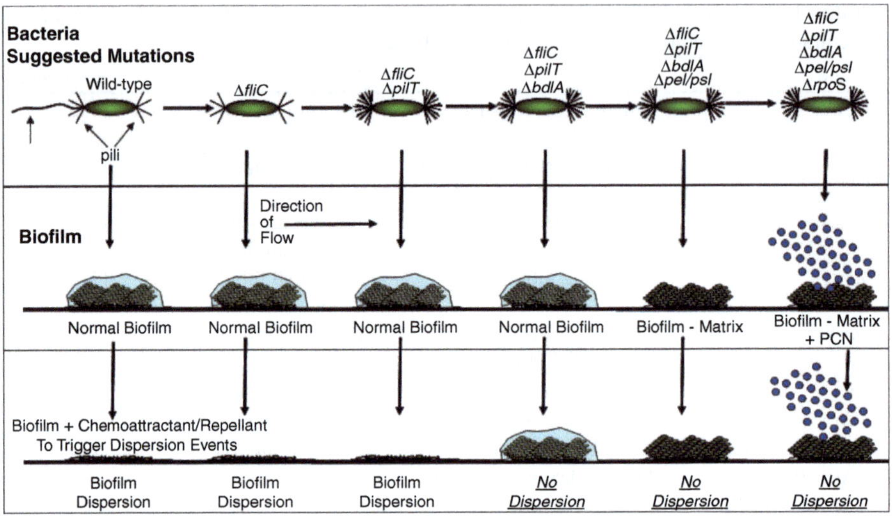

Fig. 14.5 Schematic diagram of bacterial structure, biofilm formation characteristics, and dispersion properties of *P. aeruginosa* strains with single or multiple genetic mutations to optimize electrogenic properties in an MFC

determinants involved in biofilm dispersal have been identified by S. Lory and colleagues at Harvard [132]. These involve a physiological process involving intracellular signaling via levels of bis-(3′-5′)-cyclic dimeric GMP (c-di-GMP) mediated by diguanylate cyclases (DGCs) and phosphodiesterases (PDEs).

However, in an MFC, it is advantageous to maintain bacteria in biofilms present on the anode surface, thus preventing dispersion, to ensure maximal electron transfer to anode and hence the current density. In contrast, bacteria that disperse from biofilms would decrease the level of electron transfer. Therefore, it is advantageous to utilize bacteria that have a limited ability to disperse from biofilms. This apparent limitation can be created by constructing genetic mutants lacking proteins responsible for chemotaxis. A *P. aeruginosa* gene known as *bdlA* (**b**iofilm **d**ispersion **l**ocus) was found by Morgan et al. [128] to be required for bacterial dispersion from biofilms. Thus, bacteria lacking *bdlA* causes a disruption of a cellular circuit that paralyzes the ability of attached bacteria to detach, thus maximizing sessile cell density on the anode.

3. *Limit production of nutrient-restrictive biofilm matrix.* Overproduction of the thick matrix of biofilms that is composed mostly of polysaccharide, protein, lipid, and DNA/RNA could arguably significantly impact the overall performance of an MFC. A less dense exopolymeric matrix surrounding a robust biofilm on an anode surface would have a reduced mass transfer resistance for the nutrients in the bulk fluid to reach the cells on the anode surface. Davies et al. [134] showed that *lasI* mutants of PA form thinner, more compact biofilms than the parental strain. Yet the cell density attached to the glass surface is very high,

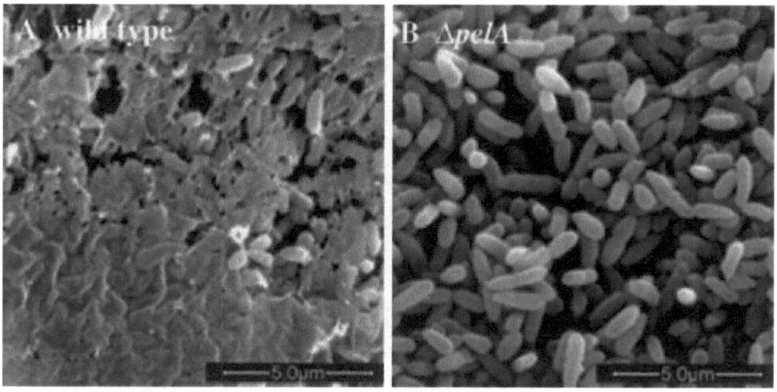

Fig. 14.6 Scanning electron micrographs of *P. aeruginosa* PA14 colonies. (**a**) PA14 wild type. (**b**) *ΔpelA*. Colonies were grown for 7 days at room temperature on 1% agar with 10 g L^{-1} tryptone plus 0.2% glycerol. Matrix material appears to surround the wild-type cells in (**a**) and is absent from the *pelA* mutant sample (Reprinted from Friedman and Kolter [142] with kind permission of © John Wiley and Sons)

thereby adding additional surface attachment properties. These organisms also display decreased polysaccharide production in the biofilm matrix. Another positive of the *lasI* mutants is that they overproduce the redox-active antibiotic pyocyanin (PCN, discussed below), which has been shown to be mediator of electron transfer in previous MFC studies using *P. aeruginosa* [122]. Other genes known as *pel* (for **pel**licle) and *psl* (**p**oly**s**accharide **l**ocus) also affect polysaccharide formation in biofilms. Specifically, Friedman and Kolter [142] were able to show that a *pelA* mutant of *P. aeruginosa* PA14 lacked the matrix that typically encases biofilm bacteria as shown in Fig. 14.6. Therefore, such mutations could also enhance MFC performance.

4. *Inhibit anaerobic respiration.* PA can undergo an alternative form of respiratory metabolism in the absence of oxygen (Fig. 14.7 for aerobic vs. anaerobic respiration scheme) known as anaerobic respiration or denitrification. NO_3^- reductase (NAR) is the first step in the eight-electron reduction of NO_3^- to N_2. The enzyme is located on the cytoplasmic side of the inner membrane. The rationale behind genetic inactivation of *narG* is that the electrons generated from oxidation of carbon molecules at the MFC anode will be directed solely to the anode and not diverted to the task of dissimilatory NO_3^- reduction even if NO_3^- is present in the wastewaters. In contrast, NO_2^- reductase (NIR) is the second step in the overall process of NO_3^- reduction to nitrogen gas during anaerobic respiration. NIR catalyzes both the one-electron reduction of NO_2^- to NO but also the four-electron reduction of O_2 to $2H_2O$. Therefore, *nirS* would also be inactivated to ensure that electron flow is maximized through the pili and not through anaerobic NO_3^- reduction. Inactivation of the *nirS* gene will also ensure that the greenhouse gas, nitrous oxide (N_2O), is not produced.

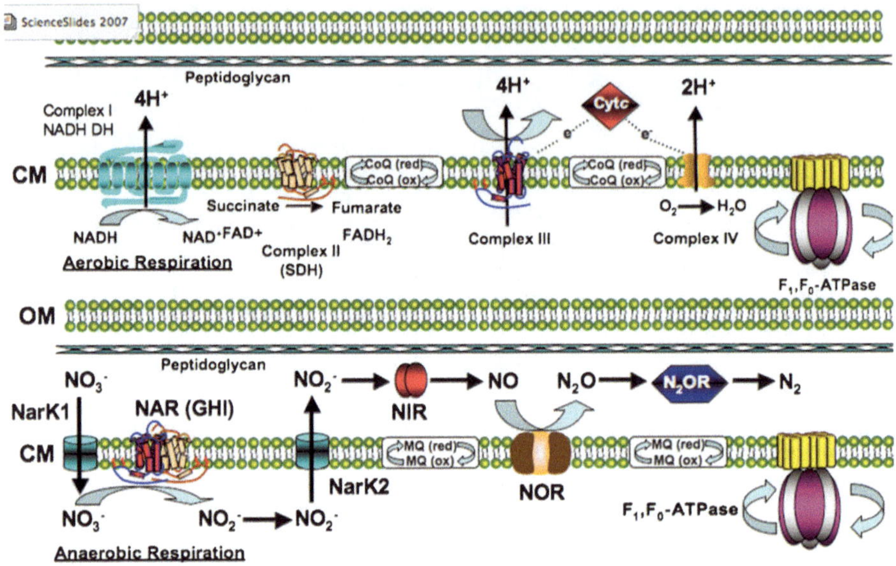

Fig. 14.7 Aerobic (*top*) vs. anaerobic (*bottom*) respiration in PA (a schematic diagram)

5. *Maximize production and secretion of electrogenic mediators. P. aeruginosa* lacking the stationary phase sigma factor RpoS overproduces the blue phenazine antibiotic PCN [143]. Rabaey and colleagues [122] have previously shown that the secreted redox-active mediator pyocyanin from *P. aeruginosa* was essential for optimal electrogenesis in an MFC. In addition, a mutant lacking PhzS overproduced aeruginosins A/B, collectively also referred to as pyorubrin (PR), a deep red pigment. This mutant was found to provide significant power generation even without added PCN [122, 144].

   Thus, because biofilm formation on inanimate surfaces are not entirely homogenous, areas upon which cells have died or sloughed off such surfaces could still provide surface area for reduced mediators such as PCN/PR in the suspensions to fill the electron reduction void left by the uncolonized anodic bacteria.
6. *Control rate of cell division.* All bacteria reproduce by binary fission and the replication rate of *P. aeruginosa* in rich medium is typically less than 30 min at 37°C. Thus, it is critical to be able to tightly control the cell division rate of these organisms at the anodic surface to prevent potential unanticipated problems associated with biofilm matrix accumulation, clogging of the MFC or premature cell death of the bacteria. Thus, yet another genetic modification could be made through the stringent control of *ftsZ*, a gene encoding a protein responsible for cell division. We fully realize that construction of an isogenic *ftsZ* mutant is not possible because it is an essential gene [145]. However, the precise control of the production of *FtsZ* can be achieved by placing the *ftsZ* gene under tight

genetic control using promoter systems that are inducible by low concentrations of non-metabolizable substrates such as the simple plant sugar, arabinose, and ultimately placing the entire construct in the nonessential *attB* site on the *PA* chromosome [146] copy with 1% arabinose [147].

7. *Increase the rate of substrate oxidation.* Uncouplers are compounds that when administered to bacteria or mitochondria separate the normally coupled processes of oxidation and phosphorylation, thereby dissipating the proton gradient from the periplasmic space to the cytoplasm. Thus, bacteria treated with the classical uncoupler, dinitrophenol, will have an equal concentration of protons in both the periplasmic space and cytoplasm. By dissipating the proton gradient, the rate of substrate oxidation and hence electron flow to the anode will likely increase substantially. However, we do not propose the use of a chemical uncoupler. Rather, we propose using eukaryotic uncoupling proteins. These could include thermogenin, UCXP-1, UCP-2, or UCP-3 under an inducible (e.g., tet/arabinose) system into an ectopic nonessential site on the chromosome. Were these uncouplers to be expressed in PA or other bacteria, the rate of substrate oxidation would likely increase dramatically and enhance the power density of MFCs.

14.3 Biocathode MFCs for Wastes and Wastewater Treatment

14.3.1 Chemical Cathode

In comparison with the extensive investigations on the overall behavior of the MFC anode in the context of MFC performance, events occurring at the cathode have only recently attracted attention. In lab investigations, ferricyanide is a typical electron acceptor used in chemical cathode MFCs due to its small overpotential loss. The highest power produced from an MFC using a potassium ferricyanide cathode reached 450 W/m^3 [148]. However, using ferricyanide is not feasible in practical applications due to its cost and toxicity. Alternatively, some oxidants like Mn(VII), Cr(VI), V(V), Cu(II), and U(VI) widely present in some wastes and wastewater have been used as electron acceptors in MFCs due to their comparatively high redox potentials [149–154] (Table 14.4). In comparison with hexacyanoferrate, using Mn(VII) increased power production 4.5-fold [153], while using Cr(VI) produced a power of 2.2 W/m^3 at a Cr(VI) removal rate of 0.67 mg/L/h [152]. At the same time, the real electroplating wastewater had been tested as an oxidant in MFCs. A maximum power density of 16 W/m^3 with a Cr(VI) removal rate of 8.1 mg/L/h had been achieved with a Cr(VI) wastewater stream containing 204 mg/L Cr(VI) [149]. Using graphite cathode coated with rutile achieved a Cr(VI) reduction rate of 0.97 mg/L/h with light irradiation in comparison with 0.77 mg/L/h without light irradiation [150].

Table 14.4 Pollutants reduced in MFCs with chemical cathodes

Type of substrate	Electron donor	Reactor and electrode	Operation mode	Electron acceptor	Substrate degradation rate[a]	Power (W/m^3)[b]	References
Nitrilotriacetic acid	Nitrilotriacetic acid, bioanode	Two-chambered, graphite felt anode and cathode	Batch	O_2	18.4	0.0000576	[155]
Acid Orange 7	Acetate	Two-chambered, granule graphite anode and cathode	Anode and cathode continuously fed	Acid Orange 7 pH 7.0	36.2–59.5	0.3–0.6	[156]
Nitrobenzene	Acetate	Two-chambered, granule graphite anode and cathode	Anolyte and catholyte re-circulated and continuously fed	Nitrobenzene pH 7.0	6.66	0.06	[157]
	Glucose	Two-chambered, carbon fiber brush anode and carbon cloth cathode with Pt catalyst	Batch anode and recirculated cathode	Nitrobenzene pH 3.0	182.5	6.3	[158]
iopromide	Acetate	Two-chambered, granular graphite anode and cathode	Anolyte and catholyte recirculated	Iopromide pH 7.0	0.28	–0.8[c]	[159]
p- chloro-phenol	Acetate	Two-chambered, carbon fiber anode and cathode	Batch	p- chlorophenol pH 7.0	1.33	0.124	[160]
Cr(VI)	Acetate	Two-chambered, graphite plate anode and cathode	Batch	Cr(VI) pH 2.0	0.67	2.2	[152]
	Acetate	Two-chambered, carbon felt anode and graphite paper cathode	Batch	Cr(VI) pH 2.5	8.1	16	[149]

Mn(VI)	Glucose	Two-chambered, carbon paper anode and carbon cloth cathode	Batch	Mn(VI) pH 3.6	5.8	2.2	[153]
	Glucose	Bushing MFC, carbon paper anode and carbon cloth cathode	Batch	Mn(VI) pH 3.6	Not provided	79.7	[153]
Se(IV)	Glucose	Single-chambered, carbon cloth anode and Pt-coated carbon cloth cathode	Batch	Se(IV) pH 7.0	1.26	25	[161]
	Acetate	Single-chambered, carbon cloth anode and Pt-coated carbon cloth cathode	Batch	Se(IV) pH 7.0	0.47	18.4	[161]
V(V)	Glucose	Two-chambered, carbon fiber anode and cathode	Batch	V(V) pH 2.0	1.76	7.3	[24, 154]
Cu(II)	Acetate	Two-chambered, graphite plate anode and graphite foil cathode	Continuous	Cu(II) pH 3.0	6.94	1.2	[151]
	Acetate	Graphite plate anode and graphite foil cathode	Continuous	Cu(II) + O_2 pH 3.0	5.90	2.2	[151]

[a]Calculated on the basis of net cathodic compartment (mg/L/h)
[b]Power output calculated on the basis of net cathodic compartment (W/m^3)
[c]Set cathodic potential vs. standard hydrogen electrode (SHE) (V)

Recently, Pandit and coworkers [162] studied four types of electron acceptors in a chemical cathode MFC and found that potassium persulfate was the most suitable electron acceptor due to its higher open circuit potential and a longer period of sustained voltage output although permanganate showed the highest power. Copper ions, often present in the mining and metallurgical industries, can be used as an oxidant at the cathode in MFCs for electricity production and recovery of metallic copper at the cathode [151]. Introducing oxygen in the system increased the maximum power density of MFC from 1.2 to 2.2 W/m^3 but decreased the copper recovery from 84% (anaerobic) to 43% (aerobic). This metallurgical MFC technology provides a potential strategy for metal recovery with simultaneous wastewater treatment [151]. Besides some oxidative heavy metals, some oxidative organics like chlorophenol, azo dyes of Acid Orange 7, nitrobenzene, nitrophenol, and iopromide can also be used as oxidants for reduction at the abiotic cathode [156, 157, 159, 160, 163] (Table 14.4). *p*-Chlorophenol can be degraded at a rate of 1.33 mg/L/h with simultaneous electricity generation of 0.124 W/m^3 [160] while with the presence of scrap iron, a higher degradation rate of 11.5 mg/L/h for *p*-nitrophenol and an electricity generation of 0.56 W/m^3 were achieved with an abiotic cathode [163]. Combined with the oxidative conditions in the anode and reductive environment in cathode, employing sulfide as an electron donor in the anode chamber and V(V) as an electron acceptor in cathode chamber can achieve sulfide oxidation at anode and V(V) reduction at cathode simultaneously [24, 154]. This abiotic-cathode MFC provided a proof-of-concept strategy for treating two kinds of wastewater streams at the same time. The efficiency of this system was influenced by the solution pH, catholyte concentrations, dissolved oxygen, and operation modes (Table 14.4) [154, 156, 157, 159].

With respect to the types of electron acceptors, oxygen is regarded as a better choice as its advantages such as sustainability, free availability, and high redox potential. However, the overpotential of interfacing electro-catalyzed reactions of oxygen at the cathode is high in the neutral pH medium in MFCs, resulting in low power generation and requirement of precious metals such as platinum as catalysts [164]. Platinum is an effective cathode catalyst, but it is prohibitively expensive. Additionally, polymer coatings and binders are also used to prepare efficient air-cathode MFCs [165]. Some more inexpensive metal materials such as pyrolysed iron(II), phthalocyanine, and cobalt tetramethoxyphenylporphyrin have been touted as alternative catalysts for oxygen reduction at air cathodes [102, 166]. Recently, manganese-based catalysts prepared by a chemical precipitation method or an electrochemical deposition method provided a cost-effective material for oxygen reduction in MFCs although its performance was not as good as that of Pt [35, 167]. Besides catalysts, electrode-based materials such as carbon also affect the performance of the electrode in the presence of heteroatoms (O, H, S, N) bonded on the surface of carbon particles. For example, carbon cloth cathode and anode both modified with carbon nanotubes improved the system performance [70].

Electrodes using carbon powder (Vulcan XC-72R) as catalyst support material showed varied performances in oxygen reduction if treated with 5% nitric acid, or 0.2N phosphoric acid, or 0.2N potassium hydroxide, or 10% H_2O_2. HNO_3-treated

Vulcan carbon-supported Pt catalyst demonstrated a higher current density of 1,115 mA/m^2 than the untreated carbon-supported Pt catalyst [168]. In contrast to using O_2 as the electron acceptor, photo-generated electrons at semiconductor conduction band can reduce protons in the catholyte, while the holes in valence band recombine with the biological electrons generated at the anode, providing the possibility of solar energy for MFCs [169]. By employing a *p*-type Cu_2O nanowire-arrayed photocathode and *S. oneidensis* MR-1, Qian et al. [169] successfully established a solar-driven, self-biased microbial photoelectrochemical cell in utilizing solar energy for microbial electricity generation, opening new opportunities for microbial/nanoelectronic hybrid devices with possible applications in energy conversion, environmental protection, and biomedical research. However, using a surface-based biofilm is not competitive compared with a volume-based suspension algal culture in the utilization of solar energy. Furthermore, solid-state photovoltaic devices are more practical.

14.3.2 Aerobic Biocathodes

MFCs used for practical applications must be inexpensive and sustainable. Using bacteria instead of the precious metal Pt or added ferricyanide as cathode catalysts, MFCs with biofilm-catalyzed biocathodes could satisfy these demands and enhance the economic viability and environmental sustainability of MFC systems. In this form of MFCs, both the anodic and the cathodic reactions are catalyzed by microbes growing as biofilms, are known as full biological MFCs. The biocathode can be both aerobic and anaerobic depending on the oxidant involved. Aerobic biocathodes using oxygen or ambient air as a terminal electron acceptor have been extensively studied [170–172]. Recently, an aerobic biocathode was used successfully to decolorize the liquid of azo, dyes and it completely degraded them with simultaneous improved electricity generation, illustrating its potential application of MFCs for recalcitrant wastewater treatment and waste remediation [173].

At an aerobic biocathode, oxygen reduction could be catalyzed directly by microorganisms or indirectly via mediators such as manganese and Fe^{2+}/Fe^{3+}. However, electron transfer at the biocathode was not likely carried out through pilus nanowires. More research on the electron-transfer mechanism is still needed. In addition, a quantitative link between the accepted electrons by microorganisms and the subsequent oxygen reduction has not yet been fully appreciated because the bacteria inside can reduce oxygen in diverse pathways [174]. Parameters including cathode potential, mass transfer, and electrode materials were also illustrated to limit the system performance significantly [175, 232]. System optimization based on these factors is therefore necessary for an efficient aerobic biocathode.

Biocathode research has attracted much attention in the past couple years. However, the implementation of biocathodes still face several challenges such as the requirement of dissolved oxygen and the transfer of oxygen from the cathode to the anode chamber if an aerobic biofilm is used on the cathode and the buildup

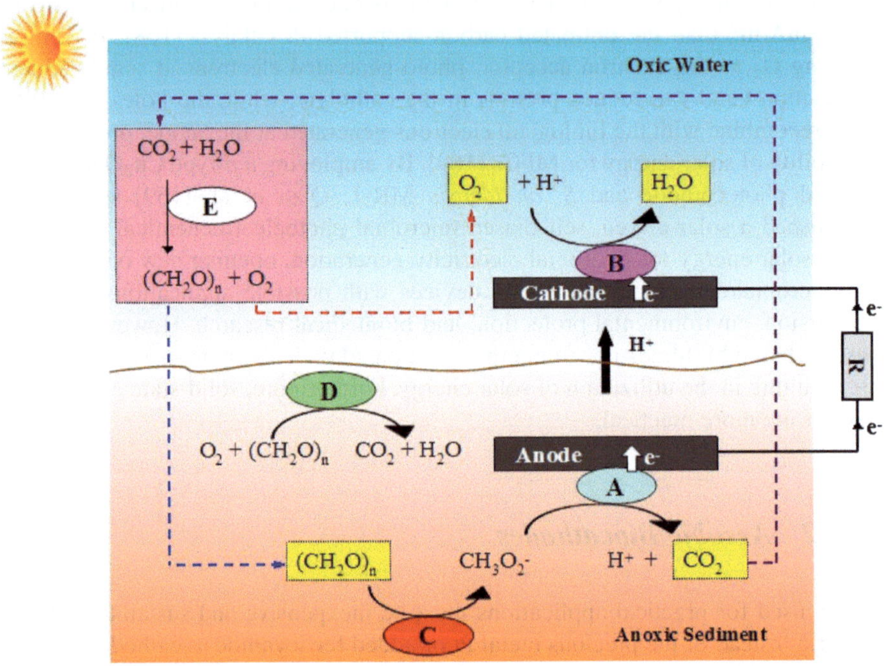

Fig. 14.8 Configuration of interdependent photosynthetic microbial fuel cells (Redrawn after original drawing of Malik et al. [180]). *A* anodic bacteria, *B* cathodic bacteria, *C* anaerobically bacterial dissimilation, *D* Aerobically bacterial dissimilation, *E* photosynthetic organisms)

of a pH gradient between anodic and cathodic chambers that adversely affects system performance [176]. A sufficiently low air supply to the cathodic chamber can prevent oxygen transfer to the anode to some extent [177]. Repeated inversion of the anode and cathode or changing the anode and cathode alternatively is an effective way to decrease the oxygen transfer [91]. Considering the requirement of relatively small quantities of dissolved oxygen to sediment MFCs, aerobic biocathodes may be better applied for sediment MFCs [178, 179]. Using sunlight and an MFC, long-term electricity generation from sunlight without replenishment of the MFCs reactants was possible because that the reactants of an MFC (organic matters and oxygen) were internally regenerated by a photosynthetic microbial consortium whose reactants (carbon dioxide and water) were the products of the MFCs (Fig. 14.8) [180, 181]. Much work is still needed to increase the rate of mass transport at the anode, provide effective microorganisms for high glucose yield, and optimize the design of MFCs with high current densities [180]. When MFCs use sunlight as the energy input for the sole purpose of generating electricity, they have to compete directly with photovoltaic technology that is cheaper and far more advanced. Thus, MFCs powered by sunlight will more likely remain as an academic curiosity or, at best, a tool to study bioelectrochemistry in MFCs.

Yet another creation used the interaction between mixed heterotrophic bacteria in sediments with plants such as reed manna grass (*Glyceria maxima*) [182] and rice plants [183, 184] formed a new type of MFC, sediment-type photoMFCs [182], providing a direct correlation between photosynthetic activity and power production. In this form of an MFC, the plants' photosynthetic process produces rhizodeposits as fuel source in MFCs for electricity generation.

14.3.3 Anaerobic Biocathodes

Because of the limitations of aerobic biocathodes for reductive removal of contaminants in wastewater, researchers quickly turned to anaerobic biocathodes. Contaminants removed at the anaerobic biocathode include nitrate, sulfate, CO_2, Cr(VI), Mn(IV), U(VI), fumarate, perchlorate, trichloroethene, tetrachloroethene, 2-chlorophenol [172, 176, 185–193, 227, 233–236], and CO_2 [188]. Nitrate has been extensively studied because the standard reduction potential of the NO_3^-/N_2 couple is relatively high ($E^{0\prime} = +0.74$ V), and nitrate is a common pollutant in agricultural runoffs. Biological nitrate reduction was initially demonstrated by removing electrons from the cathode using a potentiostat-poised half cell [190, 194]. Clauwaert and coworkers [170] succeeded in nitrate removal from catholyte in a completely biological MFC. Optimizing solution chemistry and using a flow loop operation can improve the efficiency of nitrate reduction [94, 195, 196], in which the operational conditions affected bacterial metabolism and biofilm robustness, but not the phylogenetic affiliation of dominant bacteria [197].

Most people shy away from sulfate reduction in wastewaters by biocathodes, because it has a negative standard potential and more importantly it generates hydrogen sulfide gas that is toxic and has a strong repugnant odor at even very low concentrations. However, sulfate is a major pollutant in agricultural runoffs. New heterogeneous catalysts are being developed to remove or catalytically convert hydrogen sulfide to product(s) with no odor. It is claimed (http://www.swapsol.com) that the Stenger-Wasas process reacts hydrogen sulfide gas with carbon dioxide gas on a solid heterogeneous catalyst in an exothermic and kinetically favorable (at or below 25°C to 150°C and above, 15–700 psia) reaction that reduces hydrogen sulfate to below 4 ppb in a single pass.

Most perchlorate manufactured in the USA is intended for use as an ingredient in solid fuel for missiles and rockets. However, it is now routinely detected in municipal water supplies. Using 2,6-anthraquinone disulfonate as a mediator, perchlorate was indirectly reduced by *Dechloromonas agitata* strain CKB and *Azospira suillum* strain PS [235]. H_2 utilization by conventional perchlorate-reducing bacteria probably occurred with the applied potential of −0.3 V (vs. standard hydrogen electrode, SHE) in this system. The perchlorate-reducing bacteria in a denitrifying biocathode successfully reduced perchlorate without exogenous electron shuttle supplementation [192]. The perchlorate-reducing biocathode was mainly composed of putative denitrifying betaproteobacteria that was totally different from the community of a purely denitrifying biocathode that is primarily composed of putative iron-oxidizing genera [187]. This successful development of a perchlorate-reducing microbial community in a MFC biocathode supplied with nitrate and perchlorate makes this technology potentially suitable for treating water supplies, superfund sites, or industrial wastewaters with co-contaminants of perchlorate and nitrate [187].

Inoculating with a mixture of denitrifying and anaerobic bacteria in a biocathode MFCs was successfully used to reduce Cr(VI) [233] in lab tests. Therefore, the diverse microbial consortia in wastewater are able to adaptively evolve for the degradation of various contaminants in fed streams of MFCs [78, 174]. Besides metal contaminants, chlorinated compounds like trichloroethene, tetrachloroethene, and 2-chlorophenol were successfully reduced at biocathodes [185, 193, 198]. A pure culture of *Anaeromyxobacter dehalogenans*, previously dechlorinating 2-chlorophenol in conventional biological process [199], performed not only direct electron transfer at the anode using acetate as an electron donor but also directly accepted electrons from a cathodic electrode with simultaneous reduction of 2-chlorophenol at a set biocathode potential of −300 mV (vs. SHE). This process provided a potential strategy for efficient treatment of a variety of contaminants with microbe-electrode interactions [193]. The reactor architecture and electrode surface area had substantial effects on the performance of this MFC system.

Two-chambered and H-type architectures with graphite plate cathode and anode were often used for MFCs with anaerobic biocathodes [233] (Fig. 14.9a). Huang et al. [175] reported that using graphite granules that have a high surface area can improve Cr(VI) reduction rate and power generation [175] (Fig. 14.9a). An efficient tubular MFC coupled with proper ratio of cathode surface area to anode surface area further improved performance of biocathode MFC with Cr(VI) as a terminal electron acceptor [200] (Fig. 14.9b). An increase in the cathode surface roughness of the stainless steel cathode electrode colonized by *G. sulfurreducens* led to an increase in current by a factor of 1.6 when fumarate was oxidized because the overall current density increased with an increase in biofilm coverage [201].

In the case of CO_2 reduction, a light-dependent biocathode can perform bicarbonate reduction, providing a potential method for carbon dioxide sequestration [188]. With an external potential of less than −0.7 V (vs. Ag/AgCl), carbon dioxide can be reduced to methane by hydrogenophilic methanogenesis at the biocathode via direct extracellular electron-transfer mechanisms. This indicates that methane can be produced by a biocathode in an MFC as a fuel or a value-added product [202, 203]. CO_2 discharged from the anode can be directly utilized by the algae *Chlorella vulgaris* at the cathode, increasing a maximum power from 4.1 to 5.6 W/m^3. This process demonstrated a potentially effective technology for CO_2 emission reduction with simultaneous voltage output without aeration [204]. Alternatively, *C. vulgaris* can accept electrons from the cathode via the mediator methylene blue to simultaneously remove CO_2 from the atmosphere [237]. The drawback of using MFCs for CO_2 removal is that a surface-based technology rather than a volume-based technology is used in the case of an algal suspension. However, the latter does not produce electricity in the removal process.

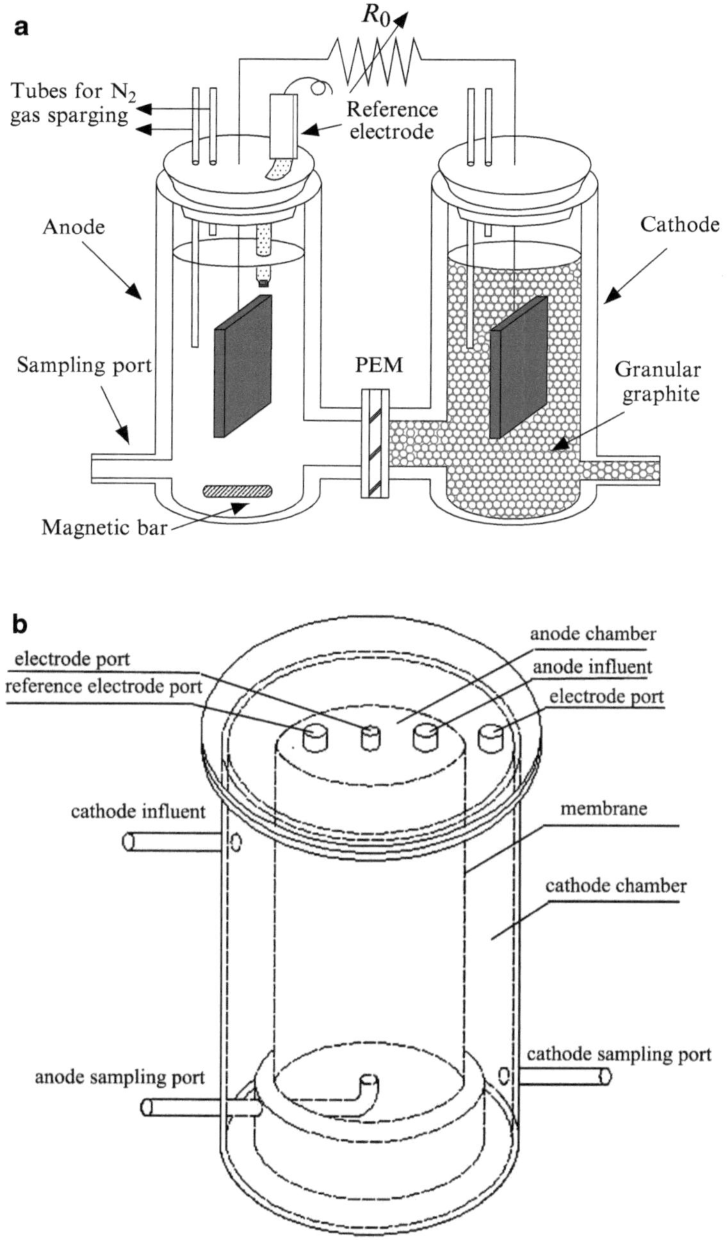

Fig. 14.9 Two-chambered anaerobic biocathode MFCs: H-type reactor with graphite plate cathode or graphite granule cathode (**a**) and tubular-type reactor with graphite fiber/graphite felt/graphite granule cathode (**b**) (Reprinted from Wang et al. [152] with kind permission of © Springer and Huang et al. [200] with kind permission of © Elsevier)

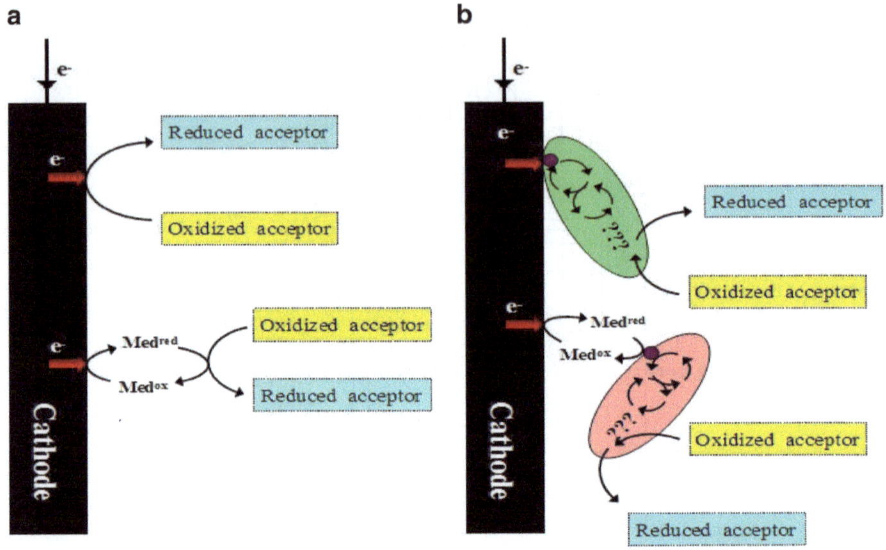

Fig. 14.10 Electron-transfer mechanisms at a chemical cathode (**a**) and a biocathode (**b**) MFCs

14.3.4 Electron-Transfer Mechanisms for Cathodes

MFCs can be classified as either chemical cathode or biocathode MFCs based on the catalysts used at the cathode. Chemical cathode MFCs use chemical catalysts on the cathode and biocathode MFCs use biofilm bacteria for catalysis on the cathode. Accordingly, electrons from a chemical cathode are transferred to electron acceptors directly or indirectly via the mediation of external added electron mediators with chemical catalysts (Fig. 14.10a). In contrast to chemical cathodes, biocathodes transfer electrons from the cathode substratum to the microorganisms directly or indirectly via mediators. But the mechanism for the subsequent electron transfer from microorganisms to final electron acceptors is still unclear ([174], Fig. 14.10b). In comparison with bioanodes, in which electrons are transferred directly or indirectly via mediators or pilus nanowires, the precise mechanism underlying the exceptional electron transfer of pili in biocathodes has not yet reported. *S. putrefaciens* can donate electrons to an anode via the extracellular electron transfer with the help of secreted flavins, menaquinone-related redox mediators, and outer membrane cytochromes [112, 205, 238, 239]. The organism can also carry out electron transfer in microbially cathodic oxygen reduction via an outer membrane-bound redox compound released into the surrounding medium [206]. The mechanism of electron transfer at a biocathode with *G. sulfurreducens* is different from that at a bioanode with the same bacteria [207, 240]. However, electron-transfer mechanisms between the electrode and *S. putrefaciens* are not fully understood. It remains unclear whether the electron-transfer mechanism from the cathode to this bacterium is similar to that at the anode. Finally, electron transfer in biocathodes can be revealed by investigating purified proteins that transfer electrons between the outer cell surface and the electrodes [123, 241, 242].

14.3.5 Outlook

While biocathodes have attracted much attention during the past few years, there are still many technological hurdles before their practical application: (1) Systematically studying the effects of parameters such as electrode potential, electrode materials, solution chemistry, mixed/pure cultures, biofilm thickness, pH, temperature, and reactor architecture on the performance of biocathodes is required. Additionally, investigation of the biocathode cost, duration, selectivity, stability, and the compatibility with operational conditions is essential for efficient waste and wastewater treatment. (2) Much work is needed to overcome the loss of organic matter in the anode chamber caused by leakage through the membrane from the anode to the cathode and on reducing pH gradients caused by transport of specific ions with higher concentrations instead of protons through membrane resulting in acidification at the anode and alkaline production at the cathode, especially when two different waste streams are treated at the anode and cathode separately [208]. Using a loop to exchange anolyte and catholyte can balance pH gradients and make the biocathodes operationally sustainable [94, 170, 243]. Another approach is to create a bioelectrode that can perform anodic and cathodic reactions alternately without the transport of protons through the membrane [91, 93]. Further investigations are needed.

14.4 Reactor Design Rationale for Practical Wastewater Treatment

In real-world wastewater treatment, there are several major concerns over the use of membranes. Membranes are a major cost factor in MFC construction. Their easy fouling makes MFC operational sustainability unlikely attainable. In an ideal MFC operation, electrons flow from anode to cathode through an external circuit, and internally, protons flow from the anode to the cathode in an unimpeded fashion. If convective flow is permitted for a wastewater stream from anode to the cathode, the internal resistance will be greatly minimized. However, this is not possible with a membrane in place. Membrane-less MFCs have been proposed by several researchers for wastewater treatment. Figure 14.11 shows an upward flow membrane-less MFC for wastewater treatment using an air cathode. In this kind of design, upward flow helps oxygen bubbles to move upward to prevent back diffusion of oxygen into the anode region at a cost of energy input for pumping the flow.

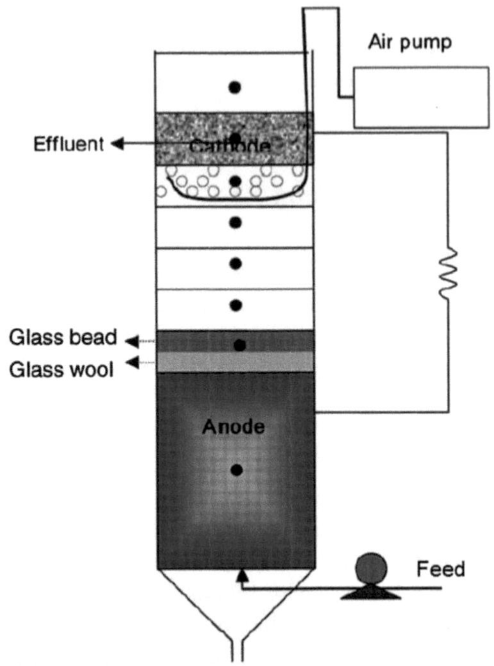

Fig. 14.11 An upflow membrane-less MFC with an air cathode. The *black dots* indicate sampling locations (Reprinted from Jang et al. [209] with kind permission of © Elsevier)

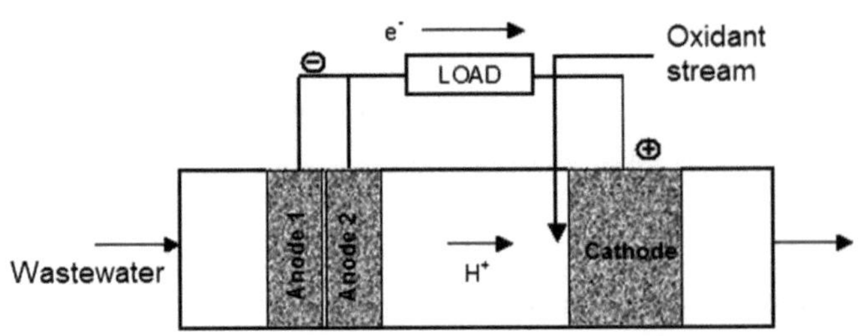

Fig. 14.12 A membrane-less MFC with dual anodes and a supplemental oxidant stream

Figure 14.12 is a simplistic membrane-less MFC reactor proposed for wastewater treatment. This kind of configuration is low cost and low maintenance. Two different anodes with different biofilm consortia are used: one for digesting simpler molecules such as volatile fatty acids and small sugar molecules and the other for digesting complicated biomass such as lignocellulosic wastes. The electrode materials should be inexpensive and have large surface areas such as graphite granules. Instead of using oxygen, a supplemental feed stream containing a dissolved oxidant such as nitrate can be used. Nitrate is readily available from agricultural runoffs. Coulombic efficiency can be greatly improved if the amount of organic carbons reaching the cathode is small or the cathode biofilm is impaired in utilizing them. Oxidant contamination of the anode area can be reduced if the wastewater's bulk-flow rate is sufficiently high. The flow direction in such a reactor can be horizontal or even downward to minimize or eliminate energy input in fluid handling as long as oxidant backflow is in check. A small amount of organic carbon reaching the cathode will reduce the CEs slightly, but it is actually beneficial for the cathode biofilm to rejuvenate itself. The electrodes in Fig. 14.12 are modularized for quick replacement and offline biofilm resetting in order to reduce maintenance costs.

A fast axial flow rate in the simplistic design in Fig. 14.12 is only possible by using highly efficient biofilms. Such biofilms will likely come from the new research efforts on "superbugs" that can efficiently digest a large array of organic carbons in wastewaters much more quickly than existing biofilm bacteria. Genetically engineered superbugs will likely be constructed before wild-type superbugs or biofilm consortia containing superbugs are found.

14.5 Value-Added Products from MXCs

Based on a life-cycle assessment [210], environmental impact is a key driver for the development of MFC technology. Application of MFC in wastes treatment and remediation of contaminated sites with simultaneous production of value added by-products is arguably the main research theme for the future. While microbial electrolysis cells (MECs) are being explored for new value-added products, the newly developed microbial desalination cells (MDCs) have also substantially broadened the applications of MFCs. MFCs and their variants are called MXCs.

14.5.1 Microbial Electrolysis Cells for Value-Added Products

An electrically assisted MFC is a device that produces H_2 at the cathode, in which the driving force is from the oxidation of organic matter by microorganisms at the anode and the external power supply [211]. In this form of MFCs, organic matter degradation occurs, and water is used to supply H^+ and OH^- with an internal voltage, a process that occurs in microbial electrolysis cells (MECs) [211, 212]. In a MEC, an external voltage must be applied to overcome the thermodynamic barrier. A voltage of 0.14–0.22 V is sufficient for H_2 production in MECs theoretically while a 0.25–0.5 V is actually needed due to electrode overpotentials and other losses. In comparison with the theoretical threshold of 1.23 V and the actual value of 1.8–2.0 V required for water electrolysis [213], the 0.25–0.5 V for MECs is much lower and results in higher electrical energy efficiency. More importantly, this technology could degrade and utilize various organic materials including the large quantity of those in wastewaters. MECs present a promising technology for wastewater treatment with concomitant bioproduct production. Diverse reactor architectures with different operation modes, as well as various anodic and cathodic materials that are the same as those used in MFCs coated with different catalysts such as platinum, Pd/Pt, and biocatalysts (biofilms), have been extensively investigated to improve volumetric efficiency [165, 191, 214–216]. MFCs can supply some, if not all, of the needed electricity for MECs. An MEC-MFC combination was successfully demonstrated, in which the power generated from an MFC was utilized to offset the electricity needed by an MEC that produced biohydrogen [217]. A major challenge for MECs producing biohydrogen is to minimize methanogensis that consumes hydrogen to produce methane while competing for carbon sources with exoelectrogenic microorganisms when single chamber MECs are used [215, 216]. In another way, the utilization of carbon dioxide for methane production via hydrogenophilic methanogenesis implied the production of a value-added product of methane by a biocathode [202, 203].

In a conventional biological system, ethanol was usually produced by microbial reduction of acetate with hydrogen as electron donor. However, in an MEC system, ethanol could be produced by the reduction of acetate at a cathode with an applied cathode potential of −0.55 V (vs. NHE) and methyl viologen as mediator [218]. MEC system can be also used to produce other value-added matters. Under laboratory conditions, a liter-scale lamellar bioelectrochemical system produced a caustic solution of 3.4 wt% with a conversion efficiency of 61% (from acetate to caustic) at the cathode (with a 1.77 V applied cell voltage, 1,015 A/m^3 anode volume). This system was tested for producing caustic using wastewater from a brewing process [219]. With a loop operation of circulating the anodic effluent to the cathode chamber, steroid hormones like dibutyl phthalate, 2-ethylformanilide, and 2,5-pyrrolidinedione can be produced at the cathode using *Dioscorea zingiberensis* processing wastewater as a fuel [220]. Crude glycerol is regarded as a waste stream from biodiesel production. Glycerol has been tested for power production in single-chambered air-cathode MFC catalyzed by a pure culture of *Bacillus subtilis* [221]. By the biofilm catalysis of a pure culture *S. oneidensis*, glycerol can be converted into ethanol at the anode of an MFC, providing a new approach for renewable bioproduct production [41].

14.5.2 From Oxygen to Hydrogen Peroxide (H_2O_2)

H_2O_2, a strong oxidant used in many applications such as in the pulp and paper industry, textile bleaching, and some chemical syntheses, is currently produced in an energy intensive process. With an external voltage of 0.5 V, a bioelectrochemical system can convert oxygen to H_2O_2 at the cathode with a rate of 1.9 ± 0.2 kg H_2O_2/m^3/day with simultaneous oxidation of organic matters in a wastewater feed stream to the anodic chamber, providing a potential application of such a system for H_2O_2 production utilizing the energy released by oxidation of organic matters [222]. This on-site-produced H_2O_2 efficiently achieved Cr(VI) reduction [223] and degradation of azo dyes amaranth [224].

14.5.3 *Microbial Desalination Cells*

A proof-of-concept microbial desalination cell (MDC) was composed of three chambers: anode chamber, cathode chamber, and center chamber between two membranes with an anion exchange membrane facing the anode side and a cation exchange membrane facing the cathode side [188]. In the MDC, current was produced through the process of organic carbon oxidization at the anode, oxidant reduction at the cathode, and water desalination at the center chamber, realizing power generation with simultaneous wastewater treatment and desalination. A maximum power density of 31 W/m^3 was achieved in this system with simultaneous 90% removal of the salt in a single desalination cycle [188]. Recently, it was demonstrated that by applying a voltage of 0.8 V, this system achieved a maximum H_2 production rate of 1.5 $m^3/m^3/d$ (1.6 mL/h) from the cathode chamber with simultaneous 98.8% removal of the 10 g/L NaCl from the middle chamber [225]. Energy balance calculations showed that it was possible that the energy produced in the form of hydrogen more than offset the energy consumption [226]. Further investigations are needed to evaluate the performances of larger systems using real-world feed streams.

14.6 Conclusion

MFC-based technologies hold promises for the production of renewable bioenergy and value-added products from waste biomass or wastewaters. They are still in infancy stages. Much work is needed on biofilm engineering, electrode materials, solution chemistry, and reactor architectures to improve overall performances. The integration of MFCs with other technologies will be surely developed for widened applications of MFCs. The involvement of multiple disciplines such as materials science, microbiology, genetic engineering, electrochemistry, environmental engineering, and chemical engineering is needed to make practical MFC-based technologies an economic reality.

Genetically engineered superbug biofilms present a unique opportunity to resolve electron-transfer bottlenecks at the anode and cathode, to allow a huge array of organic matters to be digested with increased metabolic rates, resulting in reduced residence times (i.e., faster flow) and improved CEs in the absence of a membrane. These kinds of biofilms will make the simplistic and low-cost MFC design shown in Fig. 14.12 possible. This new approach may be a game-changer in the MFC technology. To assure a high CE of the cell without a membrane, superbug biofilms for the bioanode and the biocathode are required. Arguably, the main hurdle is to obtain highly competitive and metabolically voracious bacteria that can devour hundreds of different carbon sources requiring small residence times. We touched on one such organism, PA, in this review. Being a natural soil-dwelling organism, PA has the "hard-wired" properties of being ultracompetitive. For example, if one were to place PA and *Escherichia coli* together, PA would rapidly kill this bacterium and utilize its cellular material for food, especially when starved for phosphate. Still, unlike the well-studied electrogens, *Shewanella* and *Geobacter*, the genes involved in biofilm attachment, matrix formation, micro-macrocolony differentiation, and dispersion have been well studied in PA. Thus, arguably, the most critical feature of MFC power efficiency is a robust biofilm in the anodic surface mediated, in part, by numerous conductive pili. Thus, we propose that the most desirable properties of genetically modified electrogens would be (1) robust biofilm formation on the anode; (2) a limited or an inability to disperse from the biofilm once attached; (3) inability to undergo anaerobic respiration, thereby not wasting precious electron flow to the process of denitrification; (4) reduction of the biofilm matrix, thereby increasing nutrient flow to biofilm bacteria; (5) increase mediator (e.g., PCN) secretion, allowing for yet another source of redox-based power generation as well as an antibiotic for competitive fitness; and (6) the possibility of cloning uncoupler proteins into organisms which would increase the rate of oxidation of wastewater substrates. We wish to emphasize that these are only a few strategic suggestions. However, as more genes are discovered that affect power generation in electrogenic organisms, additional light will be shed as to even greater improvements that can be made to push this technology to the next level. Furthermore, there are avenues that have to date been unexplored which include consortia of genetically modified bacteria of potentially different genera that coexist in a symbiotic relationship. Such consortia would be particularly beneficial in the event of either wastewaters with ever-changing nutrient flow or MFCs fed with multiple different wastewater streams.

Acknowledgments Huang and Cheng gratefully acknowledge financial support from the Natural Science Foundation of China (Nos. 51178077, 21077017 and 21073163), Science & Research Program of Zhejiang Province (2010 C31014), Open Project of State Key Laboratory of Clean Energy Utilization (ZJUCEU2010001), Program for Changjiang Scholars and Innovative Research Team in University (IRT0813), and "Energy + X" (2008) key program through Dalian University of Technology.

References

1. Heidrich ES, Curtis TP, Dolfing J (2011) Determination of the internal chemical energy of wastewater. Environ Sci Technol 45:827–32
2. Pant D, Bogaert GV, Diels L, Vanbroekhoven K (2010) A review of the substrates used in microbial fuel cells (MFCs) for sustainable energy production. Bioresour Technol 101:1533–1543
3. Liu H, Ramnarayanan R, Logan BE (2004) Production of electricity during wastewater treatment using a single chamber microbial fuel cell. Environ Sci Technol 38:2281–2285
4. Min B, Logan BE (2004) Continuous electricity generation from domestic wastewater and organic substrates in a flat plate microbial fuel cell. Environ Sci Technol 38:5809–5814
5. Min B, Kim J, Oh S, Regan J, Logan BE (2005) Electricity generation from swine wastewater using microbial fuel cells. Water Res 39:4961–4968
6. Lu N, Zhou S, Zhuang L, Zhang J, Ni J (2009) Electricity generation from starch processing wastewater using microbial fuel cell technology. Biochem Eng J 43:246–251

7. Feng Y, Wang X, Logan BE, Lee H (2008) Brewery wastewater treatment using air-cathode microbial fuel cells. Appl Microbiol Biotechnol 78:873–880
8. Gálvez A, Greenman J, Ieropoulos I (2009) Landfill leachate treatment with microbial fuel cells; scale-up through plurality. Bioresour Technol 100:5085–5091
9. Greenman J, Gálvez A, Giusti L, Ieropoulos I (2009) Electricity from landfill leachate using microbial fuel cells: comparison with a biological aerated filter. Enzym Microb Technol 44:112–119
10. Zhang J, Zhao Q, You S, Jiang J, Ren N (2008) Continuous electricity production from leachate in a novel upflow air-cathode membrane-free microbial fuel cell. Water Sci Technol 57:1017–1021
11. Jiang J, Zhao Q, Zhang J, Zhang G, Lee D (2009) Electricity generation from bio-treatment of sewage sludge with microbial fuel cell. Bioresour Technol 100:5808–5812
12. Jiang J, Zhao Q, Wei L, Wang K (2010) Extracellular biological organic matters in microbial fuel cell using sewage sludge as fuel. Water Res 44:2163–2170
13. Jiang XC, Hu JS, Fitzgerald LA, Biffinger JC, Xie P, Ringeisen BR, Lieber CM (2010) Probing electron transfer mechanisms in *Shewanella oneidensis* MR-1 using a nanoelectrode platform and single-cell imaging. Proc Natl Acad Sci 107:16806–16810
14. Niessen J, Schröder U, Harnisch F, Scholz F (2005) Gaining electricity from in situ oxidation of hydrogen produced by fermentative cellulose degradation. Lett Appl Microbiol 41:286–290
15. Huang L, Logan BE (2008) Electricity generation and treatment of paper recycling wastewater using a microbial fuel cell. Appl Microbiol Biotechnol 80:349–355
16. Huang L, Cheng S, Rezaei F, Logan BE (2009) Reducing organic loading in industrial effluents using microbial fuel cells. Environ Technol 30:499–504
17. Ren Z, Ward TE, Regan JM (2007) Electricity production from cellulose in a microbial fuel cell using a defined binary culture. Environ Sci Technol 41:4781–4786
18. Rezaei F, Richard TL, Brennan RA, Logan BE (2007) Substrate-enhanced microbial fuel cells for improve remote power generation from sediment-based system. Environ Sci Technol 41:4053–4058
19. Rezaei F, Xing D, Wagner R, Richard T, Logan BE (2009) Simultaneous cellulose degradation and electricity production by *Enterobacter cloacae* in a microbial fuel cell. Appl Environ Microbiol 75:3673–3678
20. Rezaei F, Richard TL, Logan BE (2009) Analysis of chitin particle size on maximum power generation, power longevity, and Coulombic efficiency in solid-substrate microbial fuel cells. J Power Sources 192:304–309
21. Rismani-Yazdi H, Christy AD, Dehority BA, Morrison M, Yu ZT, Tuovinen OH (2007) Electricity generation from cellulose by rumen microorganisms in microbial fuel cells. Biotechnol Bioeng 97:1398–1407
22. Velasquez-Orta SB, Curtis TP, Logan BE (2009) Energy from algae using microbial fuel cells. Biotechnol Bioeng 103:1068–1076
23. Rezaei F, Richard TL, Logan BE (2008) Enzymatic hydrolysis of cellulose coupled with electricity generation in a microbial fuel cell. Biotechnol Bioeng 101:1163–1169
24. Zhang B, Zhao H, Shi C, Zhou S, Ni J (2009) Simultaneous removal of sulfide and organics with vanadium(V) reduction in microbial fuel cells. J Chem Technol Biotechnol 84:1780–1786
25. Zuo Y, Maness P, Logan BE (2006) Electricity production from steam-exploded corn stover biomass. Energy Fuel 20:1716–1721
26. Wang X, Feng Y, Wang H, Ou Y, Yu Y, Ren N, Li N, Wang E, Lee H, Logan BE (2009) Bioaugmentation for electricity generation from corn stover biomass using microbial fuel cells. Environ Sci Technol 43:6088–6093
27. Zang G, Sheng G, Tong Z, Liu X, Teng S, Li W, Yu H (2010) Direct electricity recovery from *Canna indica* by an air-cathode microbial fuel cell inoculated with rumen microorganisms. Environ Sci Technol 44:2715–2720

28. Catal T, Cysneiros D, O'Flaherty V, Leech D (2011) Electricity generation in single-chamber microbial fuel cells using a carbon source sampled from anaerobic reactors utilizing grass silage. Bioresour Technol 102:404–410
29. Catal T, Xu S, Li K, Bermek H, Liu H (2008) Electricity generation from polyalcohols in single-chamber microbial fuel cells. Biosens Bioelectron 24:849–854
30. Huang L, Logan BE (2008) Electricity production from xylose in fed-batch and continuous-flow microbial fuel cells. Appl Microbiol Biotechnol 80:655–664
31. Huang L, Zeng J, Angelidaki I (2008) Electricity production from xylose using a mediator-less microbial fuel cell. Bioresour Technol 99:4178–4184
32. Catal T, Fan YZ, Li KC, Bermek H, Liu H (2008) Effects of furan derivatives and phenolic compounds on electricity generation in microbial fuel cells. J Power Sources 180:162–166
33. Luo Y, Liu G, Zhang R, Zhang C (2010) Power generation from furfural using the microbial fuel cell. J Power Sources 195:190–194
34. Cheng S, Kiely P, Logan BE (2011) Pre-acclimation of a wastewater inoculum to cellulose in an aqueous-cathode MEC improves power generation in air-cathode MFCs. Bioresour Technol 102:367–371
35. Zhang C, Li M, Liu G, Luo H, Zhang R (2009) Pyridine degradation in the microbial fuel cells. J Hazard Mater 172:465–471
36. Luo Y, Zhang R, Liu G, Li J, Li M, Zhang C (2010) Electricity generation from indole and microbial community analysis in the microbial fuel cell. J Hazard Mater 176:759–764
37. Zhang C, Liu G, Zhang R, Luo H (2010) Electricity production from and biodegradation of quinoline in the microbial fuel cell. J Environ Sci Health Part A 45:250–256
38. Huang L, Yang X, Quan X, Chen J, Yang F (2010) A microbial fuel cell-electro-oxidation system for coking wastewater treatment and bioelectricity generation. J Chem Technol Biotechnol 85:621–627
39. He Z, Kan J, Wang Y, Huang Y, Mansfeld F, Nealson KH (2009) Electricity production coupled to ammonium in a microbial fuel cell. Environ Sci Technol 43:3391–3397
40. Kim JR, Zuo Y, Regan JM, Logan BE (2008) Analysis of ammonia loss mechanisms in microbial fuel cells treating animal wastewater. Biotechnol Bioeng 99:1120–1127
41. Flynn JM, Ross DE, Hunt KA, Bond DR, Gralnick JA (2011) Enabling unbalanced fermentations by using engineered electrode-interfaced bacteria. mBio 1:e00190–10
42. Raghavulu SV, Goud RK, Sarma PN, Mohan SV (2011) *Saccharomyces cerevisiae* as anodic biocatalyst for power generation in biofuel cell: influence of redox condition and substrate load. Bioresour Technol 102:2751–2757
43. Finch AS, Mackie TD, Sund CJ, Sumner JJ (2011) Metabolite analysis of *Clostridium acetobutylicum*: fermentation in a microbial fuel cell. Bioresour Technol 102:312–315
44. Rismani-Yazdi H, Christy AD, Carver SM, Yu Z, Dehority BA, Tuovinen OH (2011) Effect of external resistance on bacterial diversity and metabolism in cellulose-fed microbial fuel cells. Bioresour Technol 102:278–283
45. Katuri KP, Scott K, Head IM, Picioreanu C, Curtis TP (2011) Microbial fuel cells meet with external resistance. Bioresour Technol 102:2758–2766
46. Picioreanu C, Head IM, Katuri KP, van Loosdrecht MCM, Scott K (2007) A computational model for biofilm-based microbial fuel cells. Water Res 41:2921–2940
47. Picioreanu C, Katuri KP, Head IM, van Loosdrecht MCM, Scott K (2008) Mathematical model for microbial fuel cells with anodic biofilms and anaerobic digestion. Water Sci Technol 57:965–971
48. Jung S, Regan JM (2011) Influence of external resistance on electrogenesis, methanogenesis, and anode prokaryotic communities in microbial fuel cells. Appl Environ Microbiol 77:564–571
49. Lee HS, Parameswaran P, Kato-Marcus A, Torres CI, Rittmann BE (2008) Evaluation of energy-conversion efficiencies in microbial fuel cells (MFCs) utilizing fermentable and non-fermentable substrates. Water Res 42:1501–1510
50. Freguia S, Rabaey K, Yuan Z, Keller J (2008) Syntrophic processes drive the conversion of glucose in microbial fuel cell anodes. Environ Sci Technol 42:7937–7943

51. Kim HJ, Park HS, Hyun MS, Chang IS, Kim M, Kim BH (2002) A mediator-less microbial fuel cell using a metal reducing bacterium, *Shewanella putrefaciens*. Enzym Microb Technol 30:145–152
52. Liu H, Cheng SA, Logan BE (2005) Power generation in fed-batch microbial fuel cells as a function of ionic strength, temperature, and reactor configuration. Environ Sci Technol 39:5488–5493
53. Wang YF, Cheng SS, Tsujimura S, Ikeda T, Kano K (2006) *Escherichia coli*-catalyzed bioelectrochemical oxidation of acetate in the presence of mediators. Bioelectrochemistry 69:74–81
54. Biffinger JC, Byrd JN, Dudley BL, Ringeisen BR (2008) Oxygen exposure promotes fuel diversity for *Shewanella oneidensis* microbial fuel cells. Biosens Bioelectron 23:820–826
55. Li SL, Freguiab S, Liu SM, Cheng SS, Tsujimura S, Shirai O, Kano K (2010) Effects of oxygen on *Shewanella decolorationis* NTOU1 electron transfer to carbon-felt electrodes. Biosens Bioelectron 25:2651–2656
56. Xie X, Hu L, Pasta M, Wells GF, Kong D, Criddle CS, Cui Y (2011) Three-dimensional carbon nanotube textile anode for high-performance microbial fuel cells. Nano Lett 11:291–296
57. Chaudhuri SK, Lovley DR (2003) Electricity generation by direct oxidation of glucose in mediatorless microbial fuel cells. Nat Biotechnol 21:1229–1232
58. He Z, Minteer SD, Angenent LT (2005) Electricity generation from artificial wastewater using an upflow microbial fuel cell. Environ Sci Technol 39:5262–5267
59. Cheng S, Liu H, Logan BE (2006) Increased performance of single-chamber microbial fuel cells using an improved cathode structure. Electrochem Commun 8:489–494
60. Cheng S, Liu H, Logan BE (2006) Increased power generation in a continuous flow MFC with advective flow through the porous anode and reduced electrode spacing. Environ Sci Technol 40:2426–2432
61. Logan BE, Cheng S, Watson V, Estadt G (2007) Graphite fiber brush anodes for increased power production in air-cathode microbial fuel cells. Environ Sci Technol 41:3341–3346
62. Liu H, Cheng S, Huang L, Logan BE (2008) Scale-up of membrane-free single-chamber microbial fuel cells. J Power Sources 179:274–279
63. Zeng L, Zhang L, Li W, Zhao S, Lei J, Zhou Z (2010) Molybdenum carbide as anodic catalyst for microbial fuel cell based on *Klebsiella pneumoniae*. Biosens Bioelectron 25:2696–2700
64. Dewan A, Beyenal H, Lewandowski Z (2008) Scaling up microbial fuel cells. Environ Sci Technol 42:7643–7648
65. Larrosa-Guerrero A, Scott K, Katuri KP, Godinez C, Head IM, Curtis T (2010) Open circuit versus closed circuit enrichment of anodic biofilms in MFC: effect on performance and anodic communities. Appl Microbiol Biotechnol 87:1699–1713
66. Larrosa-Guerrero A, Scott K, Head IM, Mateo F, Ginesta A, Godinez C (2010) Effect of temperature on the performance of microbial fuel cells. Fuel 89:3985–3994
67. Tang X, Guo K, Li H, Du Z, Tian J (2011) Electrochemical treatment of graphite to enhance electron transfer from bacteria to electrodes. Bioresour Technol 102:3558–3560
68. Lowy DA, Tender LM (2008) Harvesting energy from the marine sediment–water interface III. Kinetic activity of quinone- and antimony-based anode materials. J Power Sources 185:70–75
69. Saito T, Mehanna M, Wang X, Cusick RD, Feng Y, Hickner MA, Logan BE (2011) Effect of nitrogen addition on the performance of microbial fuel cell anodes. Bioresour Technol 102:395–398
70. Tsai HY, Wu CC, Lee CY, Shih EP (2009) Microbial fuel cell performance of multiwall carbon nanotubes on carbon cloth as electrodes. J Power Sources 194:199–205
71. Peng L, You S, Wang J (2010) Carbon nanotubes as electrode modifier promoting direct electron transfer from *Shewanella oneidensis*. Biosens Bioelectron 25:1248–1251
72. Mottaghitalab V, Spinks GM, Wallace GG (2005) The influence of carbon nanotubes on mechanical and electrical properties of polyaniline fibers. Synth Met 152:77–80
73. Qiao Y, Li CM, Bao S, Bao Q (2007) Carbon nanotube/polyaniline composite as anode material for microbial fuel cells. J Power Sources 170:79–84

74. Zou Y, Pisciotta J, Baskakov IV (2010) Nanostructured polypyrrole-coated anode for sun-powered microbial fuel cells. Bioelectrochemistry 79:50–56
75. Yuan SJ, Sheng GP, Li WW, Lin ZQ, Zeng RJ, Tong ZH, Yu HQ (2010) Degradation of organic pollutants in a photoelectrocatalytic system enhanced by a microbial fuel cell. Environ Sci Technol 44:5575–5580
76. Schröder U (2007) Anodic electron transfer mechanisms in microbial fuel cells and their energy efficiency. Phys Chem Chem Phys 9:2619–2629
77. Aelterman P, Freguia S, Keller J, Verstraete W, Rabaey K (2008) The anode potential regulates bacterial activity in microbial fuel cells. Appl Microbiol Biotechnol 78:409–418
78. Yi H, Nevin KP, Kim BC, Franks AE, Klimes A, Tender LM, Lovley DR (2009) Selection of a variant of *Geobacter sulfurreducens* with enhanced capacity for current production in microbial fuel cells. Biosens Bioelectron 24:3498–3503
79. Wang X, Feng Y, Ren N, Wang H, Lee H, Li N (2009) Accelerated start-up of two-chambered microbial fuel cells: effect of anodic positive poised potential. Electrochim Acta 54:1109–1114
80. Wei J, Liang P, Cao X, Huang X (2010) A new insight into potential regulation on growth and power generation of *Geobacter sulfurreducens* in microbial fuel cells based on energy viewpoint. Environ Sci Technol 44:3187–3191
81. Mahadevan R, Bond DR, Butler JE, Esteve-Nunez A, Coppi MV, Palsson BO, Schilling CH, Lovley DR (2006) Characterization of metabolism in the Fe (III)-reducing organism *Geobacter sulfurreducens* by constraint-based modeling. Appl Environ Microbiol 72:1558–1568
82. Straub KL, Benz M, Schink B (2001) Iron metabolism in anoxic environments at near neutral pH. FEMS Microbiol Ecol 34:181–186
83. Walczak MM, Dryer DA, Jacobson DD, Foss MG, Flynn NT (1997) pH-dependent redox couple: illustrating the Nernst equation using cyclic voltammetry. Environ Sci Technol 74:1195–1197
84. Busalmen JP, De Sanchez SR (2005) Electrochemical polarization-induced changes in the growth of individual cells and biofilms of *Pseudomonas fluorescens* (ATCC 17552). Appl Environ Microbiol 71:6235–6240
85. Luo Q, Wang H, Zhang X, Qian Y (2005) Effect of direct electric current on the cell surface properties of phenol-degrading bacteria. Appl Environ Microbiol 71:423–427
86. He Z, Huang Y, Manohar AK, Mansfeld F (2008) Effect of electrolyte pH on the rate of the anodic and cathodic reactions in an air-cathode microbial fuel cell. Bioelectrochemistry 74:78–82
87. Cheng S, Logan BE (2007) Ammonia treatment of carbon cloth anodes to enhance power generation of microbial fuel cells. Electrochem Commun 9:492–496
88. Fan YZ, Hu HQ, Liu H (2007) Sustainable power generation in microbial fuel cells using bicarbonate buffer and proton transfer mechanisms. Environ Sci Technol 41:8154–8158
89. Fornero JJ, Rosenbaum M, Cotta MA, Angenent LT (2010) Carbon dioxide addition to microbial fuel cell cathodes maintains sustainable catholyte pH and improves anolyte pH, alkalinity, and conductivity. Environ Sci Technol 44:2728–2734
90. Torres CI, Lee HS, Rittmann BE (2008) Carbonate species as OH-carriers for decreasing the pH gradient between cathode and anode in biological fuel cells. Environ Sci Technol 42:8773–8777
91. Cheng KY, Ho G, Cord-ruwisch R (2010) Anodophilic biofilm catalyzes cathodic oxygen reduction. Environ Sci Technol 44:518–525
92. Freguia S, Rabaey K, Yuan Z, Keller J (2008) Sequential anode-cathode configuration improves cathodic oxygen reduction and effluent quality of microbial fuel cells. Water Res 42:1387–1396
93. Strik DPBTB, Hamelers HVM, Buisman CJN (2010) Solar energy powered microbial fuel cell with a reversible bioelectrode. Environ Sci Technol 44:532–537
94. Virdis B, Rabaey K, Yuan Z, Keller J (2008) Microbial fuel cells for simultaneous carbon and nitrogen removal. Water Res 42:3013–3024

95. Ahn Y, Logan BE (2010) Effectiveness of domestic wastewater treatment using microbial fuel cells at ambient and mesophilic temperatures. Bioresour Technol 101:469–475
96. Cheng S, Xing D, Logan BE (2011) Electricity generation of single-chamber microbial fuel cells at low temperatures. Biosens Bioelectron 26:1913–1917
97. Jadhav GS, Ghangrekar MM (2009) Performance of microbial fuel cell subjected to variation in pH, temperature, external load and substrate concentration. Bioresour Technol 100:717–723
98. Watanabe K (2008) Recent developments in microbial fuel cell technologies for sustainable bioenergy. J Biosci Bioeng 106:528–536
99. Du Z, Li H, Gu T (2007) A state of the art review on microbial fuel cells: a promising technology for wastewater treatment and bioenergy. Biotechnol Adv 25:464–482
100. Cheng S, Logan BE (2011) Increasing power generation for scaling up single-chamber air cathode microbial fuel cells. Bioresour Technol 102:4468–4473
101. Wang X, Feng Y, Liu J, Shi X, Lee H, Li N, Ren N (2010) Power generation using adjustable Nafion/PTFE mixed binders in air-cathode microbial fuel cells. Biosens Bioelectron 26:946–948
102. Yu EH, Cheng S, Scott K, Logan BE (2007) Microbial fuel cell performance with non-Pt cathode catalysts. J Power Sources 171:275–281
103. Zuo Y, Cheng S, Logan BE (2008) Ion exchange membrane cathodes for scalable microbial fuel cells. Environ Sci Technol 42:6967–6972
104. Bullen RA, Arnot TC, Lakeman JB, Walsh FC (2006) Biofuel cells and their development. Biosens Bioelectron 21:2015–2045
105. Logan BE, Regan JM (2006) Microbial fuel cells: challenges and applications. Environ Sci Technol 40:5172–5180
106. Cournet A, Délia ML, Bergel A, Roques C, Bergé M (2010) Electrochemical reduction of oxygen catalyzed by a wide range of bacteria including Gram-positive. Electrochem Commun 12:505–508
107. Gorby YA, Yanina S, McLean JS, Rosso KM, Moyles D, Dohnalkova A, Beveridge TJ, Chang IS, Kim BH, Kim KS, Culley DE, Reed SB, Romine MF, Saffarini DA, Hill EA, Shi L, Elias DA, Kennedy DW, Pinchuk G, Watanabe K, Ishii SI, Logan BE, Nealson KH, Fredrickson JK (2006) Electrically conductive bacterial nanowires produced by *Shewanella oneidensis* strain MR-1 and other microorganisms. Proc Natl Acad Sci 103:358–363
108. Marsili E, Baron DB, Shikhare ID, Coursolle D, Gralnick JA, Bond DR (2008) *Shewanella* secretes flavins that mediate extracellular electron transfer. Proc Natl Acad Sci 105:3968–3973
109. Newton GJ, Mori S, Nakamura R, Hashimoto K, Watanabe K (2009) Analyses of current-generating mechanisms of *Shewanella loihica* PV-4 and *Shewanella oneidensis* MR-1 in microbial fuel cells. Appl Environ Microbiol 75:7674–7681
110. Pham TH, Boon N, De Maeyer K, Höfte M, Rabaey K, Verstraete W (2008) Use of *Pseudomonas* species producing phenazine-based metabolites in the anodes of microbial fuel cells to improve electricity generation. Appl Microbiol Biotechnol. doi:10.1007/s00253-008-1619-7. 80:985-993
111. von Canstein H, Ogawa J, Shimizu S, Lloyd JR (2008) Secretion of flavins by *Shewanella* species and their role in extracellular electron transfer. Appl Environ Microbiol 74:615–623
112. Kim BH, Kim HJ, Hyun MS, Park DH (1999) Direct electrode reaction of Fe(III)-reducing bacterium, *Shewanella putrefaciens*. J Microbiol Biotechnol 9:127–131
113. Myers CR, Myers JM (1992) Localization of cytochromes to the outer membrane of anaerobically grown *Shewanella putrefaciens* MR-1. J Bacteriol 174:3429–3438
114. Wu X, Zhao F, Rahunen N, Varcoe JR, Avignone-Rossa C, Thumser AE, Slade RCT (2011) A role for microbial palladium nanoparticles in extracellular electron transfer. Angew Chem Int Ed 50:427–430
115. Reguera G, McCarthy KD, Mehta T, Nicoll JS, Tuominen MT, Lovley DR (2005) Extracellular electron transfer via microbial nanowires. Nature 435:1098–1101

116. Klimes A, Franks AE, Glaven RH, Tran H, Barrett CL, Qiu Y, Zengler K, Lovley DR (2011) Production of pilus-like filaments in *Geobacter sulfurreducens* in the absence of the type IV pilin protein PilA. FEMS Microbiol Lett 310:62–68
117. Reguera G, Nevin KP, Nicoll JS, Covalla SF, Woodard TL, Lovley DR (2006) Biofilm and nanowire production leads to increased current in *Geobacter sulfurreducens* fuel cells. Appl Environ Microbiol 72:7345–7348
118. Reguera G, Pollina RB, Nicoll JS, Lovley DR (2007) Possible nonconductive role of *Geobacter sulfurreducens* pilus nanowires in biofilm formation. J Bacteriol 189:2125–2127
119. Richter H, McCarthy K, Nevin KP, Johnson J, Rotello V, Lovley DR (2008) Electricity generation by *Geobacter sulfurreducens* attached to gold electrodes. Langmuir 24:4376–4379
120. Tremblay PL, Summers ZM, Glaven RH, Nevin KP, Zengler K, Barrett CL, Qiu Y, Palsson BO, Lovley DR (2011) A c-type cytochrome and a transcriptional regulator responsible for enhanced extracellular electron transfer in *Geobacter sulfurreducens* revealed by adaptive evolution. Environ Microbiol 13:13–23
121. Chiang P, Burrows LL (2003) Biofilm formation by hyperpiliated mutants of *Pseudomonas aeruginosa*. J Bacteriol 185:2374–2378
122. Rabaey K, Boon N, Hofte M, Verstraete W (2005) Microbial phenazine production enhances electron transfer in biofuel cells. Environ Sci Technol 39:3401–3408
123. Richter H, Nevin KP, Jia H, Lowy DA, Lovley DR, Tender LM (2009) Cyclic voltammetry of biofilms of wild type and mutant *Geobacter sulfurreducens* on fuel cell anodes indicates possible roles of OmcB, OmcZ, type IV pili, and protons in extracellular electron transfer. Energy Environ Sci 2:506–516
124. Krushkal J, Juárez K, Barbe JF, Qu Y, Andrade A, Puljic M, Adkins RM, Lovley DR, Ueki T (2010) Genome-wide survey for PilR recognition sites of the metal-reducing prokaryote *Geobacter sulfurreducens*. Gene 469:31–44
125. El-Naggar MY, Wanger G, Leung KM, Yuzvinsky TD, Southam G, Yang J, Lau WM, Nealson KH, Gorby YA (2010) Electrical transport along bacterial nanowires from *Shewanella oneidensis* MR-1. Proc Natl Acad Sci 107:18127–18131
126. Hou H, Li L, Cho Y, de Figueiredo P, Han A (2009) Microfabricated microbial fuel cell arrays reveal electrochemically active microbes. PLoS One 4:e6570
127. Hassett DJ, Korfhagen TR, Irvin TR, Schurz MJ, Sauer K, Lau GW, Sutton MD, Yu H, Hoiby N (2010) *Pseudomonas aeruginosa* biofilm infections in cystic fibrosis: insights into pathogenic processes and treatment strategies. Expert Opin Ther Targets 14:117–130
128. Morgan R, Kohn S, Hwang SH, Hassett DJ, Sauer K (2006) BdlA, a chemotaxis regulator essential for biofilm dispersion in *Pseudomonas aeruginosa*. J Bacteriol 188:7335–7343
129. O'Toole GA, Gibbs KA, Hager PW, Phibbs PV Jr, Kolter R (2000) The global carbon metabolism regulator Crc is a component of a signal transduction pathway required for biofilm development by *Pseudomonas aeruginosa*. J Bacteriol 182:425–431
130. Vallet I, Olson JW, Lory S, Lazdunski A, Filloux A (2001) The chaperone/usher pathways of *Pseudomonas aeruginosa*: identification of fimbrial gene clusters (cup) and their involvement in biofilm formation. Proc Natl Acad Sci 98:6911–6916
131. Whitchurch CB, Beatson SA, Comolli JC, Jakobsen T, Sargent JL, Bertrand JJ, West J, Klausen M, Waite LL, Kang PJ, Tolker-Nielsen T, Mattick JS, Engel JN (2005) *Pseudomonas aeruginosa* fimL regulates multiple virulence functions by intersecting with Vfr-modulated pathways. Mol Microbiol 55:1357–1378
132. Kulasakara H, Lee V, Brencic A, Liberati N, Urbach J, Miyata S, Lee DG, Neely AN, Hyodo M, Hayakawa Y, Ausubel FM, Lory S (2006) Analysis of *Pseudomonas aeruginosa* diguanylate cyclases and phosphodiesterases reveals a role for bis-(3′-5′)-cyclic-GMP in virulence. Proc Natl Acad Sci 103:2839–2844
133. O'Toole GA, Kolter R (1998) Flagellar and twitching motility are necessary for *Pseudomonas aeruginosa* biofilm development. Mol Microbiol 30:295–304
134. Davies DG, Parsek MR, Pearson JP, Iglewski BH, Costerton JW, Greenberg EP (1998) The involvement of cell-to-cell signals in the development of a bacterial biofilm. Science 280:295–298

135. Choy WK, Zhou L, Syn CK, Zhang LH, Swarup S (2004) MorA defines a new class of regulators affecting flagellar development and biofilm formation in diverse *Pseudomonas* species. J Bacteriol 186:7221–7228
136. Gooderham WJ, Gellatly SL, Sanschagrin F, McPhee JB, Bains M, Cosseau C, Levesque RC, Hancock REW (2009) The sensor kinase PhoQ mediates virulence in *Pseudomonas aeruginosa*. Microbiology 155:699–711
137. Ma L, Jackson KD, Landry RM, Parsek MR, Wozniak DJ (2006) Analysis of *Pseudomonas aeruginosa* conditional *psl* variants reveals roles for the *psl* polysaccharide in adhesion and maintaining biofilm structure postattachment. J Bacteriol 188:8213–8221
138. Ma L, Lu H, Sprinkle A, Parsek MR, Wozniak DJ (2007) *Pseudomonas aeruginosa* Psl is a galactose- and mannose-rich exopolysaccharide. J Bacteriol 189:8353–8356
139. Caiazza NC, O'Toole GA (2004) SadB is required for the transition from reversible to irreversible attachment during biofilm formation by *Pseudomonas aeruginosa* PA14. J Bacteriol 186:4476–4485
140. Ueda A, Wood TK (2009) Connecting quorum sensing, c-di-GMP, pel polysaccharide, and biofilm formation in *Pseudomonas aeruginosa* through tyrosine phosphatase TpbA (PA3885). PLoS Pathog 5:e1000483
141. Debabov VG (2008) Electricity from microorganisms. Mikrobiologiia 77:149–157
142. Friedman L, Kolter R (2004) Genes involved in matrix formation in *Pseudomonas aeruginosa* PA14 biofilms. Mol Microbiol 51:675–690
143. Suh SJ, Silo-Suh L, Woods DE, Hassett DJ, West SE, Ohman DE (1999) Effect of *rpoS* mutation on the stress response and expression of virulence factors in *Pseudomonas aeruginosa*. J Bacteriol 181:3890–3897
144. Mavrodi DV, Bonsall RF, Delaney SM, Soule MJ, Phillips G, Thomashow LS (2001) Functional analysis of genes for biosynthesis of pyocyanin and phenazine-1-carboxamide from *Pseudomonas aeruginosa* PAO1. J Bacteriol 183:6454–6465
145. Paradis-Bleau C, Sanschagrin F, Levesque RC (2005) Peptide inhibitors of the essential cell division protein *FtsA*. Protein Eng Des Sel 18:85–91
146. Hoang TT, Kutchma AJ, Becher A, Schweizer HP (2000) Integration-proficient plasmids for *Pseudomonas aeruginosa*: site-specific integration and use for engineering of reporter and expression strains. Plasmid 43:59–72
147. Qiu D, Damron FH, Mima T, Schweizer HP, Yu HD (2008) PBAD-based shuttle vectors for functional analysis of toxic and highly regulated genes in *Pseudomonas* and *Burkholderia* spp. and other bacteria. Appl Environ Microbiol 74:7422–7426
148. Rabaey K, Lissens G, Siciliano SD, Verstraete W (2003) A microbial fuel cell capable of converting glucose to electricity at high rate and efficiency. Biotechnol Lett 25:1531–1535
149. Li Z, Zhang X, Lei L (2008) Electricity production during the treatment of real electroplating wastewater containing Cr^{6+} using microbial fuel cell. Process Biochem 43:1352–1358
150. Li Y, Lu A, Ding H, Jin S, Yan Y, Wang C, Zen C, Wang X (2009) Cr(VI) reduction at rutile-catalyzed cathode in microbial fuel cells. Electrochem Commun 11:1496–1499
151. Ter Heijne A, Liu F, Van der Weijden R, Weijma J, Buisman CJN, Hamelers HVM (2010) Copper recovery combined with electricity production in a microbial fuel cell. Environ Sci Technol 44:4376–4381
152. Wang G, Huang L, Zhang Y (2008) Cathodic reduction of hexavalent chromium [Cr(VI)] coupled with electricity generation in microbial fuel cells. Biotechnol Lett 30:1959–1966
153. You S, Zhao Q, Zhang J, Jiang J, Zhao S (2006) A microbial fuel cell using permanganate as the cathodic electron acceptor. J Power Sources 162:1409–1415
154. Zhang B, Zhou S, Zhao H, Shi C, Kong L, Sun J, Yang Y, Ni J (2010) Factors affecting the performance of microbial fuel cells for sulfide and vanadium (V) treatment. Bioprocess Biosyst Eng 33:187–194
155. Jang JK, Chang IS, Moon H, Kang KH, Kim BH (2006) Continuous determination of biochemical oxygen demand using microbial fuel cell type biosensor. Biotechnol Bioeng 95:772–774

156. Mu Y, Rabaey K, Rozendal R, Yuan Z, Keller J (2009) Decolorization of azo dyes in bioelectrochemical systems. Environ Sci Technol 43:5137–5143
157. Mu Y, Rozendal R, Rabaey K, Keller J (2009) Nitrobenzene removal in bioelectrochemical systems. Environ Sci Technol 43:8690–8695
158. Li J, Liu G, Zhang R, Luo Y, Zhang C, Li M (2010) Electricity generation by two types of microbial fuel cells using nitrobenzene as the anodic or cathodic reactants. Bioresour Technol 101:4013–4020
159. Mu Y, Radjenovic J, Shen JY, Rozendal RE, Rabaey K, Keller J (2011) Dehalogenation of iodinated X-ray contrast media in a bioelectrochemical system. Environ Sci Technol 45:782–788
160. Gu HY, Zhang XW, Li ZJ, Lei LC (2007) Studies on treatment of chlorophenol-containing wastewater by microbial fuel cell. Chin Sci Bull 52:3448–3451
161. Catal T, Bermek H, Liu H (2009) Removal of selenite from wastewater using microbial fuel cells. Biotechnol Lett 31:1211–1216
162. Pandit S, Sengupta A, Kale S, Das D (2011) Performance of electron acceptors in catholyte of a two-chambered microbial fuel cell using anion exchange membrane. Bioresour Technol 102:2736–2744
163. Zhu X, Ni J (2009) Simultaneous processes of electricity generation and p-nitrophenol degradation in a microbial fuel cell. Electrochem Commun 11:274–277
164. Freguia S, Rabaey K, Yuan Z, Keller J (2007) Non-catalyzed cathodic oxygen reduction at graphite granules in microbial fuel cells. Electrochim Acta 53:598–603
165. Watson VJ, Saito T, Hickner MA, Logan BE (2011) Polymer coatings as separator layers for microbial fuel cell cathodes. J Power Sources 196:3015–3025
166. Zhao F, Harnisch F, Schröder U, Scholz F, Bogdanoff P, Herrmann I (2005) Application of pyrolysed iron(II) phthalocyanine and CoTMPP based oxygen reduction catalysts as cathode materials in microbial fuel cells. Electrochem Commun 7:1405–1410
167. Liu XW, Sun XF, Huang YX, Sheng GP, Zhou K, Zeng RJ, Dong F, Wang SG, Xu AW, Tong ZH, Yu HQ (2010) Nano-structured manganese oxide as a cathodic catalyst for enhanced oxygen reduction in a microbial fuel cell fed with a synthetic wastewater. Water Res 44:5298–5305
168. Duteanu N, Erable B, Senthil Kumar SM, Ghangrekar MM, Scott K (2010) Effect of chemically modified Vulcan XC-72R on the performance of air-breathing cathode in a single-chamber microbial fuel cell. Bioresour Technol 101:5250–5255
169. Qian F, Wang G, Li Y (2010) Solar-driven microbial photoelectrochemical cells with a nanowire photocathode. Nano Lett 10:4686–4691
170. Clauwaert P, Van Der Ha D, Boon N, Verbeken K, Verhaege M, Rabaey K, Verstraete W (2007) Open air biocathode enables effective electricity generation with microbial fuel cells. Environ Sci Technol 41:7564–7569
171. Nguyen TA, Lu Y, Yang X, Shi X (2007) Carbon and steel surfaces modified by *Leptothrix discophora* SP-6: characterization and implications. Environ Sci Technol 41:7987–7996
172. Rhoads A, Beyenal H, Lewandowski Z (2005) Microbial fuel cell using anaerobic respiration as an anodic reaction and biomineralized manganese as a cathodic reactant. Environ Sci Technol 39:4666–4671
173. Sun J, Bi Z, Hou B, Cao Y, Hu Y (2011) Further treatment of decolorization liquid of azo dye coupled with increased power production using microbial fuel cell equipped with an aerobic biocathode. Water Res 45:283–291
174. Lovley DR (2008) The microbe electric: conversion of organic matter to electricity. Curr Opin Biotechnol 19:564–571
175. Huang L, Chen J, Quan X, Yang F (2010) Enhancement of hexavalent chromium reduction and electricity production from a biocathode microbial fuel cell. Bioprocess Biosyst Eng 33:937–945
176. Huang L, Regan JM, Quan X (2011) Electron transfer mechanisms, new applications, and performance of biocathode microbial fuel cells. Bioresour Technol 102:316–323

177. Rodrigo MA, Cañizares P, Lobato J (2010) Effect of the electron-acceptors on the performance of a MFC. Bioresour Technol 101:7014–7018
178. Dumas C, Mollica A, Féron D, Basséguy R, Etcheverry L, Bergel A (2007) Marine microbial fuel cell: use of stainless steel electrodes as anode and cathode materials. Electrochim Acta 53:468–473
179. Schamphelaire LD, Boeckx P, Verstraete W (2010) Evaluation of biocathodes in freshwater and brackish sediment microbial fuel cells. Appl Microbiol Biotechnol 87:1675–1687
180. Malik S, Drott E, Grisdela P, Lee J, Lee C, Lowy DA, Gray S, Tender LM (2009) A self-assembling self-repairing microbial photoelectrochemical solar cell. Energy Environ Sci 2:292–298
181. He Z, Shao H, Angenent LT (2007) Increased power production from a sediment microbial fuel cell with a rotating cathode. Biosens Bioelectron 22:3252–3255
182. Strik DPBTB, Hamelers HVM, Snel JFH, Buisman CJN (2008) Green electricity production with living plants and bacteria in a fuel cell. Int J Energy Res 32:870–876
183. De Schamphelaire L, Van den Bossche L, Dang HS, Hofte M, Boon N, Rabaey K, Verstraete W (2008) Microbial fuel cells generating electricity from rhizo deposits of rice plants. Environ Sci Technol 42:3053–3058
184. Kaku N, Yonezawa N, Kodama Y, Watanabe K (2008) Plant/microbe cooperation for electricity generation in a rice paddy field. Appl Microbiol Biotechnol 79:43–49
185. Aulenta F, Catervi A, Majone M, Panero S, Reale P, Rossetti S (2007) Electron transfer from a solid-state electrode assisted by methyl viologen sustains efficient microbial reductive dechlorination of TCE. Environ Sci Technol 41:2554–2559
186. Aulenta F, Canosa A, Reale P, Rossetti S, Panero S, Majone M (2009) Microbial reductive dechlorination of trichloroethene to ethene with electrodes serving as electron donors without the external addition of redox mediators. Biotechnol Bioeng 103:85–91
187. Butler CS, Clauwaert P, Green SJ, Verstraete W, Nerenberg R (2010) Bioelectrochemical perchlorate reduction in a microbial fuel cell. Environ Sci Technol 44:4685–4691
188. Cao X, Huang X, Liang P, Xiao K, Zhou Y, Zhang X, Logan BE (2009) A new method for water desalination using microbial desalination cells. Environ Sci Technol 43:7148–7152
189. Clauwaert P, Rabaey K, Aelterman P, Schamphelaire LD, Pham TH, Boeckx P, Boon N, Verstraete W (2007) Biological denitrification in microbial fuel cells. Environ Sci Technol 41:3354–3360
190. Park H, Kim DK, Choi YJ, Park D (2005) Nitrate reduction using an electrode as direct electron donor in a biofilm-electrode reactor. Process Biochem 40:3383–3388
191. Rozendal RA, Jeremiasse AW, Hamelers HVM, Buisman CJN (2008) Hydrogen production with a microbial biocathode. Environ Sci Technol 42:629–634
192. Shea C, Clauwaert P, Verstraete W, Nerenberg R (2008) Adapting a denitrifying biocathode for perchlorate. Water Sci Technol 58:1941–1946
193. Strycharz SM, Gannon SM, Boles AR, Franks AE, Nevin KP, Lovley DR (2010) Reductive dechlorination of 2-chlorophenol by *Anaeromyxobacter dehalogenans* with an electrode serving as the electron donor. Environ Microbiol Rep 2:289–294
194. Gregory KB, Bond DR, Lovley DR (2004) Graphite electrodes as electron donors for anaerobic respiration. Environ Microbiol 6:596–604
195. Chen GW, Choi SJ, Lee TH, Lee GY, Cha JH, Kim CW (2008) Application of biocathode in microbial fuel cells: cell performance and microbial community. Appl Microbiol Biotechnol 79:379–388
196. Clauwaert P, Desloover J, Shea C, Nerenberg R, Boon N, Verstraete W (2009) Enhanced nitrogen removal in bio-electrochemical systems by pH control. Biotechnol Lett 31:1537–1543
197. Wrighton KC, Virdis B, Clauwaert P, Read ST, Daly RA, Boon N, Piceno Y, Andersen GL, Coates JD, Rabaey K (2010) Bacterial community structure corresponds to performance during cathodic nitrate reduction. ISME J 4:1443–1455
198. Aulenta F, Reale P, Canosa A, Rossetti S, Panero S, Majone M (2010) Characterization of an electro-active biocathode capable of dechlorinating trichloroethene and cis-dichloroethene to ethane. Biosens Bioelectron 25:1796–1802

199. Sanford RA, Cole JR, Tiedje JM (2002) Characterization and description of *Anaeromyxobacter dehalogenans* gen. nov., sp. nov., an aryl-halorespiring facultative anaerobic myxobacterium. Appl Environ Microbiol 68:893–900
200. Huang L, Chai X, Cheng S, Chen G (2011) Evaluation of carbon-based materials in tubular biocathode microbial fuel cells in terms of hexavalent chromium reduction and electricity generation. Chem Eng J 166:652–661
201. Pons L, Délia ML, Bergel A (2011) Effect of surface roughness, biofilm coverage and biofilm structure on the electrochemical efficiency of microbial cathodes. Bioresour Technol 102:2678–2683
202. Cheng S, Xing D, Call D, Logan BE (2009) Direct biological conversion of electrical current into methane by electromethanogenesis. Environ Sci Technol 43:3953–3958
203. Villano M, Aulenta F, Ciucci C, Ferri T, Giuliano A, Majone M (2010) Bioelectrochemical reduction of CO_2 to CH_4 via direct and indirect extracellular electron transfer by a hydrogenophilic methanogenic culture. Bioresour Technol 101:3085–3090
204. Wang X, Feng Y, Liu J, Lee H, Li C, Li N, Ren N (2010) Sequestration of CO_2 discharged from anode by algal cathode in microbial carbon capture cells (MCCs). Biosens Bioelectron 25:2639–2643
205. Kim BH, Ikeda T, Park HS, Kim HJ, Hyun MS, Kano K, Takagi K, Tatsumi H (1999) Electrochemical activity of an Fe(III)-reducing bacterium, *Shewanella putrefaciens* IR-1, in the presence of alternative electron acceptors. Biotechnol Tech 13:475–478
206. Freguia S, Tsujimura S, Kano K (2010) Electron transfer pathways in microbial oxygen biocathodes. Electronchim Acta 55:813–818
207. Strycharz SM, Glaven RH, Coppi MV, Gannon SM, Perpetua LA, Liu A, Nevin KP, Lovley DR (2011) Gene expression and deletion analysis of mechanisms for electron transfer from electrodes to *Geobacter sulfurreducens*. Bioelectrochemistry 80:142–150
208. Sleutels THJA, Hameler HVM, Rozendal RA, Buisman CJN (2009) Ion transport resistance in microbial electrolysis cells with anion and cation exchange membranes. Int J Hydrog Energy 34:3612–3620
209. Jang JK, Pham TH, Chang IS, Kang KH, Moon H, Cho KS, Kim BH (2004) Construction and operation of a novel mediator- and membrane-less microbial fuel cell. Process Biochem 39:1007–1012
210. Foley JM, Rozendal RA, Hertle CR, Lant PA, Rabaey K (2010) Life cycle assessment of high-rate anaerobic treatment, microbial fuel cells, and microbial electrolysis cells. Environ Sci Technol 44:3629–3637
211. Rozendal RA, Hamelers HVM, Euverink GJW, Metz SJ, Buisman CJN (2006) Principle and perspectives of hydrogen production through biocatalyzed electrolysis. Int J Hydrog Energy 31:1632–1640
212. Rozendal RA, Hamelers HVM, Molenkamp RJ, Buisman CJN (2007) Performance of single chamber biocatalyzed electrolysis with different types of ion exchange membranes. Water Res 41:1984–1994
213. Cheng H, Scott K, Ramshaw C (2002) Intensification of water electrolysis in a centrifugal field. J Electrochem Soc 149:D172–D177
214. Ambler JR, Logan BE (2011) Evaluation of stainless steel cathodes and a bicarbonate buffer for hydrogen production in microbial electrolysis cells using a new method for measuring gas production. Int J Hydrog Energy 36:160–166
215. Call D, Logan BE (2008) Hydrogen production in a single chamber microbial electrolysis cell (MEC) lacking a membrane. Environ Sci Technol 42:3401–3406
216. Tartakovskya B, Manuel MF, Neburchilov V, Wang H, Guiot SR (2008) Biocatalyzed hydrogen production in a continuous flow microbial fuel cell with a gas phase cathode. J Power Sources 182:291–297
217. Sun M, Sheng G, Zhang L, Xia C, Mu Z, Liu X, Wang H, Yu H, Qi R, Yu T, Yang M (2008) An MEC-MFC-coupled system for biohydrogen production from acetate. Environ Sci Technol 42:8095–8100

218. Steinbusch KJJ, Hamelers HVM, Schaap JD, Kampman C, Buisman CJN (2010) Bioelectrochemical ethanol production through mediated acetate reduction by mixed cultures. Environ Sci Technol 44:513–517
219. Rabaey K, Bützer S, Brown S, Keller J, Rozendal R (2010) High current generation coupled to caustic production using a lamellar bioelectrochemical system. Environ Sci Technol 44:4315–4321
220. Li H, Ni J (2011) Treatment of wastewater from *Dioscorea zingiberensis* tubers used for producing steroid hormones in a microbial fuel cell. Bioresour Technol 102:2731–2735
221. Nimje VR, Chen CY, Chen CC, Chen HR, Tseng MJ, Jean JS, Chang YF (2011) Glycerol degradation in single-chamber microbial fuel cells. Bioresour Technol 102:2629–2634
222. Rozendal RA, Leone E, Keller J, Rabaey K (2009) Efficient hydrogen peroxide generation from organic matter in a bioelectrochemical system. Electrochem Commun 11:1752–1755
223. Liu L, Yuan Y, Li F, Feng C (2011) In-situ Cr(VI) reduction with electrogenerated hydrogen peroxide driven by iron-reducing bacteria. Bioresour Technol 102:2468–2473
224. Fu L, You S, Zhang G, Yang F, Fang X (2010) Degradation of azo dyes using in-situ Fenton reaction incorporated into H_2O_2-producing microbial fuel cell. Chem Eng J 160:164–169
225. Luo H, Jenkins PE, Ren Z (2011) Concurrent desalination and hydrogen generation using microbial electrolysis and desalination cells. Environ Sci Technol. doi:10.1021/es1022202. 45:340-344
226. Mehanna M, Kiely PD, Call DF, Logan BE (2010) Microbial electrodialysis cell for simultaneous water desalination and hydrogen gas production. Environ Sci Technol 44:9578–9583
227. Huang L, Cheng S, Chen G (2011) Bioelectrochemical systems for efficient recalcitrant wastes treatment. J Chem Technol Biotechnol 86:481–491
228. Zhang Y, Min B, Huang L, Angelidaki I (2009) Generation of electricity and analysis of microbial communities in wheat straw biomass-powered microbial fuel cells. Appl Environ Microbiol 75:3389–3395
229. Cheng SA, Logan BE (2007) Ammonia treatment of carbon cloth anodes to enhance power generation of microbial fuel cells. Electrochem Commun 9:492–496
230. Wagner RC, Call DI, Logan BE (2010) Optimal set anode potentials vary in bioelectrochemical systems. Environ Sci Technol 44:6036–6041
231. Cheng SA, Logan BE (2007) Ammonia treatment of carbon cloth anodes to enhance power generation of microbial fuel cells. Electrochem Commun 9:492–496
232. You SJ, Ren NQ, Zhao QL, Wang JY, Yang FL (2009) Power generation and electrochemical analysis of biocathodemicrobial fuel cell using graphite fiber brush as cathode material. Fuel Cells 9:588–596
233. Tandukar M, Huber SJ, Onodera T, Pavlostathis SG (2009) Biological chromium(VI) reduction in the cathode of a microbial fuel cell. Environ Sci Technol 43:8159–8165
234. Gregory KB, Lovley DR (2005) Remediation and recovery of uranium from contaminated subsurface environments with electrodes. Environ Sci Technol 39:8943–8947
235. Thrash JC, Trump JI, Weber KA, Miller E, Achenbach LA, Coates JD (2007) Electrochemical stimulation of microbial perchlorate reduction. Environ Sci Technol 41:1740–1746
236. Thrash JC, Coates JD (2008) Review: direct and indirect electrical stimulation of microbial metabolism. Environ Sci Technol 42:3921–3931
237. Powell EE, Mapiour ML, Evitts RW, Hill GA (2009) Growth kinetics of Chlorella vulgaris and its use as a cathodic half cell. Bioresour Technol 100:269–274
238. Fredrickson JK, Romine MF, Beliaev AS, Auchtung JM, Driscoll ME, Gardner TS, Nealson KH, Osterman AL, Pinchuk G, Reed JL, Rodionov DA, Rodrigues JLM, Saffarini DA, Serres MH, Spormann AM, Zhulin IB, Tiedje JM (2008) Towards environmental systems biology of Shewanella. Nat Rev Microbiol 6:592–603
239. Newman DK, Kolter R (2000) A role for excreted quinones in extracellular electron transfer. Nature 405:94–97
240. Dumas C, Basseguy R, Bergel A (2008) Microbial electrocatalysis with Geobacter sulfurreducens biofilm on stainless steel cathodes. Electrochim Acta 53:2494–2500

241. Voordeckers JW, Kim BC, Izallalen M, Lovley DR (2010) Role of Geobacter sulfurreducens outer-surface c-type cytochromes in the reduction of soil humic acid and anthraquinone-2, 6-disulfonate. Appl Environ Microbiol 76:2371–2375
242. Eggleston CM, Vörös J, Shi L, Lower BH, Droubay TC, Colberg PJS (2008) Binding and direct electrochemistry of OmcA, an outer-membrane cytochrome from an iron reducing bacterium with oxide electrodes: a candidate biofuel cell system. Inorg Chim Acta 361: 769–777
243. Freguia S, Rabaey K, Yuan Z, Keller J (2008) Sequential anode-cathode configuration improves cathodic oxygen reduction and effluent quality of microbial fuel cells. Water Res 42:1387–1396

Index

S.K. Sharma and R. Sanghi (eds.), *Advances in Water Treatment and Pollution Prevention*, DOI 10.1007/978-94-007-4204-8,

MIX
Papier aus verantwortungsvollen Quellen
Paper from responsible sources
FSC® C105338

If you have any concerns about our products,
you can contact us on
ProductSafety@springernature.com

In case Publisher is established outside the EU,
the EU authorized representative is:
Springer Nature Customer Service Center GmbH
Europaplatz 3, 69115 Heidelberg, Germany

Printed by Libri Plureos GmbH
in Hamburg, Germany